THIRD EDITION

PERMANENT MAGNET MOTOR TECHNOLOGY

DESIGN AND APPLICATIONS

THIRD EDITION

PERMANENT MAGNET MOTOR TECHNOLOGY

DESIGN AND APPLICATIONS

JACEK F. GIERAS

Hamilton Sundstrand Aerospace, Rockford, Illinois, U.S.A.
University of Technology and Life Sciences, Bydgoszcz, Poland

CRC Press
Taylor & Francis Group
Boca Raton   London   New York

CRC Press is an imprint of the
Taylor & Francis Group, an **Informa** business

CRC Press
Taylor & Francis Group
6000 Broken Sound Parkway NW, Suite 300
Boca Raton, FL 33487-2742

© 2010 by Taylor and Francis Group, LLC
CRC Press is an imprint of Taylor & Francis Group, an Informa business

No claim to original U.S. Government works

International Standard Book Number: 978-1-4200-6440-7 (Hardback)

Library of Congress Cataloging-in-Publication Data

Gieras, Jacek F.
   Permanent magnet motor technology : design and applications / Jacek F. Gieras. -- 3rd ed.
      p. cm.
   Includes bibliographical references and index.
   ISBN 978-1-4200-6440-7 (hardcover : alk. paper)
   1. Permanent magnet motors. I. Title.

TK2537.G54 2010
621.46--dc22
                                            2009028498

Contents

Preface to the Third Edition

The importance of permanent magnet (PM) motor technology and its impact on electromechanical drives has significantly increased since publication of the first edition of this book in 1996 and the second edition in 2002. There was 165% PM brushless motor market growth between 2005 and 2008 in comparison with 29% overall motion control market growth in the same period of time.

It is expected that the development of electric machines, mostly PM machines and associated power electronics in the next few years will be stimulated by large scale applications such as (a) computer hardware, (b) residential and public applications, (c) land, sea and air transportation and (d) renewable energy generation [124]. The development of PM machines is, however, not limited to these four major areas of application since PM machines are vital apparatus in all sectors of modern society, such as industry, services, trade, infrastructure, healthcare, defense, and domestic life. For example, worldwide demand on PM vibration motors for mobile phones increases at least 8% each year and global shipment of hard disk drives (HDD) with PM brushless motors increases about 24% each year.

In the last two decades new topologies of high torque density PM motors, high speed PM motors, integrated PM motor drives, and special PM motors have gained maturity. The largest PM brushless motor in the world rated at 36.5 MW, 127 rpm was built in 2006 by DRS Technologies, Parsippany, NJ, U.S.A.

In comparison with the 2002 edition, the 3rd edition has been thoroughly revised and updated, new chapters on high speed motors and micromotors have been written and more numerical examples and illustrative material have been added.

Any critical remarks, corrections, and suggestions for improvements are most welcome and may be sent to the author at jgieras@ieee.org.

Prof. Jacek F. Gieras, IEEE Fellow

1

Introduction

1.1 Permanent magnet versus electromagnetic excitation

The use of permanent magnets (PMs) in construction of electrical machines brings the following benefits:

- no electrical energy is absorbed by the field excitation system and thus there are no excitation losses which means substantial increase in efficiency,
- higher power density and/or torque density than when using electromagnetic excitation,
- better dynamic performance than motors with electromagnetic excitation (higher magnetic flux density in the air gap),
- simplification of construction and maintenance,
- reduction of prices for some types of machines.

The first PM excitation systems were applied to electrical machines as early as the 19th century, e.g., J. Henry (1831), H. Pixii (1832), W. Ritchie (1833), F. Watkins (1835), T. Davenport (1837), M.H. Jacobi (1839) [35]. Of course, the use of very poor quality hard magnetic materials (steel or tungsten steel) soon discouraged their use in favor of electromagnetic excitation systems. The invention of Alnico in 1932 revived PM excitation systems; however, its application was limited to small and fractional horsepower d.c. commutator machines. At the present time most PM *d.c. commutator motors* with slotted rotors use ferrite magnets. Cost effective and simple d.c. commutator motors with *barium* or *strontium ferrite PMs* mounted on the stator will still be used in the forseeable future in road vehicles, toys, and household equipment.

Cage induction motors have been the most popular electric motors in the 20th century. Recently, owing to the dynamic progress made in the field of power electronics and control technology, their application to electrical drives has increased. Their rated ouput power ranges from 70 W to 500 kW, with 75% of them running at 1500 rpm. The main advantages of cage induction motors are their simple construction, simple maintenance, no commutator or slip rings, low price and moderate reliability. The disadvantages are their small air gap, the possibility of cracking the rotor bars due to hot spots at plugging and reversal, and lower efficiency and power factor than synchronous motors.

The use of *PM brushless motors* has become a more attractive option than induction motors. *Rare earth PMs* can not only improve the motor's steady-state performance but also the power density (output power-to-mass ratio), dynamic performance, and quality. The prices of rare earth magnets are also dropping, which is making these motors more popular. The improvements made in the field of semiconductor drives have meant that the control of brushless motors has become easier and cost effective, with the possibility of operating the motor over a large range of speeds while still maintaining good efficiency.

Servo motor technology has changed in recent years from conventional d.c. or two-phase a.c. motor drives to new maintenance-free brushless three-phase vector-controlled a.c. drives for all motor applications where quick response, light weight and large continuous and peak torques are required.

A PM brushless motor has the magnets mounted on the rotor and the armature winding mounted on the stator. Thus, the armature current is not transmitted through a commutator or slip rings and brushes. These are the major parts which require maintenance. A standard maintenance routine in 90% of motors relates to the sliding contact. In a d.c. commutator motor the power losses occur mainly in the rotor which limits the heat transfer and consequently the armature winding current density. In PM brushless motors the power losses are practically all in the stator where heat can be easily transferred through the ribbed frame or, in larger machines, water cooling systems can be used [9, 24, 223]. Considerable improvements in dynamics of brushless PM motor drives can be achieved since the rotor has a lower inertia and there is a high air gap magnetic flux density and no-speed dependent current limitation.

The *PM brushless motor electromechanical drive* has become a more viable option than its induction or reluctance counterpart in motor sizes up to $10-15$ kW. There have also been successful attempts to build PM brushless motors rated above 1 MW (Germany and U.S.A.) [9, 23, 24, 124, 223, 276]. The high performance rare-earth magnets have successfully replaced ferrite and Alnico magnets in all applications where high power density, improved dynamic performance or higher efficiency are of prime interest. Typical examples where these points are key selection criteria are stepping motors for computer peripheral applications and servo motors for machine tools or robotics.

1.2 Permanent magnet motor drives

In general, all electromechanical drives can be divided into constant-speed drives, servo drives and variable-speed drives.

A *constant-speed drive* usually employs a synchronous motor alone which can keep the speed constant without an electronic converter and feedback or any other motor when there is less restriction on the speed variation tolerance.

A *servo system* is a system consisting of several devices which continuously monitor actual information (speed, position), compare these values to desired outcome and make necessary corrections to minimize the difference. A *servo motor drive* is a drive with a speed or position feedback for precise control where the response time and the accuracy with which the motor follows the speed and position commands are extremely important.

In a *variable-speed drive* (VSD) the accuracy and the response time with which the motor follows the speed command are not important, but the main requirement is to change the speed over a wide range.

In all electromechanical drives where the speed and position are controlled, a *solid state converter* interfaces the power supply and the motor. There are three types of PM motor electromechanical drives:

- d.c. commutator motor drives
- brushless motor drives (d.c. and a.c. synchronous)
- stepping motor drives

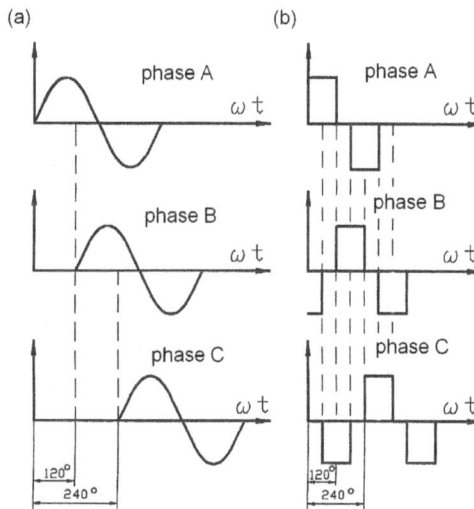

Fig. 1.1. Basic armature waveforms for three phase PM brushless motors: (a) sinusoidally excited, (b) square wave.

Brushless motor drives fall into the two principal classes of *sinusoidally excited* and *square wave* (trapezoidally excited) motors. Sinusoidally excited motors are fed with three-phase sinusoidal waveforms (Fig. 1.1a) and operate on the principle of a rotating magnetic field. They are simply called *sinewave motors* or *PM synchronous motors*. All phase windings conduct current at a time. Square wave motors are also fed with three-phase waveforms shifted by 120^0 one from another, but these waveshapes are rectangular or trapezoidal (Fig. 1.1b). Such a shape is produced when the armature current (MMF) is precisely synchronized with the rotor instantaneous position and frequency (speed). The most direct and popular method of providing the required rotor position information is to use an absolute angular position sensor mounted on the rotor shaft. Only two phase windings out of three conduct current simultaneously. Such a control scheme or *electronic commutation* is functionally equivalent to the mechanical commutation in d.c. motors. This explains why motors with square wave excitation are called *d.c. brushless motors*. An alternative name used in power electronics and motion control is *self-controlled synchronization* [166].

Although stepping motor electromechanical drives are a kind of synchronous motor drive, they are separately discussed due to their different control strategies and power electronic circuits.

1.2.1 d.c. commutator motor drives

The *d.c. commutator motor* or *d.c brush motor* is still a simple and low cost solution to variable-speed drive systems when requirements such as freedom from maintenance, operation under adverse conditions or the need to operate groups of machines in synchronism are not supreme [167]. Owing to the action of the mechanical commutator, control of a d.c. motor drive is comparatively simple and the same basic control system can satisfy the requirements of most applications. For these reasons the d.c. electromechanical drive very often turns into the cheapest alternative, in spite of the cost of the d.c. commutator motor. In many industrial applications such as agitators, extruders, kneading machines, printing machines, coating machines, some types of textile machinery, fans, blowers, simple machine tools, etc., the motor is required only to start smoothly and drive the machinery in one direction without braking or reverse running. Such a drive operates only in the first quadrant of the speed-torque characteristic and requires only one *controlled converter* (in its rectifier mode) as shown in Fig. 1.2a. At the expense of increased torque ripple and supply harmonics, a half-controlled rather than fully-controlled bridge may be used up to about 100 kW. If the motor is required to drive in both forward and reverse directions, and apply regenerative braking, a single fully controlled converter can still be used but with the possibility of reversing the armature current (Fig. 1.2a).

Electromechanical drives such as for rolling mills, cranes and mine winders are subject to rapid changes in speed or in load. Similarly, in those textile, paper or plastics machines where rapid control of tension is needed, frequent small speed adjustments may request rapid torque reversals. In these cases a *four-quadrant dual converter* comprising two semiconductor bridges in antiparallel, as in Fig. 1.2b, can be used [167]. One bridge conducts when ar-

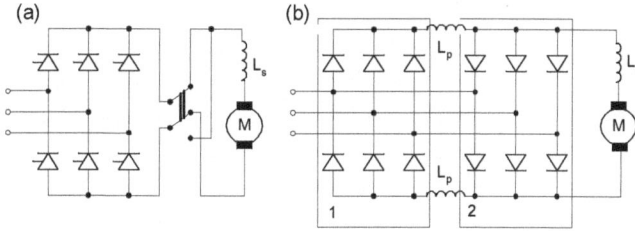

Fig. 1.2. d.c. commutator motor drives: (a) one quadrant, fully-controlled, single-converter drive, (b) four-quadrant, fully-controlled, dual converter drive; L_p are reactors limiting the currents circulating between the rectifying and inverting bridges, L_a is the armature circuit inductance.

mature current is required to be positive, and the second bridge when it is required to be negative.

Natural or *line commutation* between successively conducting power semiconductor switches takes place when the instantaneous values of consecutive phase voltages are equal and cross each other. The phase with a decreasing voltage is suppressed, while that with an increasing voltage takes over the conduction. Natural commutation may not be possible in such situations when the armature inductance is very small, e.g., control of small d.c. motors. The normal phase control would cause unacceptable torque ripple without substantial smoothing which, in turn, would impair the response of the motor [167]. Power transistors, GTO thyristors or IGBTs can be turned off by appropriate gate control signals, but conventional thyristors need to be reverse biased briefly for successful turn off. This can be accomplished by means of a forced-commutation circuit, usually consisting of capacitors, inductors and, in some designs, auxiliary thyristors. *Forced commutation* is used mostly for the frequency control of a.c. motors by variable-frequency inverters and chopper control of d.c. motors.

Fig. 1.3a shows the main components of a *one-directional chopper* circuit for controlling a PM or separately excited d.c. motor. Thyristors must be accompanied by some form of turn-off circuits or, alternatively, be replaced by GTO thyristors or IGBTs. When the mean value of the chopper output voltage is reduced below the armature EMF, the direction of the armature current cannot reverse unless T2 and D2 are added. The thyristor T2 is fired after T1 turns off, and vice versa. Now, the reversed armature current flows via T2 and increases when T1 is off. When T1 is fired, the armature current flows via D2 back to the supply. In this way regenerative braking can be achieved [167].

Full *four-quadrant operation* can be obtained with a bridge version of the chopper shown in Fig. 1.3b. Transistors or GTO thyristors allow the chopper to operate at the higher switching frequencies needed for low-inductance motors. By varying the on and off times of appropriate pairs of solid switches,

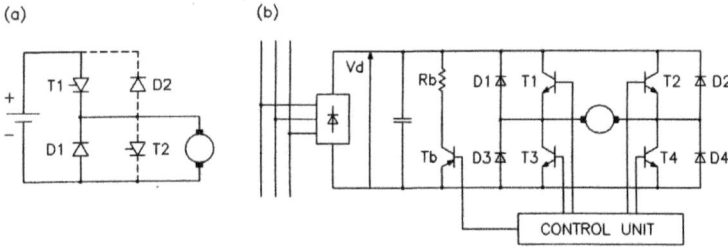

Fig. 1.3. Chopper controlled d.c. motor drives: (a) one-directional with added regenerative braking leg, (b) four-quadrant chopper controller [167].

the mean armature voltage, and hence speed, can be controlled in either direction. Typical applications are machine tools, generally with one motor and chopper for each axis, all fed from a common d.c. supply [167].

1.2.2 Synchronous motor drives

In a two-stage conversion a d.c. intermediate link is inserted between the line and motor-side converter (Fig. 1.4). For *low power PM synchronous motors* (in the range of kWs) a simple diode bridge rectifier as a line-side converter (Fig 1.4a) is used. The most widely used semiconductor switch at lower power permitting electronic turn-off is power transistor or IGBT.

Motor-side converters (inverters) have load commutation if the load, e.g., PM synchronous motor can provide the necessary reactive power for the solid state converter. Fig. 1.4b shows a basic power circuit of a *load-commutated current source thyristor converter*. The intermediate circuit energy is stored in the inductor. The inverter is a simple three-phase thyristor bridge. Load commutation is ensured by overexcitation of the synchronous motor so that it operates at a leading power factor (leading angle is approximately 30^0) [132]. This causes a decrease in the output power. The elimination of forced commutation means fewer components, simpler architecture, and consequently lower converter volume, mass, and losses. A four-quadrant operation is possible without any additional power circuitry. The motor phase EMFs required for load commutation of the inverter are not available at standstill and at very low speeds (less than 10% of the full speed). Under these conditions, the current commutation is provided by the line converter going into an inverter mode and forcing the d.c. link current to become zero, thus providing turn-off of thyristors in the load inverter [222].

The maximum output frequency of a load-commutated *current source inverter* (CSI) is limited by the time of commutation, which in turn is determined by the load. A CSI is suitable for loads of low impedance.

A *voltage source inverter* (VSI) is suitable for loads of high impedance. In a VSI the energy of the intermediate circuit is stored in the capacitor. A PWM VSI with GTOs and antiparallel diodes (Fig. 1.4c) allows a synchronous motor

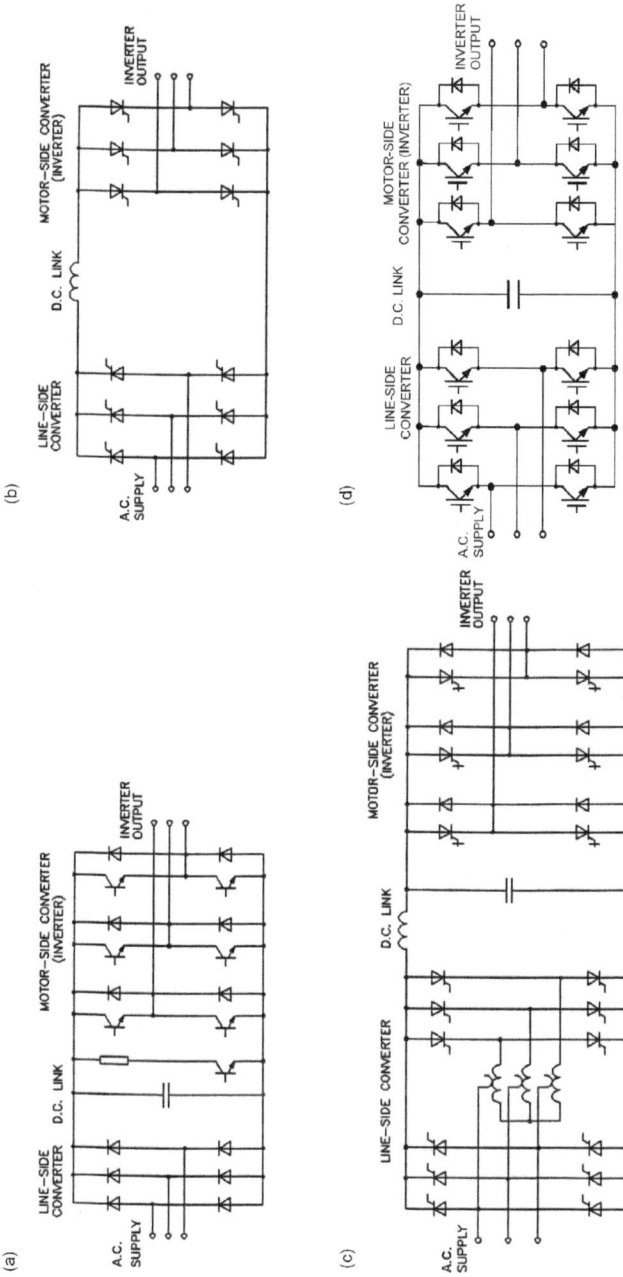

Fig. 1.4. Basic power circuits of d.c. link converters for synchronous motors with: (a) PWM transistor inverter, (b) load-commutated thyristor CSI, (c) forced-commutated GTO VSI (four-quadrant operation), (d) IGBT VSI.

to operate with unity power factor. Synchronous motors with high subtransient inductance can then be used. Four-quadrant operation is possible with a suitable power regeneration line-side converter. Replacement of thyristors by GTOs or IGBTs eliminates inverter commutation circuits and increases the pulse frequency.

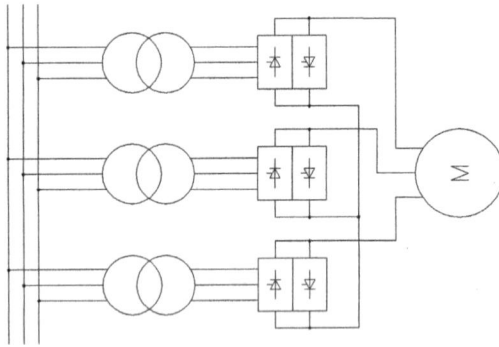

Fig. 1.5. Cycloconverter synchronous motor drive.

The maximum output frequency of the load-commutated CSI is limited to about 400 Hz even if fast thyristors are used. A higher output frequency can be achieved using an IGBT VSI with antiparallel diodes. Fig. 1.4d illustrates a typical PM brushless motor drive circuit with a three-phase PWM IGBT inverter. In the brushless d.c. mode, only two of the three phase windings are excited by properly switching the IGBTs of the inverter, resulting in ideal motor current waveforms of rectangular shape. There are six combinations of the stator winding excitation over a fundamental cycle with each combination lasting for a phase period of 60°. The corresponding two active solid state switches in each period may perform PWM to regulate the motor current. To reduce current ripple, it is often useful to have one solid state switch doing PWM while keeping the other switch conducting.

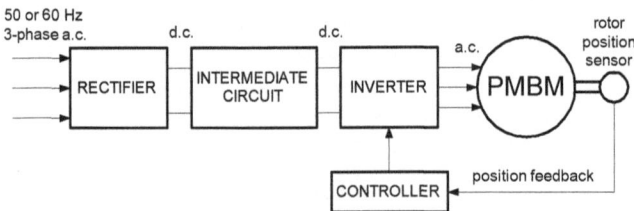

Fig. 1.6. d.c. brushless motor drive.

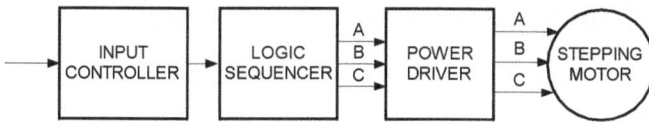

Fig. 1.7. Stepping motor drive.

A *cycloconverter* is a single stage (a.c. to a.c.) line-commutated frequency converter (Fig. 1.5). Four-quadrant operation is permitted as the power can flow in both directions between the line and the load. A cycloconverter covers narrow output frequency range, from zero to about 50% of the input frequency. Therefore, cycloconverters are usually used to supply gearless electromechanical drives with large power, low speed synchronous motors. For example, synchronous motors for ships propulsion are fed from diesel alternators mostly via cycloconverters [278]. A cycloconverter has an advantage of low torque harmonics of relatively high frequency. Drawbacks include large number of solid state switches, complex control, and poor power factor. Forced commuation can be employed to improve the power factor.

1.2.3 PM d.c. brushless motor drives

In PM d.c. brushless motors, square current waveforms are in synchronism with the rotor position angle. The basic elements of a *d.c. brushless motor drive* are: PM motor, output stage (inverter), line-side converter, shaft position sensor (encoder, resolver, Hall elements,), gate signal generator, current detector, and controller, e.g., microprocessor or computer with DSP board. A simplified block diagram is shown in Fig. 1.6.

1.2.4 Stepping motor drives

A typical *stepping motor drive* (Fig. 1.7) consists of an input controller, logic sequencer and driver. The input controller is a logic circuit that produces the required train of pulses. It can be a microprocessor or microcomputer which generates a pulse train to turn the rotor, speed up and slow down. The logic sequencer is a logic circuit that responds to step-command pulses and controls the excitation of windings sequentially [174]. Output signals of a logic sequencer are transmitted to the input terminals of a power drive which turns on and turns off the stepping motor windings. The stepping motor converts electric pulses into discrete angular displacements. The fundamental difference between a stepping motor drive and a *switched reluctance motor* (SRM) drive is that the first one operates in open loop control, without rotor position feedback.

1.3 Toward increasing the motor efficiency

Unforeseen consequences can result from problems the contemporary world currently faces, i.e.,

- fears of depletion and expected scarcities of major non-renewable energy resources over the next several decades
- increase in energy consumption
- pollution of our planet

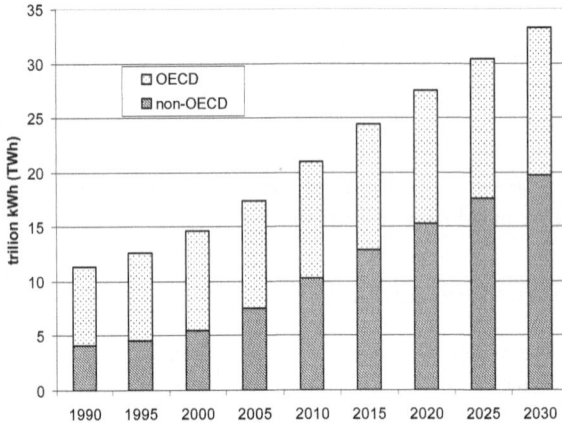

Fig. 1.8. World net electric power generation, 1990–2030 (history: 1990–2005, projections: 2005–2030) [158].

The world consumption of petroleum is about 84 million barrels per day (1 barrel = 159 l) or about 31 billion barells (about 5×10^{12} l) per year. If current laws and policies remain unchanged throughout the projection period, world marketed energy consumption is projected to grow by 50 percent over the 2005 to 2030 period [158].

About 30% of primary energy is used for generation of electricity. World net electricity generation will increase from 17,300 TWh (17.3 trillion kWh) in 2005 to 24,400 TWh in 2015 and 33,300 TWh in 2030 (Fig. 1.8). Total non-OECD [1] electricity generation increases by an average of 4.0 % per year, as compared with a projected average annual growth rate in OECD electricity generation of 1.3 % from 2005 to 2030 [158].

The mix of energy sources in the world is illustrated in Fig. 1.9. The 3.1 % projected annual growth rate for coal-fired electricity generation worldwide is

---

[1] The Organisation for Economic Co-operation and Development (OECD) is an international organisation of thirty countries that accept the principles of representative democracy and free-market economy.

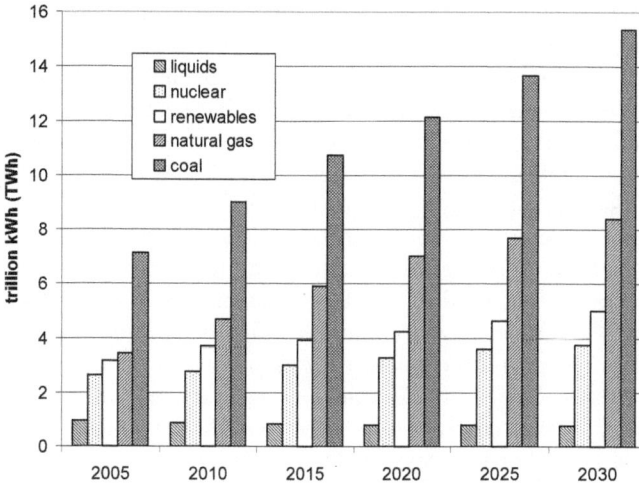

Fig. 1.9. World energy generation by fuel, 2005–2030 [158].

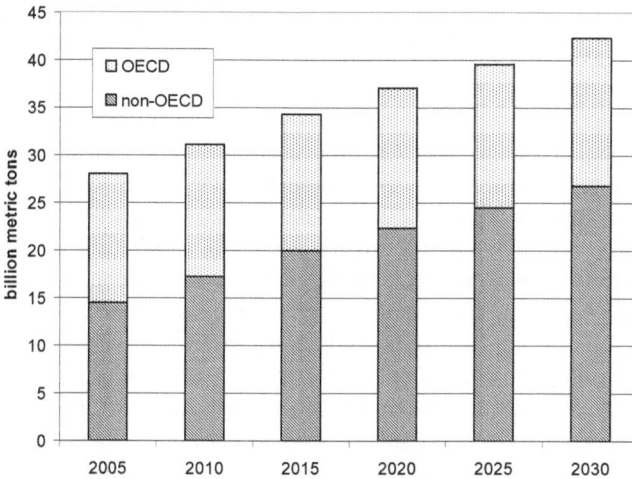

Fig. 1.10. World energy–related carbon dioxide emissions, 2005–2030 [158].

exceeded only by the 3.7-percent growth rate projected for natural-gas-fired generation [158].

The industrial sector, in developed countries, uses more than 30% of the electrical energy. More than 65% of this electrical energy is consumed by electric motor drives. The number of installed electrical machines can be estimated on the basis of their world production.

The increasing electrical energy demand causes tremendous concern for environmental pollution (Fig. 1.10). The power plants using fossil and nuclear fuel and road vehicles with combustion engines are main contributors to *air pollution, acid rain,* and the *greenhouse effect.* There is no doubt that electric propulsion and energy savings can improve these side effects considerably. For example, the population of Japan is about 50% of that of the U.S. However, carbon emmission is four times less (400 million metric ton in Japan versus over 1550 million metric ton in the U.S. in 2005). Mass public transport in Japan based on modern electrical commuter and long distance trains network plays an important part in reduction of carbon emission. It has been estimated that in developed industrialized countries, roughly 10% of electrical energy can be saved by using more efficient control strategies for electromechanical drives. This means that *electrical machines have an enormous influence on the reduction of energy consumption.* Electrical energy consumption can be saved in one of the following ways [212]:

- good housekeeping
- use of variable-speed drives
- construction of electric motors with better efficiency

Good housekeeping measures are inexpensive, quick and easy to implement. The simplest way to save energy costs is to switch idling motors off. Motors can be switched off manually or automatically. Devices exist that use either the input current to the motor or limit switches to detect an idling motor. When larger motors are being switched off and on, the high starting current drawn by the motor could cause supply interference and mechanical problems with couplings, gearboxes, belts, etc., which deteriorate from repeated starting. These problems can be avoided by using electronic *solid state converters.*

Fan and pump drives employ over 50% of motors used in industry. Most fans and pumps use some form of flow control in an attempt to match supply with demand. Traditionally, mechanical means have been used to restrict the flow, such as a damper on a fan or a throttle valve on a pump. Such methods waste energy by increasing the resistance to flow and by running the fan or pump away from its most efficient point. A much better method is to use a VSD to alter the speed of the motor. For centrifugal fans and pumps the power input is proportional to the cube of the speed, while the flow is proportional to the speed. Hence, a reduction to 80% of maximum speed (flow) will give a potential reduction in power consumption of 50% [212].

The application of PMs to electrical machines improves their efficiency by eliminating the excitation losses. The air gap magnetic flux density increases, which means greater output power for the same main dimensions.

A 3% increase in motor efficiency can save 2% of energy used [212]. Most energy is consumed by three-phase induction motors rated at below 10 kW. Consider a small three-phase, four-pole, 1.5-kW, 50-Hz cage induction motor. The full load efficiency of such a motor is usually 78%. By replacing this motor with a rare-earth PM brushless motor the efficiency can be increased to 89%. This means that the three-phase PM brushless motor draws from the mains only 1685 W instead of 1923 W drawn by the three phase cage induction motor. The power saving is 238 W per motor. If in a country, say, one million such motors are installed, the reduction in power consumption will be 238 MW, or one quite large turboalternator can be disconnected from the power system. It also means a reduction in CO_2 and NO_x emitted into the atmosphere if the energy is generated by thermal power plants.

1.4 Classification of permanent magnet electric motors

In general, rotary PM motors for continuous operation are classified into:

- d.c. brush commutator motors
- d.c. brushless motors
- a.c. synchronous motors

The construction of a PM d.c. commutator motor is similar to a d.c. motor with the electromagnetic excitation system replaced by PMs. PM d.c. brushless and a.c. synchronous motor designs are practically the same: with a polyphase stator and PMs located on the rotor. The only difference is in the control and shape of the excitation voltage: an a.c. synchronous motor is fed with more or less sinusoidal waveforms which in turn produce a rotating magnetic field. In PM d.c. brushless motors the armature current has a shape of a square (trapezoidal) waveform, only two phase windings (for Y connection) conduct the current at the same time, and the switching pattern is synchronized with the rotor angular position (electronic commutation).

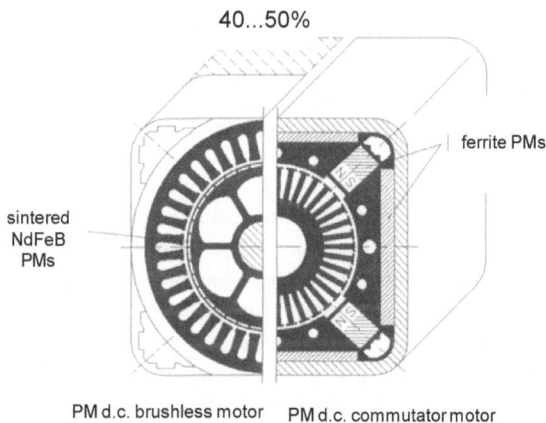

PM d.c. brushless motor   PM d.c. commutator motor

Fig. 1.11. Comparison of PM brushless and PM d.c. commutator motors.

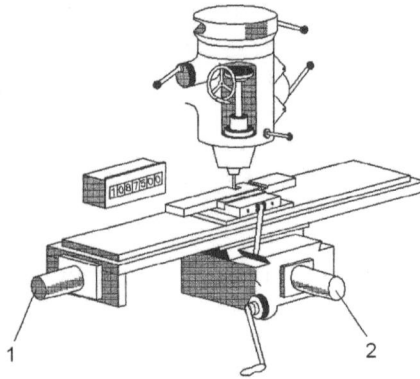

Fig. 1.12. Machine tool for milling grooves across a steel bar stock: 1 — PM brushless servo motor for indexing the stock, 2 — stepping motor for controlling the mill stroke. Courtesy of *Parker Hannifin Corporation*, Rohnert Park, CA, U.S.A.

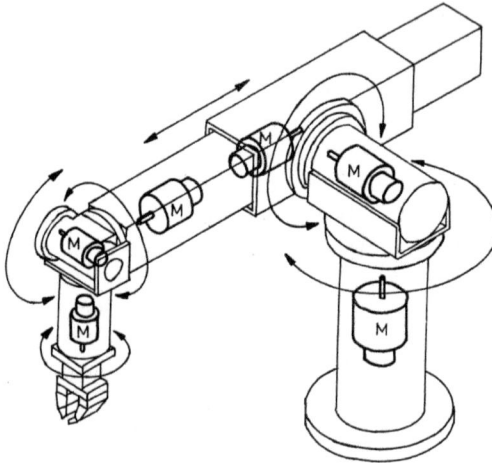

Fig. 1.13. An industrial robot (M — electric motor).

The armature current of synchronous and d.c. brushless motors is not transmitted through brushes, which are subject to wear and require maintenance. Another advantage of the brushless motor is the fact that the power losses occur in the stator, where heat transfer conditions are good. Consequently the power density can be increased as compared with a d.c. commutator motor. In addition, considerable improvements in dynamics can be achieved because the air gap magnetic flux density is high, the rotor has a lower inertia and there are no speed-dependent current limitations. Thus, the volume of a brushless PM motor can be reduced by 40 to 50% while still keeping the same rating as that of a PM commutator motor [75] (Fig. 1.11).

The following constructions of PM d.c. comutator motors have been developed:

- motors with conventional slotted rotors
- motors with slotless (surface-wound) rotors
- motors with moving coil rotors:
    (a) outside field type
    – cylindrical
    – wound disk rotor
    – printed circuit disk rotor
    (b) inside field type with cylindrical rotor
    – honeycomb armature winding
    – rhombic armature winding
    – bell armature winding
    – ball armature winding

The PM a.c. synchronous and d.c. brushless motors (moving magnet rotor) are designed as:

- motors with conventional slotted stators,
- motors with slotless (surface-wound) stators,
- cylindrical type:
    – surface magnet rotor (uniform thickness PMs, bread loaf PMs)
    – inset magnet rotor
    – interior magnet rotor (single layer PMs, double layer PMs)
    – rotor with buried magnets symmetrically distributed
    – rotor with buried magnets asymmetrically distributed
- disk type:
    (a) single-sided
    (b) double-sided
    –   with internal rotor
    –   with internal stator (armature)

The stator (armature) winding of PM brushless motors can be made of coils distributed in slots, concentrated non-overlapping coils or slotless coils.

1.5 Trends in permanent magnet motors and drives industry

The electromechanical drives market analysis shows that the d.c. commutator motor drive sales increase only slightly each year while the demand for a.c. motor drives increases substantially [288]. A similar tendency is seen in the PM d.c. commutator motor drives and PM brushless motor drives.

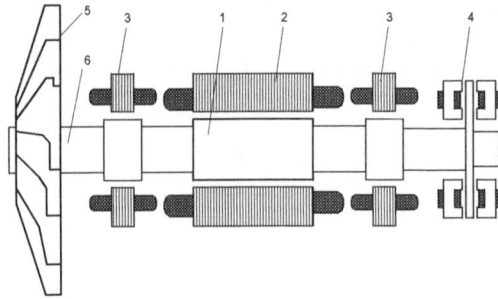

Fig. 1.14. Centrifugal chiller compressor with high speed PM brushless motor and magnetic bearings: 1 — rotor with surface PMs and nonmagnetic can, 2 — stator, 3 — radial magnetic bearing, 4 — axial magnetic bearing, 5 — impeller, 6 — shaft.

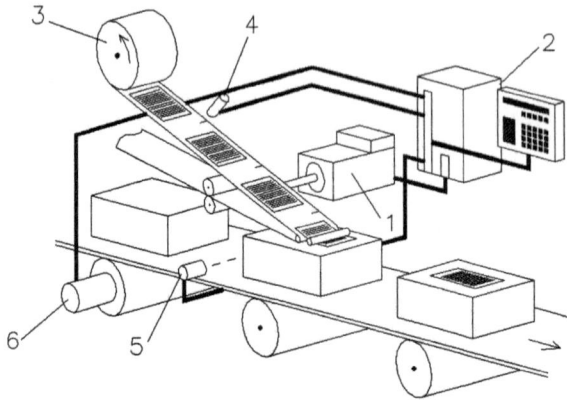

Fig. 1.15. Automatic labelling systems with a PM brushless motor: 1 — PM brushless motor servo drive, 2 — controller, 3 — spool of self-adhesive labels, 4 — registration mark sensor, 5 — box position sensor, 6 — conveyor speed encoder. Courtesy of *Parker Hannifin Corporation*, Rohnert Park, CA, U.S.A.

Small PM motors are especially demanded by manufacturers of computer hardware, automobiles, office equipment, medical equipment, instrumentation for measurements and control, robots, and handling systems. The 2002 world production of PM motors was estimated to be 4.68 billion units with a total value of U.S.$ 38.9 billion. Commutator motors account for 74.8% (3,500 million units), brushless motors account for 11.5% (540 million units) and stepping motors account for 13.7% (640 million units). From today's perspective, the Far East (Japan, China, and South Korea), America and Europe will remain the largest market area.

Advances in electronics and PM quality have outpaced similar improvements in associated mechanical transmission systems, making ball lead screws and gearing the limiting factors in motion control. For the small motor busi-

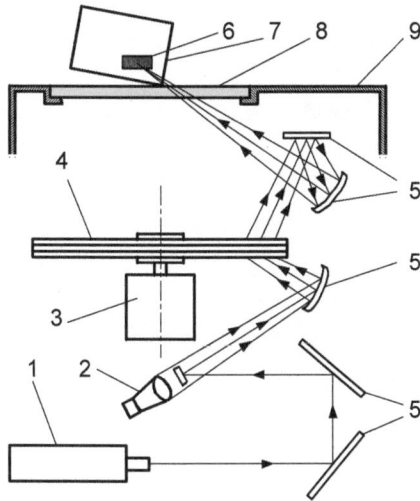

Fig. 1.16. Bar code scanner: 1 — laser, 2 — photodecoder converting laser beam into electric signal, 3 — PM brushless motor, 4 — holographic three-layer disk, 5 — mirror, 6 — bar code, 7 — scanned object, 8 — scan window, 9 — housing.

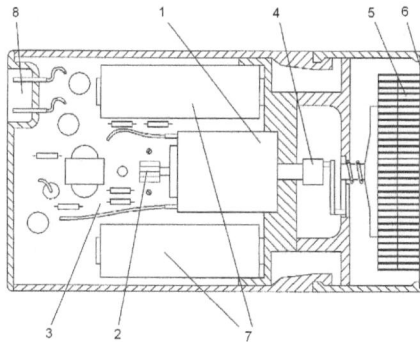

Fig. 1.17. Electric shaver with a PM brushless motor: 1 — PM brushless motor, 2 — position sensor, 3 — printed circuit board, 4 — cam-shaft, 5 — twizzer head, 6 — platinum-coated shaving foil, 7 — rechargeable battery, 8 — input terminals 110/220 V.

ness, a substantially higher integration of motor components will increasingly help to bridge this gap in the future [211]. However, there is always the question of cost analysis, which ultimately is the key factor for specific customer needs.

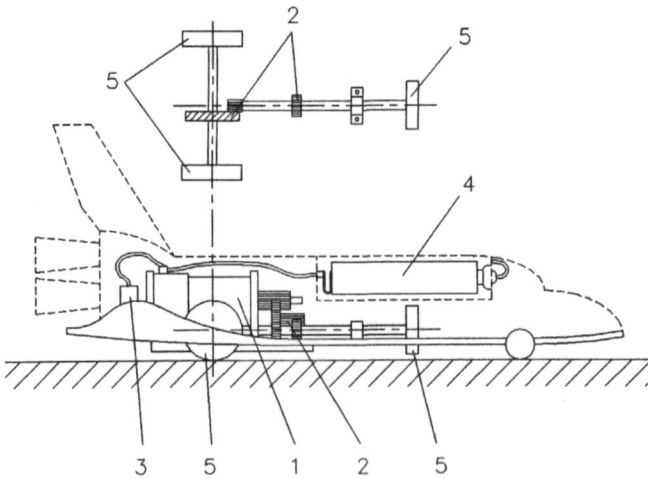

Fig. 1.18. Toy space shuttle: 1 — PM d.c. commutator motor, 2 — transmission, 3 — on–off switch, 4 — 1.5-V battery, 5 — driven wheels.

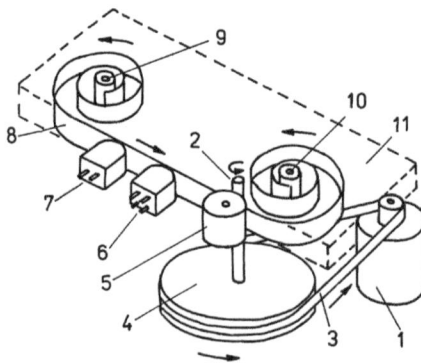

Fig. 1.19. Cassette deck: 1 — PM motor, 2 — capstan, 3 — belt, 4 — flywheel, 5 — pressure roller, 6 — rec/play head, 7 — erase head, 8 — tape, 9 — supply reel table, 10 — take-up reel table, 11 — cassette.

1.6 Applications of permanent magnet motors

PM motors are used in a broad power range from mWs to hundreds kWs. There are also attempts to apply PMs to large motors rated at minimum 1 MW. Thus, PM motors cover a wide variety of application fields, from stepping motors for wrist watches, through industrial drives for machine tools to large PM synchronous motors for ship propulsion (navy frigates, cruise ships, medium size cargo vessels and ice breakers) [109, 124, 278]. The application of PM electric motors includes:

- Industry (Figs 1.12, 1.14 1.13 and 1.15):
  - industrial drives, e.g., pumps, fans, blowers, compressors (Fig. 1.14), centrifuges, mills, hoists, handling systems, etc.
  - machine tools
  - servo drives
  - automation processes
  - internal transportation systems
  - robots
- Public life:
  - heating, ventilating and air conditioning (HVAC) systems
  - catering equipment
  - coin laundry machines
  - autobank machines
  - automatic vending machines
  - money changing machines
  - ticketing machines
  - bar code scanners at supermarkets (Fig. 1.16)
  - environmental control systems
  - clocks
  - amusement park equipment
- Domestic life (Fig. 1.17, 1.18 and 1.19):
  - kitchen equipment (refrigerators, microwave ovens, in-sink garbage disposers, dishwashers, mixers, grills, etc.)
  - bathroom equipment (shavers, hair dryers, tooth brushes, massage apparatus)
  - washing machines and clothes dryers
  - HVAC systems, humidifiers and dehumidifiers
  - vacuum cleaners
  - lawn mowers
  - pumps (wells, swimming pools, jacuzzi whirlpool tubs)
  - toys
  - vision and sound equipment
  - cameras
  - cellular phones
  - security systems (automatic garage doors, automatic gates)
- Information and office equipment (Figs. 1.20 and 1.21):
  - computers [161, 163]
  - printers
  - plotters
  - scanners
  - facsimile machines
  - photocopiers
  - audiovisual aids
- Automobiles with combustion engines (Fig. 1.22);

- Transportation (Figs. 1.23, 1.24, 1.25 and 1.27):
  - elevators and escalators
  - people movers
  - light railways and streetcars (trams)
  - electric road vehicles
  - aircraft flight control surface actuation
  - electric ships
  - electric boats
  - electric aicrafts (Fig. 1.27
- Defense forces (1.26):
  - tanks
  - missiles
  - radar systems
  - submarines
  - torpedos
- Aerospace:
  - rockets
  - space shuttles
  - satellites
- Medical and healthcare equipment:
  - dental handpieces (dentist's drills)
  - electric wheelchairs
  - air compressors
  - trotters
  - rehabilitation equipment
  - artificial heart motors
- Power tools (Fig. 1.28):
  - drills
  - hammers
  - screwdrivers
  - grinders
  - polishers
  - saws
  - sanders
  - sheep shearing handpieces [256]
- Renewable energy systems (Fig. 1.29)
- Research and exploration equipment (Fig. 1.30)

The automotive industry is the biggest user of PM d.c. commutator motors. The number of auxiliary d.c. PM commutator motors can vary from a few in an inexpensive car to about one hundred in a luxury car [175].

Small PM brushless motors are first of all used in computer hard disk drives (HDDs) and cooling fans. The 2002 worldwide production of computers is estimated to be 200 million units and production of HDDs approximately 250 million units.

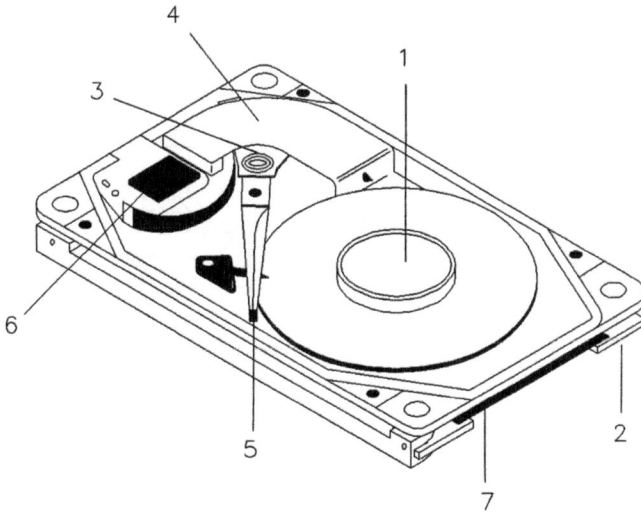

Fig. 1.20. Computer hard disk drive (HDD) with a PM brushless motor: 1 — in-hub PM brushless motor mounted in 2.5-inch disk, 2 — integrated interface/drive controller, 3 — balanced moving-coil rotary actuator, 4 — actuator PM, 5 — read/write heads, 6 — read/write preamplifier, 7 — 44-pin connector.

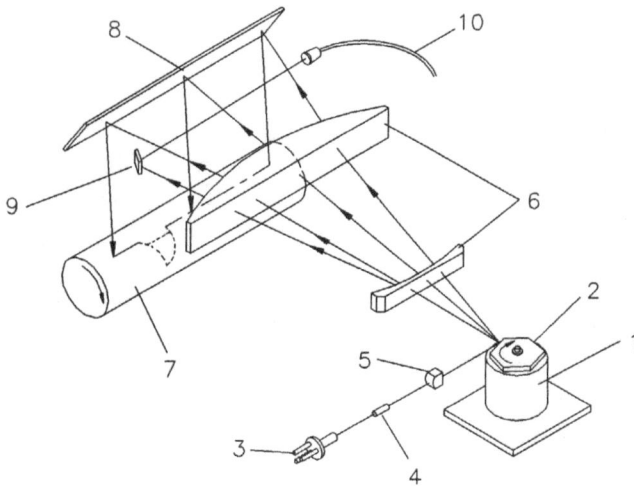

Fig. 1.21. Laser beam printer: 1 — stepping motor, 2 — scanner mirror, 3 — semiconductor laser, 4 — collimator lens, 5 — cylindrical lens, 6 — focusing lenses, 7 — photosensitive drum, 8 — mirror, 9 — beam detect mirror, 10 — optical fiber.

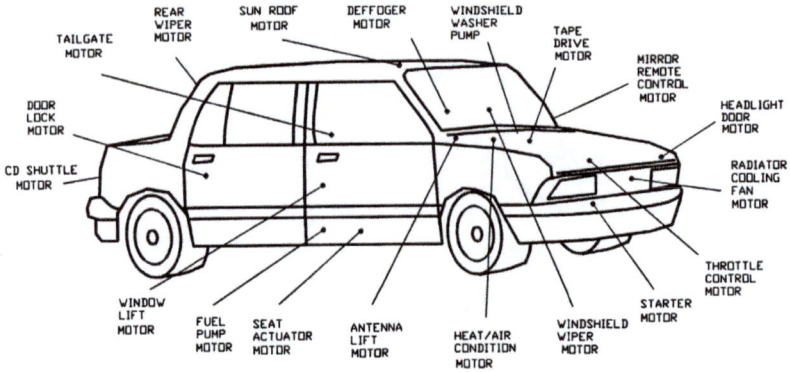

Fig. 1.22. PM motors installed in a car.

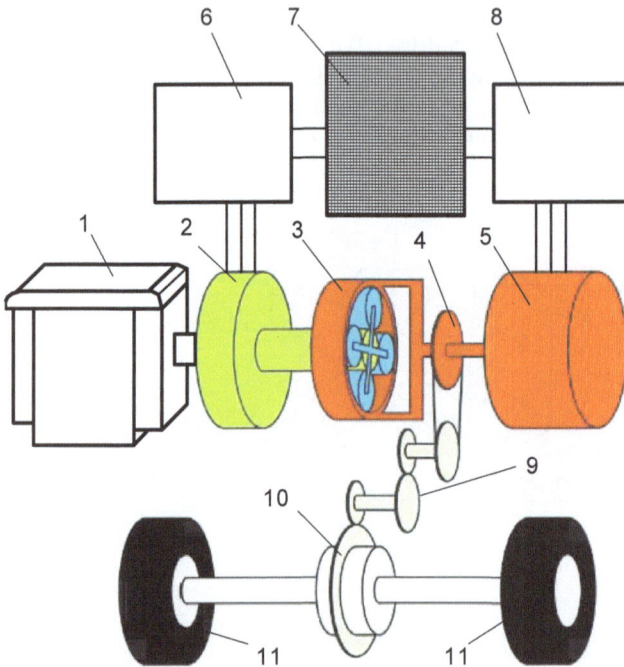

Fig. 1.23. Power train of Toyota Prius hybrid electric vehicle (HEV):1 — gasoline engine, 2 — PM brushless generator/starter (GS), 3 — power split device (PSD), 4 — silent chain, 5 — PM brushless motor/generator (MG), 6 — GS solid state converter, 7 — battery, 8 — MG solid state converter, 9 — reduction gears, 10 — differential, 11 — front wheels.

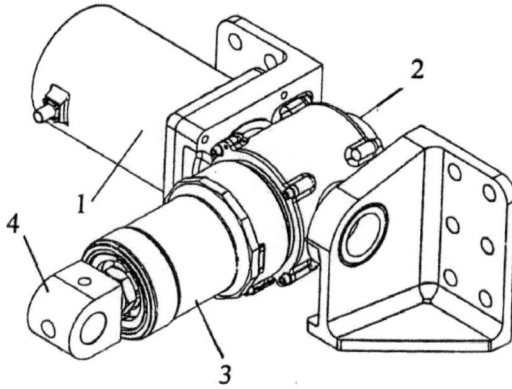

Fig. 1.24. Electromechanical actuator for flight control surfaces: 1 — brushless motor, 2 — gearbox, 3 — ball lead screw, 4 — clevis or spherical joint end.

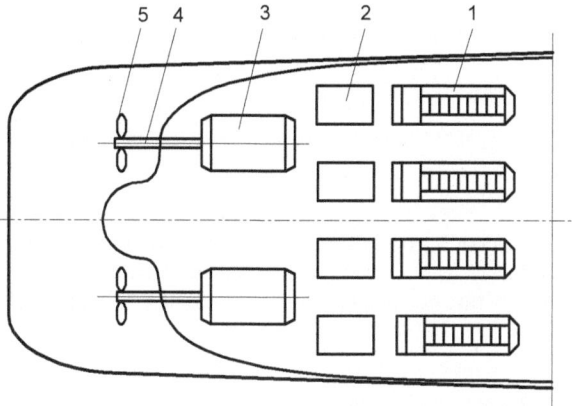

Fig. 1.25. Ship propulsion system with a PM brushless motor: 1 — diesel engine and synchronous generator, 2 — converter, 3 — large PM brushless motor, 4 — propeller shaft, 5 — propeller.

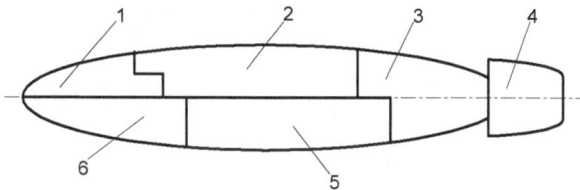

Fig. 1.26. Stealth torpedo: 1 — guiding system, 2 — conformal acoustic arrays, 3 — advanced rechargeable batteries, 4 — integrated PM brushless motor propulsor, 5 — active and passive noise control, 6 — synergic drag reduction.

Fig. 1.27. *Taurus Electro* two-seat self launching glider with a 30 kW, 1800 rpm, 15.8 kg PM brushless motor. Photo courtesy of Pipistrel, Ajdovscina, Slovenia.

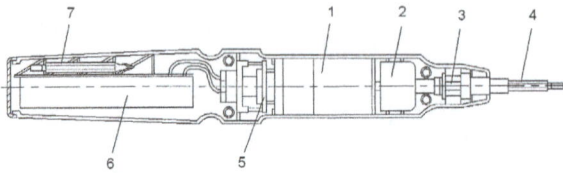

Fig. 1.28. Cordless electric screwdriver: 1 — PM d.c. commutator motor 3.6 V/240 mA, 2 — speed reducer, 3 — locker for manual screwing, 4 — screwdriver bit, 5 — forward-reverse rotation switch, 6 — rechargeable Ni-Cd battery, 7 — bit compartment.

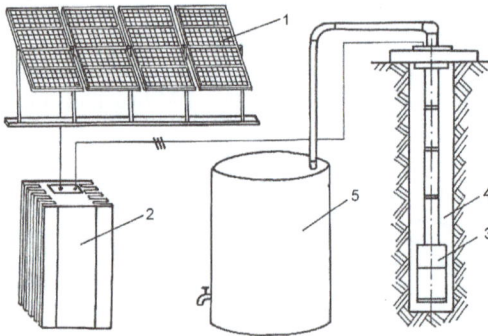

Fig. 1.29. Water pumping system for a remote population center: 1 — solar panels, 2 — inverter, 3 — submersible PM brushless motor–pump unit, 4 — well, 5 — water storage tank.

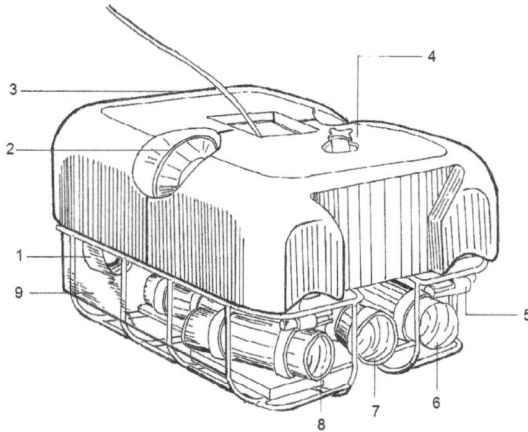

Fig. 1.30. Underwater robotic vehicle for remote inspection of hydro power plants: 1 — two thruster sets for forward and reverse (200-W d.c. brushless motors), 2 — two thruster sets for lateral and vertical (200-W d.c. brushless motors), 3 — buoyancy material, 4 — transponder, 5 — flood light, 6 — still camera, 7 — video camera, 8 — electric flash, 9 — cover. Courtesy of Mitsui Ocean Development and Eng. Co., Tokyo, Japan.

PM brushless motors rated from 50 to 100 kW seem to be the best propulsion motors for electric and hybrid road vehicles.

Given below are some typical applications of PM motors in industry, manufacturing processes, factory automation systems, domestic life, computers, transportation and clinical engineering:

- *Industrial robots* and x, y-*axis coordinate machines*: PM brushless motors
- *Indexing rotary tables*: PM stepping motors
- *X–Y tables*, e.g. for milling grooves across steel bars: PM brushless servo motors
- *Linear actuators with ball or roller screws*: PM brushless and stepping motors
- *Transfer machines for drilling a number of holes*: ball lead screw drives with PM brushless motors
- *Monofilament nylon winders*: PM d.c. commutator motor as a torque motor and PM brushless motor as a traverse motor (ball screw drive)
- *Mobile phones*: PM d.c. commutator or brushless vibration motors
- *Bathroom equipment*: PM commutator or brushless motors
- *Toys*: PM d.c. commutator motors
- *Computer hard disk drives* (HDD): PM brushless motors
- *Computer printers*: PM stepping motors
- *Cooling fans for computers and instruments*: PM brushless motors

- *Auxiliary motors for automobiles*: PM d.c. commutator and PM brushless motors
- *Gearless elevators*: PM brushless motors
- *Electric and hybrid electric vehicles* (EV and HEV): PM brushless motors of cylindrical or disk type
- *Ship propulsion*: large PM brushless motors or transverse flux motors (above 1 MW)
- *Submarine periscope drives*: direct-drive PM d.c. brushless torque motors
- *More electric aircraft* (MEE): PM brushless motors
- *Dental and surgical handpieces*: slotless PM brushless motors
- *Implantable blood pumps*: PM brushless motors integrated with impellers.

1.7 Mechatronics

A new technology called *mechatronics* emerged in the late 1970s. Mechatronics is the intelligent integration of mechanical engineering with microelectronics and computer control in product design and manufacture to give improved performance and cost saving. Applications of mechatronics can be found in the aerospace and defense industries, in intelligent machines such as industrial robots, automatic guided vehicles, computer-controlled manufacturing machines and in consumer products such as computer hard disk drives (HDD), video cassette players and recorders, cameras, CD players and quartz watches.

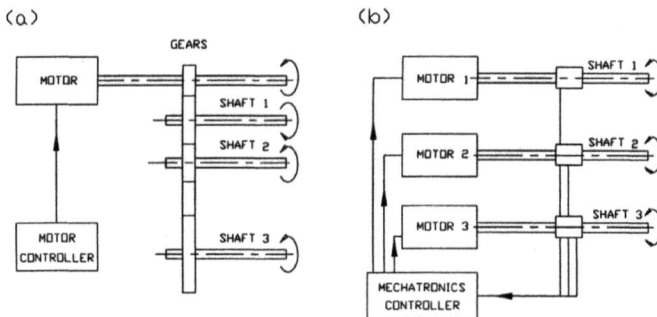

Fig. 1.31. Gear train design: (a) conventional system, (b) mechatronics system.

A typical example of a novel mechatronics application is in the control of multi-shaft motion. A gear train has traditionally been employed with the performance, i.e. speed, torque and direction of rotation determined by the motor and gear rated parameters as shown in Fig. 1.31a. Such a configuration is acceptable for constant speed of each shaft but where variable speeds are required, a different set of gears is needed for each gear ratio. In the mechatronics solution (Fig. 1.31b) each shaft is driven by an electronically controlled motor, e.g. a PM brushless motor with feedback which provides more flexibility than can be obtained from mechanical gear trains. By adding a microprocessor or microcomputer, any required motion of the mechanism can be programmed by software. The common term for this type of control system is *mechatronics control system* or *mechatronics controller*. The "electronic gearbox" is more flexible, versatile and reliable than the mechanical gearbox. It also reduces acoustic noise and does not require maintenance.

1.8 Fundamentals of mechanics of machines

1.8.1 Torque and power

The shaft torque T as a function of mechanical power P is expressed as

$$T = F\frac{D}{2} = \frac{P}{\Omega} = \frac{P}{2\pi n} \tag{1.1}$$

where $\Omega = 2\pi n$ is the angular speed and n is the rotational speed in rev/s.

Table 1.1. Basic formulae for linear and rotational motions

Linear motion			Rotational motion		
Quantity	Formula	Unit	Quantity	Formula	Unit
Linear displacement	$s = \theta r$	m	Angular displacement	θ	rad
Linear velocity	$v = ds/dt$ $v = \Omega r$	m/s	Angular velocity	$\Omega = d\theta/dt$	rad/s
Linear acceleration	$a = dv/dt$ $a_t = \alpha r$ $a_r = \Omega^2 r$	m/s²	Angular acceleration	$\alpha = d\Omega/dt$	rad/s²
Mass	m	kg	Moment of inertia	J	kgm²
Force	$F = mdv/dt$ $= ma$	N	Torque	$T = Jd\Omega/dt$ $= J\alpha$	Nm
Friction force	Dds/dt $= Dv$	N	Friction torque	$Dd\theta/dt$ $= D\Omega$	Nm
Spring force	Ks	N	Spring torque	$K\theta$	Nm
Work	$dW = Fds$	Nm	Work	$dW = Td\theta$	Nm
Kinetic energy	$E_k = 0.5mv^2$	J or Nm	Kinetic energy	$E_k = 0.5J\Omega^2$	J
Power	$P = dW/dt$ $= Fv$	W	Power	$P = dW/dt$ $= T\Omega$	W

1.8.2 Simple gear trains

In the simple trains shown in Fig. 1.32, let n_1, n_2 = speeds of 1 and 2, z_1 and z_2 = numbers of teeth on 1 and 2, D_1, D_2 = pitch circle diameters of 1 and 2.

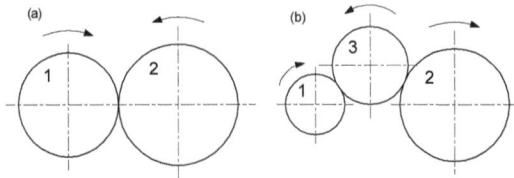

Fig. 1.32. Simple trains.

- in train according to Fig. 1.32a

$$\gamma = \frac{n_1}{n_2} = -\frac{z_2}{z_1} = -\frac{D_2}{D_1} \tag{1.2}$$

- in train according to Fig. 1.32b

$$\gamma = \frac{n_1}{n_2} = \frac{z_2}{z_1} = \frac{D_2}{D_1} \tag{1.3}$$

The negative sign signifies that 1 and 2 rotate in opposite directions. The idler, 3, Fig. 1.32b, does not affect the velocity ratio of 1 to 2 but decides on the directions of 2. The ratio $\gamma = z_2/z_1$ is called the *gear ratio*.

1.8.3 Efficiency of a gear train

Allowing for friction, the efficiency of a gear train is

$$\eta = \frac{\text{output power}}{\text{input power}} = \frac{P_2}{P_1} \tag{1.4}$$

Thus,

$$\eta = \frac{P_2}{P_1} = \frac{T_2(2\pi n_2)}{T_1(2\pi n_1)} = \frac{T_2 n_2}{T_1 n_1} \tag{1.5}$$

According to eqn (1.2) $n_2/n_1 = |z_1/z_2|$, so that eqn (1.5) becomes

$$\frac{T_2 n_2}{T_1 n_1} = \frac{T_2 z_1}{T_1 z_2}$$

The torque on wheel 1

$$T_1 = T_2 \frac{z_1}{z_2} \frac{1}{\eta} \tag{1.6}$$

1.8.4 Equivalent moment of inertia

In the simple trains shown in Fig. 1.32a, let J_1, J_2 = moments of inertia of rotating masses of 1 and 2, Ω_1 and Ω_2 = angular speed of 1 and 2, D_1, D_2 = pitch circle diameters of 1 and 2, $0.5J_1\Omega_1^2$, $0.5J_2\Omega_2^2$ = kinetic energy of 1 and 2, respectively.

The net energy supplied to a system in unit time is equal to the rate of change of its kinetic energy E_k (Table 1.1), i.e.

$$P = \frac{dE_k}{dt} = T\Omega_1$$

$$T\Omega_1 = \frac{d}{dt}[0.5J_1\Omega_1^2 + 0.5J_2\Omega_2^2] = 0.5\left(J_1 + \frac{\Omega_2^2}{\Omega_1^2}J_2\right) \times \frac{d}{dt}\Omega_1^2 \qquad (1.7)$$

$$= 0.5\left(J_1 + \frac{\Omega_2^2}{\Omega_1^2}J_2\right) \times 2\Omega_1\frac{d\Omega_1}{dt} \qquad (1.8)$$

The quantity $J_1 + (\Omega_2/\Omega_1)^2 J_2$ may be regarded as the equivalent moment of inertia of the gears referred to wheel 1. The moments of inertia of the various gears may be reduced to an equivalent moment of inertia of the motor shaft, i.e.

$$T = \left(J_1 + \frac{\Omega_2^2}{\Omega_1^2}J_2\right)\frac{d\Omega_1}{dt} = \left(J_1 + \frac{z_1^2}{z_2^2}J_2\right)\frac{d\Omega_1}{dt} \qquad (1.9)$$

The equivalent moment of inertia is equal to the moment of inertia of each wheel in the train being multiplied by the square of its gear ratio relative to the reference wheel.

1.8.5 Rotor dynamics

All spinning shafts, even in the absence of external load, deflect during rotation. Fig. 1.33 shows a shaft with two rotating masses m_1 and m_2. The mass m_1 can represent a cylindrical rotor of an electric machine while the mass m_2 can represent a load. The mass of the shaft is m_{sh}. The combined mass of the rotor, load and shaft can cause deflection of the shaft that will create resonant vibration at a certain speed called *whirling* or *critical speed*. The frequency when the shaft reaches its critical speed can be found by calculating the frequency at which transverse vibration occurs. The critical speed in rev/s of the ith rotating mass can be found as [258]

$$n_{cri} = \frac{1}{2\pi}\sqrt{\frac{K_i}{m_i}} = \frac{1}{2\pi}\sqrt{\frac{g}{\sigma_i}} \qquad (1.10)$$

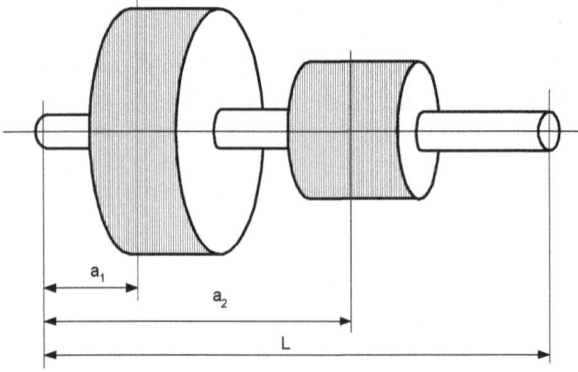

Fig. 1.33. Solid cylindrical shaft loaded with two masses m_1 and m_2.

where K_i is the stiffness of the ith rotor in N/m, m_i is the mass of the ith rotor in kg, $g = 9.81$ m/s^2 is the acceleration due to gravity and δ_i is the static deflection at the ith position of the rotor due to the ith rotor only, i.e.,

$$\sigma_i = \frac{m_i g a_i^2 (L - a_i)^2}{3 E_i I_i L} \tag{1.11}$$

In the above equation E_i is the modulus of elasticity (for steel $E = 200 \times 10^9$ Pa), I_i is the area moment of inertia of cross-sectional area, L is the length of the shaft and a_i is the location of the ith rotor from the left end of the shaft (Fig. 1.33). The area moment of inertia can be found as

$$I_i = \frac{\pi D_i^4}{64} \tag{1.12}$$

The resultant angular critical speed $\Omega_{cr} = 2\pi n_{cr}$ for the shaft loaded with i number of rotors ($\Omega_{cri} = 2\pi n_{cri}$) can be found on the basis of Dunkerley's equation [88] as

$$\frac{1}{\Omega_{cr}^2} = \sum_i \frac{1}{\Omega_{cri}^2} \tag{1.13}$$

or Rayleigh's equation [258]

$$\Omega_{cr} = \sqrt{\frac{g \sum_i (m_i \sigma_i)}{\sum_i (m_i \sigma_i^2)}} \tag{1.14}$$

The shaft is also considered a rotor with mass m_{sh} concentrated at $0.5L$ where L is the length of the shaft (bearing–to–bearing).

Dunkerley's empirical method uses the frequencies that each individual load creates when each load acts alone and then combines them to give an approximation for the whole system [88]. Thus eqn (1.13) is an approximation to the *first natural frequency* of vibration of the system which is assumed nearly equal to the critical speed of rotation. Rayleigh's method is based on the fact that the maximum kinetic energy must be equal to maximum potential energy for a conservative system under free vibration [258].

1.8.6 Mechanical characteristics of machines

In general, the mechanical characteristic $T = f(\Omega)$ of a machine driven by an electric motor can be described by the following equation:

$$T = T_r \left(\frac{\Omega}{\Omega_r}\right)^\beta \tag{1.15}$$

where T_r is the resisting torque of the machine at rated angular speed Ω_r, $\beta = 0$ for hoists, belt conveyors, rotating machines and vehicles (constant torque machines), $\beta = 1$ for mills, callanders, paper machines and textile machines, $\beta = 2$ for rotary pumps, fans, turbocompressors and blowers.

1.9 Torque balance equation

An electromechanical system can simply be described with the aid of the following torque balance equation,

$$J\frac{d^2\theta}{dt^2} + D\frac{d\theta}{dt} + K\theta = \pm T_d \mp T_{sh} \tag{1.16}$$

where J is the *moment of inertia* of the system in kgm^2 assumed as constant ($dJ/dt = 0$), D is the *damping coefficient* in Nm s/rad, K is the *stiffness coefficient* or *spring constant* in Nm/rad, T_d is the instantaneous electromagnetic torque developed by the motor, T_{sh} is the instantaneous external (shaft) passive load torque, θ is the rotor angular displacement, $T_d > T_{sh}$ for acceleration, $T_d < T_{sh}$ for deceleration. Assuming $D = 0$ and $K = 0$ the torque balance equation (1.16) becomes

$$J\frac{d^2\theta}{dt^2} \approx \pm T_d \mp T_{sh} \tag{1.17}$$

1.10 Evaluation of cost of a permanent magnet motor

The cost of an electrical machine is a function of a large number of variables. The cost can be evaluated only approximately because it depends on:

- number of electrical machines of the same type manufactured per year
- manufacturing equipment (how modern is the equipment, level of automation, production capacity per year, necessary investment, etc.)

- organization of production process (engineering staff–to–administrative and supporting staff ratio, qualification and experience of technical management, overhead costs, productivity of employees, small company or large corporation, company culture, etc.)
- cost of labor (low in third world countries, high in North America, Europe and Japan)
- quality of materials (good quality materials cost more) and many other aspects

It is impossible to take into account all these factors in a general mathematical model of costs. A logical approach is to select the most important components of the total cost and express them as functions of dimensions of the machine [186].

The most important costs of an electrical machine can be expressed by the following approximate equation:

$$C = k_N(C_w + C_c + C_{PM} + C_{sh} + C_0) \qquad (1.18)$$

where $k_N \leq 1$ is the coefficient depending on the number of manufactured machines per annum, C_w is the cost of winding, C_c is the cost of ferromagnetic core and components dependent on the size of core (frame, end disks, bearings, etc), C_{PM} is the cost of PMs, C_{sh} is the cost of shaft and C_0 is the cost of all other components independent of the shape of the machine, e.g. nameplate, encoder, terminal board, terminal leads, commutator in d.c. brush machine, etc.

The cost of winding is [186]

$$C_w = k_{sp}k_{ii}k_{sr}\rho_{Cu}c_{Cu}V_{sp} \qquad (1.19)$$

where $k_{sp} < 1$ is the slot space (fill) factor, $k_{ii} > 1$ is the coefficient of the cost of fabrication of coils including placing in slots, insulation, impregnation, etc., $k_{sr} \geq 1$ is the cost of the stator and rotor winding–to–the cost of the stator winding ratio (if the rotor winding exists, e.g., damper), ρ_{Cu} is the specific mass density of the conductor material (copper) in kg/m^3, c_{Cu} is the cost of conductor per kilogram and V_{sp} is the space designed for the winding and insulation in m^3.

The cost of a ferromagnetic core consists of the cost of laminated parts C_{cl} and other material parts, e.g., sintered powder parts C_{csp}, i.e.,

$$C_c = k_p(C_{cl} + C_{csp}) \qquad (1.20)$$

where $k_p > 1$ is the coefficient accounting for the cost $\sum C_{ci}$ of all machine parts dependent on the dimensions of the stator core (frame, end plates, bearings, etc.) expressed as

$$k_p = 1 + \frac{\sum C_{ci}}{C_c} \qquad (1.21)$$

The cost of laminated core

$$C_{cl} = k_u k_i k_{ss} \rho_{Fe} c_{Fe} \frac{\pi D_{out}^2}{4} \sum L \qquad (1.22)$$

where $k_u > 1$ is the coefficient of utilization of electrotechnical steel sheet or strip (total surface of sheet/strip corresponding to single lamination to the surface of the lamination), $k_i < 1$ is the stacking (insulation) factor, $k_{ss} > 1$ is the coefficient accounting for the cost of stamping, stacking and other operations, ρ_{Fe} is the specific mass density of electrotechnical steel, c_{Fe} is the cost of electrotechnical sheet steel per kilogram, D_{out} is the outer diameter of the core and $\sum L$ is the total length of laminated stacks (the stack can be divided into segments).

The cost of sintered powder core

$$C_{csp} = k_{sh} \rho_{sp} c_{sp} V_{sp} \qquad (1.23)$$

where $k_{sh} > 1$ is the coefficient accounting for the increase of the cost of sintered powder part dependent on the complexity of its shape (related to a simple shape, e.g., a cube), ρ_{sp} is the specific mass density of the sintered powder material, c_{sp} is the cost of sintered powder material per kilogram, V_{sp} is the volume of the sintered powder part. The cost of the solid steel core can be calculated in a similar way.

The cost of PMs is

$$C_{PM} = k_{shPM} k_{magn} \rho_{PM} c_{PM} V_M \qquad (1.24)$$

where $k_{shPM} > 1$ is the coefficient accounting for the increase of the cost of PMs due complexity of their shape (related to a simple shape, e.g., a cube), $k_{magn} > 1$ is the coefficient taking into account the cost of magnetization of PMs, ρ_{PM} is the specific mass density of the PM material, c_{PM} is the cost of PM material per kilogram and V_M is the volume of PMs.

The cost of the shaft

$$C_{sh} = k_{ush} k_m \rho_{steel} c_{steel} V_{sh} \qquad (1.25)$$

where $k_{ush} > 1$ is the coefficient of utilization of the round steel bar (total volume of the steel bar to the volume of the shaft), $k_m > 1$ is the coefficient accounting for the cost of machining, ρ_{steel} is the specific mass density of steel, c_{steel} is the cost of steel bar per kilogram, V_{sh} is the shaft volume.

Numerical examples

Numerical example 1.1

Find the steady-state torque, output power, and shaft moment of inertia of an electric motor propelling a rolling mill as in Fig. 1.34. The speed of the

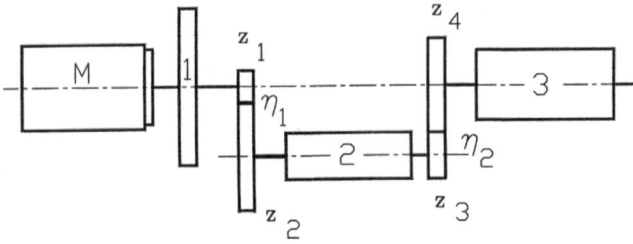

Fig. 1.34. Electric motor driven rolling mill. *Numerical example 1.1.*

motor is $n = 730$ rpm. The flywheel and rollers are made of steel with the specific mass density $\rho = 7800$ kg/m^3.

Solid steel flywheel 1: diameter $D_1 = 1.5$ m, thickness $l_1 = 0.2$ m.

First roller 2: diameter $D_2 = 0.4$ m, length $l_2 = 0.8$ m, circumferential force $F_2 = 20$ kN, number of teeth of the first gear $z_1 = 15$, $z_2 = 35$, efficiency of the first gear $\eta_1 = 0.87$.

Second roller 3: diameter $D_3 = 0.5$ m, length $l_3 = 1.2$ m, circumferential force $F_3 = 14$ kN, number of teeth of the second gear $z_3 = 20$, $z_4 = 45$, efficiency of the second gear $\eta_2 = 0.9$.

Solution

The shaft (load) torque according to eqn (1.6)

$$T_{sh} = F_2 \frac{D_2}{2} \frac{z_1}{z_2} \frac{1}{\eta_1} + F_3 \frac{D_3}{2} \frac{z_1}{z_2} \frac{z_3}{z_4} \frac{1}{\eta_1} \frac{1}{\eta_2} = 2.82 \text{ kNm}$$

where the gear ratio is $\gamma = z_2/z_1$.
    Output power of the motor

$$P_{out} = 2\pi n T_{sh} = 2\pi \times (730/60) \times 2820 = 216 \text{ kW}$$

The mass of flywheel

$$m_1 = \rho \frac{\pi D_1^2}{4} l_1 = 2757 \text{ kg}$$

The mass of the first roller

$$m_2 = \rho \frac{\pi D_2^2}{4} l_2 = 785 \text{ kg}$$

The mass of the second roller

$$m_3 = \rho \frac{\pi D_3^2}{4} l_3 = 1840 \text{ kg}$$

Moment of inertia of the flywheel

$$J_1 = m_1 \frac{D_1^2}{8} = 776 \text{ kgm}^2$$

Moment of inertia of the first roller

$$J_2 = m_2 \frac{D_2^2}{8} = 15.7 \text{ kgm}^2$$

Moment of inertia of the second roller

$$J_3 = m_3 \frac{D_3^2}{8} = 57.5 \text{ kgm}^2$$

The total moment of inertia of the system with respect to the motor shaft according to eqn (1.9)

$$J = J_1 + J_2 \left(\frac{z_1}{z_2} \right)^2 + J_3 \left(\frac{z_1 z_3}{z_2 z_4} \right)^2 = 781 \text{ kgm}^2$$

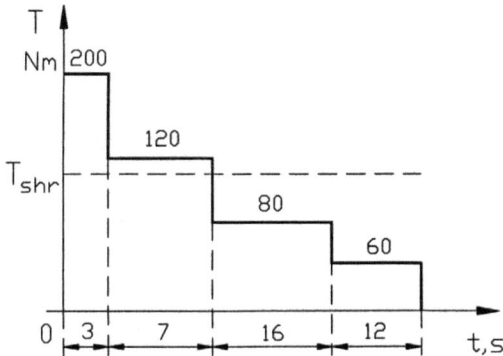

Fig. 1.35. Torque profile of the electric motor. *Numerical example 1.2.*

Numerical example 1.2

A 12-kW, 1000-rpm electric motor operates with almost constant speed according to the torque profile given in Fig. 1.35. The overload capacity factor $k_{ocf} = T_{max}/T_{shr} = 2$. Find the thermal utilization coefficient of the motor.

Solution

The required rated shaft torque

$$T_{shr} = \frac{P_{out}}{2\pi n} = \frac{12,000}{2\pi \times (1000/60)} = 114.6 \text{ Nm}$$

The rms torque based on the given duty cycle

$$T_{rms}^2(t_1 + t_2 + \ldots + t_n) = T_1^2 t_1 + T_2^2 t_2 + \ldots + T_n^2 t_n \quad \text{or} \quad T_{rms}^2 \sum t_i = \sum T_i^2 t_i$$

Thus

$$T_{rms} = \sqrt{\frac{\sum T_i^2 t_i}{\sum t_i}} = \sqrt{\frac{200^2 \times 3 + 120^2 \times 7 + 80^2 \times 16 + 60^2 \times 12}{3 + 7 + 16 + 12}}$$

$$= 95.5 \text{ Nm}$$

Note that in electric circuits the rms or effective current is

$$I_{rms} = \sqrt{\frac{1}{T} \int_0^T i^2 dt}$$

since the average power delivered to the resistor is $P = RI_{rms}^2$.

The maximum torque in Fig. 1.35 cannot exceed the rated shaft torque times overload capacity factor $k_{ocf} \times T_{shr}$. Also, the required T_{shr} should be greater than or equal to T_{rms}.

The coefficient of thermal utilization of the motor

$$\frac{T_{rms}}{T_{sh}} \times 100\% = \frac{95.5}{114.6} \times 100\% = 83.3\%$$

Numerical example 1.3

The required torque and speed profiles of a servo drive are given in Fig. 1.36. At constant speed 2500 rpm the load shaft torque is $T_{sh} = 1.5$ Nm. The load intertia subjected to the motor axis is $J_L = 0.004$ kgm^2. Assuming the servomotor intertia $J_M = 0.5 J_L$, select a PM brushless servomotor.

Solution

The mechanical balance according to eqn (1.17) is expressed as

$$2\pi J \frac{\Delta n}{\Delta t} = T_d \pm T_{sh}$$

where T_d is the electromagnetic torque developed by the motor for acceleration or braking, the "$-$" sign is for acceleleration, and the "$+$" sign is for deceleration. The motor torque required for acceleration

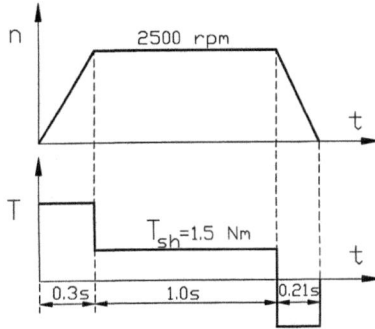

Fig. 1.36. Speed and torque profiles of a servo drive. *Numerical example 1.3.*

$$T_d = 2\pi(J_M + J_L)\frac{\Delta n}{\Delta t} + T_{sh} = \frac{2\pi}{60}(0.004 + 0.002)\frac{2500 - 0}{0.3 - 0} + 1.5 \approx 6.74 \text{ Nm}$$

The motor torque required for braking

$$T_d = 2\pi(J_M + J_L)\frac{\Delta n}{\Delta t} - T_{sh} = \frac{2\pi}{60}(0.004 + 0.002)\frac{2500}{0.21} - 1.5 \approx 5.98 \text{ Nm}$$

The *rms* torque

$$T_{rms} = \sqrt{\frac{6.74^2 \times 0.3 + 1.5^2 \times 1.0 + 5.98^2 \times 0.21}{0.3 + 1.0 + 0.21}} = 3.93 \text{ Nm}$$

The output power calculated for the *rms* torque

$$P_{out} = T_{rms}(2\pi n) = 3.93 \times 2\pi \times \frac{2500}{60} = 1030 \text{ W}$$

The overload capacity factor for $T_{dmax} = 6.74$ Nm

$$\frac{T_{dmax}}{T_{rms}} = \frac{6.74}{3.93} = 1.715$$

A PM brushless motor rated at 1.1 kW with minimum 1.8 overload capacity factor is recommended.

Numerical example 1.4

A 10 kW, 1450 rpm electric motor has been used to drive the following machines: (a) a hoist ($\beta = 0$), (b) a mill ($\beta = 1$) and (c) a fan ($\beta = 2$). The load torque in each case is 60 Nm. Find the drop in mechanical power if the speed is reduced to $n = 1200$ rpm.

Solution

The output power delivered by the motor at the resisting torque $T_r = 60$ Nm and $n_r = 1450$ rpm

$$P_{outr} = T_r \Omega_r = T_r(2\pi n_r) = 60(2\pi \frac{1450}{60}) = 9111 \text{ kW}$$

As the speed is reduced to $n = 1200$ rpm, the load torque is subject to a change according to eqn (1.15), i.e.,

(a) for the hoist

$$T = 60 \left(\frac{1200}{1450} \right)^0 = 60 \text{ Nm}$$

$$P_{out} = T(2\pi n) = 60 \times \left(2\pi \frac{1200}{60} \right) = 7540 \text{ W}$$

(b) for the mill

$$T = 60 \left(\frac{1200}{1450} \right)^1 = 49.7 \text{ Nm}$$

$$P_{out} = T(2\pi n) = 49.7 \times \left(2\pi \frac{1200}{60} \right) = 6245 \text{ W}$$

(c) for the fan

$$T = 60 \left(\frac{1200}{1450} \right)^2 = 41.1 \text{ Nm}$$

$$P_{out} = T(2\pi n) = 41.1 \times \left(2\pi \frac{1200}{60} \right) = 5165 \text{ W}$$

The mechanical power at reduced speed and referred to the rated power is

(a) for the hoist

$$\frac{7540}{9111} \times 10\% = 82.7\%$$

(b) for the mill

$$\frac{6245}{9111} \times 10\% = 68.5\%$$

(c) for the fan

$$\frac{5165}{9111} \times 10\% = 56.7\%$$

Numerical example 1.5

A 25-kg disk is pin-supported at its center. The radius of the disk is $r = 0.2$ m. It is acted upon by a constant force $F = 20$ N which is applied to a cord wrapped around its periphery. Determine the number of revolutions it must make to attain an angular velocity of 25 rad/s starting from rest. Neglect the mass of the cord.

Solution

According to Table 1.1 the kinetic energy $E_k = 0.5J\Omega^2$ where the moment of inertia is $J = 0.5mr^2$. Initially, the disk is at rest ($\Omega_1 = 0$), so that

$$E_{k1} = 0$$

The kinetic energy at $\Omega_2 = 25$ rad/s

$$E_{k2} = \frac{1}{2}J\Omega_2^2 = \frac{1}{4}mr^2\Omega_2^2 = \frac{1}{4}25 \times 0.2^2 \times 25^2 = 156.25 \text{ J}$$

The constant force F does positive work $W = Fs$ as the cord moves downward where $s = \theta r$ (Table 1.1). Thus, the *principle of work and energy* may be written as

$$E_{k1} + Fs = E_{k2} \qquad \text{or} \qquad E_{k1} + Fr\theta = E_{k2}$$

Therefore

$$\theta = \frac{E_{k2} - E_{k1}}{Fr} = \frac{156.25 - 0}{20 \times 0.2} = 39.06 \text{ rad}$$

The number of revolutions is

$$39.06 \text{ rad} \left(\frac{1\text{rev}}{2\pi \text{ rad}}\right) = 6.22 \text{ rad}$$

Numerical Example 1.6

Find the critical speed of rotation of the system consisting of a cylindrical rotor (laminated stack with PMs), steel shaft and driven wheel. Moduli of elasticity, specific mass densities, diameters and widths (lengths) are as follows:

(a) $E_1 = 200 \times 10^9$ Pa, $\rho_1 = 7600$ kg/m^3, $D_1 = 0.24$ m, $w_1 = 0.24$ m for the rotor;

(b) $E_2 = 200 \times 10^9$ Pa, $\rho_2 = 7650$ kg/m^3, $D_2 = 0.4$, $w_2 = 0.15$ m for the driven wheel;

(c) $E_{sh} = 210 \times 10^9$ Pa, $\rho_{sh} = 7700$ kg/m^3, $D_{sh} = 0.0508$ m, $L = 0.76$ m for the shaft.

The location of the rotor from the left end of the shaft is $a_1 = 0.28$ m and the location of the driven wheel from the same end of the shaft is $a_2 = 0.6$ m (Fig. 1.33). The acceleration of gravity is 9.81 m/s^2.

Solution

Mass of rotor

$$m_1 = \rho_1 \frac{\pi D_1^2}{4} w_1 = 7600 \frac{\pi 0.24^2}{4} \times 0.24 = 82.52 \text{ kg}$$

Mass of driven wheel

$$m_2 = \rho_2 \frac{\pi D_2^2}{4} w_2 = 7650 \frac{\pi 0.4^2}{4} \times 0.15 = 144.2 \text{ kg}$$

Mass of shaft

$$m_{sh} = \rho_{sh} \frac{\pi D_{sh}^2}{4} L = 7700 \frac{\pi 0.0508^2}{4} \times 0.76 = 11.86 \text{ kg}$$

Area moment of inertia of the rotor according to eqn (1.12)

$$I_1 = \frac{\pi 0.24^4}{64} = 1.629 \times 10^{-4} \text{ m}^4$$

Area moment of inertia of the driven wheel to eqn (1.12)

$$I_2 = \frac{\pi 0.4^4}{64} = 12.57 \times 10^{-4} \text{ m}^4$$

Area moment of inertia of the shaft according to eqn (1.12)

$$I_{sh} = \frac{\pi 0.0508^4}{64} = 3.269 \times 10^{-7} \text{ m}^4$$

Static deflection of the shaft at position of rotor due to rotor only as given by eqn (1.11)

$$\sigma_1 = \frac{82.52 \times 9.81 \times 0.28^2 (0.76 - 0.28)^2}{3 \times 200 \times 10^9 \times 1.629 \times 10^{-4} \times 0.76} = 1.969 \times 10^{-7} \text{ m}$$

Static deflection of the shaft at position of driven wheel due to driven wheel only as given by eqn (1.11)

$$\sigma_2 = \frac{144.2 \times 9.81 \times 0.6^2 (0.76 - 0.6)^2}{3 \times 200 \times 10^9 \times 12.57 \times 10^{-4} \times 0.76} = 2.275 \times 10^{-8} \text{ m}$$

Static deflection of shaft due to shaft only as given by eqn (1.11)

$$\sigma_{sh} = \frac{11.86 \times 9.81 \times 0.38^2 (0.76 - 0.38)^2}{3 \times 210 \times 10^9 \times 3.269 \times 10^{-7} \times 0.76} = 1.55 \times 10^{-5} \text{ m}$$

where the midpoint of the shaft is $0.5L = 0.5 \times 0.76 = 0.38$ m. Thus, critical speeds according to eqn (1.10) are

- Critical speed of the rotor

$$n_{cr1} = \frac{1}{2\pi} \sqrt{\frac{9.81}{1.969 \times 10^{-7}}} = 1123.4 \text{ rev/s} = 67\ 405.2 \text{ rpm}$$

- Critical speed of the driven wheel

$$n_{cr2} = \frac{1}{2\pi} \sqrt{\frac{9.81}{2.275 \times 10^{-8}}} = 3304.9 \text{ rev/s} = 198\ 292.5 \text{ rpm}$$

- Critical speed of the shaft

$$n_{crsh} = \frac{1}{2\pi} \sqrt{\frac{9.81}{1.55 \times 10^{-5}}} = 126.6 \text{ rev/s} = 7596.8 \text{ rpm}$$

Critical angular speeds for the rotor is $\Omega_{cr1} = 2\pi 1123.4 = 7058.7$ rad/s, for the driven wheel is $\Omega_{cr2} = 2\pi 3304.9 = 20\ 765.1$ rad/s and for the shaft is $\Omega_{crsh} = 2\pi 126.6 = 795.5$ rad/s. According to Dunkerley equation

$$x = \frac{1}{\Omega_{cr1}^2} + \frac{1}{\Omega_{cr2}^2} + \frac{1}{\Omega_{crsh}^2} = \frac{1}{7058.7^2} + \frac{1}{20\ 765.1^2} + \frac{1}{795.5^2} = 1.602 \times 10^{-6} \text{ s}^2/\text{rad}^2$$

Critical angular speed of the system as given by eqn (1.13)

$$\Omega_{cr} = \frac{1}{\sqrt{x}} = \frac{1}{\sqrt{1.602 \times 10^{-6}}} = 790 \text{ rad/s}$$

Critical speed of rotation of the system according to Dunkerley equation

$$n_{cr} = \frac{\Omega_{cr}}{2\pi} = \frac{790}{2\pi} = 125.7 \text{ rev/s} = 7543.6 \text{ rpm}$$

Critical speed of rotation of the system according to Rayleigh's method - eqn (1.14)

$$n_{cr} =$$

$$\frac{1}{2\pi} \sqrt{\frac{9.81(82.52 \times 1.969 \times 10^{-7} + 144.2 \times 2.275 \times 10^{-8} + 11.86 \times 1.55 \times 10^{-5})}{82.52 \times (1.969 \times 10^{-7})^2 + 144.2 \times (2.275 \times 10^{-8})^2 + 11.86 \times (1.55 \times 10^{-5})^2}}$$

$$= 133.1 \text{ rev/s} = 7985.5 \text{ rpm}$$

The results according to Dunkerley [88] and Rayleigh [258] are not the same.

Numerical example 1.7

Estimate the cost of a 3-phase, 7.5 kW PM brushless servo motor with sintered NdFeB PMs and laminated stator and rotor cores. The mass of stator copper conductors is $m_{Cu} = 7.8$ kg, mass of stack $m_{Fe} = 28.5$ kg (stator and rotor), mass of PMs $m_{PM} = 2.10$ kg and mass of shaft $m_{sh} = 6.2$ kg. The cost of materials in U.S. dollars per kilogram is: copper conductor $c_{Cu} = 5.55$, steel laminations $c_{Fe} = 2.75$, NdFeB magnets $c_{PM} = 54.50$ and shaft steel $c_{steel} = 0.65$. The cost of components independent of the machine geometry (nameplate, encoder, terminal leads, terminal board) is $C_0 = \$146.72$.

Coefficients taking into account manufacturing, utilization, complexity and economic factors of PM brushless motors are as follows:

- coefficient dependent of the number of machines manufactured per annum, $k_N = 0.85$ (10,000 machines manufactured per year)
- coefficient taking into account the cost of frame, end bells and bearings, $k_p = 1.62$
- coefficient of the cost of fabrication of coils (insulation, assembly, impregnation), $k_{ii} = 2.00$
- coefficient taking into account the cost of the rotor winding, $k_{sr} = 1.0$ (no rotor winding)
- coefficient of utilization of electrotechnical steel, $k_u = 1.3$
- stacking (insulation) factor, $k_i = 0.96$
- coefficient including the cost of stamping, stacking and other operations, $k_{ss} = 1.4$
- coefficient accounting for the increase of the cost of PMs due to complexity of their shape, $k_{shPM} = 1.15$
- coefficient including the cost of magnetization of PMs, $k_{magn} = 1.1$
- total volume of the steel bar to the volume of the shaft, $k_{ush} = 1.94$
- coefficient taking into account the cost of machining of the shaft, $k_m = 3.15$

It has been assumed for cost analysis that 10,000 machines are manufactured per year.

<u>Solution</u>

The cost of the laminated stack with frame (enclosure), end bells and bearings

$$C_{cl} = k_p k_u k_i k_{ss} m_{Fe} c_{Fe} = 1.62 \times 1.3 \times 0.96 \times 1.4 \times 28.5 \times 2.75) = \$221.84$$

The cost of the copper winding

$$C_w = k_{ii} k_{sr} m_{Cu} c_{Cu} = 2.00 \times 1.0 \times 7.8 \times 5.55 = \$86.58$$

The cost of PMs

$$C_{PM} = k_{shPM}k_{magn}m_{PM}c_{PM} = 1.15 \times 1.1 \times 2.10 \times 54.50 = \$144.78$$

The cost of the shaft

$$C_{sh} = k_{ush}k_m m_{sh}c_{steel} = 1.94 \times 3.15 \times 6.2 \times 0.65 = \$24.63$$

Total cost of the motor

$$C = k_N(C_{cl} + C_w + C_{PM} + C_{sh} + C_0)$$

$$= 0.85(221.84 + 86.58 + 144.78 + 24.63 + 146.72) = \$530.87$$

The evaluated cost of 10,000 machines (annual production) is $10,000 \times 530.87 = \$5,308,700$.

2

Permanent Magnet Materials and Circuits

A *permanent magnet* (PM) can produce a magnetic field in an air gap with no field excition winding and no dissipation of electric power. External energy is involved only in changing the energy of the magnetic field, not in maintaining it. As any other ferromagnetic material, a PM can be described by its B—H hysteresis loop. PMs are also called *hard magnetic materials*, meaning ferromagnetic materials with a wide hysteresis loop.

2.1 Demagnetization curve and magnetic parameters

The basis for the evaluation of a PM is the portion of its hysteresis loop located in the upper left-hand quadrant, called the *demagnetization curve* (Fig. 2.1). If a reverse magnetic field intensity is applied to a previously magnetized, say, toroidal specimen, the magnetic flux density drops down to the magnitude determined by the point K. When the reversal magnetic flux density is removed, the flux density returns to the point L according to a minor hysteresis loop. Thus, the application of a reverse field has reduced the *remanence*, or *remanent magnetism*. Reapplying a magnetic field intensity will again reduce the flux density, completing the minor hysteresis loop by returning the core to approximately the same value of flux density at the point K as before. The minor hysteresis loop may usually be replaced with little error by a straight line called the *recoil line*. This line has a slope called the *recoil permeability* μ_{rec}.

As long as the negative value of applied magnetic field intensity does not exceed the maximum value corresponding to the point K, the PM may be regarded as being reasonably permanent. If, however, a greater negative field intensity H is applied, the magnetic flux density will be reduced to a value lower than that at point K. On the removal of H, a new and lower recoil line will be established.

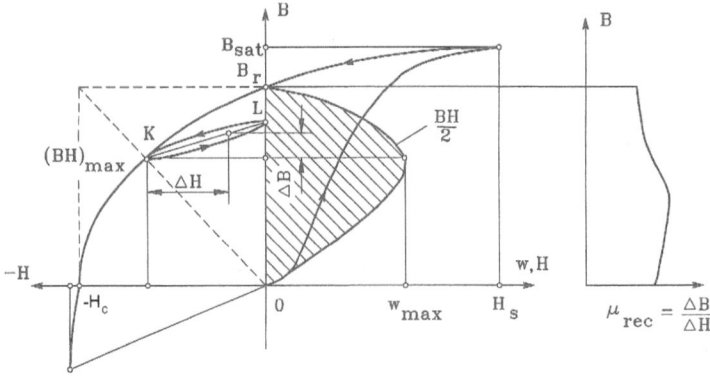

Fig. 2.1. Demagnetization curve, recoil loop, energy of a PM, and recoil magnetic permeability.

The general relationship between the magnetic flux density B, intrinsic magnetization B_i due to the presence of the ferromagnetic material, and magnetic field intensity H may be expressed as [208, 236]

$$B = \mu_0 H + B_i = \mu_0(H + M) = \mu_0(1 + \chi)H = \mu_0 \mu_r H \qquad (2.1)$$

in which $\mathbf{B}$, $\mathbf{H}$, $\mathbf{B}_i$ and $\mathbf{M} = \mathbf{B}_i/\mu_0$ are parallel or antiparallel vectors, so that eqn (2.1) can be written in a scalar form. The magnetic permeability of free space $\mu_0 = 0.4\pi \times 10^{-6}$ H/m. The relative magnetic permeability of ferromagnetic materials $\mu_r = 1 + \chi \gg 1$. The magnetization vector $\mathbf{M}$ is proportional to the magnetic susceptibility χ of the material, i.e., $\mathbf{M} = \chi\mathbf{H}$. The flux density $\mu_0 H$ would be present within, say, a toroid if the ferromagnetic core was not in place. The flux density B_i is the contribution of the ferromagnetic core.

A PM is inherently different than an electromagnet. If an external field H_a is applied to the PM, as was necessary to obtain the hysteresis loop of Fig. 2.1, the resultant magnetic field is

$$H = H_a + H_d \qquad (2.2)$$

where $-H_d$ is a potential existing between the poles, 180^0 opposed to B_i, proportional to the intrinsic magnetization B_i. In a closed magnetic circuit, e.g., toroidal, the magnetic field intensity resulting from the intrinsic magnetization $H_d = 0$. If the PM is removed from the magnetic circuit

$$H_d = -\frac{M_b B_i}{\mu_o} \qquad (2.3)$$

where M_b is the coefficient of demagnetization dependent on geometry of a specimen. Usually $M_b < 1$, see Section 2.6.3. Putting $B_i = B_d - \mu_o H_d$ to eqn (2.3) the equation relating the magnetic flux density B_d and the self-demagnetizing field H_d to the magnet geometry is [236]

$$\frac{B_d}{\mu_0 H_d} = 1 - \frac{1}{M_b} \qquad (2.4)$$

The coefficient $(1 - 1/M_b)$ is proportional to the permeance of the external magnetic circuit.

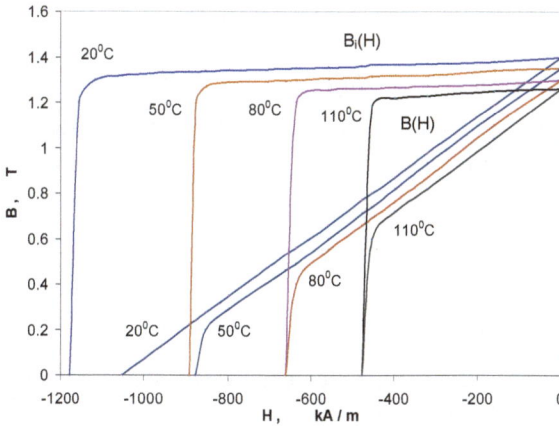

Fig. 2.2. Comparison of $B - H$ and $B_i - H$ demagnetization curves and their variations with the temperature for sintered N48M NdFeB PMs. Courtesy of *ShinEtsu*, Japan.

PMs are characterized by the parameters listed below.

Saturation magnetic flux density B_{sat} and corresponding saturation magnetic field intensity H_{sat}. At this point the alignment of all the *magnetic moments of domains* is in the direction of the external applied magnetic field.

Remanent magnetic flux density B_r, or **remanence**, is the magnetic flux density corresponding to zero magnetic field intensity. High remanence means the magnet can support higher magnetic flux density in the air gap of the magnetic circuit.

Coercive field strength H_c, or **coercivity**, is the value of demagnetizing field intensity necessary to bring the magnetic flux density to zero in a material previously magnetized (in a symmetrically cyclically magnetized condition). High coercivity means that a thinner magnet can be used to withstand the demagnetization field.

Intrinsic demagnetization curve (Fig. 2.2) is the portion of the $B_i = f(H)$ hysteresis loop located in the upper left-hand quadrant, where $B_i = B - \mu_0 H$

is according to eqn (2.1). For $H = 0$ the instrinsic magnetic flux density $B_i = B_r$.

Intrinsic coercivity $_iH_c$ is the magnetic field strength required to bring to zero the intrinsic magnetic flux density B_i of a magnetic material described by the $B_i = f(H)$ curve. For PM materials $_iH_c > H_c$.

Recoil magnetic permeability μ_{rec} is the ratio of the magnetic flux density to magnetic field intensity at any point on the demagnetization curve, i.e.,

$$\mu_{rec} = \mu_0 \mu_{rrec} = \frac{\Delta B}{\Delta H} \tag{2.5}$$

where the *relative recoil permeability* $\mu_{rrec} = 1 \ldots 3.5$.

Maximum magnetic energy per unit produced by a PM in the external space is equal to the maximum magnetic energy density per volume, i.e.

$$w_{max} = \frac{(BH)_{max}}{2} \qquad \text{J/m}^3 \tag{2.6}$$

where the product $(BH)_{max}$ corresponds to the maximum energy density point on the demagnetization curve with coordinates B_{max} and H_{max}.

Form factor of the demagnetization curve characterizes the concave shape of the demagnetization curve, i.e.,

$$\gamma = \frac{(BH)_{max}}{B_r H_c} = \frac{B_{max} H_{max}}{B_r H_c} \tag{2.7}$$

For a square demagnetization curve $\gamma = 1$ and for a straight line (rare-earth PM) $\gamma = 0.25$.

Owing to the leakage fluxes, PMs used in electrical machines are subject to nonuniform demagnetization. Therefore, the demagnetization curve is not the same for the whole volume of a PM. To simplify the calculation, in general, it is assumed that the whole volume of a PM is described by one demagnetization curve with B_r and H_c about 5 to 10% lower than those for uniform magnetization.

The leakage flux causes magnetic flux to be distributed nonuniformly along the height $2h_M$ of a PM. As a result, the MMF produced by the PM is not constant. The magnetic flux is higher in the neutral cross section and lower at the ends, but the behavior of the MMF distribution is the opposite (Fig. 2.3).

The PM surface is not equipotential. The magnetic potential at each point on the surface is a function of the distance to the neutral zone. To simplify the calculation the magnetic flux, which is a function of the MMF distribution along the height h_M per pole, is replaced by an equivalent flux. This equivalent flux goes through the whole height h_M and exits from the surface of the poles.

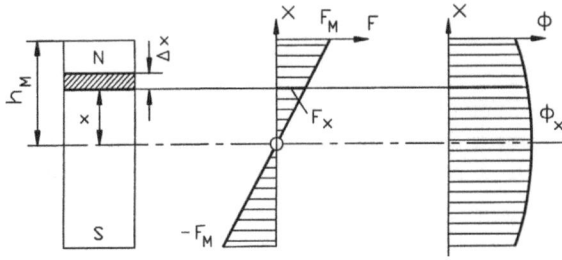

Fig. 2.3. Distribution of the MMF and magnetic flux along the height h_M of a rectangular PM.

To find the equivalent leakage flux and the whole flux of a PM, the equivalent magnetic field intensity has to be found, i.e.,

$$H = \frac{1}{h_M} \int_0^{h_M} H_x dx = \frac{F_M}{h_M} \tag{2.8}$$

where H_x is the magnetic field intensity at a distance x from the neutral cross section and F_M is the MMF of the PM per pole (MMF $= 2F_M$ per pole pair).

The equivalent magnetic field intensity (2.8) allows the equivalent leakage flux of the PM to be found, i.e.,

$$\Phi_{lM} = \Phi_M - \Phi_g \tag{2.9}$$

where Φ_M is the full equivalent flux of the PM and Φ_g is the air gap magnetic flux. The *coefficient of leakage flux* of the PM,

$$\sigma_{lM} = \frac{\Phi_M}{\Phi_g} = 1 + \frac{\Phi_{lM}}{\Phi_g} > 1 \tag{2.10}$$

simply allows the air gap magnetic flux to be expressed as $\Phi_g = \Phi_M / \sigma_{lM}$.

The following leakage permeance expressed in the flux Φ—MMF coordinate system corresponds to the equivalent leakage flux of the PM:

$$G_{lM} = \frac{\Phi_{lM}}{F_M} \tag{2.11}$$

An accurate estimation of the leakage permeance G_{lM} is the most difficult task in calculating magnetic circuits with PMs. Of course, the accuracy problem exists only in the circuital approach since using the field approach e.g., the finite element method (FEM), the leakage permeance can be found fairly accurately.

The average equivalent magnetic flux and equivalent MMF mean that the magnetic flux density and magnetic field intensity are assumed to be the same in the whole volume of a PM. The full energy produced by the magnet in the outer space is

$$W = \frac{BH}{2}V_M \quad \text{J} \quad (2.12)$$

where V_M is the volume of the PM or a system of PMs.

For PMs with linear demagnetization curve, i.e., NdFeB magnets, the coercive field strength at room temperature can simply be calculated on the basis of B_r and μ_{rrec} as

$$H_c = \frac{B_r}{\mu_0\mu_{rrec}} \quad (2.13)$$

The magnetic flux density produced in the air gap g by a PM with linear demagnetization curve and its height h_M placed in a magnetic circuit with infinitely large magnetic permeability and air gap g is, approximately,

$$B_g \approx \frac{B_r}{1 + \mu_{rrec}g/h_M} \quad (2.14)$$

The above eqn (2.14) has been derived assuming $H_c h_M \approx H_M h_M + H_g g$, neglecting magnetic voltage drop (MVD) in the mild steel portion of the magnetic circuit, putting H_c according to eqn (2.13), $H_M = B_g/(\mu_0\mu_{rrec})$ and $H_g = B_g/\mu_0$. More explanation is given in Section 8.8.5.

For a simple PM circuit with rectangular cross section consisting of a PM with height per pole h_M, width w_M, length l_M, two mild steel yokes with average length $2l_{Fe}$ and an air gap of thickness g the Ampère's circuital law can be written as

$$2H_M h_M = H_g g + 2H_{Fe}l_{Fe} = H_g g \left(1 + \frac{2H_{Fe}l_{Fe}}{H_g g}\right) \quad (2.15)$$

where H_g, H_{Fe}, and H_M are the magnetic field intensities in the air gap, mild steel yoke, and PM, respectively. Since $\Phi_g \sigma_{lM} = \Phi_M$ or $B_g S_g = B_M S_M/\sigma_{lM}$, where B_g is the air gap magnetic flux density, B_M is the PM magnetic flux density, S_g is the cross section area of the air gap, and $S_M = w_M l_M$ is the cross section area of the PM, the magnetic flux balance equation is

$$\frac{V_M}{2h_M}\frac{1}{\sigma_{lM}}B_M = \mu_0 H_g \frac{V_g}{g} \quad (2.16)$$

where $V_g = S_g g$ is the volume of the air gap and $V_M = 2h_M S_M$ is the volume of the PM. The fringing flux in the air gap has been neglected. Multiplying through eqns (2.15) for magnetic voltage drops and (2.16) for magnetic flux, the air gap magnetic field intensity is found as

$$H_g = \sqrt{\frac{1}{\mu_0}\frac{1}{\sigma_{lM}}\left(1 + \frac{2H_{Fe}l_{Fe}}{H_g g}\right)^{-1}\frac{V_M}{V_g}B_M H_M} \approx \sqrt{\frac{1}{\mu_0}\frac{V_M}{V_g}B_M H_M} \quad (2.17)$$

For a PM circuit the magnetic field strength H_g in a given air gap volume V_g is directly proportional to the square root of the energy product $(B_M H_M)$ and the volume of magnet $V_M = 2h_M w_M l_M$.

Following the trend to smaller packaging, smaller mass and higher efficiency, the material research in the field of PMs has focused on finding materials with high values of the maximum energy product $(BH)_{max}$.

2.2 Early history of permanent magnets

A hard magnetic material called *loadstone* was mentioned by Greek philosopher Thales of Miletus as early as ca. 600 B.C. [236]. This was a natural magnetic mineral, a form of magnetitie Fe_3O_4. Loadstone was given the name *magnes* because it was found in Magnesia, a district in Thessaly.

The first artificial magnets were iron needles magnetized by touching the loadstone. Man's first practical use of magnetism may have been the compass. Around 1200 A.D. there are references in a French poem written by Guyot de Provins to a touched needle of iron supported by a floating straw [236]. Other references suggest that good magnet steel was available from China in about 500 A.D.

The earliest systematic reporting of magnets was a scientific work by William Gilbert in 1600 [125]. Gilbert described how to arm loadstones with soft iron pole tips to increase attractive force on contact and how to magnetize pieces of iron or steel [236]. The next great advance in magnetism came with the invention of the electromagnet by J. Henry and W. Sturgeon in 1825.

By 1867, German handbooks recorded that ferromagnetic alloys could be made from nonferromagnetic materials and nonferromagnetic alloys of ferromagnetic materials, mainly iron. For example in 1901, *Heusler alloys* (Cu_2MnAl), which had outstanding properties compared to previous magnets, were reported[1]. The composition of a typical Heusler alloy was 10 to 30% manganese and 15 to 19% aluminum, the balance being copper.

In 1917 *cobalt steel alloys* and in 1931 *Alnico* (Al, Ni, Co, Fe) were discovered in Japan. In 1938, also in Japan, Kato and Takei developed magnets made of *powdered oxides*. This development was the forerunner of the modern *ferrite*.

2.3 Properties of permanent magnets

There are three classes of PMs currently used for electric motors:

- Alnicos (Al, Ni, Co, Fe);
- Ceramics (ferrites), e.g., barium ferrite $BaO \times 6Fe_2O_3$ and strontium ferrite $SrO \times 6Fe_2O_3$;

---

[1] Heusler F, Stark W, Haupt E. Verh. Deutsche Phys., Gessel 5(219), 1903.

- Rare-earth materials, i.e., samarium-cobalt SmCo and neodymium-iron-boron NdFeB.

Demagnetization curves of the above PM materials are given in Fig. 2.4. Demagnetization curves are sensitive to the temperature (Fig. 2.2). Both B_r and H_c decrease as the magnet temperature increases, i.e.,

$$B_r = B_{r20}[1 + \frac{\alpha_B}{100}(\vartheta_{PM} - 20)] \qquad (2.18)$$

$$H_c = H_{c20}[1 + \frac{\alpha_H}{100}(\vartheta_{PM} - 20)] \qquad (2.19)$$

where ϑ_{PM} is the temperature of PM, B_{r20} and H_{c20} are the remanent magnetic flux density and coercive force at 20^0C and $\alpha_B < 0$ and $\alpha_H < 0$ are temperature coefficients for B_r and H_c in $\%/^0$C, respectively.

Table 2.1. Temperature coefficients and Curie temperature for common PM materials acccording to Arnold Magnetic Technologies, Rochester, NY, U.S.A.

Material	Reversible temperature coefficient for B_r $\%/^0$C	Reversible temperature coefficient for $_iH_c$ $\%/^0$C	Curie temperature ^{0}C
Alnico 5	−0.02	−0.01	900
Alnico 8	−0.02	−0.01	860
Sm$_2$Co$_{17}$	−0.03	−0.20	800
SmCo$_5$	−0.045	−0.40	700
Bonded NdFeB MQP-C (15% Co)	−0.07	−0.40	470
Sintered NdFeB 40 MGOe (0% Co)	−0.10	−0.60	310
Ferrite 8	−0.20	+0.27	450
Plastiform 2401 Ferrite-Neo hybrid	−0.14	−0.04	−

2.3.1 Alnico

The main advantages of *Alnico* are its high magnetic remanent flux density and low temperature coefficients. The temperature coefficient of B_r is $-0.02\%/^0$C and maximum service temperature is 520^0C. These advantages allow a high air gap magnetic flux density at high magnet temperature. Unfortunately, coercive force is very low and the demagnetization curve is extremely nonlinear. Therefore, it is very easy not only to magnetize but also to demagnetize Alnico. Alnico has been used in PM d.c. commutator motors of disk type with relatively large air gaps. This results in a negligible armature

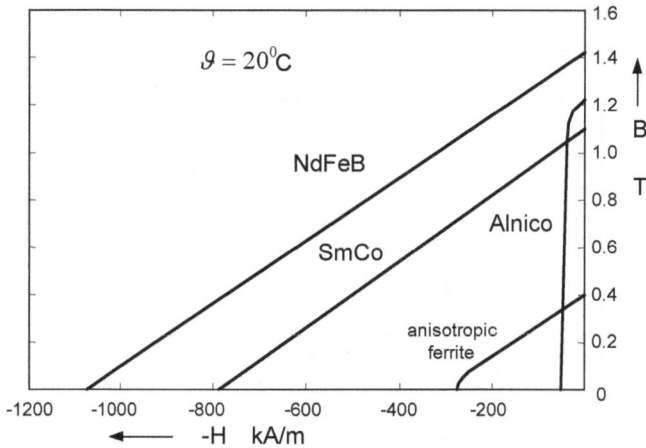

Fig. 2.4. Demagnetization curves for different permanent magnet materials.

reaction magnetic flux acting on the PMs. Sometimes, Alnico PMs are protected from the armature flux, and consequently from demagnetization, using additional mild steel pole shoes. Alnicos dominated the PM motors market in the range from a few watts to 150 kW between the mid 1940s and the late 1960s when ferrites became the most widely used materials [236].

2.3.2 Ferrites

Barium and strontium *ferrites* were invented in the 1950s. A ferrite has a higher coercive force than Alnico, but at the same time has a lower remanent magnetic flux density. Temperature coefficients are relatively high, i.e. the coefficient of B_r is $-0.20\%/^0$C and the coefficient of H_c is $-0.27\%/^0$C. The maximum service temperature is 400^0C. The main advantages of ferrites are their low cost and very high electric resistance, which means no eddy-current losses in the PM volume. Ferrite magnets are most economical in fractional horsepower motors and may show an economic advantage over Alnico up to about 7.5 kW. Barium ferrite PMs (Fig. 2.5) are commonly used in small d.c. commutator motors for automobiles (blowers, fans, windscreen wipers, pumps, etc.) and electric toys.

Ferrites are produced by powder metallurgy. Their chemical formulation may be expressed as $MO \times 6(Fe_2O_3)$, where M is Ba, Sr, or Pb. Strontium ferrite has a higher coercive force than barium ferrite. Lead ferrite has a production disadvantage from an environmental point of view. Ferrite magnets are available in *isotropic* and *anisotropic* grades.

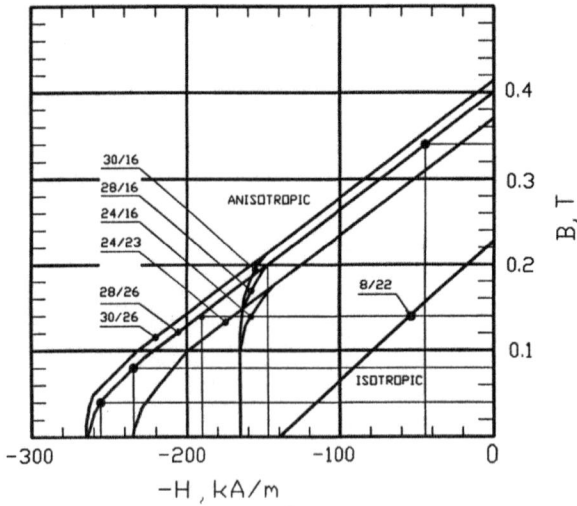

Fig. 2.5. Demagnetization curves of *Hardferrite* barium ferrites manufactured by Magnetfabrik Schramberg GmbH & Co., Schramberg, Germany.

2.3.3 Rare-earth permanent magnets

During the last three decades great progress regarding available energy density $(BH)_{max}$ has been achieved with the development of *rare-earth PMs*. The rare-earth elements are in general not rare at all, but their natural minerals are widely mixed compounds. To produce one particular rare-earth metal, several others, for which no commercial application exists, have to be refined. This limits the availability of these metals. The first generation of these new alloys based on the composition $SmCo_5$ and invented in the 1960s has been commercially produced since the early 1970s. Today it is a well-established hard magnetic material. $SmCo_5$ has the advantage of high remanent flux density, high coercive force, high energy product, linear demagnetization curve and low temperature coefficient (Table 2.2). The temperature coefficient of B_r is −0.02 to −0.045%/^{0}C and the temperature coefficient of H_c is −0.14 to −0.40%/^{0}C. Maximum service temperature is 300 to 350^{0}C. It is well suited to build motors with low volume, high power density and class of insulation F or H. The cost is the only drawback. Both Sm and Co are relatively expensive due to their supply restrictions.

Generally, SmCo magnets do not need any coatings or platings after processing. Occasionally advanced coatings such as phenolic resin is desirable if cleanliness is an issue. The coating in this case acts much like a sealant. Sometimes SmCo magnets are plated with Nickel.

With the discovery in the recent years of a second generation of rare-earth magnets on the basis of inexpensive neodymium (Nd), remarkable progress with regard to lowering raw material costs has been achieved. This new gener-

Table 2.2. Physical properties of Vacomax sintered Sm_2Co_{17} PM materials at room temperature 20^0C manufactured by Vacuumschmelze GmbH, Hanau, Germany

Property	Vacomax 240 HR	Vacomax 225 HR	Vacomax 240
Remanent flux density, B_r, T	1.05 to 1.12	1.03 to 1.10	0.98 to 1.05
Coercivity, H_c, kA/m	600 to 730	720 to 820	580 to 720
Intrinsic coercivity, iH_c, kA/m	640 to 800	1590 to 2070	640 to 800
$(BH)_{max}$, kJ/m^3	200 to 240	190 to 225	180 to 210
Relative recoil magnetic permeability	1.22 to 1.39	1.06 to 1.34	1.16 to 1.34
Temperature coefficient α_B of B_r at 20 to 100^0C, %/^0C		−0.030	
Temperature coefficient α_{iH} of iH_c at 20 to 100^0C, %/^0C	−0.15	−0.18	−0.15
Temperature coefficient α_B of B_r at 20 to 150^0C, %/^0C		−0.035	
Temperature coefficient α_{iH} of iH_c at 20 to 150^0C, %/^0C	−0.16	−0.19	−0.16
Curie temperature, ^0C		approximately 800	
Maximum continuous service temperature, ^0C	300	350	300
Thermal conductivity, W/(m ^0C)		approximately 12	
Specific mass density, ρ_{PM}, kg/m^3		8400	
Electric conductivity, $\times10^6$ S/m		1.18 to 1.33	
Coefficient of thermal expansion at 20 to 100^0C, $\times10^{-6}$/^0C		10	
Young's modulus, $\times10^6$ MPa		0.150	
Bending stress, MPa		90 to 150	
Vicker's hardness		approximately 640	

ation of rare-earth PMs was announced by Sumitomo Special Metals, Japan, in 1983 at the 29th Annual Conference of Magnetism and Magnetic Materials held in Pittsburgh, PA, U.S.A. The Nd is a much more abundant rare-earth element than Sm. NdFeB magnets, which are now produced in increasing quantities, have better magnetic properties than those of SmCo, but unfortunately only at room temperaure. The demagnetization curves, especially the coercive force, are strongly temperature dependent. The temperature coefficient of B_r is −0.09 to −0.15%/^0C and the temperature coefficient of H_c is −0.40 to −0.80%/^0C. The maximum service temperature is 250^0C and Curie temperature is 350^0C). The NdFeB is also susceptible to corrosion. NdFeB magnets have great potential for considerably improving the *performance-to-cost* ratio for many applications. For this reason they will have a major impact on the development and application of PM apparatus in the future.

The latest grades of NdFeB have a higher remanent magnetic flux density and better thermal stability (Table 2.3). Metallic or resin coating are employed to improve resistance to corrosion.

Table 2.3. Physical properties of Vacodym sintered NdFeB PM materials at room temperature 20^0C manufactured by Vacuumschmelze GmbH, Hanau, Germany

Property	Vacodym 633 HR	Vacodym 362 TP	Vacodym 633 AP
Remanent flux density, B_r, T	1.29 to 1.35	1.25 to 1.30	1.22 to 1.26
Coercivity, H_c, kA/m	980 to 1040	950 to 1005	915 to 965
Intrinsic coercivity, iH_c, kA/m	1275 to 1430	1195 to 1355	1355 to 1510
$(BH)_{max}$, kJ/m^3	315 to 350	295 to 325	280 to 305
Relative recoil magnetic permeability	1.03 to 1.05		1.04 to 1.06
Temperature coefficient α_B of B_r at 20 to 100^0C, %/^{0}C	−0.095	−0.115	−0.095
Temperature coefficient α_{iH} of iH_c at 20 to 100^0C, %/^{0}C	−0.65	−0.72	−0.64
Temperature coefficient α_B of B_r at 20 to 150^0C, %/^{0}C	−0.105	−0.130	−0.105
Temperature coefficient α_{iH} of iH_c at 20 to 150^0C, %/^{0}C	−0.55	−0.61	−0.54
Curie temperature, ^{0}C	approximately 330		
Maximum continuous service temperature, ^{0}C	110	100	120
Thermal conductivity, W/(m ^{0}C)	approximately 9		
Specific mass density, ρ_{PM}, kg/m^3	7700	7600	7700
Electric conductivity, $\times 10^6$ S/m	0.62 to 0.83		
Coefficient of thermal expansion at 20 to 100^0C, $\times 10^{-6}/^0$C	5		
Young's modulus, $\times 10^6$ MPa	0.150		
Bending stress, MPa	270		
Vicker's hardness	approximately 570		

Nowadays, for the industrial production of rare-earth PMs the powder metallurgical route is mainly used [236]. Neglecting some material specific parameters, this processing technology is, in general, the same for all rare-earth magnet materials [75]. The alloys are produced by vacuum induction melting or by a calciothermic reduction of the oxides. The material is then size-reduced by crushing and milling to a single crystalline powder with particle sizes less than 10 μm.

In order to obtain anisotropic PMs with the highest possible $(BH)_{max}$ value, the powders are then aligned in an external magnetic field, pressed

Table 2.4. Physical properties of Neoquench-DR neodymium based PM materials at room temperature 20^0C manufactured by Daido Steel Co. Ltd

Property	ND-31HR	ND-31SHR	ND-35R
Remanent flux density, B_r, T	1.14 to 1.24	1.08 to 1.18	1.22 to 1.32
Coercivity, H_c, kA/m	828 to 907	820 to 899	875 to 955
Intrinsic coercivity, iH_c, kA/m	1114 to 1432	1592 to 1989	1035 to 1353
$(BH)_{max}$, kJ/m^3	239 to 279	231 to 163	179 to 318
Relative recoil magnetic permeability	1.05		
Temperature coefficient α_B of B_r, %/^{0}C	-0.10		
Temperature coefficient α_H of H_c, %/^{0}C	-0.50		
Curie temperature ^{0}C	360		
Thermal conductivity, W/(m ^{0}C)	4.756		
Specific mass density, ρ_{PM}, kg/m^3	7600	7700	7600
Electric conductivity, $\times10^6$ S/m	1.35		
Coefficient of thermal expansion at 20 to 200^{0}C, $\times10^{-6}/^0$C	1 to 2 in radial direction -1 to 0 in axial direction		
Young's modulus, $\times10^6$ MPa	0.152		
Bending stress, MPa	196.2		
Vicker's hardness	approximately 750		

and densified to nearly theoretical density by sintering. The most economical method for mass production of simple shaped parts like blocks, rings or arc segments of the mass in the range of a few grams up to about 100 g is a die pressing of the powders in an approximate final shape. Larger parts or smaller quantities can be produced from isostatically pressed bigger blocks by cutting and slicing.

Sintering and the heat treatment that follows are done under vacuum or under an inert gas atmosphere. Sintering temperatures are in the range of 1000 to 1200^{0}C depending on the PM material with sintering times ranging from 30 to 60 min. During annealing after sintering the microstructure of the material is optimized, which increases the intrinsic coercivity $_iH_c$ of the magnets considerably. After machining to get dimensional tolerances the last step in the manufacturing process is magnetizing. The magnetization fields to reach complete saturation are in the range of 1000 to 4000 kA/m, depending on material composition.

Researchers at General Motors, MI, U.S.A., have developed a fabrication method based on the melt-spinning casting system originally invented for production of amorphous metal alloys. In this technology molten stream of

NdFeCoB material is first formed into ribbons 30 to 50-μm thick by rapid quenching, then cold pressed, extruded and hot pressed into bulk. Hot pressing and hot working are carried out while maintaining the fine grain to provide high density close to 100% which eliminates the possibility of internal corrosion. The standard electro-deposited epoxy resin coating provides excellent corrosion resistance. Parameters of Neoquench-DR PM materials manufactured by Daido Steel Co. Ltd, are given in Table 2.4. Those PM materials are suitable for fabrication of large diameter (over 200 mm) radially oriented ring magnets.

NdFeB magnets are mechanically very strong, not as brittle as SmCo. Surface treatment is required (nickel, aluminum chromate or polymer coatings).

Table 2.5. Hydrogen exposure test on rare-earth PMs at 30 MPa pressure and 25^0C temperature. Courtesy of Shin Etsu Chemical Co., Ltd., Takefu, Japan

rare-earth PM	before test	after test		
		1 day	3 days	7 days
NdFeB 20 μm Ni coating				
Sm_2Co_{17}				

2.3.4 Corrosion and chemical reactivity

In comparison with SmCo magnets, NdFeB magnets have low corrosion resistance. Their chemical reactivity is similar to that of alkaline earth metals like magnesium. Under normal conditions the NdFeB reacts slowly. The reaction is faster at higher temperatures and in the presence of humidity. Most PMs are assembled using adhesives. Adhesives with acid content must not be used since they lead to rapid decomposition of the PM material. NdFeB magnets cannot be used under the following conditions:

- in an acidic, alkaline, or organic solvent (unless hermetically sealed inside a can)
- in water or oil (unless hermetically sealed)
- in an electrically conductive liquid, such as electrolyte containing water
- in a hydrogen-containing atmosphere, especially at elevated temperatures since hydrogenation causes the magnet material to disintegrate (Table 2.5)

- in corrosive gasses, such as Cl, NH_3, N_{ox}, etc.
- in the presence of radioactive rays (NdFeB magnets can be damaged by radiation, mainly gamma ray)

Motors used for ship propulsion are exposed to salty atmospheric moisture and air humidity can exceed 90%. Hot unprotected NdFeB PMs are very vulnerable to salty mist and after a few days the penetration of corrosion reaches 0.1 mm [280]. Compare with grinding tolerance of ±0.05 mm and cutting from blocks tolerance of ±0.1 mm. Protection of NdFeB magnets by fiberglass bandages impregnated in resin is not sufficient [280]. Metallic (Sn or Ni) or organic (electro-painting) are the best methods of protection against corrosion. Air-drying varnishes are the only cost-effective coatings and partially effective to protect PMs against corrosion.

2.3.5 Market issues

Today for large-scale production the prices of NdFeB magnets, depending on the grade, are about US$55 per kg. This is about 60 to 70% of the cost of SmCo magnets. Assuming constant raw material prices, it is expected that with the increasing production, and obviously with the decreasing manufacturing costs, the prices will still be coming down.

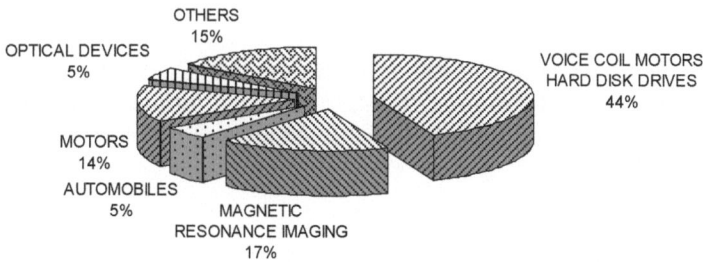

Fig. 2.6. Fields of application of rare-earth PMs according to Shin-Etsu Chemical Co., Ltd., Takefu, Japan.

A breakdown in the sales of rare-earth magnets according to application fields is given in Fig. 2.6. The most important application area of rare-earth magnets with about 40 to 45% of the sales are voice coil motors (VCMs) and hard disk drives (HDDs) for computers. Other major areas of application are magnetic resonance imaging (MRI), electric motors, optical devices, automobiles, acoustic devices, and magnetomechanical devices (holding devices, couplings, magnetic separators, magnetic bearings). PM motors are used in a broad power range from a few mW to more than 1 MW, covering a wide variety of applications from stepping motors for wristwatches via industrial servo-drives for machine tools (up to 15 kW) to large synchronous motors. High performance rare-earth magnets have successfully replaced Alnico and ferrite magnets in all applications where high power density, improved dynamic performance or higher efficiency are of prime interest.

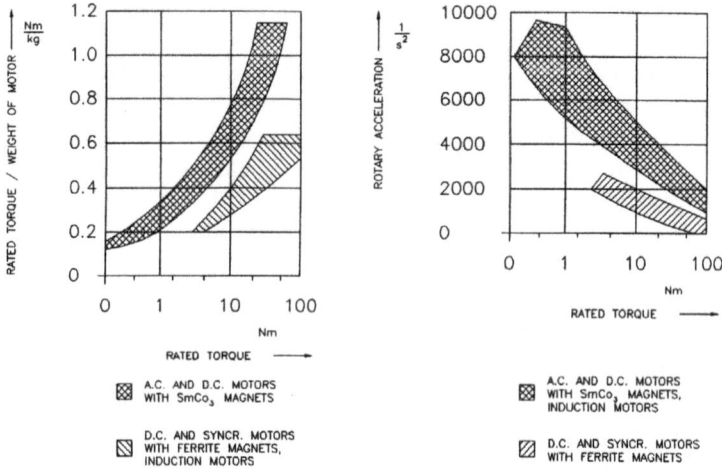

Fig. 2.7. Torque–mass ratio and rotary acceleration of different types of servo motors.

A comparison of the *torque–to–mass* ratio and rotary acceleration of different types of servo motors in Fig. 2.7 shows that the torque and acceleration of servo motors with rare-earth PMs can be increased by more than double [75].

2.4 Approximation of demagnetization curve and recoil line

The most widely used approximation of the demagnetization curve is the approximation using a hyperbola, i.e.,

$$B = B_r \frac{H_c - H}{H_c - a_0 H} \tag{2.20}$$

where B and H are coordinates, and a_0 is the constant coefficient, which can be evaluated as [20]

$$a_0 = \frac{B_r}{B_{sat}} = \frac{2\sqrt{\gamma} - 1}{\gamma} \tag{2.21}$$

or on the basis of the demagnetization curve

$$a_0 = \frac{1}{n} \sum_{i=1}^{n} \left(\frac{H_c}{H_i} + \frac{B_r}{B_i} - \frac{H_c}{H_i} \frac{B_r}{B_i} \right) \tag{2.22}$$

where (B_i, H_i) are coordinates of points $i = 1, 2, \ldots n$ on the demagnetization curve, arbitrarily chosen, and n is the number of points on the demagnetization curve.

The recoil magnetic permeability is assumed to be constant and equal to [233]

$$\mu_{rec} = \frac{B_r}{H_c}(1 - a_0) \tag{2.23}$$

The above equations give good accuracy between calculated and measured demagnetization curves for Alnicos and isotropic ferrites with low magnetic energy. Application to anisotropic ferrites with high coercivity can in some cases cause errors.

For rare-earth PMs at room temperature 20^0C the approximation is simple due to their practically linear demagnetization curves, i.e.,

$$B = B_r \left(1 - \frac{H}{H_c} \right) \tag{2.24}$$

This means that putting $a_0 = 0$ or $\gamma = 0.25$, eqn (2.20) takes the form of eqn (2.24). Eqn (2.24) cannot be used for NdFeB magnets at the temperature higher that 20^0C since their demagnetization curves become nonlinear.

2.5 Operating diagram

2.5.1 Construction of the operating diagram

The energy of a PM in the external space exists only if the reluctance of the external magnetic circuit is higher than zero. If a previously magnetized PM is placed inside a closed ideal ferromagnetic circuit, i.e. toroid, this PM does not show any magnetic properties in the external space, in spite of the fact that there is magnetic flux Φ_r corresponding to the remanent flux density B_r inside the PM.

A PM previously magnetized and placed alone in an open space, as in Fig. 2.8a, generates a magnetic field. To sustain a magnetic flux in the external open space, an MMF developed by the magnet is necessary. The state of the PM is characterized by the point K on the demagnetization curve (Fig. 2.9). The location of the point K is at the intersection of the demagnetization curve with a straight line representing the permeance of the external magnetic circuit (open space):

$$G_{ext} = \frac{\Phi_K}{F_K}, \qquad \tan \alpha_{ext} = \frac{\Phi_K / \Phi_r}{F_K / F_c} = G_{ext} \frac{F_c}{\Phi_r} \tag{2.25}$$

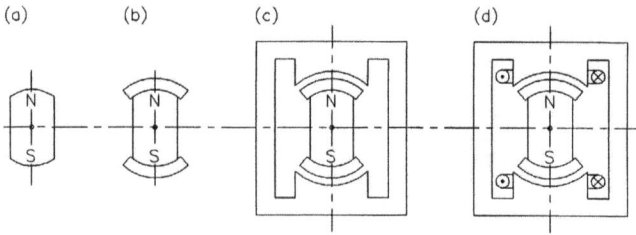

Fig. 2.8. Stabilization of a PM: (a) PM alone, (b) PM with pole shoes, (c) PM inside an external magnetic circuit, (d) PM with a complete external armature system.

The permeance G_{ext} corresponds to a flux Φ—MMF coordinate system and is referred to as MMF at the ends of the PM. The magnetic energy per unit produced by the PM in the external space is $w_K = B_K H_K/2$. This energy is proportional to the rectangle limited by the coordinate system and lines perpendicular to the Φ and F coordinates projected from the point K. It is obvious that the maximum magnetic energy is for $B_K = B_{max}$ and $H_K = H_{max}$.

If the poles are furnished with pole shoes (Fig. 2.8b) the permeance of the external space increases. The point which characterizes a new state of the PM in Fig. 2.9 moves along the recoil line from the point K to the point A. The recoil line KG_M is the same as the internal permeance of the PM, i.e.,

$$G_M = \mu_{rec}\frac{w_M l_M}{h_M} = \mu_{rec}\frac{S_M}{h_M} \tag{2.26}$$

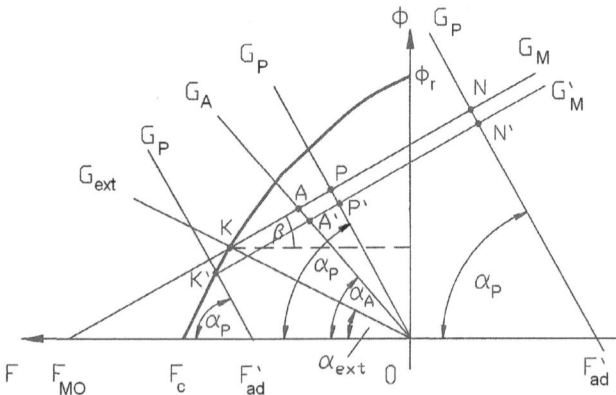

Fig. 2.9. Diagram of a PM for finding the origin of the recoil line and operating point.

The point A is the intersection of the recoil line KG_M and the straight line OG_A representing the leakage permeance of the PM with pole shoes, i.e.,

$$G_A = \frac{\Phi_A}{F_A}, \qquad \tan \alpha_A = G_A \frac{F_c}{\Phi_r} \qquad (2.27)$$

The energy produced by the PM in the external space decreases as compared with the previous case, i.e., $w_A = B_A H_A / 2$.

The next stage is to place the PM in an external ferromagnetic circuit as shown in Fig. 2.8c. The resultant permeance of this system is

$$G_P = \frac{\Phi_P}{F_P}, \qquad \tan \alpha_P = G_P \frac{F_c}{\Phi_r} \qquad (2.28)$$

which meets the condition $G_P > G_A > G_{ext}$. For an external magnetic circuit without any electric circuit carrying the armature current, the magnetic state of the PM is characterized by the point P (Fig. 2.9), i.e., the intersection of the recoil line KG_M and the permeance line OG_P.

When the external magnetic circuit is furnished with an armature winding and when this winding is fed with a current that produces an MMF magnetizing the PM (Fig. 2.8d), the magnetic flux in the PM increases to the value Φ_N. The d-axis MMF F'_{ad} of the external (armature) field acting directly on the PM corresponds to Φ_N. The magnetic state of the PM is described by the point N located on the recoil line on the right-hand side of the origin of the coordinate system. To obtain this point it is necessary to lay off the distance OF'_{ad} and to draw a line G_P from the point F'_{ad} inclined by the angle α_P to the F-axis. The intersection of the recoil line and the permeance line G_P gives the point N. If the exciting current in the external armature winding is increased further, the point N will move further along the recoil line to the right, up to the saturation of the PM.

When the excitation current is reversed, the external armature magnetic field will demagnetize the PM. For this case it is necessary to lay off the distance OF'_{ad} from the origin of the coordinate system to the left (Fig. 2.9). The line G_P drawn from the point F'_{ad} with the slope α_P intersects the demagnetization curve at the point K'. This point can be above or below the point K (for the PM alone in the open space). The point K' is the origin of a new recoil line $K'G'_M$. Now if the armature exciting current decreases, the operating point will move along the new recoil line $K'G'_M$ to the right. If the armature current drops down to zero, the operating point takes the position P' (intersection of the new recoil line $K'G'_M$ with the permeance line G_P drawn from the origin of the coordinate system).

On the basis of Fig. 2.9 the energies $w_{P'} = B_{P'} H_{P'}/2$, $w_P = B_P H_P/2$, and $w_{P'} < w_P$. The location of the origin of the recoil line, as well as the location of the operating point, determines the level of utilization of the energy produced by the PM. A PM behaves in a different way than a d.c. electromagnet; the energy of a PM is not constant if the permeance and exciting current of the external armature change.

The location of the origin of the recoil line is determined by the minimum value of the permeance of the external magnetic circuit or the demagnetization action of the external field.

To improve the properties of PMs independent of the external fields PMs are stabilized. *Stabilization* means the PM is demagnetized up to a value that is slightly higher than the most dangerous demagnetization field during the operation of a system where the PM is installed. In magnetic circuits with stabilized PMs the operating point describing the state of the PM is located on the recoil line.

2.5.2 Operating point for magnetization without armature

Let us assume that the PM has been magnetized without the armature and has then been placed in the armature system, e.g., the same as that for an electrical machine with an air gap. The beginning of the recoil line is determined by the leakage permeance G_{ext} of the PM alone located in open space (Fig. 2.10). In order to obtain the point K, the set of eqns (2.20), (2.25) in flux Φ—MMF coordinate system is to be solved; this results in the following second order equation:

$$a_0 G_{ext} F_K^2 - (G_{ext} F_c + \Phi_r) F_K + \Phi_r F_c = 0$$

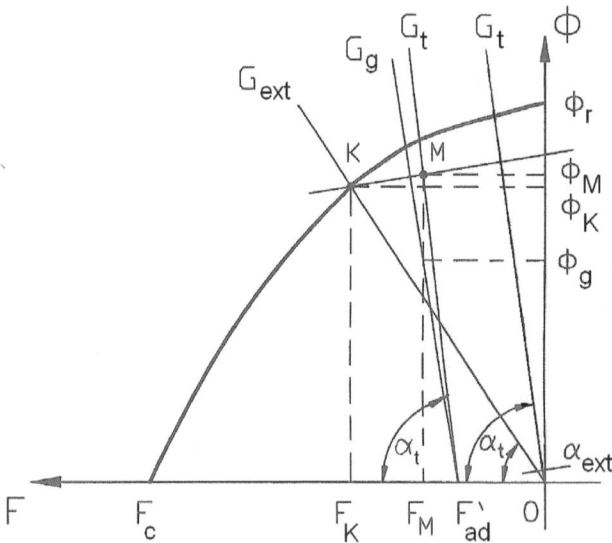

Fig. 2.10. Location of the operating point for the magnetization without the armature.

If $a_0 > 0$, the MMF corresponding to the point K is

$$F_K = \frac{F_c}{2a_0} + \frac{\Phi_r}{2a_0 G_{ext}} \pm \sqrt{\left(\frac{F_c}{2a_0} + \frac{\Phi_r}{2a_0 G_{ext}}\right)^2 - \frac{\Phi_r F_c}{a_0 G_{ext}}}$$

$$= b_0 \pm \sqrt{b_0^2 - c_0} \qquad (2.29)$$

where

$$b_0 = \frac{F_c}{2a_0} + \frac{\Phi_r}{2a_0 G_{ext}} \quad \text{and} \quad c_0 = \frac{\Phi_r F_c}{a_0 G_{ext}}$$

If $a_0 = 0$ (for rare-earth PMs), the MMF F_K is

$$F_K = \frac{\Phi_r}{G_{ext} + \Phi_r/F_c} \qquad (2.30)$$

The magnetic flux Φ_K can be found on the basis of eqn (2.25).

The equation of the recoil line for $a_0 > 0$ is

(i) in the B—H coordinate system

$$B = B_K + (H_K - H)\mu_{rec} \qquad (2.31)$$

(ii) in the flux Φ—MMF coordinate system

$$\Phi = \Phi_K + (F_K - F)\mu_{rec}\frac{S_M}{h_M} \qquad (2.32)$$

For rare-earth PMs with $a_0 = 0$ the recoil permeability $\mu_{rec} = (h_M/S_M)(\Phi_r - \Phi_K)/F_K = \Phi_r h_M/(F_c S_M)$ and the equation of the recoil line is the same as that for the demagnetization line, i.e.,

$$\Phi = \Phi_r\left(1 - \frac{F}{F_c}\right) \qquad (2.33)$$

The d-axis armature MMF F'_{ad} acting directly on the magnet usually demagnetizes the PM, so that the line of the resultant magnetic permeance,

$$G_t = \frac{\Phi_M}{F_M - F'_{ad}} \qquad (2.34)$$

intersects the recoil line between the point K and the magnetic flux axis. Solving eqn (2.32) in which $\Phi = \Phi_M$ and $F = F_M$ and eqn (2.34) the MMF of the PM is given by the equation

$$F_M = \frac{\Phi_K + F_K\mu_{rec}(S_M/h_M) + G_t F'_{ad}}{G_t + \mu_{rec}(S_M/h_M)} \qquad (2.35)$$

For rare-earth PMs eqns (2.33) and (2.34) should be solved to obtain

$$F_M = \frac{\Phi_r + G_t F'_{ad}}{G_t + \Phi_r/F_c} \quad (2.36)$$

The magnetic flux $\Phi_M = G_t(F_M - F'_{ad})$ in the PM is according to eqn (2.34). The useful flux density in the air gap can be found using the coefficient of leakage flux (2.10), i.e.,

$$B_g = \frac{\Phi_M}{S_g \sigma_{lM}} = \frac{G_t(F_M - F'_{ad})}{S_g \sigma_{lM}}$$

$$= \frac{G_t}{S_g \sigma_{lM}} \left[\frac{\Phi_K + F_K \mu_{rec}(S_M/h_M) + G_t F'_{ad}}{G_t + \mu_{rec}(S_M/h_M)} - F'_{ad} \right] \quad (2.37)$$

where S_g is the surface of the air gap. With the fringing effect being neglected the corresponding magnetic field intensity is

$$H_g = H_M = \frac{F_M}{h_M} = \frac{\Phi_K + F_K \mu_{rec}(S_M/h_M) + G_t F'_{ad}}{h_M[G_t + \mu_{rec}(S_M/h_M)]} \quad (2.38)$$

The external magnetic energy per volume is

$$w_{ext} = \frac{B_M H_M}{2}$$

$$= \frac{G_t}{V_M} \left[\frac{\Phi_K + F_K \mu_{rec}(S_M/h_M) + G_t F'_{ad}}{G_t + \mu_{rec}(S_M/h_M)} - F'_{ad} \right]$$

$$\times \frac{\Phi_K + F_K \mu_{rec}(S_M/h_M) + G_t F'_{ad}}{G_t + \mu_{rec}(S_M/h_M)} \quad (2.39)$$

where $B_M = B_g \sigma_{lM}$ and $V_M = 2h_M S_g$. The above equation is correct if $G_{ext} < G_t$. A PM produces the maximum energy in the external space if the operating point is located on the demagnetization curve and the absolute maximum of this energy is for $\tan \alpha_{ext} = G_{ext}(F_c/\Phi_r) \approx 1.0$.

In the general case, the resultant permeance G_t of the external magnetic circuit consists of the useful permeance G_g of the air gap and the leakage permeance G_{lM} of the PM, i.e.,

$$G_t = G_g + G_{lM} = \sigma_{lM} G_g \quad (2.40)$$

The useful permeance G_g corresponds to the useful flux in the active portion of the magnetic circuit. The leakage permeance G_{lM} is the referred leakage permeance of a single PM or PM with armature. Consequently, the external energy w_{ext} can be divided into the useful energy w_g and leakage energy w_{lM}. The useful energy per volume in the external space is

$$w_g = \frac{B_g H_g}{2}$$

$$= \frac{G_g}{V_M} \left[\frac{\Phi_K + F_K \mu_{rec}(S_M/h_M) + G_t F'_{ad}}{G_t + \mu_{rec}(S_M/h_M)} - F'_{ad} \right]$$

$$\times \frac{\Phi_K + F_K \mu_{rec}(S_M/h_M) + G_t F'_{ad}}{G_t + \mu_{rec}(S_M/h_M)} \tag{2.41}$$

The leakage energy

$$w_{lM} = w_{ext} - w_g = (\sigma_{lM} - 1)w_g \tag{2.42}$$

2.5.3 Operating point for magnetization with armature

If the magnet is placed in the external armature circuit and then magnetized by the armature field or magnetized in a magnetizer and then the poles of the magnetizer are in a continuous way replaced by the poles of the armature, the origin K of the recoil line is determined by the resultant magnetic permeance G_t drawn from the point F'_{admax} at the F-coordinate. The MMF F'_{admax} corresponds to the maximum demagnetizing d-axis field acting directly on the magnet which can appear during the machine operation. In Fig. 2.11 this is the intersection point K of the demagnetization curve and the line G_t:

$$G_t = \frac{\Phi_K}{F_K - F'_{admax}} \tag{2.43}$$

The maximum armature demagnetizng MMF F'_{admax} can be determined for the reversal or locked-rotor condition.

The beginning of the recoil line permanent magnet!recoil line is determined by the resultant permeance G_t of the PM mounted in the armature (Fig. 2.11). In order to obtain the point K, the set of eqns (2.20), (2.43) in the flux Φ—MMF coordinate system is solved; this results in the following second order equation:

$$a_0 G_t F_K^2 - (G_t F_c + a_0 G_t F'_{admax} + \Phi_r)F_K + (G_t F'_{admax} + \Phi_r)F_c = 0$$

If $a_0 > 0$, the MMF corresponding to the point K is

$$F_K = b_0 \pm \sqrt{b_0^2 - c_0} \tag{2.44}$$

where

$$b_0 = 0.5 \left(\frac{F_c}{a_o} + F'_{admax} + \frac{\Phi_r}{a_0 G_t} \right) \quad \text{and} \quad c_0 = \frac{(G_t F'_{admax} + \Phi_r)F_c}{a_0 G_t}$$

If $a_0 = 0$ (for rare-earth PMs), the MMF F_K is

$$F_K = \frac{\Phi_r + G_t F'_{admax}}{G_t + \Phi_r/F_c} \tag{2.45}$$

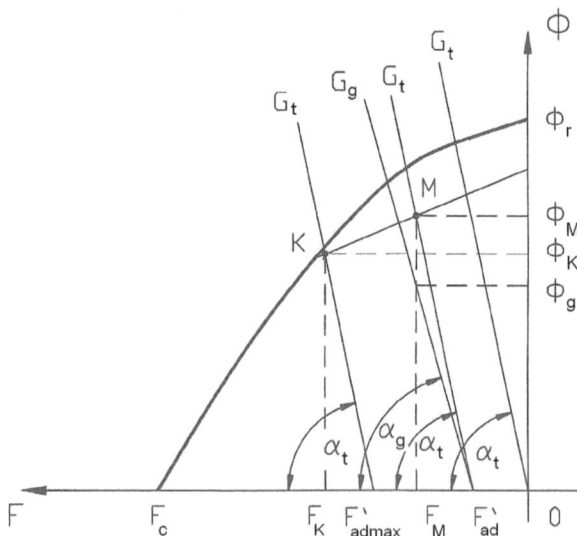

Fig. 2.11. Location of the operating point for the magnetization with the armature.

The magnetic flux Φ_K can be found on the basis of eqn (2.43).

The rest of the construction is similar to that shown in Fig. 2.10 for the demagnetization action of the armature winding (point M). The coordinates of the point M are expressed by eqns (2.34), (2.35) and (2.36).

2.5.4 Magnets with different demagnetization curves

In practical electrical machines, except for very small motors, the excitation magnetic flux is produced by more than one magnet. The operating point is found assuming that all magnets have the same demagnetization curves. Owing to nonhomogeneous materials, deviations of dimensions and differences in magnetization, PMs of the same type can have different demagnetization curves. Manufacturers of PMs give both maximum and minimum B_r and H_c of magnets of the same type (Tables 2.2, 2.3 and 2.4) or permissible deviations $\pm \Delta B_r$ and $\pm \Delta H_c$.

Permissible disparities of properties of PMs assembled in series are dependent on the permeance of the external magnetic circuit. The higher the permeance of the external magnetic circuit, the more homogeneous PMs are required [78]. In an unfavorable case, the stronger magnet magnetizes the weaker magnet and is subject to faster demagnetization than a case in which all magnets have the same B_r and H_c [78].

2.6 Permeances for main and leakage fluxes

2.6.1 Permeance evaluation by flux plotting

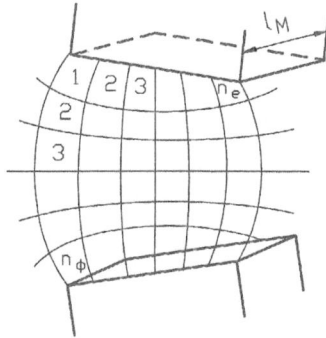

Fig. 2.12. Permeance evaluation by flux plotting.

The procedure to be followed in field plotting is simple. On a diagram of the magnetic circuit, several equipotential lines are drawn. Flux lines connecting the surfaces of opposite polarity are then added in such a manner so as to fulfill the following requirements:

- all flux lines and equipotential lines must be mutually perpendicular at each point of intersection,
- each figure bonded by two adjacent flux lines and two adjacent equipotential lines must be a curvilinear square,
- the ratio of the average width to average height of each square should be equal to unity.

When the full plot has been completed, the magnetic permeance can be found by dividing the number of curvilinear squares between any two adjacent equipotential lines, designated as n_e, by the number of curvilinear squares between any two adjacent flux lines, n_Φ, and multiplying by the length l_M of the field perpendicular to the plane of the flux plot (Fig. 2.12), i.e.,

$$G = \mu_0 \frac{n_e}{n_\Phi} l_M \tag{2.46}$$

Table 2.6 shows equations for calculating the permeances of air gaps between poles of different configurations.

2.6.2 Permeance evaluation by dividing the magnetic field into simple solids

Fig. 2.13 shows a flat model of an electrical machine with smooth armature core (without slots) and salient-pole PM excitation system. The armature is

Table 2.6. Permeances of air gaps between poles of different configurations

System	Configuration of poles	Permeance
1		Rectangular poles (neglecting fringing flux paths) $$G = \mu_0 \frac{w_M l_M}{g}$$ where $g/w_M < 0.1$ and $g/l_M < 0.1$
2		Halfspace and a rectangular pole $$G = \mu_0 \frac{1}{g}(w_M + \frac{0.614g}{\pi})(l_M + \frac{0.614g}{\pi})$$
3		Fringe paths originating on lateral flat surfaces $$G = \mu_0 \frac{x w_M}{0.17g + 0.4x}$$ or $$G = \mu_0 \frac{w_M}{\pi} \ln\left[1 + 2\sqrt{\frac{x + (x^2 + xg)}{g}}\right]$$
4		Cylindrical poles (neglecting fringing flux) $$G = \mu_0 \frac{\pi d_M^2}{4g}$$ More accurate formula for $g/d_M < 0.2$ is $$G = \mu_0 d_M \left[\frac{\pi d_M}{4g} + \frac{0.36 d_M}{2.4 d_M + g} + 0.48\right]$$
5	as above	Fringe paths originating on lateral cylindrical surfaces $$G = \mu_0 \frac{x d_M}{0.22g + 0.4x}$$

Table 2.6. Permeances of air gaps between poles of different configurations (continued)

System	Configuration of poles	Permeance
6		Between lateral surfaces inclined by an angle Θ $G = \mu_0 \frac{l_M}{\Theta} \int_{R_1}^{R_2} \frac{dx}{x} = \mu_0 \frac{l_M}{\Theta} \ln \frac{R_2}{R_1}$
7		Between rectangles lying on the same surface $G = \mu_0 \frac{1}{2\pi} \ln[2m^2 - 1 + 2m\sqrt{m^2 - 1}] l_M$ or $G = \mu_0 \frac{1}{\pi} \ln \left(1 + \frac{2w_M}{g}\right) l_M$
8		Between two rectangles of different area lying in the same plane $G = \mu_0 \frac{1}{\pi} \ln \left[\frac{\Delta^2 - (\epsilon + x)^2}{\Delta(g - x)} - \frac{\epsilon + x}{\Delta} \right] l_M$ where $\epsilon = \frac{w_2 - w_1}{2}$, $2\Delta = w_1 + w_2 + 2g$, $x = \frac{1}{2\epsilon}[\Delta^2 - g^2 - \epsilon^2$ $- \sqrt{\Delta^2 - g^2 - \epsilon^2 - 4\epsilon^2 g^2}]$
9		Single cylindrical airgap of a salient-pole electrical machine $G = \mu_0 \frac{l_M \Theta}{\ln(1 + g/h_M)}$ When $g/h_M \leq 0.02$ then the formula reduces to $G = \mu_0 \frac{l_M h_M \Theta}{g}$

Table 2.6. Permeances of air gaps between poles of different configurations (continued)

System	Configuration of poles	Permeance
10		Cylindrical space between two salient poles without a rotor $$G = \mu_0 l_M \int_0^\Theta \frac{\tan \alpha}{\alpha} \, d\alpha$$ To take into account the fringing flux the permeance G should be increased by 10 to 15%
11		Between a cylinder parallel to the salient pole with rectangular cross section at $w_M > 4h$ $$G = \mu_0 \frac{\pi}{\ln(2n + \sqrt{4n^2 - 1})} l_M$$ where $n = h/(2r)$. For $w_M = (1.25 \ldots 2.5)h$ the permeance G should be multiplied by the correction factor $0.85 \ldots 0.92$
12		A cylinder located parallelly and symmetrically between two salient poles of rectangular cross section $$G = \mu_0 \frac{(1.25 \ldots 1.40)\pi}{\ln(2n + \sqrt{4n^2 - 1})} l_M$$ where $n = h/(2r)$
13		Between two parallel cylinders of different diameters $$G = \mu_0 \frac{2\pi}{\ln(u + \sqrt{u^2 - 1})}$$ where $u = \frac{h^2 - r_1^2 - r_2^2}{2r_1 r_2}$

in the form of a cylinder made of steel laminations. The PMs are fixed to the mild steel ferromagnetic rotor yoke.

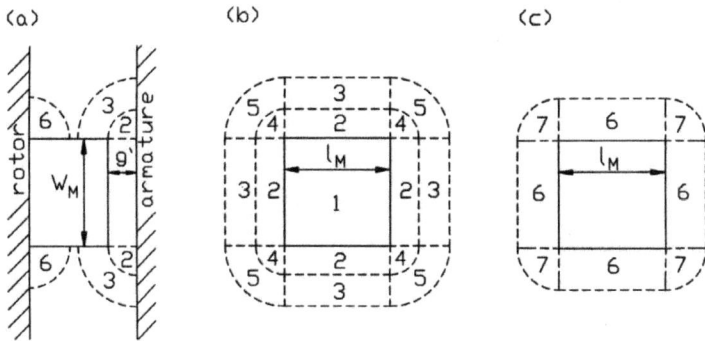

Fig. 2.13. Flat model of a simple PM electric machine — division of the space occupied by the magnetic field into simple solids: (a) longitudinal section, (b) air gap field, (c) leakage field. The width of the PM w_M is equal to its length l_M.

The pole pitch (circumference per pole) is τ, the width of each PM is w_M, and its length is l_M. It is assumed that the width of a PM is equal to its length, i.e., $w_M = l_M$. The space between the pole face and the armature core is divided into a prism (1), four quarters of a cylinder (2), four quarters of a ring (3), four pieces of 1/8 of a sphere (4), and four pieces of 1/8 of a shell (5). Formulae for the permeance calculations are found on the assumption that the permeance of a solid is equal to its average cross section area to the average length of the flux line. Neglecting the fringing flux, the permeance of a rectangular air gap per pole (prism 1 in Fig. 2.13) is

$$G_{g1} = \mu_0 \frac{w_M l_M}{g'} \tag{2.47}$$

The equivalent air gap g' is only equal to the nonferromagnetic gap (mechanical clearance) g for a slotless and unsaturated armature core. To take into account slots (if they exist) and magnetic saturation, the air gap g is increased to $g' = g k_C k_{sat}$, where $k_C > 1$ is Carter's coefficient taking into account slots (Appendix A, eqn (A.22)), and $k_{sat} > 1$ is the saturation factor of the magnetic circuit defined as the ratio of the sum of magnetic voltage drops (MVDs) per pole in the iron and air gap pair to the air gap MVD taken twice, i.e.,

$$k_{sat} = \frac{2(V_g + V_{1t} + V_{2t}) + V_{1y} + V_{2y}}{2V_g}$$

$$= 1 + \frac{2(V_{1t} + V_{2t}) + V_{1y} + V_{2y}}{2V_g} \tag{2.48}$$

where V_g is the MVD across the air gap, V_{1t} is the MVD along the armature teeth (if they exist), V_{2t} is the MVD along the PM pole shoe teeth (if there is a cage winding), V_{1y} is the MVD along the armature yoke, and V_{2y} is the MVD along the excitation system yoke.

To take into account the fringing flux it is necessary to include all paths for the magnetic flux coming from the excitation system through the air gap to the armature system (Fig. 2.13), i.e.,

$$G_g = G_{g1} + 4(G_{g2} + G_{g3} + G_{g4} + G_{g5}) \qquad (2.49)$$

where G_{g1} is the air gap permeance according to eqn (2.47) and G_{g2} to G_{g5} are the air gap permeances for fringing fluxes. The permeances G_{g2} to G_{g5} can be found using the formulae for calculating the permeances of simple solids given in Table 2.7.

In a similar way, the resultant permeance for the leakage flux of the PM can be found, i.e.,

$$G_{lM} = 4(G_{l6} + G_{l7}) \qquad (2.50)$$

where G_{l6} and G_{l7} are the permeances for leakage fluxes between the PM and rotor yoke according to Fig. 2.13c.

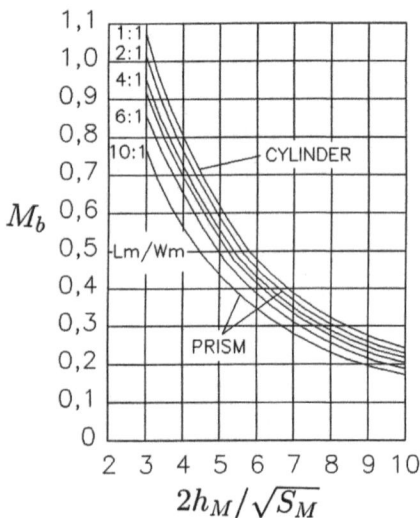

Fig. 2.14. Ballistic coefficient of demagnetization M_b for cylinders and prisms with different ratios of l_M/w_M (experimental curves).

Table 2.7. Equations for calculating the permeances of simple solids

System	Auxiliary sketch (configuration)	Permeance
1		Rectangular prism $G = \mu_0 \frac{w_M l_M}{g}$
2		Cylinder $G = \mu_0 \frac{\pi d_M^2}{4g}$
3		Half-cylinder $G = 0.26\mu_0 l_M$ $g_{av} = 1.22g, \ S_{av} = 0.322g l_M$
4		One-quarter of a cylinder $G = 0.52\mu_0 l_M$
5		Half-ring $G = \mu_0 \frac{2l_M}{\pi(g/c+1)}$ For $g < 3c$, $\ G = \mu_0 \frac{l_M}{\pi} \ln\left(1 + \frac{2c}{g}\right)$
6		One-quarter of a ring $G = \mu_0 \frac{2l_M}{\pi(g/c+0.5)}$ For $g < 3c$, $\ G = \mu_0 \frac{2l_M}{\pi} \ln\left(1 + \frac{c}{g}\right)$

Table 2.7. Equations for calculating the permeances of simple solids (continued)

System	Auxiliary sketch (configuration)	Permeance
7		One-quarter of a sphere $G = 0.077\mu_0 g$
8		One-eighth of a sphere $G = 0.308\mu_0 g$
9		One-quarter of a shell $G = \mu_0 \frac{c}{4}$
10		One-eighth of a shell $G = \mu_0 \frac{c}{2}$
11		Ring with a semi-circular cross section $G = 1.63\mu_0 \left(r + \frac{g}{4}\right)$
12		Hollow ring with a semi-circular cross section $G = \mu_0 \frac{4(r+0.5g)}{1+0.5g}$ For $g < 3c$, $G = \mu_o(2r + g)\ln\left(1 + \frac{2c}{g}\right)$

2.6.3 Calculation of leakage permeances for prisms and cylinders located in an open space

In the case of simple shaped PMs, the permeance for leakage fluxes of a PM alone can be found as

$$G_{ext} = \mu_0 \frac{2\pi}{M_b} \frac{S_M}{h_M} \qquad (2.51)$$

where M_b is the ballistic coefficient of demagnetization. This coefficient can be estimated with the aid of graphs [20] as shown in Fig. 2.14. The cross section area of a cylindrical PM is $S_M = \pi d_M^2/4$ where d_M is the PM diameter and $S_M = w_M l_M$ for a rectangular PM. In the case of hollow cylinders (rings) the coefficient M_b is practically the same as that for solid cylinders.

For cylindrical PMs with small h_M and large cross sections $\pi d_M^2/4$ (button-shaped PMs), the leakage permeance can be calculated using the following equation

$$G_{ext} \approx 0.716 \mu_o \frac{d_M^2}{h_M} \qquad (2.52)$$

Eqns (2.51) and (2.52) can be used for finding the origin K of the recoil line for PMs magnetized without the armature (Fig. 2.10).

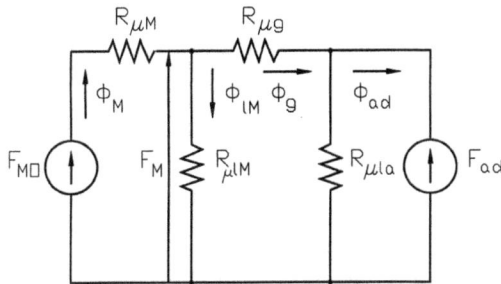

Fig. 2.15. Equivalent circuit of a PM system with armature.

2.7 Calculation of magnetic circuits with permanent magnets

Fig. 2.15 shows the equivalent magnetic circuit of a PM system with armature. The reluctances of pole shoes (mild steel) and armature stack (electrotechnical laminated steel) are much smaller than those of the air gap and PM and have been neglected. The "open circuit" MMF acting along the internal magnet permeance $G_M = 1/R_{\mu M}$ is $F_{M0} = H_{M0}h_M$ (Fig. 2.9). For a linear demagnetization curve $H_{M0} = H_c$. The d-axis armature reaction MMF is F_{ad}, the total magnetic flux of the permanent magnet is Φ_M, the leakage flux of the PM is Φ_{lM}, the useful air gap magnetic flux is Φ_g, the leakage flux of the external armature system is Φ_{la}, the flux produced by the armature is Φ_{ad} (demagnetizing or magnetizing), the reluctance for the PM leakage flux is $R_{\mu lM} = 1/G_{lM}$, the air gap reluctance is $R_{\mu g} = 1/G_g$, and the external armature leakage reactance is $R_{\mu la} = 1/G_{gla}$. The following Kirchhoff's equations can be written on the basis of the equivalent circuit shown in Fig. 2.15

$$\Phi_M = \Phi_{lM} + \Phi_g$$

$$\Phi_{la} = \frac{\pm F_{ad}}{R_{\mu la}}$$

$$F_{M0} - \Phi_M R_{\mu M} - \Phi_{lM} R_{\mu lM} = 0$$

$$\Phi_{lM} R_{lM} - \Phi_g R_{\mu g} \mp F_{ad} = 0$$

The solution to the above equation system gives the air gap magnetic flux:

$$\Phi_g = \left[F_{M0} \mp F_{ad} \frac{G_g}{G_g + G_{lM}} \frac{(G_g + G_{lM})(G_M + G_{lM})}{G_g G_M} \right] \frac{G_g G_M}{G_g + G_{lM} + G_M}$$

or

$$\Phi_g = \left[F_{M0} \mp F'_{ad} \frac{G_t(G_M + G_{lM})}{G_g G_M} \right] \frac{G_g G_M}{G_t + G_M} \tag{2.53}$$

where the total resultant permeance G_t for the flux of the PM is according to eqn (2.40) and the direct-axis armature MMF acting directly on the PM is

$$F'_{ad} = F_{ad} \frac{G_g}{G_g + G_{lM}} = F_{ad} \left(1 + \frac{G_{lM}}{G_g} \right)^{-1} = \frac{F_{ad}}{\sigma_{lM}} \tag{2.54}$$

The upper sign in eqn (2.53) is for the demagnetizing armature flux and the lower sign is for the magnetizing armature flux.

The general expression for the coefficient of the PM leakage flux is

$$\sigma_{lM} = 1 + \frac{\Phi_{lM}}{\Phi_g} = 1 + \frac{G_{lM}}{G_g} \tag{2.55}$$

2.8 Mallinson–Halbach array and Halbach cylinder

In 1972 J.C. Mallinson of Ampex Corporation, Redwood City, CA, U.S.A. discovered a "magnetic curiosity" of a PM configuration that concentrates magnetic flux on one side of the array and cancels it to near zero on the other [206]. Another interesting and unique quality of this configuration is that the array of PMs is stronger than its individual components, i.e., a single PM, because of superposition of field lines. This effect was rediscovered in the late 1970s by K. Halbach of Lawrence Berkeley National Laboratory, Berkeley Hills, CA, U.S.A., applied to particle accelerators and expanded upon cylindrical configurations [135, 136, 137]. The polarities of individual PMs in the array are arranged such that the magnetization vector rotates as a function of distance along the array.

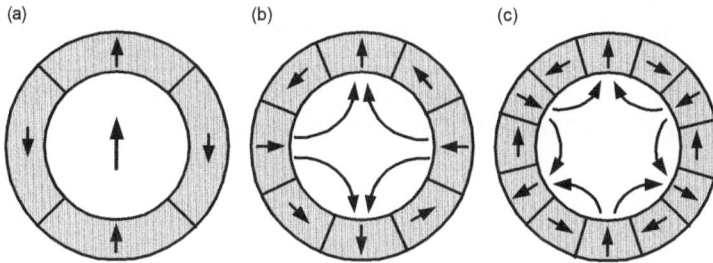

Fig. 2.16. Halbach cylinders: (a) $\lambda = 2\pi$, $2p = 2$; (b) $\lambda = \pi$, $2p = 4$; $\lambda = 2\pi/3$, $2p = 6$.

A Halbach cylinder is a cylinder composed of rare earth PMs producing an intense magnetic field confined entirely inside or outside the cylinder with zero field on the other cylindrical surface (Fig. 2.16). The magnetization vector can be described as [135, 317]

$$\mathbf{M} = M_r \cos(p\theta)\mathbf{1_r} + M_r \sin(\pm p\theta)\mathbf{1_\theta} \qquad (2.56)$$

where M_r is the remanent magnetization, $\mathbf{1_r}$, $\mathbf{1_\theta}$ are unit vectors in radial and tangential directions, repectively, r, θ are cylindrical coordinates, p is the number of pole pairs (wave number), the sign $+$ is for internal field and the sign $-$ is for external field. The number of poles expressed with the aid of wavelength (spatial period of the array) λ is

$$p = \frac{2\pi}{\lambda} \qquad (2.57)$$

The uniform magnetic flux density inside the cylinder is described by the following equation [135]

$$B = B_r \ln \left(\frac{D_{out}}{D_{in}} \right) \left(\frac{\sin \pi/n_M}{\pi/n_M} \right) \tag{2.58}$$

where B_r is the remanent magnetic flux density, D_{out} is the outer diameter, D_{in} is the inner diameter and n_M is the number of PM pieces per wavelength λ. For $\lambda = 2\pi$ the number of pole pairs is $p = 2\pi/2\pi = 1$, for $\lambda = \pi$ the number of pole pairs $p = 2$ and for $\lambda = 2\pi/3$ the number of poles $p = 3$ (Fig. 2.16). If the ratio of outer to inner diameters is greater than the base of the natural logarithm e, the magnetic flux density inside the bore exceeds the remanence flux density of the PM. Magnetic fields of over 5 T in a 2 mm gap in Mallinson–Halbach type PM dipole at room temperature has been achieved [188]. Mallinson–Halbach array has the following advantages:

- the fundamental field is stronger by a factor of 1.4 than in a conventional PM array, and thus the power efficiency of the machine is doubled;
- the array of PMs does not require any backing steel magnetic circuit and PMs can be bonded directly to a non-ferromagnetic supporting structure (aluminum, plastics);
- the magnetic field is more sinusoidal than that of a conventional PM array;
- Mallinson–Halbach array has very low back-side fields.

Numerical examples

Numerical example 2.1

A simple stationary magnetic circuit is shown in Fig. 2.17. There are two Vacodym 362TP NdFeB PMs (Table 2.3) with minimum value of $B_r = 1.25$ T, minimum value of $H_c = 950$ kA/m, temperature coefficients $\alpha_B = -0.13\%/^0$C and $\alpha_H = -0.61\%/^0$C at $20 \leq \vartheta_{PM} \leq 150^0$C. The height of the PM per pole is $h_M = 6$ mm and the air gap thickness $g = 1$ mm. The U-shaped and I-shaped (top) ferromagnetic cores are made of a laminated electrotechnical steel. The width of the magnets and cores is 17 mm. Calculate the air gap magnetic flux density, air gap magnetic field strength, the useful energy of PMs and normal attraction force per two poles at: (a) $\vartheta_{PM} = 20^0$C and (b) $\vartheta_{PM} = 100^0$C. The MVD in the laminated core, leakage and fringing magnetic flux can be neglected.

<u>Solution:</u>

(a) Magnet temperature $\vartheta_{PM} = 20^0$C

The relative recoil magnetic permeability according to eqn (2.5) for a straight line demagnetization curve

$$\mu_{rrec} = \frac{1}{\mu_0} \frac{\Delta B}{\Delta H} = \frac{1}{0.4\pi \times 10^{-6}} \frac{1.25 - 0}{950,000 - 0} \approx 1.05$$

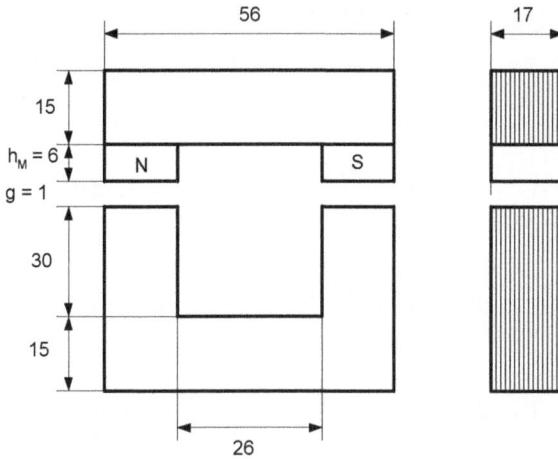

Fig. 2.17. A simple stationary magnetic circuit with PMs and air gap. *Numerical example 2.1.*

The air gap magnetic flux density according to eqn (2.14)

$$B_g \approx \frac{1.25}{1 + 1.05 \times 1.0/6.0} = 1.064 \text{ T}$$

The air gap magnetic field strength according to eqn (2.24) in which $H = H_g$ and $B = B_g$

$$H_g = H_c \left(1 - \frac{B_g}{B_r} \right) = 950 \times 10^3 \left(1 - \frac{1.064}{1.25} \right) = 141.15 \times 10^3 \text{ A/m}$$

The useful energy per magnet volume

$$w_g = \frac{B_g H_g}{2} = \frac{1.064 \times 141,150}{2} = 75,112.8 \text{ J/m}^3$$

The useful energy per pole pair according to eqn (2.12)

$$W_g = w_g V_M = 75,112.8(2 \times 6 \times 15 \times 17 \times 10^{-9}) = 0.23 \text{ J}$$

The normal attraction force per 2 poles

$$F = \frac{B_g^2}{2\mu_0}(2S_M) = \frac{1.064^2}{0.4\pi \times 10^{-6}}(15 \times 17 \times 10^{-6}) = 229.8 \text{ N}$$

(b) Magnet temperature $\vartheta_{PM} = 100^0$C

The remanent magnetic flux density and coercivity at 100^0C according to eqns (2.18) and (2.19)

$$B_r = 1.25 \left[1 + \frac{-0.13}{100}(100 - 20) \right] = 1.12 \text{ T}$$

$$H_c = 950 \times 10^3 \left[1 + \frac{-0.61}{100}(100 - 20) \right] = 486.4 \times 10^3 \text{ A/m}$$

At $\vartheta_{PM} = 100^0$C the demagnetization curve is nonlinear. Its linear part is only between 0.6 T and B_r parallel to the demagnetization curve at 20^0C.
  The air gap magnetic flux density according to eqn (2.14)

$$B_g \approx \frac{1.12}{1 + 1.05 \times 1.0/6.0} = 0.954 \text{ T}$$

The air gap magnetic field strength according to eqn (2.24) in which $H = H_g$ and $B = B_g$

$$H_g = H_c \left(1 - \frac{B_g}{B_r} \right) = 486.4 \times 10^3 \left(1 - \frac{0.954}{1.12} \right) = 72.27 \times 10^3 \text{ A/m}$$

The useful energy per magnet volume

$$w_g \approx \frac{0.954 \times 72,27}{2} = 34,458.2 \text{ J/m}^3$$

The useful energy per pole pair according to eqn (2.12)

$$W_g = 34,458.2(2 \times 6 \times 15 \times 17 \times 10^{-9}) = 0.105 \text{ J}$$

The normal attraction force per 2 poles

$$F = \frac{0.954^2}{0.4\pi \times 10^{-6}}(15 \times 17 \times 10^{-6}) = 184.5 \text{ N}$$

Numerical example 2.2

A primitive 2-pole electrical machine is shown in Fig. 2.18. The anisotropic barium ferrite PM *Hardferrite 28/26* (Fig. 2.5) [204] with the dimensions $h_M = 15$ mm, $w_M = 25$ mm, and $l_M = 20$ mm and demagnetization curve as shown in Fig. 2.5, has been magnetized after being assembled in the armature. The remanent magnetic flux density is $B_r = 0.4$ T, the coercive force is $H_c = 265,000$ A/m, and the recoil magnetic permeability is $\mu_{rec} = 1.35 \times 10^{-6}$ H/m. The cross section area of the armature magnetic circuit is 25×20 mm^2

and its average length is $l_{Fe} = 70 + 2 \times 50 + 2 \times 20 = 210$ mm. It is made of a cold-rolled laminated steel with its thickness 0.5 mm and stacking factor $k_i = 0.96$. The air gap across one pole (mechanical clearance) $g = 0.6$ mm. The armature winding is wound with $N = 1100$ turns of copper wire, the diameter of which is $d_a = 0.75$ mm (without insulation). The machine is fed from a d.c. voltage source $V = 24$ V via a rheostat. Find the magnetic flux density in the air gap when the armature winding is fed with current $I_a = 1.25$ A.

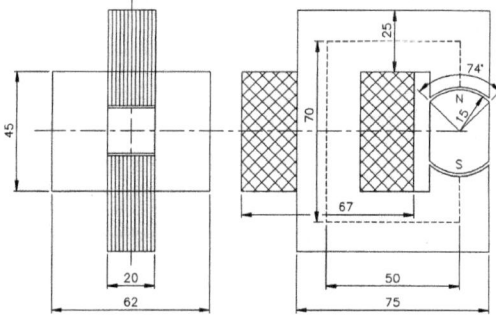

Fig. 2.18. A primitive 2-pole PM machine. *Numerical example 2.2.*

Solution:

1. The armature winding is distributed in 22 layers, each of them consisting of 50 turns. Without insulation, the height of the winding is $0.75 \times 50 = 37.5$ mm and its thickness $0.75 \times 22 = 16.5$ mm.
2. The average length of each armature turn

$$l_{av} \approx 2(46 + 41) = 174 \ \text{mm}$$

3. The cross section area of the armature conductor

$$s_a = 0.25\pi d_a^2 = 0.25\pi \times (0.75 \times 10^{-3})^2 = 0.4418 \times 10^{-6} \ \text{m}^2$$

4. The resistance of the winding at 75^0C

$$R_a = \frac{N l_{av}}{\sigma s_a} = \frac{1100 \times 0.174}{47 \times 10^6 \times 0.4418 \times 10^{-6}} = 9.218 \ \Omega$$

5. The maximum armature current is for the external resistance $R_{rhe} = 0$, i.e.,

$$I_{amax} = \frac{V}{R_a} = \frac{24.0}{9.218} = 2.6 \ \text{A}$$

6. Current density corresponding to the maximum armature current

$$J_{amax} = \frac{2.6}{0.4418 \times 10^{-6}} = 5.885 \times 10^{-6} \ \text{A/m}^2$$

The winding can withstand this current density without forced ventilation for a short period of time.

7. Maximum armature demagnetizing MMF acting directly on the PM and referred to one pole

$$F'_{admax} \approx F_{admax} = \frac{I_{amax}N}{2p} = \frac{2.6 \times 1100}{2} = 1430 \ \text{A} \approx 1.43 \ \text{kA}$$

8. Maximum armature demagnetizing magnetic field intensity in d-axis

$$H'_{admax} = \frac{F'_{admax}}{h_M} = \frac{1430}{0.015} = 95,333.3 \ \text{A/m}$$

9. The MMF corresponding to the *Hardferrite 28/26* coercive force $H_c = 265,000 \ \text{A}$

$$F_c = H_c h_M = 26,5000 \times 0.015 = 3975 \ \text{A} = 3.975 \ \text{kA}$$

10. At the rated armature current $I_a = 1.25 \ \text{A}$ the current density is only $J_a = 1.25/(0.4418 \times 10^{-6}) = 2.83 \times 10^6 \ \text{A/m}^2$ and the winding can withstand the continuous current $I_a = 1.25 \ \text{A}$ without any forced ventilation system.

11. The d-axis MMF and magnetic field intensity acting directly on the PM at $I_a = 1.25 \ \text{A}$ are

$$F'_{ad} = \frac{1.25 \times 1100}{2} = 687.5 \ \text{A}, \qquad H'_{ad} = \frac{687.5}{0.015} = 45,833.3 \ \text{A/m}$$

12. The permeance of the air gap according to Table 2.6, configuration 9

$$G_g = \mu_o \frac{l_M \Theta}{\ln(1 + g/h_M)} = 0.4\pi \times 10^{-6} \frac{0.02 \times 1.2915}{\ln(1 + 0.0006/0.015)}$$

$$= 0.876 \times 10^{-6} \ \text{H}$$

The angle $\Theta = 74^0 = 1.2915$ rad according to Fig. 2.18.

13. Assuming the leakage factor $\sigma_{lM} = 1.15$, the total permeance is

$$G_t = \sigma_{lM} G_g = 1.15 \times 0.876 \times 10^{-6} = 1.0074 \times 10^{-6} \ \text{H}$$

14. The demagnetization curve of *Hardferrite 28/26* shown in Fig. 2.5 can be approximated with the aid of eqns (2.20) and (2.22). Five points ($n = 5$) with coordinates (257,000; 0.04), (235,000; 0.08), (188,000; 0.14), (146,000;

0.2), and (40,000; 0.34) have been selected to estimate the coefficient a_0, i.e.,

$$a_0 = \frac{1}{5}\Big(\frac{265,000}{257,000} + \frac{0.4}{0.04} - \frac{265,000}{257,000}\frac{0.4}{0.04} + \frac{265,000}{235,000} + \frac{0.4}{0.08} - \frac{265,000}{235,000}\frac{0.4}{0.08}$$

$$+\frac{265,000}{188,000} + \frac{0.4}{0.14} - \frac{265,000}{188,000}\frac{0.4}{0.14} + \frac{265,000}{146,000} + \frac{0.4}{0.20} - \frac{265,000}{146,000}\frac{0.4}{0.20}$$

$$+\frac{265,000}{40,000} + \frac{0.4}{0.34} - \frac{265,000}{40,000}\frac{0.4}{0.34}\Big)$$

$$= \frac{1}{5}(0.7201 + 0.4896 + 0.2393 + 0.185 + 0.0104) = 0.329$$

Demagnetization curves according to the manufacturer [204] and according to eqn (2.20), i.e.,

$$B = 0.4\frac{265,000 - H}{265,000 - 0.329H}$$

are plotted in Fig. 2.19.

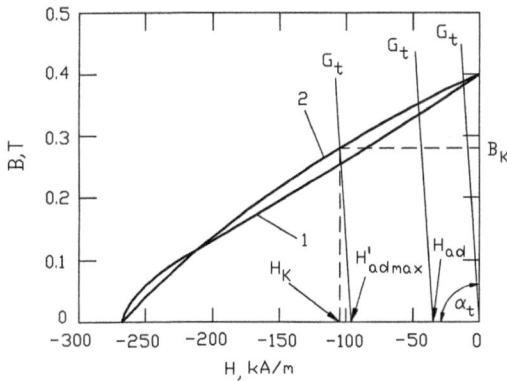

Fig. 2.19. Demagnetization curves for *Hardferrite* 28/26 anisotropic barium ferrite: 1 — according to the manufacturer, 2 — approximated curve according to eqn (2.20) for $a_0 = 0.329$. *Numerical example 2.2.*

15. The saturation magnetic flux density according to eqn (2.21)

$$B_{sat} = \frac{B_r}{a_0} = \frac{0.4}{0.329} = 1.216 \text{ T}$$

16. The magnetic flux corresponding to the remanent magnetic flux density

$$\Phi_r = B_r w_M l_M = 0.4 \times 0.2 \times 0.025 = 2.0 \times 10^{-4} \text{ Wb}$$

17. The MMF corresponding to the point K in Fig. 2.11 is calculated using eqn (2.44)

$$b_0 = 0.5 \left(\frac{F_c}{a_0} + F'_{admax} + \frac{\Phi_r}{a_0 G_t} \right)$$

$$= 0.5 \left(\frac{3975}{0.329} + 1430 + \frac{2.0 \times 10^{-4}}{0.329 \times 1.0074 \times 10^{-6}} \right) = 7057.75 \text{ A}$$

$$c_0 = \frac{(G_t F'_{admax} + \Phi_r) F_c}{a_0 G_t}$$

$$= \frac{(1.0074 \times 10^{-6} \times 1430 + 2.0 \times 10^{-4}) 3975}{0.329 \times 1.0074 \times 10^{-6}} = 19.675 \times 10^6 \text{ A}^2$$

$$F_{K1,2} = b_o \pm \sqrt{b_o^2 - c_o}$$

$$= 7057.75 \pm \sqrt{7057.75^2 - 19.676 \times 10^6} = 1568.14 \quad \text{or} \quad 12,547.36 \text{ A}$$

There are two solutions $F_{K1} = 1568.14$ A and $F_{K2} = 12,547.36$ A. Of course, $| F_K | < | F_c |$ so the true solution is $F_K = 1568.14$ A/m.

18. The magnetic field intensity at the point K is

$$H_K = \frac{F_K}{h_M} = \frac{1568.14}{0.015} = 104,542.67 \text{ A/m}$$

19. The magnetic flux at point K can be calculated with the aid of eqn (2.43), i.e.,

$$\Phi_K = G_t(F_K - F'_{admax})$$

$$= 1.0074 \times 10^{-6}(1568.14 - 1430.00) = 1.3916 \times 10^{-4} \text{ Wb}$$

20. The magnetic flux density at the point K

$$B_K = \frac{\Phi_K}{w_M l_M} = \frac{1.39 \times 10^{-4}}{0.025 \times 0.02} = 0.27832 \text{ T}$$

or

$$B_K = 0.4 \frac{265,000 - H_K}{265,000 - 0.329 H_K}$$

$$= 0.4 \frac{265,000 - 104,542}{265,000 - 0.329 \times 104,542} = 0.27832 \text{ T}$$

21. The angle between the Φ coordinate and G_t line (slope of G_t)

$$\alpha_t = \arctan\left(G_t\frac{F_c}{\Phi_r}\right) = \arctan\left(1.0074 \times 10^{-6}\frac{3975}{2.0 \times 10^{-4}}\right) = 87.14^0$$

22. The d-axis armature reaction MMF and magnetic field intensity acting directly on the PM

$$F'_{ad} = \frac{F_{ad}}{\sigma_{lM}} = \frac{687.5}{1.15} = 598 \ A$$

$$H'_{ad} = \frac{F'_{ad}}{h_M} = \frac{598}{0.015} = 39,866.7 \ A/m$$

23. The MMF corresponding to the point M in Fig. 2.11 is calculated with the aid of eqn (2.35)

$$F_M = \frac{1.3916 \times 10^{-4} + 1568.14 \times 1.35 \times 10^{-6} \times (0.025 \times 0.02/0.015)}{1.0074 \times 10^{-6} + 1.35 \times 10^{-6} \times (0.025 \times 0.02/0.015)}$$

$$+\frac{1.0074 \times 10^{-6} \times 598}{1.0074 \times 10^{-6} + 1.35 \times 10^{-6} \times (0.025 \times 0.02/0.015)} = 771.71 \ A$$

24. The magnetic field intensity at the point M

$$H_M = \frac{F_M}{h_M} = \frac{771.71}{0.015} = 51,447.3 \ A/m$$

25. The magnetic flux from eqn (2.34) is the total flux produced by the PM including the leakage flux, i.e.,

$$\Phi_M = G_t(F_M - F'_{ad})$$

$$= 1.0074 \times 10^{-6}(771.71 - 598.0) = 1.74995 \times 10^{-4} \ Wb$$

26. The total magnetic flux density produced by the PM

$$B_M = \frac{\Phi_M}{w_M l_M} = \frac{1.74995 \times 10^{-4}}{0.025 \times 0.02} = 0.35 \ T$$

which is smaller than $B_r = 0.4$ T.

27. The useful magnetic flux density in the air gap

$$B_g = \frac{B_M}{\sigma_{lM}} = \frac{0.35}{1.15} = 0.304 \ T$$

28. The magnetic flux density in the laminated core

$$B_{Fe} = \frac{B_g}{k_i} = \frac{0.304}{0.96} = 0.317 \ T$$

Numerical example 2.3

A Halbach NdFeB cylinder has the outer diameter $D_{out} = 52$ mm and inner diameter $D_{in} = 30$ mm. The remanent magnetic flux density of NdFeB PMs is $B_r = 1.26$ T. Find how the number n_M of PM pieces per wavelength (per two poles) affects the uniform magnetic flux density inside the cylinder.

<u>Solution:</u>

The maximum magnetic flux density is for the number n_M of PM pieces per wavelength tending to infinity. Using eqn (2.58)

$$\lim_{n_M \to \infty} \left[B_r \ln \left(\frac{D_{out}}{D_{in}} \right) \left(\frac{\sin \pi / n_M}{\pi / n_M} \right) \right] = B_r \ln \left(\frac{D_{out}}{D_{in}} \right)$$

$$= 1.26 \ln \left(\frac{52}{30} \right) == 0.693 \text{ T}$$

The peak value of the magnetic flux density as a function of n_M obtained from eqn (2.58) is plotted in Fig. 2.20. For $n_M = 8$ the flux density is practically the same as for $n_M \to \infty$.

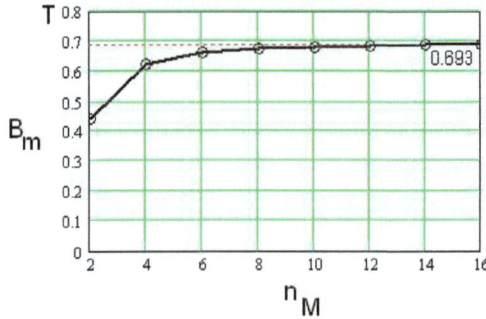

Fig. 2.20. Peak value of magnetic flux density as a function of number n_M of PMs per wavelength. *Numerical example 2.3.*

3

Finite Element Analysis

The finite element method (FEM) has proved to be particularly flexible, reliable and effective in the analysis and synthesis of power-frequency electromagnetic and electromechanical devices. Even in the hands of nonspecialists, modern FEM packages are user friendly and allow for calculating the electromagnetic field distribution and integral parameters without detailed knowledge of applied mathematics.

The FEM can analyze PM circuits of any shape and material. There is no need to calculate reluctances, leakage factors or the operating point on the recoil line. The PM demagnetization curve is input into the finite element program which can calculate the variation of the magnetic flux density throughout the PM system. An important advantage of finite element analysis over the analytical approach to PM motors is the inherent ability to calculate accurately armature reaction effects, inductances and the electromagnetic torque variation with rotor position (cogging torque). To understand and use FEM packages efficiently, the user must have the fundamental knowledge of electromagnetic field theory. It is assumed that the reader of this book is familiar with vector algebra, dot and cross products, coordinate systems, and the physical senses of *gradient*, *divergence* and *curl*.

3.1 Gradient, divergence and curl

To reduce the length of partial differential equations an operator ∇, called *del* or *nabla*, is used. In rectangular coordinates it is defined as

$$\nabla = 1_x \frac{\partial}{\partial x} + 1_y \frac{\partial}{\partial y} + 1_z \frac{\partial}{\partial z} \tag{3.1}$$

where 1_x, 1_y, and 1_z are unit vectors. The ∇ operator is a vector operator that has no physical meaning or vector direction by itself. The following operations involve the ∇ operator with scalars and vectors:

$$(Gradient\ of\ f) = \nabla f = 1_x\frac{\partial f}{\partial x} + 1_y\frac{\partial f}{\partial y} + 1_z\frac{\partial f}{\partial z} \tag{3.2}$$

$$(Divergence\ of\ \mathbf{A}) = \nabla \cdot \mathbf{A} = \frac{\partial A_x}{\partial x} + \frac{\partial A_y}{\partial y} + \frac{\partial A_z}{\partial z} \tag{3.3}$$

$$(Curl\ of\ \mathbf{A}) = \nabla \times \mathbf{A}$$

$$= 1_x\left(\frac{\partial A_z}{\partial y} - \frac{\partial A_y}{\partial z}\right) + 1_y\left(\frac{\partial A_x}{\partial z} - \frac{\partial A_z}{\partial x}\right) + 1_z\left(\frac{\partial A_y}{\partial x} - \frac{\partial A_x}{\partial y}\right) \tag{3.4}$$

$$(Laplacian\ of\ f) = \nabla \cdot \nabla f = \nabla^2 f = \frac{\partial^2 f}{\partial x^2} + \frac{\partial^2 f}{\partial y^2} + \frac{\partial^2 f}{\partial z^2} \tag{3.5}$$

It should be noted that some of the ∇ operations yield scalars while others yield vectors. In the last equation $\nabla \cdot \nabla = \nabla^2$ is called a *Laplacian operator* and is also used to operate on a vector

$$(Laplacian\ of\ \mathbf{A}) = \nabla^2\mathbf{A} = 1_x\nabla^2 A_x + 1_y\nabla^2 A_y + 1_z\nabla^2 A_z \tag{3.6}$$

Vector identities, as for example

$$div\ curl\mathbf{A} = 0 \tag{3.7}$$

$$curl\ curl\ \mathbf{A} = grad\ div\ \mathbf{A} - \nabla^2\mathbf{A} \tag{3.8}$$

can be brought to the following simpler forms

$$\nabla \cdot \nabla \times \mathbf{A} = 0 \tag{3.9}$$

$$\nabla \times (\nabla \times \mathbf{A}) = \nabla(\nabla \cdot \mathbf{A}) - \nabla^2\mathbf{A} \tag{3.10}$$

using the *nabla* operator.

3.2 Biot–Savart, Faraday's, and Gauss's laws

Maxwell's equations were derived in 1864—1865 from the earlier *Biot–Savart law* (1820), *Faraday's law* (1831), and *Gauss's law* (1840).

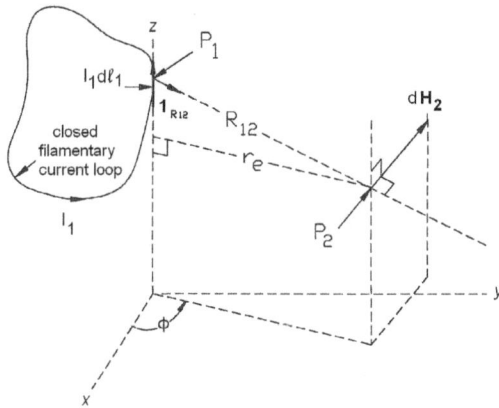

Fig. 3.1. Graphical display of the vector magnetostatic field intensity $\mathbf{dH}_2$ at P_2 produced by a current element $I_1\mathbf{dl}_1$ at P_1.

3.2.1 Biot–Savart law

The *Biot-Savart law* gives the differential magnetic field intensity $\mathbf{dH}_2$ at a point P_2, produced by a current element $I_1\mathbf{dl}_1$ at point P_1, which is filamentary and differential in length, as shown in Fig. 3.1 [208]. This law can best be stated in vector form as

$$\mathbf{dH}_2 = \frac{I_1\mathbf{dl}_1 \times \mathbf{1}_{R12}}{4\pi R_{12}^2} \tag{3.11}$$

where the subscripts indicate the point to which the quantities refer, I_1 is the filamentary current at P_1, $\mathbf{dl}_1$ is the vector length of current path (vector direction same as conventional current) at P_1, $\mathbf{1}_{R12}$ is the unit vector directed from the current element $I_1\mathbf{dl}_1$ to the location of $\mathbf{dH}_2$, from P_1 to P_2, R_{12} is the scalar distance between the current element $I_1\mathbf{dl}_1$ to the location of $\mathbf{dH}_2$, the distance between P_1 and P_2, and $\mathbf{dH}_2$ is the vector magnetostatic field intensity at P_2.

Fig. 3.1 shows graphically the relationship between the quantities found in the Biot-Savart law (3.11) when a current element $I_1\mathbf{dl}_1$ is singled out from a closed loop of filamentary current I_1. The direction of $\mathbf{dH}_2$ comes from $\mathbf{dl}_1 \times \mathbf{1}_{R12}$ and thus is perpendicular to $\mathbf{dl}_1$ and $\mathbf{1}_{R12}$. The direction of $\mathbf{dH}_2$ is also governed by the right-hand rule and is in the direction of the fingers of the right hand when one grasps the current element so that the thumb points in the direction of $\mathbf{dl}_1$. Biot-Savart law is similar to Coulomb's law of magnetostatics.

3.2.2 Faraday's law

Faraday's law says that a time-varying or space-varying magnetic field induces an EMF in a closed loop linked by that field:

$$e = -N\frac{d\Phi(x,t)}{dt} = -N\left(\frac{\partial\Phi}{\partial t} + \frac{\partial\Phi}{\partial x}\frac{\partial x}{\partial t}\right) \tag{3.12}$$

where e is the instantaneous EMF induced in a coil with N turns and Φ is the magnetic flux (the same in each turn).

3.2.3 Gauss's law

The total electric flux passing through any closed imaginary surface enclosing the charge Q is equal to Q (in SI units). The charge Q is enclosed by the closed surface and is called Q enclosed, or Q_{en}. The total flux Ψ_E is thus equal to

$$\Psi_E = \oint_S d\Psi_E = \oint_S \mathbf{D}_S \cdot \mathbf{dS} = Q_{en}$$

where $\oint_S$ indicates a double integral over the closed surface S and $\mathbf{D}_s$ is the electric flux density through the surface S. The mathematical formulation obtained from the above equation

$$\oint_S \mathbf{D}_S \cdot \mathbf{dS} = Q_{en} \tag{3.13}$$

is called *Gauss's law* after K. F. Gauss. The Q_{en} enclosed by surface S, due to a volume charge density ρ_V distribution, becomes

$$Q_{en} = \int_V \rho_V dV \tag{3.14}$$

where V is the volume.

3.3 Gauss's theorem

Gauss's theorem also called the *divergence theorem*, relates a closed surface integral of $\mathbf{D}_S \cdot \mathbf{dS}$ to a volume integral of $\nabla \cdot \mathbf{D}dV$ involving the same vector, i.e.,

$$\oint_S \mathbf{D}_S \cdot \mathbf{dS} = \int_V \nabla \cdot \mathbf{D}dV \tag{3.15}$$

It should be noted that the closed surface S encloses the volume V.

3.4 Stokes' theorem

Stokes' theorem relates the closed loop integral of $\mathbf{H} \cdot \mathbf{dl}$ to a surface integral of $\nabla \times \mathbf{H} \cdot \mathbf{dS}$, i.e.,

$$\oint_l \mathbf{H} \cdot \mathbf{dl} = \int_S (\nabla \times \mathbf{H}) \cdot \mathbf{dS} \tag{3.16}$$

where the loop l encloses the surface S.

3.5 Maxwell's equations

3.5.1 First Maxwell's equation

Maxwell introduced so-called *displacement current*, the density of which is $\partial \mathbf{D}/\partial t$, where $\mathbf{D}$ is the electric flux density (displacement) vector. There is a continuity of the displacement current and electric current $\mathbf{J}$, e.g., in a circuit with a capacitor. The differential form of the *first Maxwell's equation* is

$$curl\mathbf{H} = \mathbf{J} + \frac{\partial \mathbf{D}}{\partial t} + curl(\mathbf{D} \times \mathbf{v}) + \mathbf{v}\, div\mathbf{D} \tag{3.17}$$

or

$$\nabla \times \mathbf{H} = \mathbf{J} + \frac{\partial \mathbf{D}}{\partial t} + \nabla \times (\mathbf{D} \times \mathbf{v}) + \mathbf{v}\, \nabla \cdot \mathbf{D}$$

where $\mathbf{J}$ is the density of the electric current, $\partial \mathbf{D}/\partial t$ is the density of the displacement current, $curl(\mathbf{D} \times \mathbf{v})$ is the density of the current due to the motion of a polarized dielectric material, and $\mathbf{v}div\mathbf{D}$ is the density of the convection current. For $\mathbf{v} = 0$

$$curl\mathbf{H} = \mathbf{J} + \frac{\partial \mathbf{D}}{\partial t} \tag{3.18}$$

According to eqn (3.4) the last equation has the following scalar form:

$$\frac{\partial H_z}{\partial y} - \frac{\partial H_y}{\partial z} = J_x + \frac{\partial D_x}{\partial t}$$

$$\frac{\partial H_x}{\partial z} - \frac{\partial H_z}{\partial x} = J_y + \frac{\partial D_y}{\partial t} \tag{3.19}$$

$$\frac{\partial H_y}{\partial x} - \frac{\partial H_x}{\partial y} = J_z + \frac{\partial D_z}{\partial t}$$

3.5.2 Second Maxwell's equation

The *second Maxwell's equation* in the differential form is

$$curl\mathbf{E} = -\frac{\partial \mathbf{B}}{\partial t} - curl(\mathbf{B} \times \mathbf{v}) \tag{3.20}$$

or

$$\nabla \times \mathbf{E} = -\frac{\partial \mathbf{B}}{\partial t} - \nabla \times (\mathbf{B} \times \mathbf{v})$$

For $\mathbf{v} = 0$

$$curl\mathbf{E} = -\frac{\partial \mathbf{B}}{\partial t} \tag{3.21}$$

The scalar form of the last equation is

$$\frac{\partial E_z}{\partial y} - \frac{\partial E_y}{\partial z} = -\frac{\partial B_x}{\partial t}$$

$$\frac{\partial E_x}{\partial z} - \frac{\partial E_z}{\partial x} = -\frac{\partial B_y}{\partial t} \tag{3.22}$$

$$\frac{\partial E_y}{\partial x} - \frac{\partial E_x}{\partial y} = -\frac{\partial B_z}{\partial t}$$

For magnetically isotropic bodies

$$\mathbf{B} = \mu_0 \mu_r \mathbf{H} \tag{3.23}$$

where $\mu_0 = 0.4\pi \times 10^{-6}$ H/m is the magnetic permeability of free space, and μ_r is the relative magnetic permeability. For ferromagnetic materials (iron, steel, nickel, cobalt) $\mu_r \gg 1$, for paramagnetic materials (aluminum) $\mu_r > 1$, and for diamagnetic materials (copper) $\mu_r < 1$.

For magnetically anisotropic materials, e.g., cold-rolled electrotechnical steel sheets

$$\begin{bmatrix} B_x \\ B_y \\ B_z \end{bmatrix} = \begin{bmatrix} \mu_{11} & \mu_{12} & \mu_{13} \\ \mu_{21} & \mu_{22} & \mu_{23} \\ \mu_{31} & \mu_{32} & \mu_{33} \end{bmatrix} \begin{bmatrix} H_x \\ H_y \\ H_z \end{bmatrix}$$

If the coordinate system $0, x, y, z$ is the same as the axes of anisotropy

$$\begin{bmatrix} B_x \\ B_y \\ B_z \end{bmatrix} = \begin{bmatrix} \mu_{11} & 0 & 0 \\ 0 & \mu_{22} & 0 \\ 0 & 0 & \mu_{33} \end{bmatrix} \begin{bmatrix} H_x \\ H_y \\ H_z \end{bmatrix}$$

Since $\mu_{11} = \mu_0 \mu_{rx}$, $\mu_{22} = \mu_0 \mu_{ry}$, and $\mu_{33} = \mu_0 \mu_{rz}$

$$B_x = \mu_0 \mu_{rx} H_x, \qquad B_y = \mu_0 \mu_{ry} H_y, \qquad B_z = \mu_0 \mu_{rz} H_z \tag{3.24}$$

3.5.3 Third Maxwell's equation

From Gauss's law (3.13) for the volume charge density ρ_V and through the use of Gauss's theorem (3.15), the *third Maxwell's equation* in differential form is

$$div\mathbf{D} = \rho_V \tag{3.25}$$

or

$$\nabla \cdot \mathbf{D} = \rho_V$$

In scalar form

$$\frac{\partial D_x}{\partial x} + \frac{\partial D_y}{\partial y} + \frac{\partial D_z}{\partial z} = \rho_V \tag{3.26}$$

3.5.4 Fourth Maxwell's equation

The physical meaning of the equation

$$\oint_S \mathbf{B} \cdot \mathbf{dS} = 0$$

is that there are no magnetic charges. By means of the use of Gauss's theorem (3.15) the *fourth Maxwell equation* in differential form is

$$div\mathbf{B} = 0 \tag{3.27}$$

or

$$\nabla \cdot \mathbf{B} = 0$$

In scalar form

$$\frac{\partial B_x}{\partial x} + \frac{\partial B_y}{\partial y} + \frac{\partial B_z}{\partial z} = 0 \tag{3.28}$$

3.6 Magnetic vector potential

Through the use of the identity shown in eqn (3.7) and the fourth Maxwell's equation (3.27), which states that the divergence of $\mathbf{B}$ is always zero everywhere, the following equation can be written

$$curl\mathbf{A} = \mathbf{B} \qquad \text{or} \qquad \nabla \times \mathbf{A} = \mathbf{B} \tag{3.29}$$

On the assumptions that $\mu = const$, $\epsilon = const$, $\sigma = const$, $\mathbf{v} = 0$, and $div\mathbf{D} = 0$, the first Maxwell's equation (3.17) can be written in the form

$$curl\mathbf{B} = \mu\sigma\mathbf{E} + \mu\epsilon\frac{\partial\mathbf{E}}{\partial t}$$

Putting the magnetic vector potential and using the identity shown in eqn (3.8) the above equation takes the form

$$grad\ div\mathbf{A} - \nabla^2\mathbf{A} = \mu\sigma\mathbf{E} + \mu\epsilon\frac{\partial\mathbf{E}}{\partial t}$$

Since $div\mathbf{A} = 0$ and for power frequencies 50 or 60 Hz $\sigma\mathbf{E} >> j\omega\epsilon\mathbf{E}$, the magnetic vector potential for sinusoidal fields can be expressed with the aid of Poisson's equation

$$\nabla^2\mathbf{A} = -\mu\mathbf{J} \tag{3.30}$$

In scalar form

$$\nabla^2 A_x = -\mu J_x \qquad \nabla^2 A_y = -\mu J_y \qquad \nabla^2 A_z = -\mu J_z \tag{3.31}$$

3.7 Energy functionals

The FEM is based on the conservation of energy. The law of conservation of energy in electrical machines can be derived from Maxwell's equations. The net electrical input active power is [41, 273]

$$P = \int_V \sigma\mathbf{E}^2 dV = \int_V \mathbf{E}\cdot\mathbf{J}dV = \int_l\int_S(\mathbf{J}\cdot\mathbf{dS})\mathbf{E}\cdot\mathbf{dl} \tag{3.32}$$

The EMF from Faraday's law (3.12) for $N = 1$ is

$$e = \int_l \mathbf{E}\cdot\mathbf{dl} = -\frac{\partial}{\partial t}\int_S\mathbf{B}\cdot\mathbf{dS} \tag{3.33}$$

and the current enclosed from Ampère's circuital law is

$$I_{en} = \int_S\mathbf{J}\cdot\mathbf{dS} = \oint_l\mathbf{H}\cdot\mathbf{dl} \tag{3.34}$$

Putting $\int_l\mathbf{E}\cdot\mathbf{dl}$ (3.33) and $\int_S\mathbf{J}\cdot\mathbf{dS}$ (3.34) into eqn (3.32) for the net electrical input power

$$P = \oint_l\mathbf{H}\cdot\mathbf{dl}\left[-\frac{\partial}{\partial t}\int_S\mathbf{B}\cdot\mathbf{dS}\right] = -\int_V\mathbf{H}\cdot\frac{\partial\mathbf{B}}{dt}dV$$

or

$$\int_V\mathbf{E}\cdot\mathbf{J}dV = -\int_V\mathbf{H}\cdot\frac{\partial\mathbf{B}}{dt}dV$$

The right-hand side can be rewritten to obtain

$$P = \int_V \mathbf{E} \cdot \mathbf{J} dV = -\frac{\partial}{\partial t} \int_V \left[\int_0^B \mathbf{H} \cdot \mathbf{dB} \right] dV \qquad (3.35)$$

The term on the right-hand side is the rate of increase of the stored magnetic energy, i.e.,

$$W = \int_V \left[\int_0^B \mathbf{H} \cdot \mathbf{dB} \right] dV \quad \text{J} \qquad (3.36)$$

The input power P can also be expressed in terms of the magnetic vector potential $\mathbf{A}$. In accordance with eqns (3.21) and (3.29) the electric field intensity $\mathbf{E} = -\partial \mathbf{A}/\partial t$. Thus, the input electrical power (3.32) can be written as

$$P = -\int_V \mathbf{J} \cdot \frac{\partial \mathbf{A}}{\partial t} dV = -\frac{\partial}{\partial t} \int_V \left[\int_0^A \mathbf{J} \cdot \mathbf{dA} \right] dV \qquad (3.37)$$

and then comparing eqns (3.35) and (3.37)

$$\int_V \left[\int_0^B \mathbf{H} \cdot \mathbf{dB} \right] dV = \int_V \left[\int_0^A \mathbf{J} \cdot \mathbf{dA} \right] dV \qquad (3.38)$$

Eqn (3.38) shows that for lossless electromagnetic devices the *stored magnetic energy* equals the *input electric energy*.

Variational techniques obtain solutions to field problems *by minimizing an energy functional F* that is the difference between the stored energy and the input (applied) energy in the system volume. Thus, for magnetic systems

$$F = \int_V \left[\int_0^B \mathbf{H} \cdot \mathbf{dB} - \int_0^A \mathbf{J} \cdot \mathbf{dA} \right] dV \qquad (3.39)$$

The first term on the right-hand side is the magnetic stored energy and the second is the electric input. The energy functional F is minimized when

$$\frac{\partial F}{\partial A} = 0 \qquad (3.40)$$

Thus

$$\int_V \left[\frac{\partial}{\partial A} \int_0^B \mathbf{H} \cdot \mathbf{dB} - \mathbf{J} \right] dV = 0 \qquad (3.41)$$

Including losses $0.5 j \omega \sigma \mathbf{A}^2$ due to induced currents where ω is the angular frequency and σ is the electric conductivity, the functional according to eqn (3.39) for linear electromagnetic problems F becomes

$$F = \int_V \left[\frac{B^2}{2\mu} - \mathbf{J} \cdot \mathbf{A} + jw\frac{1}{2}\sigma\mathbf{A}^2 \right] dV \tag{3.42}$$

where

$$\int_V \left[\int_0^B \mathbf{H} \cdot d\mathbf{B} \right] dV = \int_V \frac{B^2}{2\mu} dV = \frac{1}{2\mu}(B_x^2 + B_y^2) \tag{3.43}$$

For two dimensional (planar) problems

$$F = \int_S \left[\frac{B^2}{2\mu} - \mathbf{J} \cdot \mathbf{A} + jw\frac{1}{2}\sigma\mathbf{A}^2 \right] dS \tag{3.44}$$

The functional (3.39) is further altered if an additional magnetization vector $\mathbf{B}_r$ is included in eqn (2.1) to model a hard ferromagnetic material. In problems with PMs

$$\mathbf{B} = \mu_0\mathbf{H} + \mu_0\chi\mathbf{H} + \mathbf{B_r} = \mu_0\mu_r\mathbf{H} + \mathbf{B_r} \tag{3.45}$$

The functional is

$$F = \int_V \left[\frac{B^2}{2\mu} - \frac{\mathbf{B}\mathbf{B_r}}{\mu} - \mathbf{J} \cdot \mathbf{A} + jw\frac{1}{2}\sigma\mathbf{A}^2 \right] dV \tag{3.46}$$

where $\mathbf{B_r}$ is the contribution of the PM to eqn (2.1) equivalent to the remanent magnetization of the PM material (equal to the remanent magnetic flux density).

Although most FEM packages include the magnetization vector in their code, PMs can also be simulated using a sheet current equivalent, as discussed in [108].

3.8 Finite element formulation

The 2D sinusoidally time varying field can be described with the aid of the magnetic vector potential, i.e.,

$$\frac{\partial}{\partial x}\left(\frac{1}{\mu}\frac{\partial\mathbf{A}}{\partial x}\right) + \frac{\partial}{\partial y}\left(\frac{1}{\mu}\frac{\partial\mathbf{A}}{\partial y}\right) = -\mathbf{J} + jw\sigma\mathbf{A} \tag{3.47}$$

where the magnetic vector potential $\mathbf{A}$ and excitation current density vector $\mathbf{J}$ are directed out or into the flat model along the z-axis, i.e.,

$$\mathbf{A} = \mathbf{1_z}A_z, \qquad \mathbf{J} = \mathbf{1_z}J_z \tag{3.48}$$

The magnetic flux density vector has two components in the xy plane perpendicular to $\mathbf{A}$ and $\mathbf{J}$, i.e.,

$$\mathbf{B} = \mathbf{1_x}B_x + \mathbf{1_y}B_y \tag{3.49}$$

Using the definition of the magnetic vector potential (3.29), i.e., $\nabla \times \mathbf{A} = (\nabla \times \mathbf{1}_z)A = \mathbf{B}$ and definition (3.4)

$$B_x = \frac{\partial A_z}{\partial y} \qquad B_y = -\frac{\partial A_z}{\partial x} \qquad \mathbf{1}_x \frac{\partial A_z}{\partial y} - \mathbf{1}_y \frac{\partial A_z}{\partial x} = \mathbf{B}(x, y) \qquad (3.50)$$

For the energy density

$$w = \frac{1}{2\mu}B^2 = \frac{1}{2\mu}(B_x^2 + B_y^2) = \frac{1}{2\mu}\left(\left|\frac{\partial A_z}{\partial x}\right|^2 + \left|\frac{\partial A_z}{\partial y}\right|^2\right) \qquad (3.51)$$

An equation similar to (3.47) can be written for the 2D magnetostatic field

$$\frac{\partial}{\partial x}\left(\frac{1}{\mu}\frac{\partial \mathbf{A}}{\partial x}\right) + \frac{\partial}{\partial y}\left(\frac{1}{\mu}\frac{\partial \mathbf{A}}{\partial y}\right) = -J \qquad (3.52)$$

and electrostatic field

$$\frac{\partial^2 \psi}{\partial x^2} + \frac{\partial^2 \psi}{\partial y^2} = -\frac{\rho}{\epsilon} \qquad (3.53)$$

where ψ is the electrostatic scalar potential, ρ is charge density dependent on ψ and ϵ is the permittivity.

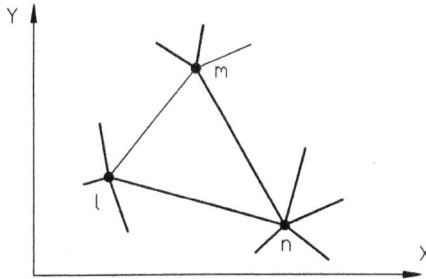

Fig. 3.2. Typical triangular finite element connected to other elements

Further considerations relate to the 2D sinusoidally time varying electromagnetic field described by eqn (3.47).

Minimization of the magnetic energy functional over a set of elements (called the mesh) leads to a matrix equation that has to solve for the magnetic vector potential $\mathbf{A}$ [41, 273]. It is in fact the minimization of $\mathbf{A}$ throughout the mesh. Fig. 3.2 shows the coordinate system for planar problems, along with part of a typical finite element mesh. The entire planar mesh may represent, for example, the stator and rotor laminations and air gap of a motor. A similar two-dimensional derivation may be made for axisymmetric problems, such as for cylindrical solenoids. In either case, the device being analyzed must be subdivided (discretized) into triangles or quadrilaterals called *elements*. Each element has at least three vertices called *nodes* or grid points. The number of nodes corresponding to each element depends on the shape of the element and also on the type of function used to model the potential within the element. This function is called the *shape function* and can be of any order depending on the desired complexity. Usually linear or second-order shape functions are used. Assuming a linear shape function of **A** within an element the vector potential inside each element is described as

$$A = \alpha_1 + \alpha_2 x + \alpha_3 y \qquad (3.54)$$

The values of A at each node are

$$A_l = \alpha_1 + \alpha_2 x_l + \alpha_3 y_l$$

$$A_m = \alpha_1 + \alpha_2 x_m + \alpha_3 y_m \qquad (3.55)$$

$$A_n = \alpha_1 + \alpha_2 x_n + \alpha_3 y_n$$

where $A_l = A(x_l, y_l)$, $A_m = A(x_m, y_m)$ and $A_n = A(x_n, y_n)$. In matrix form

$$\begin{bmatrix} A_l \\ A_m \\ A_n \end{bmatrix} = \begin{bmatrix} 1 & x_l & y_l \\ 1 & x_m & y_m \\ 1 & x_n & y_n \end{bmatrix} \begin{bmatrix} \alpha_1 \\ \alpha_2 \\ \alpha_3 \end{bmatrix} \qquad (3.56)$$

Solution to equation (3.56) yields

$$\alpha_1 = \frac{1}{2\Delta} \begin{vmatrix} A_l & x_l & y_l \\ A_m & x_m & y_m \\ A_n & x_n & y_n \end{vmatrix} \qquad \alpha_2 = \frac{1}{2\Delta} \begin{vmatrix} 1 & A_l & y_l \\ 1 & A_m & y_m \\ 1 & A_n & y_n \end{vmatrix}$$

$$\alpha_3 = \frac{1}{2\Delta} \begin{vmatrix} 1 & x_l & A_l \\ 1 & x_m & A_m \\ 1 & x_n & A_n \end{vmatrix} \qquad 2\Delta = \begin{vmatrix} 1 & x_l & y_l \\ 1 & x_m & y_m \\ 1 & x_n & y_n \end{vmatrix} \qquad (3.57)$$

where Δ is the surface area of a triangle with nodes l, m, n. Putting eqns (3.57) into eqn (3.54) the linear interpolation polynomial function is

$$A = \frac{1}{2\Delta} \sum_{k=l,m,n} [a_k + b_k x + c_k y] A_k = [N_l \ N_m \ N_n] \begin{bmatrix} A_l \\ A_m \\ A_n \end{bmatrix} \qquad (3.58)$$

where $N_l = (a_l + b_l x + c_l y)/(2\Delta)$, $N_m = (a_m + b_m x + c_m y)/(2\Delta)$ and $N_n = (a_n + b_n x + c_n y)/(2\Delta)$. Comparing eqns (3.50) and (3.58),

$$\mathbf{B}(x, y) = \frac{1}{2\Delta} \sum_{k=l,m,n} (\mathbf{1}_x c_k - \mathbf{1}_y b_k) A_k \qquad (3.59)$$

The magnetic field is constant within a particular triangular finite element.

The node point potentials A_k can be calculated by minimizing the energy functional (3.44) according to eqn (3.40). Considering a single triangular finite element yields

$$\int_S \frac{\partial}{\partial A_k} \left[\frac{B^2}{2\mu} - \mathbf{J} \cdot \mathbf{A} + j\omega \frac{1}{2} \sigma (\mathbf{A})^2 \right] dS = 0 \tag{3.60}$$

or

$$\int_S \left\{ \frac{1}{2\mu} \frac{\partial}{\partial A_k} \left[\left(\frac{\partial A}{\partial x} \right)^2 + \left(\frac{\partial A}{\partial y} \right)^2 \right] - \frac{\partial}{\partial A_k} (\mathbf{J} \cdot \mathbf{A}) \right\} dx dy$$

$$+ j \int_S \left[\omega \frac{1}{2} \sigma \frac{\partial}{\partial A_k} (\mathbf{A})^2 \right] dx dy = 0 \tag{3.61}$$

where $dS = dx dy$. The minimization of the functional with respect to the magnetic vector potential can be approximated by the following set of equations [98]

$$[S][A] = [I] \tag{3.62}$$

where $[S]$ is the global coefficient matrix, $[A]$ is the the matrix of nodal magnetic vector potentials and $[I]$ matrix represents nodal currents (forcing functions). Elements of $[S]$ and $[I]$ are expressed as [98]

$$[S] = \frac{1}{4\mu\Delta} \begin{bmatrix} b_l b_l + c_l c_l & b_l b_m + c_l c_m & b_l b_n + c_l c_n \\ b_m b_l + c_m c_l & b_m b_m + c_m c_m & b_m b_n + c_m c_n \\ b_n b_l + c_n c_l & b_n b_m + c_n c_m & b_n b_n + c_n c_n \end{bmatrix}$$

$$+ j \frac{\omega \sigma \Delta}{12} \begin{bmatrix} 2 & 1 & 1 \\ 1 & 2 & 1 \\ 1 & 1 & 2 \end{bmatrix} ; \tag{3.63}$$

$$[I] = J \frac{\Delta}{3} \begin{bmatrix} 1 \\ 1 \\ 1 \end{bmatrix} \tag{3.64}$$

The above eqns (3.62), (3.63) and (3.64) solve for the potential $\mathbf{A}$ in a region containing the triangle with nodes l, m, and n. For practical problems with K nodes, the preceding process is repeated for each element, obtaining the matrix $[S]$ with K rows and columns, $[A]$ and $[I]$ are then column matrices containing K rows of complex terms [98].

Three dimensional (3D) finite element models are built on the ideas presented in the foregoing, but with the added complexity of an additional coordinate. This means that each element has a minimum of four nodes.

3.9 Boundary conditions

Natural boundaries very rarely exist in electromagnetic field problems. In most types of applications the electromagnetic field is an infinitely extending space. In these applications boundaries are used to simplify the finite element model and approximate the magnetic vector potential at node points. Rotary electrical machines have identical pole pitches or sometimes half-pole pitches. Boundaries due to symmetry greatly reduce the size of the finite element model.

Boundary conditions can be classified into three types [41], which will be looked at more closely.

3.9.1 Dirichlet boundary conditions

Dirichlet boundary conditions require the magnetic vector potential, at a particular point, to take on a prescribed value, i.e.,

$$\mathbf{A} = m \tag{3.65}$$

where m is the specified value. Dirichlet boundaries force the flux lines to be parallel to the boundary's edge.

In 2D problems a flux line is a line of $\mathbf{A} = const$. By using the boundary condition $\mathbf{A} = 0$, flux lines are constrained to follow the boundary. The outer edge of the stator yoke, for example, could have a Dirichlet boundary of $\mathbf{A} = 0$. This is a simplification since any leakage flux which would extend beyond the stator's yoke is now neglected. The high relative permeability of the ferromagnetic yoke would ensure that most of the flux remains inside the yoke and in most motor designs this boundary condition is a reasonable simplification to make.

3.9.2 Neumann boundary conditions

Neumann boundary conditions require the normal derivative of the magnetic vector potential be zero, i.e.,

$$\frac{\partial \mathbf{A}}{\partial n} = 0 \tag{3.66}$$

This type of boundary is not satisfied exactly by the finite element solution but only the mean value of the boundary. Neumann boundaries are *natural* boundaries in the FEM since they do not have to be specified explicitly. Flux lines cross a Neumann boundary orthogonally.

Neumann boundaries are used mainly in symmetry problems where the flux is orthogonal to a plane. This occurs only in no-load operation in most electrical motors.

3.9.3 Interconnection boundary conditions

Interconnection boundary conditions set the constraint between two nodes. This could be between two geometrically adjacent nodes or between two nodes at a particular interval apart. This type of boundary has only one of the two potentials independently specified and is still exactly satisfied in the finite element solution. The relationship between the two nodes is

$$\mathbf{A_m} = a\mathbf{A_n} + b \tag{3.67}$$

where a and b are factors which link the two nodes.

The general use of the interconnection boundary condition in electrical machines is in relating two nodes that are one pole pitch or a multiple of a pole pitch apart. This type of constraint is usually called a *periodic constraint* in which $\mathbf{A_m} = \mathbf{A_n}$ or $\mathbf{A_m} = -\mathbf{A_n}$, depending on the number of pole pitches the nodes are apart, e.g.,

$$A\left(r, \Theta_0 + \frac{2\pi}{2p}\right) = -A(r, \Theta_0) \tag{3.68}$$

3.10 Mesh generation

The accuracy of the finite element solution is dependent on the mesh topology. The mesh is thus an important part of any finite element model and attention should be placed on creating it. Essentially, there are two types of mesh generators. The first being an *analytical* mesh generator that defines the problem geometry using large global elements. These global elements are subsequently refined according to the user, usually automatically. The other type of mesh generator is a *synthetic* generator where the user designs a mesh region at a node–by–node level and the model is the union of a number of different mesh regions.

Modern finite element packages can generate a mesh automatically from the geometric outline of the problem drawn in a CAD type package. These mesh generators usually construct the mesh using a *Delaunay triangulation method*. Automating the mesh generation drastically reduces the manpower costs.

Fully automated mesh generation can be achieved only if the errors which arise from the mesh discretization are taken into account. This is called *self-adaptive* meshing, and relies on an accurate and reliable method of estimating the discretization errors in the mesh. Modern finite element packages that use self-adaptive meshing normally calculate the discretization error estimates from a finite element solution. These packages usually create a crude mesh which is solved. Error estimates are made from this solution and the mesh is refined in the approximate places. This process is repeated until the required level of accuracy is obtained for the model. The error estimates used in the

program depend generally on the application, but general finite element pro-
grams usually calculate their error estimates from the change in flux density
across element edges.

3.11 Forces and torques in an electromagnetic field

The calculation of forces and torques using the FEM is one of the most important functions of this method. In electrical machine problems four methods of
calculating forces or torques are used: the *Maxwell stress tensor*, the *co-energy
method*, the *Lorentz force equation* ($\mathbf{J} \times \mathbf{B}$), and the *rate of change of field
energy method* ($\mathbf{B} \, \partial B / \partial x$). The most appropriate method is usually problem-
dependent, although the most frequently used methods are the Maxwell stress
tensor and co-energy methods.

3.11.1 Maxwell stress tensor

The use of the Maxwell stress tensor is simple from a computational perspec-
tive since it requires only the local flux density distribution along a specific
line or contour.

Using the definition of Maxwell stress tensor, the electromagnetic forces
can be determined on the basis of the magnetic flux density, i.e.,

- the total force

$$\mathbf{F} = \int \int \left[\frac{1}{\mu_0} \mathbf{B}(\mathbf{B} \cdot \mathbf{n}) - \frac{1}{2\mu_0} B^2 \mathbf{n} \right] dS \qquad (3.69)$$

- the normal force

$$F_n = \frac{L_i}{2\mu_o} \int [B_n^2 - B_t^2] dl \qquad (3.70)$$

- the tangential force

$$F_t = \frac{L_i}{\mu_o} \int B_n B_t dl \qquad (3.71)$$

where $\mathbf{n}$, L_i, l, B_n and B_t are the normal vector to the surface S, stack length,
integration contour, radial (normal) component of the magnetic flux density
and tangential component of the magnetic flux density, respectively.

The torque $\mathbf{T} = \mathbf{r} \times \mathbf{F}$ in connection with eqn (3.71) is

$$T = \frac{L_i}{\mu_0} \oint_l r B_n B_t dl \qquad (3.72)$$

where r is the radius of the circumference which lies in the air gap.

Since a finite grid is being used the above equations can be written for element i. The torque shown below in cylindrical coordinates is a sum of torques for each element i, i.e.,

$$T = \frac{L_i}{\mu_0} \sum_i r^2 \int_{\theta_i}^{\theta_{i+1}} B_{ri} B_{\theta i} d\theta \tag{3.73}$$

The accuracy of this method is markedly dependent on the model discretization and on the selection of the integration line or contour. The Maxwell stress tensor line integration necessitates a precise solution in the air gap, demanding a fine discretization of the model in the air gap since the flux density is not continuous at the nodes and across boundaries of first-order elements.

3.11.2 Co-energy method

The force or torque is calculated as the derivative of the stored magnetic co-energy W' with respect to a small displacement.

The finite difference approximation approximates the derivative of co-energy by the change in co-energy for a displacement s, called the co-energy finite difference method. The component of instantaneous force F_s in the direction of the displacement s is [41, 202]

$$F_s = \frac{dW'}{ds} \approx \frac{\Delta W'}{\Delta s} \tag{3.74}$$

or, for an instantaneous torque T with a small angular rotation displacement θ (mechanical angle),

$$T = \frac{dW'}{d\theta} \approx \frac{\Delta W'}{\Delta \theta} \tag{3.75}$$

The problem with the finite difference approach is that two finite element models have to be calculated, doubling the calculation time, and the most suitable value of angular increment $\Delta \theta$ is unknown and has to be found using a trial–and–error procedure. If $\Delta \theta$ is too small, rounding-off errors in $\Delta W'$ will dominate. If $\Delta \theta$ is too large, the calculated torque will no longer be accurate for the specific rotor position.

The instantaneous torque can also be expressed in the following form

$$T = \frac{\partial W'(i, \theta)}{\partial \theta}\Big|_{i=const} = -\frac{\partial W(\Psi, \theta)}{\partial \theta}\Big|_{\Psi=const} \tag{3.76}$$

where W, Ψ and i are the magnetic energy, flux linkage vector and current vector, respectively.

3.11.3 Lorentz force theorem

Using the Lorentz force theorem, the instantaneous torque is expressed as a function of phase EMFs and phase currents, i.e.,

$$
T = \sum_{l=A,B,C} i_l(t) \left[2pN \int_{-\pi/(2m_1 p)}^{\pi/(2m_1 p)} rL_i B(\theta, t) d\theta \right]
$$

$$
= \frac{1}{2\pi n} [e_A(t)i_A(t) + e_B(t)i_B(t) + e_C(t)i_C(t)] \tag{3.77}
$$

where p, θ, n and N are the number of pole pairs, mechanical degree, rotational speed and conductor number in phase belt, repectively.

3.12 Inductances

3.12.1 Definitions

FEM computations use the following definitions of the steady-state inductances:

- the number of flux linkages of the coil, divided by the current in the coil,
- the energy stored in the coil divided by one-half the current squared.

Both definitions give identical results for linear inductances but not for non-linear inductances [202].

If the incorrect potential distribution does not differ significantly from the correct potential, the error in energy computation is much smaller than that in potential [273]. Therefore, the steady-state inductances are often very accurately approximated even if the potential solution contains substantial errors.

3.12.2 Dynamic inductances

In order to predict accurately the dynamic behavior and performance of a solid state converter-fed PM brushless motor one needs to know the self- and mutual-winding dynamic inductance $d\Psi/di$ rather than the steady-state value Ψ/I [227].

The *current/energy perturbation method* is based upon consideration of the total energy stored in the magnetic field of a given device comprising l windings [83, 84, 227]. The voltage across the terminals of the jth winding is

$$
v_j = R_j i_j + \frac{\partial \Psi_j}{\partial i_1} \frac{di_1}{dt} + \frac{\partial \Psi_j}{\partial i_2} \frac{di_2}{dt} + \cdots
$$

$$+\frac{\partial\Psi_j}{\partial i_j}\frac{di_j}{dt}+\ldots+\frac{\partial\Psi_j}{\partial i_l}\frac{di_l}{dt}+\frac{\partial\Psi_j}{\partial\theta}\frac{d\theta}{dt} \tag{3.78}$$

For a fixed rotor position the last term in eqn (3.78) is equal to zero, because the rotor speed $d\theta/dt = 0$. The partial derivative of the flux linkage Ψ_j with respect to a winding current i_k $(k = 1, 2, \ldots, j, \ldots, l)$ in eqn (3.78) is the so-called *incremental inductance* L_{jk}^{inc} [83, 84, 227]. Therefore, the total stored global energy associated with the system of l coupled windings can be written as [83, 84, 227]:

$$w = \sum_{j=1}^{l} w_j = \sum_{j=1}^{l}\left[\sum_{k=1}^{l}\int_{i_k(0)}^{i_k(t)}(L_{jk}^{inc}i_j)di_k\right] \tag{3.79}$$

In the papers [84, 227] the self- and mutual-inductance terms of the various l windings have been expressed as the partial derivatives of the global stored energy density, w, with respect to various winding current perturbations, Δi_j. These derivatives can, in turn, be expanded around a "quiescent" magnetic field solution obtained for a given set of winding currents, in terms of various current perturbations $\pm\Delta i_j$ and $\pm\Delta i_k$ in the jth and kth windings, and the resulting change in the global energy. In a machine where $L_{jk} = L_{kj}$, this process yields the following for the self- and mutual-inductance terms:

$$L_{jj} = \frac{\partial^2 w}{\partial(\Delta i_j)^2} \approx [w(i_j - \Delta i_j) - 2w + w(i_j + \Delta i_j)]/(\Delta i_j)^2 \tag{3.80}$$

$$L_{jk} = \frac{\partial^2 w}{\partial(\Delta i_j)\partial(\Delta i_k)} \approx [w(i_j + \Delta i_j, i_k + \Delta i_k) - w(i_j - \Delta i_j, i_k + \Delta i_k)$$

$$-w(i_j + \Delta i_j, i_k - \Delta i_k) + w(i_j - \Delta i_j, i_k - \Delta i_k]/(4\Delta i_j\Delta i_k) \tag{3.81}$$

The details are given in [83, 84, 227]. For the two-dimensional field distribution the current/energy perturbation method does not take into account the end connection leakage.

For steady-state problems a similar accuracy can be obtained by first calculating the synchronous reactance, then mutual reactance and finally the slot and differential leakage reactance as a difference between synchronous and mutual reactances.

3.12.3 Steady-state inductance

In design calculations it is enough to calculate the steady-state inductance through the use of the flux linkage Ψ [53], Stokes' theorem (3.16), and magnetic vector potential (3.29), i.e.:

$$L = \frac{\Psi}{I} = \frac{\int_S \nabla \times \mathbf{A} \cdot d\mathbf{S}}{I} = \frac{\oint \mathbf{A} \cdot d\mathbf{l}}{I} \qquad (3.82)$$

where $\mathbf{A}$ is the magnetic vector potential around the contour l.

3.12.4 Reactances of synchronous machines

The *synchronous reactance* is when the total flux Ψ_{sd} or Ψ_{sq} includes both mutual and leakage fluxes, i.e., the armature slot, tooth top, and differential leakage flux. A flux plot through the air gap does not include the stator leakage, but simply the d-axis and q-axis linkage flux Ψ_{ad} or Ψ_{aq}. The first harmonics of these main fluxes give the *armature reaction* or *magnetizing reactances* [252, 253]. A combination of the total fluxes Ψ_{sd}, Ψ_{sq} and linkage fluxes Ψ_{ad} and Ψ_{aq} will give the *armature leakage reactance* (excluding the end connection leakage reactance).

This method is easy to implement using any FEM software since it can be automated which is important when trying to find the performance characteristics at a particular input voltage using an iterative loop.

If the armature current $I_a = 0$, then it follows that the normal component of the rotor magnetic flux density, B_z, determines the d-axis. A line integral through the air gap gives the distribution of the magnetic vector potential. The values of constant vector potential represent flux lines. Numerical Fourier analysis of this vector potential yields an analytical expression for the first harmonic, i.e.,

$$A_z(p\alpha) = a_1 \cos(p\alpha) + b_1 \sin(p\alpha) = A_{o1} \sin(p\alpha + \alpha_d) \qquad (3.83)$$

where $A_{o1} = \sqrt{a_1^2 + b_1^2}$ and $\alpha_d = \arctan(b_1/a_1)$. The angle α_d relates to the d-axis since it shows the angle of zero crossing of the magnetic vector potential through the air gap line contour. This angle is usually found to be zero due to the symmetry in the machine. The q-axis is related to the d-axis by a shift of $\pi/(2p)$, thus

$$\alpha_q = \alpha_d + \frac{\pi}{2p} \qquad (3.84)$$

3.12.5 Synchronous reactances

In a two-dimensional FEM model the d-axis and q-axis synchronous reactances, respectively, excluding the end connection leakage flux are

$$X_{sd} = 2\pi f \frac{\Psi_{sd}}{I_{ad}} \qquad\qquad X_{sq} = 2\pi f \frac{\Psi_{sq}}{I_{aq}} \qquad (3.85)$$

where Ψ_{sd} and Ψ_{sq} are the total fluxes in the d and q axis, I_{ad} and I_{aq} are the d-axis and q-axis armature currents, respectively, and f is the armature supply (input) frequency.

The d-axis and q-axis fluxes are obtained from the combination of the phase belt linkages [59, 60, 82, 239]. The real and imaginary components of the flux represent the d-axis and q-axis flux linkages. The phasor diagram for a synchronous motor shows that the rotor excitation flux and d-axis armature flux are in the same direction while the q-axis armature flux is perpendicular.

The calculation of the synchronous reactances using eqns (3.85) are sensitive to the values of I_{ad} and I_{aq}, respectively. This is due to the fact that in the full-load analysis of a PM synchronous motor the values of I_{ad} and I_{aq} both approach zero at different load angles. The rounding off errors that occur in the FEM amplify the error in the synchronous reactance as the armature current components tend toward zero. The use of a constant current disturbance, in both I_{ad} and I_{aq}, solves this problem. The synchronous reactances are then

$$X'_{sd} = 2\pi f \frac{\Delta \Psi_{sd}}{\Delta I_{ad}} \qquad\qquad X'_{sq} = 2\pi f \frac{\Delta \Psi_{sq}}{\Delta I_{aq}} \qquad (3.86)$$

where $\Delta \Psi_{sd}$ and $\Delta \Psi_{sq}$ are the changes in magnetic fluxes at the load point in the d-axis and q-axis, respectively, and ΔI_{ad} and ΔI_{aq} are the values of the d-axis and q-axis stator current disturbances, respectively.

The magnetic saturation that occurs in the loaded FEM solution should be transferred to the problem with current disturbance. This is done by storing the permeability of every element obtained from the loaded nonlinear magnetic field computation. These permeabilities can then be used for the linear calculation with current disturbance. This ensures that the saturation effect, which occurs in the loaded model is not ignored in the disturbed results.

3.12.6 Armature reaction reactances

The d-axis and q-axis fundamental components of magnetic flux in the air gap can be derived by performing a Fourier analysis on the vector potentials **A** around the inner surface of the armature core. In Fourier series the cosine term coefficient a_1 expresses the quantity of half the q-axis flux per pole and the sine term coefficient b_1 expresses the quantity of half the d-axis flux per pole [253]. Thus, the fundamental harmonic of the resultant air gap flux per pole and the inner torque angle δ_i (Fig. 5.5, Chapter 5), are

$$\Phi_g = 2L_i \sqrt{a_1^2 + b_1^2} \qquad\qquad \delta_i = \arctan\left(\frac{b_1}{a_1}\right) \qquad (3.87)$$

This magnetic flux rotates with the synchronous speed $n_s = f/p$ and induces in each phase winding the following EMF

$$E_i = \pi \sqrt{2} f N_1 k_{w1} \Phi_g \qquad (3.88)$$

The phasor diagrams shown in Fig. 5.5 (Chapter 5) give the following expressions for the d-axis and q-axis armature reaction (mutual) reactances:

- in the case of an underexcited motor

$$X_{ad} = \frac{E_i \cos \delta_i - E_f}{I_{ad}}; \qquad X_{aq} = \frac{E_i \sin \delta_i}{I_{aq}} \qquad (3.89)$$

- in the case of an overexcited motor

$$X_{ad} = \frac{E_f - E_i \cos \delta_i}{I_{ad}}; \qquad X_{aq} = \frac{E_i \sin \delta_i}{I_{aq}} \qquad (3.90)$$

The armature reaction affects the saturation of magnetic circuit in proportion to the load torque. To take into account this effect another set of equations similar to eqns (3.89) or (3.90) is needed [253].

3.12.7 Leakage reactance

The armature leakage reactance can be obtained in two ways:

(a) from numerical evaluation of the energy stored in the slots and in the end connections [214],
(b) as the difference between the synchronous reactance and armature reaction reactance, i.e.,

$$X_{1d} = X_{sd} - X_{ad} \qquad X_{1q} = X_{sq} - X_{aq} \qquad (3.91)$$

The first method allows for calculating the slot leakage reactance, including the magnetic saturation effect and the end connection leakage reactance. It can be performed provided that each slot and each tooth is defined as a volume V in which the local contribution to the leakage energy can be calculated [214]. The leakage flux is practically independent of the relative stator–rotor position.

The result of the second method is the sum of the slot, differential, and tooth-top leakage reactance obtained as the difference between synchronous and armature reaction reactances. In the case of PM brushless motors this method is preferred since it uses only synchronous and armature reaction reactances. Note that according to the second method, leakage reactances in the d and q-axis differ slightly.

In general, the calculation of the end connection leakage reactance is rather difficult using the two-dimensional FEM program and does not bring satisfactory results. For this purpose three-dimensional FEM software package is recommended [302].

3.13 Interactive FEM programs

Modern finite element packages use advanced graphical displays which are usually menu-driven to make the process of problem-solving an easy task. All packages have three main components, although they may be integrated into one; they are the pre-processor, processor (solver), and post-processor. The most popular commercial packages used in the electromagnetic analysis of electrical machines and electromagnetic devices are: (a) MagNet from Infolytica Co., Montreal, Canada, (b) Maxwell from Ansoft Co., Pittsburgh, PA, U.S.A., (c) Flux from Magsoft Co., Troy, NY, U.S.A., (d) JMAG from JMAG Group, Tokyo, Japan and (e) Opera from Vector Fields Ltd., Oxford, U.K. All these packages have 2D and 3D solvers for electrostatic, magnetostatic and eddy-current problems. A transient and motion analysis is also included to some extent.

3.13.1 Pre-processor

The *pre-processor* is a module where the finite element model is created by the user. This module allows new models to be created and old models to be altered. The different parts of any finite element model are:

- **Drawing**: Drawing the *geometric* outline of the model using graphical drawing tools much like any CAD package including mirroring and copying features.

- **Materials**: The different regions of the geometry model are assigned magnetic material properties. The materials can have linear or nonlinear magnetic characteristics. For each material a particular conductivity is defined. For rare-earth PM materials only the remanence and coercivity is defined instead of the whole demagnetization curve.

- **Electric circuit**: Regions that contain coils are linked to current or voltage sources. The user specifies the number of turns per coil and the current magnitude.

- **Constraints**: The edges of the model usually need constraints and this is done by defining constraints graphically. Periodic constraints are easily defined, by the user, using graphical mouse clicks.

Commercial packages also allow the user to enter new material curves. This is done by specifying a number of points on the B—H curve of the particular material and then the program creates a smooth curve from these points. The curve is a continuous function with a nondecreasing first derivative, if sufficient points are entered.

Once all the components of the model are described the finite element model can be solved.

3.13.2 Solver

The *solver* module solves numerically the field equations. The pre-processor module sets which solver (electrostatic, magnetostatic, eddy-current) should be used. Mostly, adaptive meshing is implemented to ensure efficient mesh discretization. The solver starts by creating a coarse finite element mesh and solving it. An error estimate is produced from this solution, and the mesh is refined and solved again. This is repeated until the mesh is refined sufficiently to produce an accurate result.

3.13.3 Post-processor

The *post-processor* is an interactive module that displays field quantities such as magnetic vector potential, flux density, field intensity and permeability. It also gives the user access to a vast amount of information regarding the finite element solution such as energy, force, torque and inductance, which are all built into the module.

Numerical examples

Numerical example 3.1

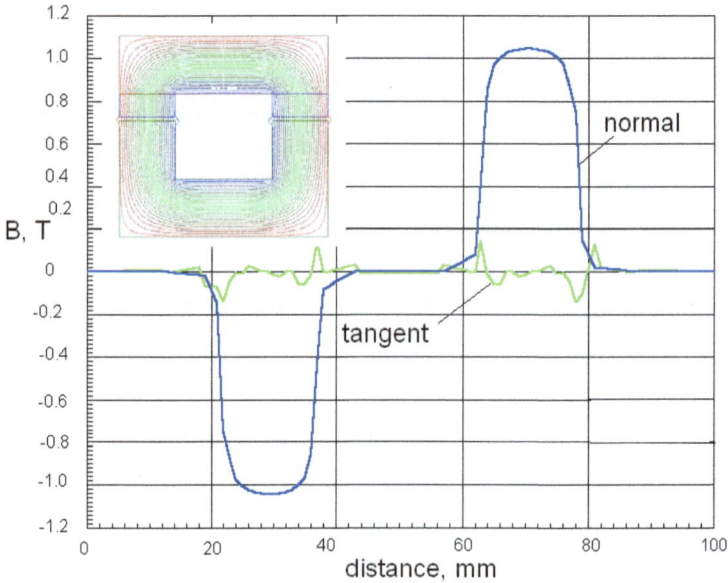

Fig. 3.3. Magnetic flux distribution, normal and tangential component of the magnetic flux density for the magnetic circuit shown in Fig. 2.17.

Find the magnetic flux distribution, magnetic flux density in the air gap and attraction force for the simple magnetic circuit shown in Fig. 2.17.

Solution:

This problem has been solved using the 2D *Maxwell* commercial FEM package from Ansoft, Pittsburgh, PA, U.S.A.

The magnetic flux distribution, normal and tangential component of the magnetic flux density in the air gap are plotted in Fig. 3.3. The normal component of magnetic flux density (over 1 T) is very close to that obtained from analytical solution (Numerical example 2.1).

The normal attraction force as obtained from the 2D FEM is $F = 198.9$ N at 20°C (compare with 229.7 N from the analytical approach). The analytical approach gives a higher attractive force because it has been assumed that the normal component of the magnetic flux density in the air gap is uniform, fringing and leakage fluxes are neglected and the MVDs in the laminated cores are zero.

Numerical example 3.2

The primitive 2-pole PM electrical machine shown in Fig. 2.18 is used in this example. The dimensions and material properties are as shown in Example 2.1. The problem is essentially the same as was set out in Chapter 2: to find the magnetic flux density in the air gap when the armature winding is fed with current $I_a = 1.25$ A.

Using the FEM more information can be obtained about this machine with very little extra effort. Important design details such as the winding inductance and rotor output torque will also be calculated.

Solution:

1. *Finite element model*

    A two-dimensional finite element model of the primitive machine is built using *MagNet* 2D commercial package from Infolytica, Montreal, Canada. The outline of the model is first drawn, as shown in Fig. 3.4. The material properties are then assigned to the different regions. The materials' B—H curves are entered, including the PM demagnetization curve, using the *MagNet* curve module.

    The Dirichlet boundary condition is applied to the area around the model. This boundary is used since most of the flux is expected to remain within the model. Any leakage flux will be seen, although it may not distribute as widely away from the core due to the boundary. The air region around the model should be increased if the leakage flux is of interest.

    The current source circuit is then defined with the current going in opposite directions for the two coil regions. A current of $I_a = 1.25$ A is set with each coil having $N = 1100$ turns. The problem can now be solved.

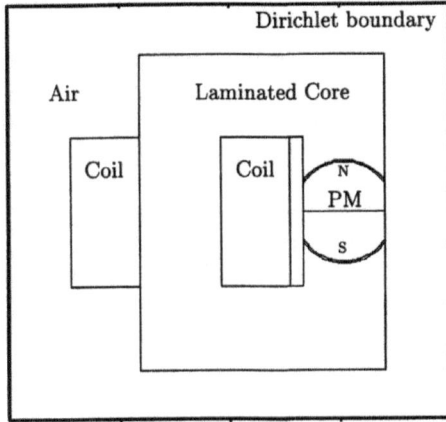

Fig. 3.4. Geometric outline used in FEM analysis of a primitive electrical machine according to Fig. 2.18.

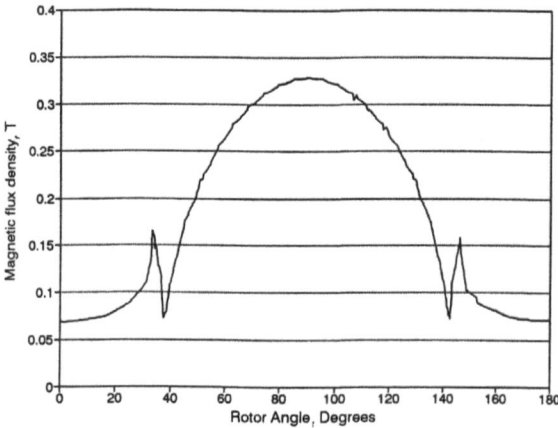

Fig. 3.5. The magnitude of the air gap magnetic flux density.

2. *Air gap magnetic flux density*

A line contour through the air gap is used to plot the magnitude of the air gap magnetic flux density, as shown in Fig. 3.5. The plot is made for an input current $I_a = 1.25$ A. From Fig. 3.5 it can be seen that the maximum magnetic flux density in the center of the pole is 0.318 T. From the classical approach the value of 0.304 T has been obtained as the average value over the magnet pole face width w_M. The flux distributions for an input current $I_a = 1.25$ A and $I_a = 2.6$ A are shown in Fig. 3.6.

(a)   (b)

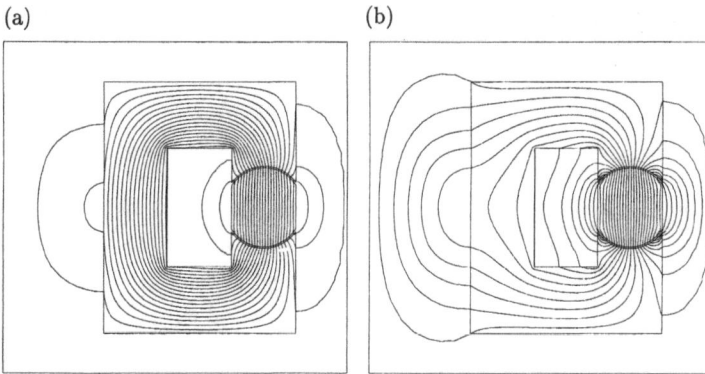

Fig. 3.6. Flux distribution with: (a) $I_a = 1.25$ A, (b) $I_a = 2.6$ A.

(a)   (b)

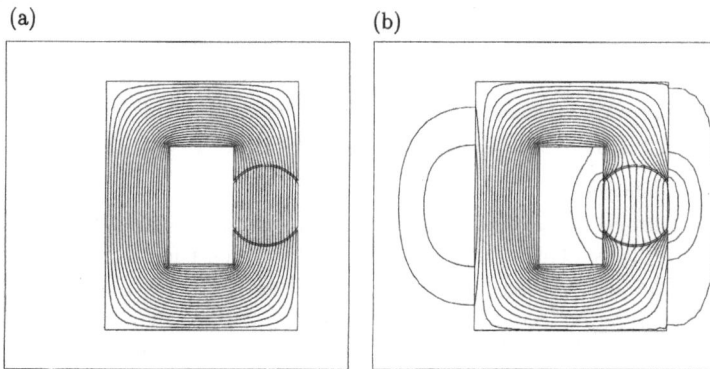

Fig. 3.7. Flux distribution with: (a) no armature current $I_a = 0$ A, (b) $I_a = 1.25$ A and the magnet unmagnetized.

Fig. 3.7 shows the flux distribution of the PM with no armature current and the flux distribution for $I_a = 1.25$ A with the PM unmagnetized.

3. *Winding inductance*

The winding inductance is calculated using the energy stored in the coil divided by half the current squared and by the current/energy perturbation method. The first method is easily calculated from a single finite element solution. The winding inductance, using this method, is calculated by the *MagNet* post-processor. The perturbation method is a little more complicated and requires three finite element solutions. The inductance is then calculated using eqn (3.80). The winding inductance obtained from the stored energy method is 0.1502 H and from the energy perturbation method is 0.1054 H. The difference in the inductance values are due to

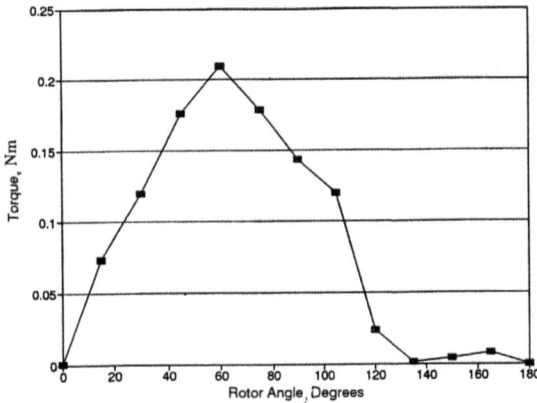

Fig. 3.8. Developed torque versus rotor position angle. In the region from 0 to 135° the force is attractive and from 135° to 180° the force is repulsive.

the oversimplification made in the stored energy form. The energy perturbation result is seen as being a more accurate result.

4. *Developed torque*

The developed torque is calculated using Maxwell's stress tensor. A line contour is drawn through the air gap along which the calculation is done. The torque is obtained for a number of different rotor positions. This is done by changing the geometric rotation of the rotor and resolving the model. Fig 3.8 shows the torque versus rotor angle. The attractive forces are in the region from 0° to 135° and the repulsive forces from 135° to 180°.

Numerical example 3.3

A 24-pole ring-shaped PM has its outer diameter of 208 mm, width of 17 mm and radial thickness of 6 mm. The magnet is located in a free space and made of a sintered NdFeB with $B_r = 1.16$ T and $H_c = 876$ kA/m. Assuming that the magnetization vector is radially aligned, create a 3D plot of the magnetic flux density vector.

Solution:

The problem requires a 3D magnetostatic field simulator, e.g., the commercial 3D *Maxwell* FEM package from Ansoft, Pittsburgh, PA, U.S.A.

The distribution of magnetization of the multipole ring magnet magnetized radially can be approximated with the aid of a square wave. It is enough to take into account only the three most significant components of Fourier series of a square wave, i.e.,

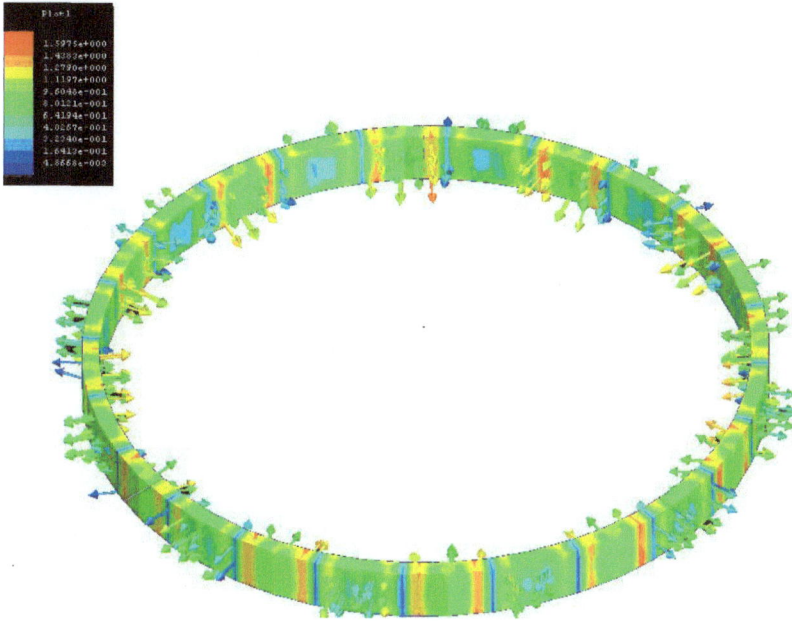

Fig. 3.9. Distribution of the magnetic flux density vector about a 24-pole PM ring.

$$f_m(\alpha) \approx \frac{4}{\pi} M_p \left[\sin(p\alpha) + \frac{1}{3}\sin(3p\alpha) + \frac{1}{5}\sin(5p\alpha) \right]$$

where $M_p = \mu_{rrec} H_c = 1.053 \times 876,000 = 922,428$ A/m, $\mu_{rrec} = B_r/(\mu_0 H_c)$ $= 1.16/(0.4\pi \times 10^{-6} \times 876,000) = 1.053$, $p = 12$ is the number of pole pairs and α is the space angle which denotes the radial position of magnetization. The function $f_m(\alpha)$ describes the variation of the magnetization with the angle α. In a Cartesian coordinate system the ring magnet lies in the xy plane and the z axis perpendicular to the xy plane is the center axis of the ring. Thus, the components for a directional vector are:

$$m_x(\alpha) = f_m(\alpha)\cos\alpha \qquad m_y(\alpha) = f_m(\alpha)\sin\alpha \qquad m_z(\alpha) = 0$$

Once the material properties are assigned to the PM, a solution for the fields can be generated and the magnetic flux density vector can be plotted in the post-processor. The 3D plot is shown in Fig. 3.9.

4

Permanent Magnet d.c. Commutator Motors

4.1 Construction

A d.c. PM commutator (brush) motor can be compared with a d.c. separately excited motor. The only difference is in the excitation flux Φ_g in the air gap: for a PM motor $\Phi_g = const$ whilst for a separately excited motor Φ_g can be controlled. This means that the speed of a standard d.c. PM commutator motor can normally be controlled only by changing the armature input voltage or armature current. A typical d.c. PM commutator motor is shown in Fig. 4.1. By adding an additional field excitation winding, the flux Φ_g as well as the speed can be changed in a certain limited range.

Alnico PMs used to be common in motors having ratings in the range of 0.5 to 150 kW. Ceramic magnets are now most popular and economical in fractional horsepower motors and may have an economic advantage over Alnico up to about 7.5 kW. Rare-earth magnet materials are costly, but are the best economic choice in small motors.

Magnetic circuit configurations of different types of PM d.c. commutator motors are shown in Figs 4.2 to 4.4. There are four fundamental armature (rotor) structures:

- conventional slotted rotor (Fig. 4.2a and 4.3),
- slotless (surface wound) rotor (Fig. 4.2b),
- moving-coil cylindrical rotor (Fig. 4.4a),
- moving-coil disk (pancake) rotor (Fig. 4.4b and c).

The slotted- and slotless-rotor PM commutator motors have armature windings fixed to the laminated core. The armature winding, armature core and shaft create one integral part.

The moving-coil d.c. motor has armature windings fixed to an insulating cylinder or disk rotating between PMs or PMs and a laminated core. The types of moving-coil motors are listed in Table 4.1. In a moving-coil motor the moment of inertia of the rotor is very small since all ferromagnetic cores are

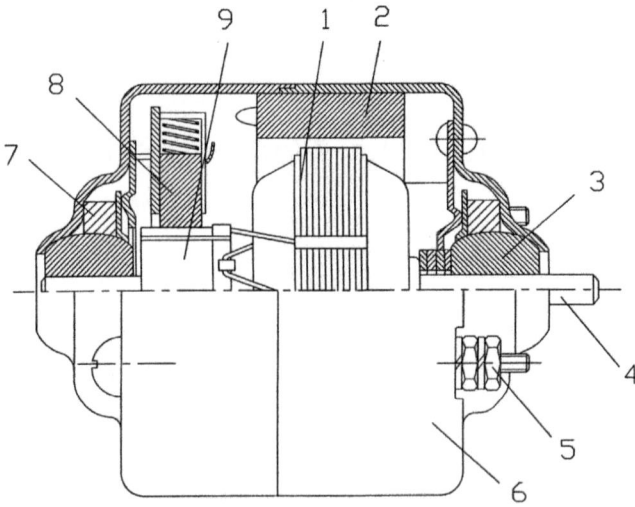

Fig. 4.1. d.c. commutator motor with segmental PMs. 1 — armature, 2 — barium ferrite PM, 3 — porous metal bearing, 4 — shaft, 5 — terminal, 6 — steel frame, 7 — oil soaked felt pad, 8 — brush, 9 — commutator.

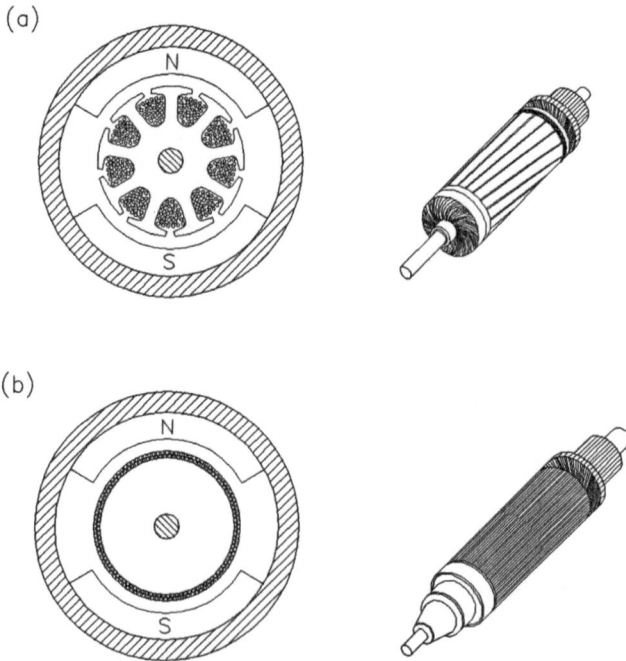

Fig. 4.2. Construction of d.c. PM commutator motors with laminated-core rotors: (a) slotted rotor, (b) slotless rotor.

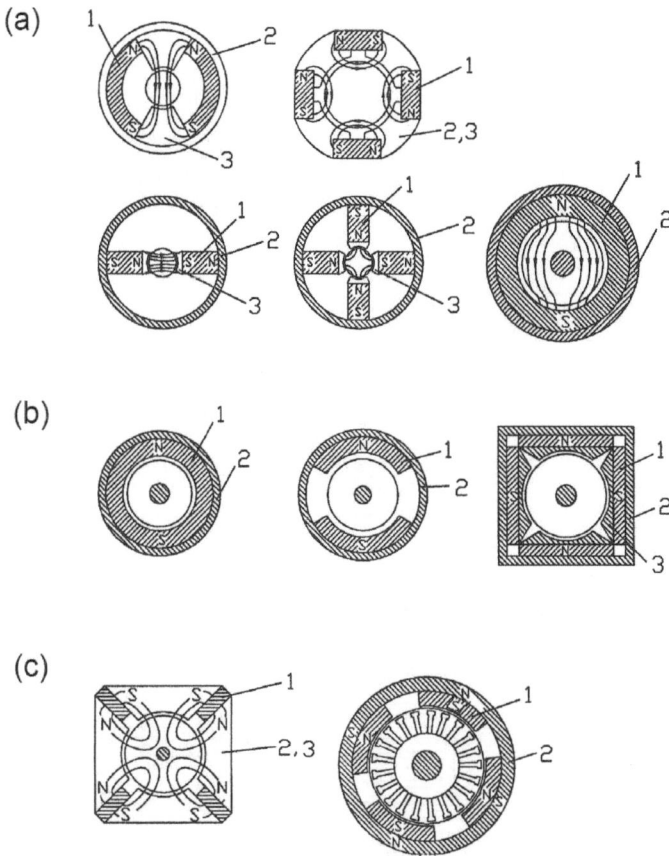

Fig. 4.3. Excitation systems of d.c. commutator motors with laminated-core rotors using different types of PMs: (a) Alnico, (b) ferrites, (c) rare-earth. 1 — PM, 2 — mild steel yoke, 3 — pole shoe.

Table 4.1. Classification of moving-coil motors

Cylindrical type		Disk (pancake) type		
Outside-field type	Inside-field type • honeycomb winding • rhombic winding • bell winding • ball winding	Wound armature	Printed armature	Three-coil armature

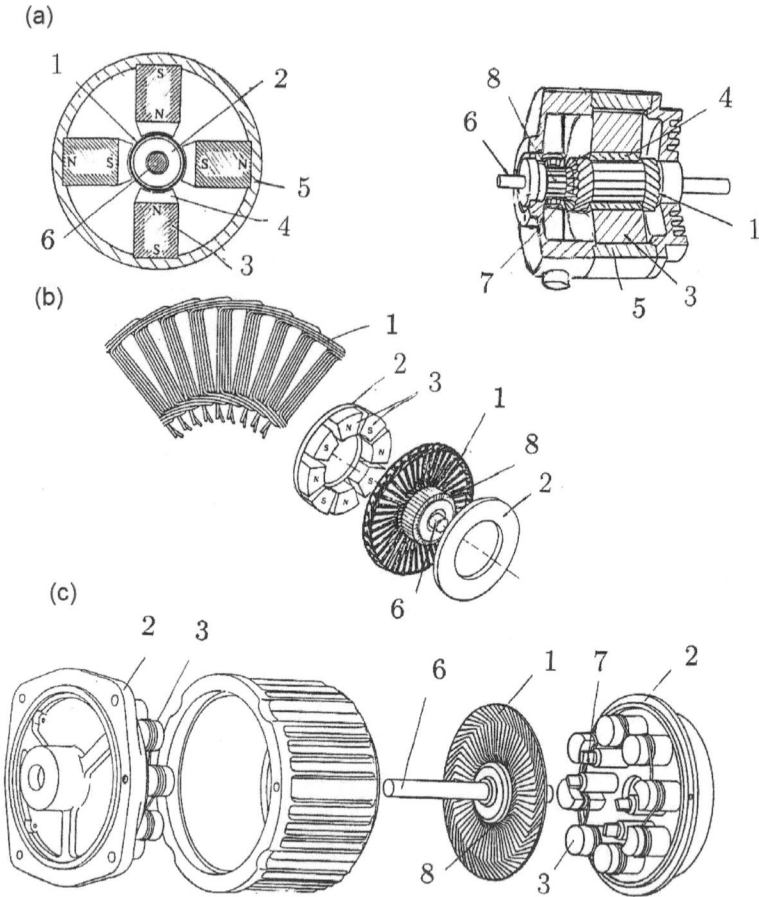

Fig. 4.4. Outside-field type moving coil PM commutator motors: (a) cylindrical motor, (b) disk motor with wound rotor, (c) disk motor with printed rotor winding and hybrid excitation system. 1 — moving coil armature winding, 2 — mild steel yoke, 3 — PM, 4 — pole shoe, 5 — mild steel frame, 6 — shaft, 7 — brush, 8 — commutator.

stationary; i.e., they do not move in the magnetic field and no eddy-currents or hysteresis losses are produced in them. The efficiency of a moving-coil motor is better than that of a slotted rotor motor.

Owing to low moment of inertia the mechanical time constants of moving-coil motors are much smaller than those of steel-core armature motors.

4.1.1 Slotted-rotor PM d.c. motors

The core of a slotted rotor is a lamination of a silicon steel sheet or carbon steel sheet. The armature winding is located in the rotor slots. The torque acts on the conductors secured in the slots and reinforced by the slot insulation and epoxy resin. Thus a slotted rotor is more durable and reliable than a slotless rotor. A core having many slots is usually desirable because the greater the number of slots, the less the cogging torque and electromagnetic noise. Cores having even numbers of slots are usually used for the motors manufactured by an automated mass production process because of the ease of production. From the motor quality point of view, ferromagnetic cores with odd numbers of slots are preferred due to low cogging torque.

Table 4.2. Data of small PM d.c. commutator motors with slotted rotors manufactured by Buehler Motors GmbH, Nuremberg, Germany

Specifications	1.13.044.235	1.13.044.413	1.13.044.236	1.13.044.414
Rated voltage, V	12	12	24	24
Rated torque, Nm$\times 10^{-3}$	150	180	150	180
Rated speed, rpm	3000			
Rated current, A	6.2	7.3	3.1	3.5
Diameter of frame, mm	51.6			
Length of frame, mm	88.6	103.6	88.6	103.6
Diameter of shaft, mm	6			

Skewed slots in Fig. 4.2a reduce the cogging torque that is produced by interaction between the rotor teeth and PM pole shoes (change in the air gap reluctance).

Specifications of small PM commutator motors with slotted rotors manufactured by Buehler Motors GmbH, Nuremberg, Germany are given in Table 4.2.

4.1.2 Slotless-rotor PM motors

Extremely low cogging torque can be produced by fixing the windings on a cylindrical steel core without any slots (Fig. 4.2b). In this case the torque is exerted on the conductors uniformly distributed on the rotor surface. However, the flux decreases in comparison with the slotted rotor since the gap between the rotor core and the pole shoes is larger. Therefore, larger volume of PMs must be used to get sufficient magnetic flux.

4.1.3 Moving-coil cylindrical motors

Cylindrical outside-field type

This type of motor (Fig. 4.4a) has a very small mechanical time constant T_m, sometimes $T_m < 1$ ms. In order to obtain a small T_m, the ratio Φ_g/J must be as large as possible where Φ_g is the air gap magnetic flux and J is the moment of inertia of the rotor. More flux is produced from an Alnico or rare-earth PMs, which have a high remanence B_r than from a ferrite PM. Since *Alnico* PMs are easy to demagnetize, a long Alnico magnet, magnetized lengthwise, is used in order to avoid demagnetization. A modern design is a stator magnetic circuit with rare-earth PMs.

Cylindrical inside-field type

Moving-coil motors of the inner-field type, which are also known as *coreless motors*, are often used for applications of less than 10 W, very rarely up to 225 W. In this type of motor the PM is inside the moving-coil armature. Though the moment of inertia of this rotor is low, the mechanical time constant is not always low, because the magnetic flux produced by a small size PM placed inside the armature is low. Coreless PM d.c. brush motors were extensively used for driving capstans of audio cassette players (Fig. 1.19), VCRs, zoom lenses of cameras, etc. due to their outstanding characteristics: (a) high power per volume, (b) high efficiency (no core losses), (c) zero cogging torque, (d) low damping coefficient D — eqn (1.16). Today applications of inside-field type PM d.c. brush motors include medical and laboratory equipment, robotics and automation, optics, technical instruments, office equipment, public life, prototyping, etc.

The mechanical component of the damping coefficient is partly due to friction caused by bearing lubrication components and bearing seals. Another mechanical effect which may manifest itself at high speed is the windage element due to rapidly rotating parts [179]. The electromagnetic components of the damping effect consist of [179]: (a) circulating current caused by currents flowing in commutated coils in the presence of stray fields and (b) eddy currents induced in armature conductors moving in the magnetic field.

The inside-field type cylindrical moving coil motors are shown in Fig. 4.5. The armature windings are classified as follows: (a) winding with skew-wound coils, (b) rhombic winding, (c) bell winding, (d) ball winding.

The *winding with skew-wound coils* or *honeycomb winding*, which is also known as *Faulhaber winding* (U.S. patent 3360668), was the first type of winding (1965) to be employed in the widely used coreless motors. Fig. 4.5a shows

(a)

(b)

(1)

(c)

(2)

(d)

(3)

PLASTIC
BASE

WINDING   PM

(4)

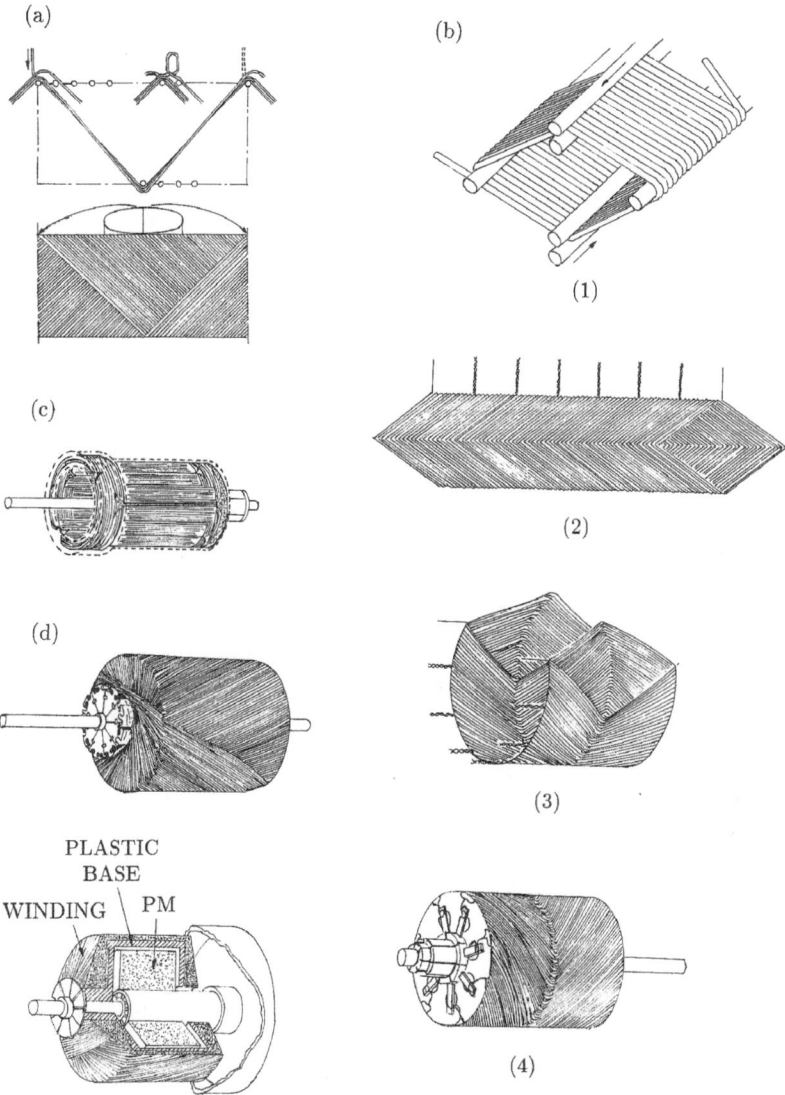

Fig. 4.5. Inside-field type cylindrical moving coil rotors with: (a) winding with skew-wound coils (honeycomb winding) according to U.S. patent 3360668, (b) rhombic winding according to U.S. patent application publication 2007/0103025, (c) bell winding according to US patent 3467847, (d) ball winding.

Table 4.3. Moving coil cylindrical coreless PM d.c. commutator motors (inside–field type) manufactured by Minimotor SA, Faulhaber Group, Croglio, Switzerland

Series	Brushes	Output power W	Outer diameter mm	Length mm	Shaft diameter mm	Rated voltage V	No-load speed rpm	Stall torque 10^{-3} Nm
0615...S	PrM	0.12	6	15	1.5	1,5 to 4,5	20,200	0.24
0816...S	PrM	0.18	8	16	1.5	3 to 8	16,500	0.41
1016...G	PrM	0.42	10	16	1.5	3 to 12	18,400	0.87
1024...S	PrM	1.11	10	24	1.5	3 to12	14,700	2.89
1219...G	PrM	0.50	12	19	1.5	4,5 to 15	16,200	1.19
1224...S	PrM	1.3	12	24	1.5	6 to 15	13,100	3.69
1224...SR	PrM	1.95	12	24	1.5	6 to 15	13,800	5.43
1319...SR	PrM	1.10	13	19	1.5	6 to 24	14,600	2.91
1331...SR	PrM	3.11	13	31	1.5	6 to 24	10,600	11.20
1336...C	Gr	2.02	13	36	2	6 to 24	9200	8.40
1516...S	PrM	0.42	15	16	1.5	1,5 to 12	16,200	1.04
1516...SR	PrM	0.54	15	16	1.5	6 to 12	12,900	1.61
1524...SR	PrM	1.92	15	24	1.5	3 to 24	10,800	7.12
1624...S	PrM	1.87	16	24	1.5	3 to 24	14,400	5.16
1717...SR	PrM	1.97	17	17	1.5	3 to 24	14,000	5.38
1724...SR	PrM	2.83	17	24	1.5	3 to 24	8600	13.20
1727...C	Gr	2.37	17	27	2	6 to 24	7800	11.6
2224...SR	PrM	4.55	22	24	2	3 to 36	8200	21.40
2230...S	PrM	3.69	22	30	1.5/2	3 to 40	9600	14.70
2232...SR	PrM	11	22	32	2	6 to 24	7400	59.2
2233...S	PrM	3.85	22	33	1.5/2	4,5 to 30	9300	18.40
2342...CR	Gr	20.50	23	42	3	6 to 48	9000	91.40
2642...CR	Gr	23.2	26	42	4	12 to 48	6400	139
2657...CR	Gr	47.9	26	57	4	12 to 48	6400	286
3242...CR	Gr	27.3	32	42	5	12 to 48	5400	193
3257...CR	Gr	84.5	32	57	5	12 to 48	5900	547
3557...C	Gr	15	35	57	4	6 to 32	5000	122
3557...CS	Gr	28.1	35	57	4	9 to 48	5700	188
3863...C	Gr	226	38	63	6	12 to 48	6700	1290
PrM = precious metal brushes,				Gr = graphite brushes				

the armature winding with coil terminals invented by F. Faulhaber. The production of the winding takes place on a cylindrical coil form, which is provided with holding devices, such as pins, at the ends of the armature for forming reversing points for the wire from which the winding is produced. Such a "distributed" winding consists of two layers, which together have a thickness equal to the thickness of two conductors. Originally, this motor type used an *Alnico* magnet in order to obtain a high magnetic flux. The housing, which also serves as the return path for magnetic flux, is made of mild carbon steel. Two bearings are usually placed in the center hole of the PM to support the shaft.

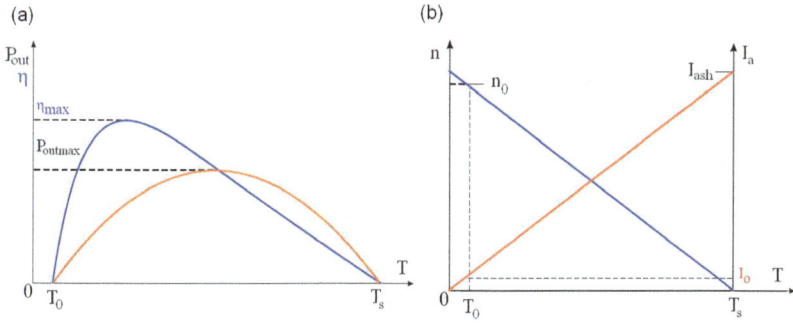

Fig. 4.6. Steady-state characteristics of coreless d.c. commutator motors with Faulhaber skew wound coils: (a) efficiency η and output power P_{out} versus torque T; (b) Speed n and armature current I_a versus torque T. Symbols: n_0 is the no-load speed, T_0 is the friction loss torque, I_{ash} is the "short circuit" armature current at $n = 0$ ($E = 0$) and T_s is the stall torque torque corresponding to I_{ash}.

For Faulhaber winding motors, as well as other coreless motors, commutators are made small for the following reasons:

(a) Both commutator and brushes use precious metals (gold, silver, platinum, and/or palladium), which are resistant to electrochemical processes during operation. Because precious metals are expensive, the size of the brushes must be as small as possible;

(b) The linear surface speed (4.13) of the commutator for stable commutation must be low;

(c) The machine size must be as small as possible.

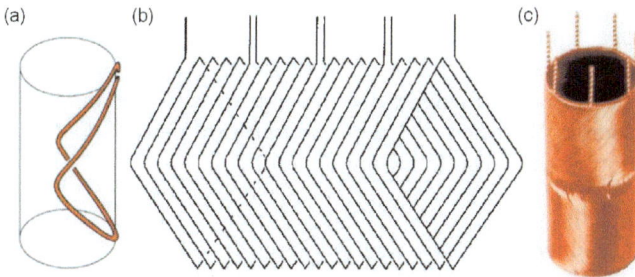

Fig. 4.7. Moving coil cylindrical coreless stator winding of rhombic type according to U.S. patent application publication 2007/0103025: (a) loop shape, (b)arrangement of layers, (c) complete cylindrical rhombic winding.

Characteristics of coreless d.c. commutator motors with Faulhaber skew wound coils are shown in Fig. 4.6. The maximum efficiency depends on the

Table 4.4. Moving coil cylindrical coreless two-pole PM d.c. commutator motors (inside–field type) with precious metal brushes manufactured by Maxon, Sachseln, Switzerland

Specifications	RE 16		RE-max 21		RE 25	
Outer diameter, mm	16	16	21	21	25	25
Power rating, W	3.2	3.2	5.0	5.0	10	10
Nominal voltage, V	24	30	24	36	24	32
No-load speed, rpm	7250	6460	9110	10,300	5190	5510
Max. permissible speed, rpm	7600	7600	16,000	16,000	5500	5500
No-load current, mA	3	2	5.71	4.69	14.3	11.6
Starting current, A	0.561	0.341	1.13	0.943	3.28	2.76
Max. continuous current, mA	172	120	249	185	668	529
Stall torque, mNm	17.6	15.0	28.3	31.4	144	152
Max. continuous torque, mNm	5.40	5.28	6.11	6.03	28.8	28.6
Max. output power at nominal voltage, W	3.34	2.53	−	−	−	−
Max. efficiency, %	86	85	87	87	87	88
Torque constant, mNm/A	31.4	44.1	25.0	33.3	44.0	55.2
Speed constant, $\times 10^{-3}$ V/rpm	3.289	4.608	2.625	3.484	4.608	5.780
Mechanical time constant, ms	5	5	7.06	7.08	3.97	3.97
Rotor moment of inertia, gcm^2	1.15	1.10	2.08	2.05	10.5	10.4
Terminal resistance Ω	42.8	88.0	21.3	38.2	7.31	11.6
Terminal inductance, mH	1.75	3.44	0.784	1.38	0.832	1.31
Thermal time constant (winding), s	9	9	8.77	8.77	12.4	12.4
Mass, g	38	38	42	42	130	130
Power density, W/kg	842	842	1190	1190	769	769

stall torque–to–friction torque ratio and is a function of the armature terminal voltage.

The *bell-type winding* (Fig. 4.5c) invented also by F. Faulhaber (U.S. patent 3467847) uses straight rectangular or skewed rectangular coils. A bell-type winding has a toothed carrier disk assembled with a commutator.

In comparison with other PM d.c. commutator motors, Faulhaber type coreless PM motors show the following advantages:

- high power density
- low starting voltage (very low friction losses)
- low inertia of the rotor
- very rapid starting
- high efficiency
- linear voltage–speed characteristics
- linear current–torque characteristics
- no cogging torque
- able to bear high overloads for short period of time
- high precision assures long life

Inside-field type d.c. motors with rhombic windings manufactured by Maxon, Sachseln, Switzerland have similar characteristics. The rhombic winding construction is shown in Figs 4.5b and 4.7. Specifications of small motors are listed in Table 4.4.

The *ball winding* method makes the rotor shape like a ball (Fig. 4.5d). The PM producing the magnetic flux is placed inside the ball winding. There is a plastic cylinder between the magnet and the winding, as illustrated in the cutaway view in Fig. 4.5d.

4.1.4 Disk motors

There are three main types of disk (pancake) motors: wound-rotor motor, printed rotor winding motor, and the three-coil motor.

In the *pancake wound-rotor motor* the winding is made of copper wires and moulded with resin (Fig. 4.4b). The commutator is similar to that of the conventional type. One application of these motors can be found in radiator fans.

The *disk-type printed armature winding motor* is shown in Fig. 4.4c. The coils are stamped from pieces of sheet copper and then welded, forming a wave winding. When this motor was invented by J. Henry Baudot [22], the armature was made using a similar method to that by which printed circuit boards are fabricated. Hence, this is called the printed winding motor. The magnetic flux of the printed motor can be produced using either Alnico or ferrite magnets.

The *disk-type three-coil motor* has three flat armature coils on the rotor and a four-pole PM system on the stator. The coil connection is different from that of an ordinary lap or wave winding [172]. Three-coil motors are usually designed as micromotors.

4.2 Fundamental equations

4.2.1 Terminal voltage

From the Kirchhoff's voltage law, the terminal (input) voltage is

$$V = E + I_a \sum R_a + \Delta V_{br} \tag{4.1}$$

where E is the voltage induced in the armature winding (back EMF), I_a is the armature current, $\sum R_a$ is the resistance of the armature circuit and ΔV_{br} is the brush voltage drop. The brush voltage drop is approximately constant and for most typical d.c. motors is practically independent of the armature current. For carbon (graphite) brushes $\Delta V_{br} \approx 2$ V; for other materials ΔV_{br} is given in Table 4.5. For d.c. PM motors with interpoles, $\sum R_a = R_a + R_{int}$, where R_a is the resistance of the armature winding and R_{int} is the resistance of the interpole winding.

4.2.2 Armature winding EMF

The EMF induced in the armature winding by the main flux Φ_g in the air gap is

$$E = \frac{N}{a} p n \Phi_g = c_E n \Phi_g \tag{4.2}$$

where N is the number of armature conductors, a is the number of pairs of armature current parallel paths, p is the number of pole pairs, Φ_g is the air gap (useful) magnetic flux, and

$$c_E = \frac{Np}{a} \tag{4.3}$$

is the *EMF constant* or *armature constant*. For a PM excitation $k_E = c_E \Phi_g = const$, thus

$$E = k_E n \tag{4.4}$$

The following relationship exists between the number of armature conductors N and the number of commutator segments C:

$$N = 2CN_c \tag{4.5}$$

where N_c is the number of turns per armature coil.

Table 4.5. Brushes for d.c. commutator motors

Material	Maximum current density, A/cm^2	Voltage drop (2 brushes) ΔV_{br}, V	Maximum commutator speed v_C, m/s	Friction coefficient at $v = 15$ m/s	Pressure, N/cm^2
Carbon –graphite	6 to 8	$2^{\pm 0.5}$	10 to 15	0.25 to 0.30	1.96 to 2.35
Graphite	7 to 11	1.9 to $2.2^{\pm 0.5}$	12 to 25	0.25 to 0.30	1.96 to 2.35
Electro –graphite	10	2.4 to $2.7^{\pm 0.6}$	25 to 40	0.20 to 0.75	1.96 to 3.92
Copper –graphite	12 to 20	0.2 to $1.8^{\pm 0.5}$	20 to 25	0.20 to 0.25	1.47 to 2.35
Bronze –graphite	20	$0.3^{\pm 0.1}$	20	0.25	1.68 to 2.16

4.2.3 Electromagnetic (developed) torque

The electromagnetic torque developed by the d.c. commutator motor is

$$T_d = \frac{N}{a}\frac{p}{2\pi}\Phi_g I_a = c_T \Phi_g I_a \qquad (4.6)$$

where

$$c_T = \frac{Np}{2\pi a} = \frac{c_E}{2\pi} \qquad (4.7)$$

is the *torque constant*. The electromagnetic torque is proportional to the armature current.

PMs produce a constant field flux $\Phi_g = const$ (neglecting the armature reaction). The developed torque is

$$T_d = k_T I_a \qquad (4.8)$$

where $k_T = c_T \Phi_g$.

When the brushes are shifted from the geometrical neutral zone by an angle Ψ, the developed torque is proportional to the $\cos\Psi$, i.e., $T_d = (N/a)[p/(2\pi)]I_a \Phi_g \cos\Psi$. The angle Ψ is between the q-axis and magnetic field axis of the rotor. If the brushes are in the neutral zone, $\cos\Psi = 1$.

4.2.4 Electromagnetic power

The electromagnetic power developed by the motor is

$$P_{elm} = \Omega T_d \qquad (4.9)$$

where the rotor angular speed is

$$\Omega = 2\pi n \qquad (4.10)$$

The electromagnetic power is also a product of the EMF and armature current, i.e.,

$$P_{elm} = E I_a \qquad (4.11)$$

4.2.5 Rotor and commutator linear speed

The rotor (armature) linear surface speed is

$$v = \pi D n \qquad (4.12)$$

where D is the outer diameter of the rotor (armature). Similarly, the commutator linear surface speed is

$$v_C = \pi D_C n \qquad (4.13)$$

where D_C is the outer diameter of the commutator.

4.2.6 Input and output power

A motor converts an electrical input power

$$P_{in} = V I_a \tag{4.14}$$

into a mechanical output power

$$P_{out} = \Omega T_{sh} = \eta P_{in} \tag{4.15}$$

where T_{sh} is the shaft (output) torque, and η is the efficiency.

4.2.7 Losses

The d.c. PM motor losses are

$$\sum \Delta P = \Delta P_a + \Delta P_{Fe} + \Delta P_{br} + \Delta P_{rot} + \Delta P_{str} \tag{4.16}$$

where

- the armature winding losses

$$\Delta P_a = I_a^2 \sum R_a \tag{4.17}$$

- the armature core loss

$$\Delta P_{Fe} = \Delta P_{ht} + \Delta P_{et} + \Delta P_{hy} + \Delta P_{ey} + \Delta P_{ad}$$

$$\propto f^{4/3} [B_t^2 m_t + B_y^2 m_y] \tag{4.18}$$

- the brush–drop loss

$$\Delta P_{br} = I_a \Delta V_{br} \approx 2 I_a \tag{4.19}$$

- the rotational losses

$$\Delta P_{rot} = \Delta P_{fr} + \Delta P_{wind} + \Delta P_{vent} \tag{4.20}$$

- the stray load losses

$$\Delta P_{str} \approx 0.01 P_{out} \tag{4.21}$$

and $\Delta P_{ht} \propto f B_t^2$ are the hysteresis losses in the armature teeth, $\Delta P_{et} \propto f^2 B_t^2$ are the eddy-current losses in the armature teeth, $\Delta P_{hy} \propto f B_y^2$ are the hysteresis losses in the armature yoke, $\Delta P_{ey} \propto f^2 B_y^2$ are the eddy-current losses in the armature yoke, P_{ad} are the additional losses in the armature core, $\Delta p_{1/50}$ is the specific core loss in W/kg at 1 T and 50 Hz, B_t is the magnetic flux density in the armature tooth, B_y is the magnetic flux density in the armature yoke, m_t is the mass of the armature teeth, m_y is the mass of the armature yoke, ΔP_{fr} are the friction losses (bearings and commutator–brushes), ΔP_{wind} are the windage losses, and ΔP_{vent} are the ventilation losses. For calculating the armature core losses, eqn (B.19) given in Appendix B for a.c. motors can be used.

The frequency of the armature current is

$$f = pn \tag{4.22}$$

Rotational losses can be calculated on the basis of eqns (B.31) to (B.33) given in Appendix B. Stray losses ΔP_{str} due to flux pulsation in pole shoes and steel rotor bandages are important only in medium and large power motors. Calculation of stray losses is a difficult problem with no guarantee of obtaining accurate results. It is better to assume that the stray losses are approximately equal to 1% of the ouput power.

It is sometimes convenient to express the motor losses as a function of the motor's efficiency, i.e.,

$$\sum \Delta P = P_{in} - P_{out} = \frac{P_{out}}{\eta} - P_{out} = P_{out}\frac{1 - \eta}{\eta} \tag{4.23}$$

The electromagnetic power can also be found by multiplying eqn (4.1) by the armature current I_a as given by eqn (4.9), i.e.,

$$P_{elm} = EI_a = P_{in} - I_a^2 \sum R_a - \Delta V_{br}I_a = \frac{P_{out}}{\eta} - (\Delta P_a + \Delta P_{br})$$

According to experimental tests the armature winding and brush-drop losses $(\Delta P_a + \Delta P_{br})$ in motors rated up to 1 kW are on average about 2/3 of the total losses. Thus the electromagnetic power of small d.c. commutator motors is

$$P_{elm} \approx \frac{P_{out}}{\eta} - \frac{2}{3}P_{out}\frac{1 - \eta}{\eta} = \frac{1 + 2\eta}{3\eta}P_{out} \tag{4.24}$$

The above equation is used in calculating the main dimensions of continuous-duty small d.c. commutator motors. For short-time or short-time intermittent duty, $(\Delta P_a + \Delta P_{br})$ amounts to 3/4 of the total losses and

$$P_{elm} \approx \frac{1 + 3\eta}{4\eta}P_{out} \tag{4.25}$$

For linear torque–speed and torque–current characteristics (Fig. 4.6) the maximum efficiency can be estimated as [279]

$$\eta_{max} \approx \left(1 - \sqrt{\frac{T_0}{T_s}}\right)^2 = \left(1 - \sqrt{\frac{I_{a0}}{I_{sh}}}\right)^2 \tag{4.26}$$

where I_{a0} is the no-load armature current, I_{ash} is the "short circuit" armature current at zero speed given by eqn (4.61), T_0 is the friction loss torque and T_s is the stall torque given by eqn (4.62).

4.2.8 Pole pitch

The pole pitch is defined as the armature circumference πD divided by the number of poles $2p$, i.e.,

$$\tau = \frac{\pi D}{2p} \tag{4.27}$$

The pole pitch can also be expressed in slots as a number of armature slots per number of poles. The ratio

$$\alpha_i = \frac{b_p}{\tau} = 0.55\ldots0.75 \tag{4.28}$$

is called the *effective pole arc coefficient* in which b_p is the pole shoe width.

4.2.9 Air gap magnetic flux density

The *air gap magnetic flux density*, or the *specific magnetic loading*, is

$$B_g = \frac{\Phi_g}{\alpha_i \tau L_i} = \frac{2p\Phi_g}{\pi \alpha_i L_i D} \tag{4.29}$$

where Φ_g is the air gap magnetic flux.

4.2.10 Armature line current density

The *armature line current density*, or the *specific electric loading*, is defined as the number of armature conductors N times the current in one parallel current path $I_a/(2a)$ divided by the armature circumference πD, i.e.,

$$A = \frac{N}{\pi D} \frac{I_a}{2a} \tag{4.30}$$

4.2.11 Armature winding current density

The *armature winding current density* is defined as the current in one parallel path $I_a/(2a)$ divided by the total cross section s_a of conductor (bar)

$$J_a = \frac{I_a}{2a s_a} \tag{4.31}$$

For a conductor consisting of parallel wires

$$s_a = s_{as} a_w \tag{4.32}$$

where s_{as} is the cross section of a single conductor and a_w is the number of parallel conductors.

The permissible current density as a function of armature linear speed $v = \pi D n$ is shown in Fig. 4.8. The large range of current density is due to variety of cooling systems, class of insulation and type of enclosure.

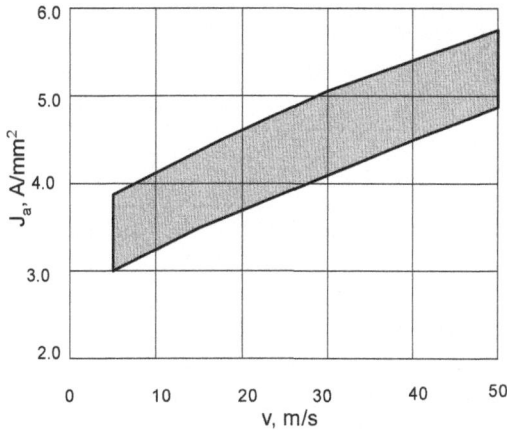

Fig. 4.8. Armature winding current density as a function of linear speed v — eqn (4.12) for air cooling system.

4.2.12 Armature winding resistance

The resistance of the armature winding is expressed as follows:

$$R_a = \frac{Nl_{av}}{\sigma s_a} \frac{1}{(2a)^2} \tag{4.33}$$

where N is the number of armature conductors, σ is the electric conductivity of the winding, s_a is the cross section area of armature conductor and $2a$ is the number of parallel current paths. The average length of the armature conductor (half of a coil) is $l_{av} = L_i + 1.2D$ for $p = 1$ and $l_{av} = L_i + 0.8D$ for $p > 1$.

Skin effect in the armature conductors or large commutator motors can be included by multiplying eqn (4.33) by coefficient k_{1R} derived in Chapter 8, Section 8.6.3, for high power density brushless motors.

4.2.13 Armature winding inductance

For a d.c. motor without compoles, the air gap armature winding inductance is expressed as [128]

$$L_a = \mu_0 \frac{\pi}{12} \frac{\alpha_i^3 DL_i}{g'} \left(\frac{N}{4pa}\right)^2 \tag{4.34}$$

For most PM machines the air gap $g' \approx k_C k_{sat} g + h_M/\mu_{rrec}$, where g is the mechanical clearance, $k_C \geq 1$ is the Carter's coefficient and $k_{sat} \geq 1$ is saturation factor of the magnetic circuit according to eqn (2.48). The electrical time constant of the armature winding is $T_{ae} = R_a/L_a$.

4.2.14 Mechanical time constant

The mechanical time constant

$$T_m = \frac{2\pi n_0 J}{T_{st}} = \frac{2\pi n_o J}{c_T \Phi I_{ast}} \tag{4.35}$$

is proportional to the no-load angular speed $2\pi n_0$ and the moment of inertia J of the rotor, and inversely proportional to the starting torque T_{st}.

4.3 Sizing procedure

For a cylindrical rotor d.c. motor the electromagnetic (internal) power as a function of specific magnetic (4.29) and electric (4.30) loading is

$$P_{elm} = EI_a = \frac{N}{a}pn\Phi_g\frac{2\pi aDA}{N} = \frac{N}{a}pn\alpha_i B_g L_i\frac{\pi D}{2p}\frac{2\pi aDA}{N} = \alpha_i\pi^2 D^2 L_i n B_g A$$

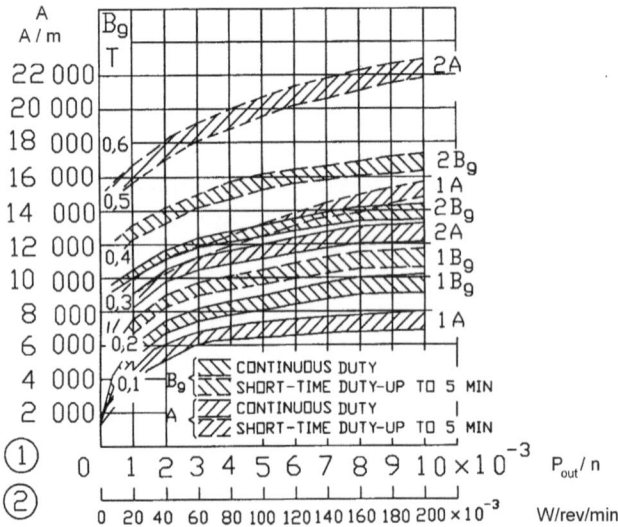

Fig. 4.9. Magnetic and electric loadings of d.c. commutator motors with Alnico or ferrite magnets as functions of the *output power–to–speed* ratio.

The ratio

$$\sigma_p = \frac{P_{elm}}{D^2 L_i n} = \alpha_i\pi^2 B_g A \tag{4.36}$$

is termed *the output coefficient*. It is expressed in N/m^2 or VAs/m^3.

The specification of the machine will decide the output power P_{out}, the efficiency η and the speed n. The electromagnetic power can be estimated using eqns (4.24) or (4.25). With guidance to suitable values for B_g and A (Fig. 4.9), the $D^2 L_i$ product is determined.

The electromagnetic torque of a d.c. machine can be expressed with the aid of electric and magnetic loadings, i.e.,

$$T_d = \frac{P_{elm}}{2\pi n} = \alpha_i \frac{\pi}{2} D^2 L_i B_g A \qquad (4.37)$$

Electric A and magnetic B_g loadings express the so called *shear stress*, i.e., electromagnetic force per unit of the whole rotor surface:

$$p_{sh} = \alpha_i B_g A \qquad (4.38)$$

The losses in the armature winding per surface area of the rotor are

$$\frac{\Delta P_a}{\pi D L_i} = \frac{1}{\pi D L_i} \frac{N l_{av}}{\sigma s_a} \frac{1}{(2a)^2} I_a^2 = \frac{l_{av}}{\sigma L_i} J_a A \qquad (4.39)$$

where J_a is according to eqn (4.31) and A is according to eqn (4.30). The product $J_a A$ must not exceed certain permissible values. Approximately, for motors up to 10 kW the product $J_a A \leq 12 \times 10^{10}$ A^2/m^3 [128].

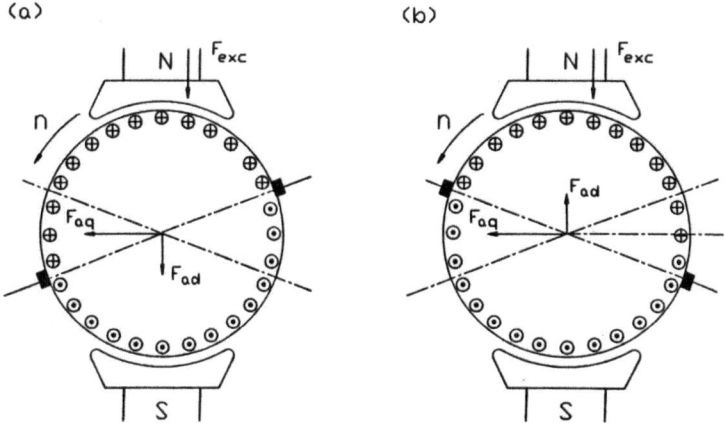

Fig. 4.10. Direct-axis F_{ad} and cross F_{aq} magnetizing force of armature reaction when brushes are shifted from the neutral line: (a) ahead, (b) back.

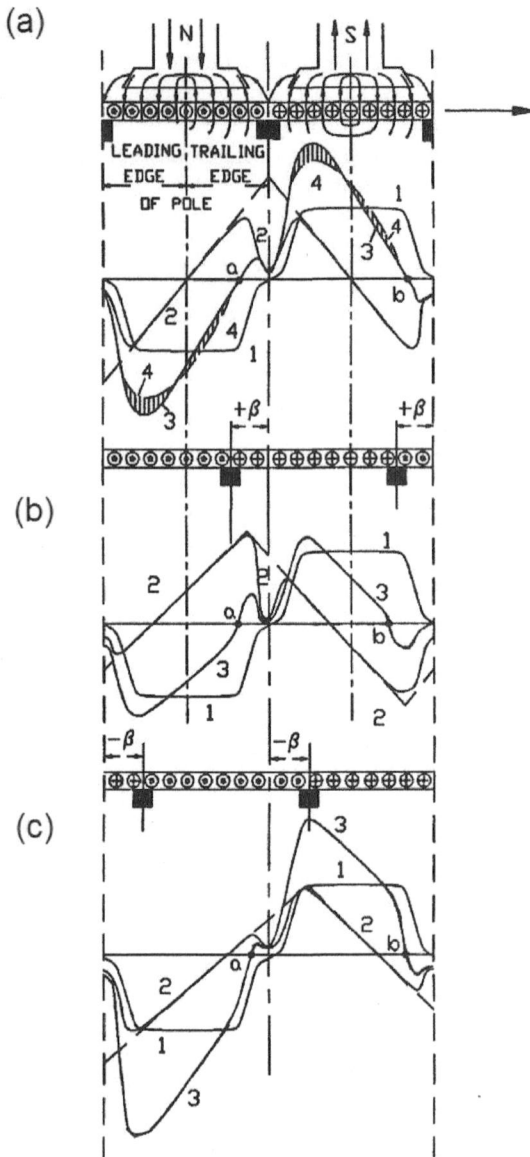

Fig. 4.11. Curves of d.c. motor fields for different position of brushes: (a) brushes in geometrical neutral line, (b) brushes shifted back from the neutral line, (c) brushes shifted ahead of the neutral line. 1 — MMF of the field winding, 2 — MMF of the armature winding, 3 — resultant MMF with magnetic saturation being neglected, 4 — resultant MMF with magnetic saturation taken into account.

4.4 Armature reaction

The action of the armature MMF on the MMF of the field winding is termed *armature reaction*.

If the brushes are set along the geometrical neutral line, the armature field is directed across the axis of the main poles, i.e., at 90^0. Such an armature field is termed the *cross* or *quadrature* MMF of the armature. The quadrature MMF per pole pair is

$$2F_a = 2F_{aq} = A\tau \qquad (4.40)$$

where the line current density A is according to eqn (4.30).

In general, the brushes can be shifted from the neutral line by an angle Ψ or over the corresponding arc b_{br} on the armature periphery. The armature can be considered as two superposed electromagnets, one of which, created by the position of the winding with the double angle 2Ψ ($2b_{br}$), forms the armature *direct-axis MMF* F_{ad} and the other, created by the remainder of the winding over the arc $(\tau - 2b_{br})$, produces the armature *quadrature MMF* F_{aq} per pole pair, i.e.,

$$2F_{ad} = Ab_{br} \qquad (4.41)$$

where the brush shift for small motors rated below 100 W is negligible, i.e., $b_{br} = 0.15\ldots0.3$ mm [100], whilst for medium power d.c. motors it can exceed 3 mm [185].

When the brushes of a motor are shifted ahead (in the direction of rotation) from the neutral line there appears a direct axis-armature reaction of a magnetizing nature (Fig. 4.10a). When the brushes are shifted back (against the speed) from the neutral line there appears a direct-axis armature reaction of a demagnetizing nature (Fig. 4.10b).

The curves of armature and field winding MMFs for different positions of the brushes are shown in Fig. 4.11. The saturation of the magnetic circuit modulates the resultant curve as well.

The armature MMF per pole pair

$$2F_a = 2h_M H_a = 2F_{aq} \pm 2F_{ad} \pm 2F_{aK} \qquad (4.42)$$

holds three components: the cross (quadrature) MMF F_{aq}, the direct-axis MMF F_{ad}, and the direct axis MMF F_{aK} of the coil sections being commutated. The '+' sign is for a generator, the '−' sign is for a motor.

The MMF induced by the currents of the coil sections being commutated can be estimated as [20]

$$2F_{aK} = \frac{b_K N^2 n}{C \sum R_c} \lambda_c A \qquad (4.43)$$

where the width of the commutation zone $b_K = 0.8\tau(1 - \alpha_i)$, C is according to eqn (4.5), $\sum R_c$ is the resistance of the short-circuited coil during its commutation, λ_c is the leakage permeance corresponding to a short-circuited coil section during its commutation and n is the rotor speed in rev/s.

The total resistance of a short circuited coil section during its commutation is

$$\sum R_c = R_c + 2R_{br} = \frac{R_a}{2C} + p\frac{\Delta V_{br}}{2I_a} \qquad (4.44)$$

where $R_c = R_a/(2C)$ is the resistance of the coil section alone, R_a is according to eqn (4.33), and C is according to eqn (4.5). The resistance of the contact layer brush–commutator $R_{br} = p\Delta V_{br}/(2I_a) \approx p/I_a$.

The leakage permeance corresponding to a short-circuited section during its commutation consists of two terms: the slot leakage permeance $2\lambda_s$ and end connection leakage permeance $\lambda_e l_e/L_i$ [185], i.e.,

$$\lambda_c = 2\mu_0 L_i \left(2\lambda_s + \lambda_e \frac{l_e}{L_i} \right) \qquad \text{H} \qquad (4.45)$$

and the resultant self-inductance for a full pitch winding is

$$L_c = N_c^2 \lambda_c = 2\mu_0 N_c^2 L_i \left(2\lambda_{sl} + \lambda_e \frac{l_e}{L_i} \right) \qquad \text{H}$$

The coefficient of the slot leakage permeance or specific slot leakage permeance λ_s depends on the slot shape and is given in the literature, e.g., [185] (see also Appendix A). The coefficient of the end connection leakage permeance for bands made of ferromagnetic materials is $\lambda_e = 0.75$ and for bands made of non-ferromagnetic materials is $\lambda_e = 0.5$. The length of a single end connection (overhang) is $l_e = l_{av} - L_i$, where l_{av} is according to eqn (4.33).

The MMF of a PM per pole pair $2F_M = 2H_M h_M$ must counterbalance the MVD per pole pair

$$2F_p = \frac{\Phi_g}{G_g} \qquad (4.46)$$

where Φ_g is the air gap magnetic flux per pole, G_g is the permeance of the air gap with saturation of the magnetic circuit being included, and the MMF of the armature F_a, i.e.,

$$F_M = F_p + F_a \qquad (4.47)$$

The point K (Fig. 2.11), which is given by the intersection of the demagnetization curve and the total permeance line, is usually determined for plugging when

$$F_{amax} = H_{amax} h_M = F_{aq}(I_{amax}) + F_{ad}(I_{amax}) + F_{aK}(I_{amax}) \qquad (4.48)$$

For a PM d.c. commutator motor F'_{admax} in Fig. 2.11 is equal to F_{amax}/σ_{lM}. The armature current for plugging can reach the following value [20]:

$$I_{amax} = (0.6\ldots0.9)\frac{V+E}{\sum R_a} \tag{4.49}$$

If the brushes are set along the geometrical neutral axis and the magnetic circuit is unsaturated, the cross MMF of the armature winding per pole pair is simply $2F_{aq} = A\tau$, where A is according to eqn (4.30). The armature cross MMF distorts the main field in motors, weakening it under the trailing edge. If the magnetic circuit is saturated, the reluctance of the leading edges of the poles increases more quickly than it reduces under the trailing edges.

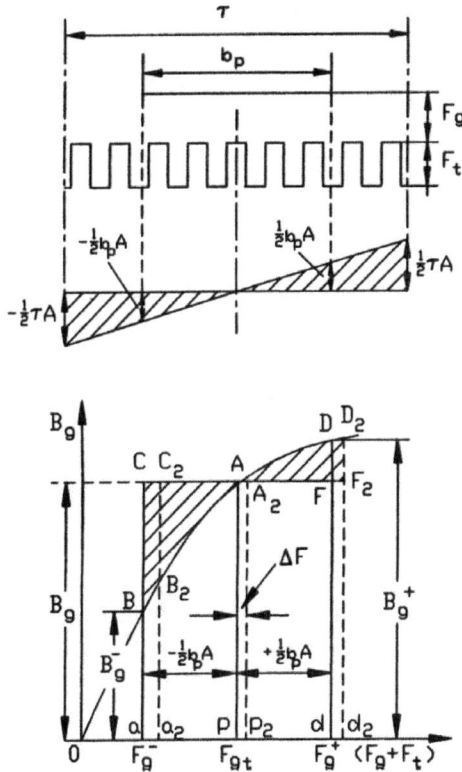

Fig. 4.12. Determination of demagnetizing effect of cross armature reaction.

The curve B_g plotted against $(F_g + F_t)$ (Fig. 4.12), where B_g is the air gap magnetic flux density, F_g is the MVD across the air gap, and F_t is the MVD drop along the armature tooth, is used for evaluation of the demagnetizing effect of cross armature reaction. Since the rectangle $aCFd$ has its base proportional to the pole arc b_p and its height equal to B_g, its area can serve as a measure of the no-load flux. In the same manner the area of the curvilinear tetragon $aBDd$ serves as a measure of the flux under load. If the machine is saturated, then the triangles $ACB > AFD$. To obtain the same magnetic flux Φ_g and EMF E under load as with no-load, the MMF of the excitation system must be somewhat higher, say, by ΔF, i.e., the MMF per pole pair of PMs of a loaded machine must be

$$2F_M = 2F_{Mo} + 2\Delta F \tag{4.50}$$

where F_{Mo} is the MMF of the PM for no-load. To determine this increase it is sufficient to move rectangle $aCFd$ to the right so that area $AC_2B_2 = AF_2D_2$. The areas of rectangles $aCFd$ and $a_2C_2F_2d_2$, and of the curvilinear tetragon $a_2B_2D_2d_2$, are equal, and hence the air gap flux and also the EMF E recover their initial value owing to the increase ΔF in the MMF of the PM. The increase ΔF compensates for the effect of cross armature reaction $\Delta F = F_{aq}$, since F_g and F_t are for a single air gap and a single tooth, respectively.

The methods of evaluating the F_{aq} published so far for d.c. machines with electromagnetic excitation, e.g., by Gogolewski and Gabryś [128]

$$2F_{aq} \approx \frac{b_p A}{b_g}$$

or by Voldek [299]

$$2F_{aq} \approx \frac{b_p A}{3b_g}$$

where

$$b_g = \frac{B_g^+ - B_g^-}{2B_g - B_g^+ - B_g^-} \tag{4.51}$$

are rather rough, not in agreement with each other and cannot be used for small PM motors. The problem can be solved in a simple way if the portions AB and AD_2 of the characteristic $B_g = f(F_g + F_t)$ shown in Fig. 4.12 are approximated by line segments, as in Fig. 4.13. Since [116]

$$(0.5b_p A - \Delta F)b_l = (0.5b_p A + \Delta F)b_r$$

and

$$\tan \delta_l = \frac{b_l}{0.5b_p A - \Delta F} = \frac{B_g - B_g^-}{0.5b_p A}$$

$$\tan \delta_r = \frac{b_r}{0.5 b_p A + \Delta F} = \frac{B_g^+ - B_g}{0.5 b_p A}$$

the MMF of cross armature reaction is expressed by the following equation

$$2 F_{aq} = b_p A (b_g - \sqrt{b_g^2 - 1}) \tag{4.52}$$

where b_g is according to eqn (4.51).

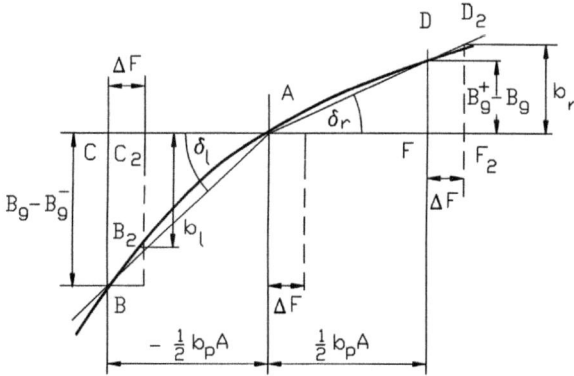

Fig. 4.13. Approximation of the $B_g = f(F_g + F_t)$ curve by two line segments.

A more accurate equation,

$$2 F_{aq} = 0.5 b_p A \frac{1 + \exp(-\alpha)}{1 - \exp(\alpha)} - F_{gt} - \frac{1}{b} \tag{4.53}$$

where

$$\alpha = b b_p A (1 - \frac{b}{a} B_g) \tag{4.54}$$

can be obtained using Froelich's equation

$$B(F) = \frac{aF}{1 + bF}$$

in which a and b are constants depending on the shape of $B(F)$ curve [116]. Another method is to use a numerical approach [114, 116].

Fig. 4.14 shows the calculation results of the cross armature reaction F_{aq} versus the shaft torque of a small 8 W low voltage d.c. commutator motor with segmental PMs. The Froelich's approximation and the numerical method show a good correlation over the whole range of torque values.

It is seen from Fig. 4.11 that the armature reaction magnetic field demagnetizes the PM at its trailing edge. An irreversible demagnetization of the

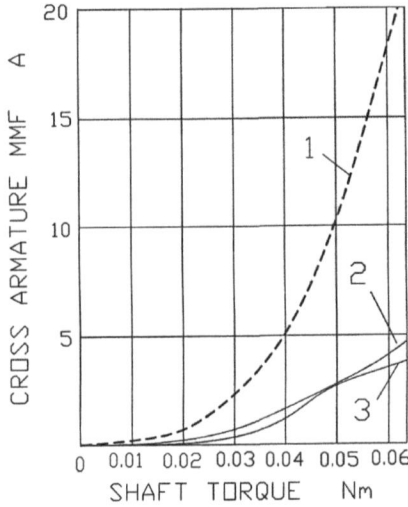

Fig. 4.14. Cross MMF of armature winding as a function of the shaft torque for 8-W d.c. commutator motor. Computation results. 1 — linear approximation, 2 — numerical method, 3 — Froelich's approximation.

trailing edges of PMs can occur if the armature reaction field is larger than the coercive force. This problem can be avoided by designing segmental PMs with the trailing edge (or edges) made of a material with high coercivity and the remaining portions made of high remanent flux density as shown in Fig. 4.15 [232]. The application advantages of two-component PMs are especially important with the increasing size of d.c. motors.

4.5 Commutation

Commutation is a group of phenomena related to current reversal in the conductors of an armature winding when conductors pass through the zone where they are short-circuited by the brushes placed on the commutator [185].

Assuming that (a) the width of the brush is equal to that of a commutator segment, (b) with a simplex lap winding the brush short-circuits only one coil section of the armature winding, and (c) the resistivity of the contact between a brush and a commutator does not depend on the current density, the time-dependent current in the short-circuited coil section is

$$i(t) = \frac{I_a}{2a} \frac{1 - 2\frac{t}{T}}{1 + [(\sum R_c + 2R_r)/R_{br}]\frac{t}{T}\left(1 - \frac{t}{T}\right)} \tag{4.55}$$

where $2a$ is the number of parallel armature current paths, T is the total duration of commutation, i.e., time during which a coil section passes through

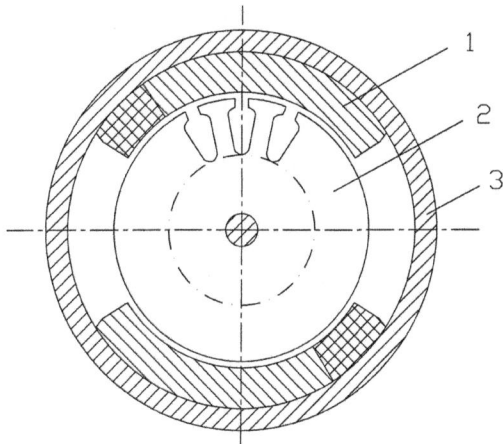

Fig. 4.15. d.c. commutator motor with two-component segmental PMs. 1 — PM, 2 — armature, 3 — frame.

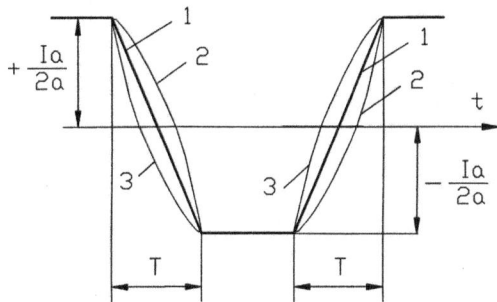

Fig. 4.16. Current variation curves in armature winding coil section with commutation taken into account: 1 — linear commutation, 2 — retarded commutation, 3 — accelerated commutation.

the short circuit (Fig. 4.16), R_{br} is the resistance of contact layer of brushes, $\sum R_c$ is the resistance of short-circuited coil section according to eqn (4.44), and R_r is the resistance of commutator risers ($\sum R_c + 2R_r > R_{br}$). For $t = 0$, the current in the short-circuited coil section is $i = +I_a/(2a)$ and for $t = T$ the current $i = -I_a/(2a)$. The current variation in the armature winding coil is shown in Fig. 4.16.

The resultant reactive EMF of self-induction and mutual induction of a short-circuited coil section is [185]

$$E_r = \mu_0 \frac{N}{C} v A L_i \left(2\lambda_s + \lambda_e \frac{l_e}{L_i} \right) \tag{4.56}$$

where N is the number of armature conductors, C is the number of commutator segments, $N/(2C)$ is the number of turns N_c per one armature coil according to eqn (4.5), L_i is the effective length of the armature core, l_e is the length of the one-sided end connection of the armature coil, λ_s is the armature slot specific permeance, λ_e is the specific permeance of the armature end connections, A is according to eqn (4.30), and v is according to eqn (4.12).

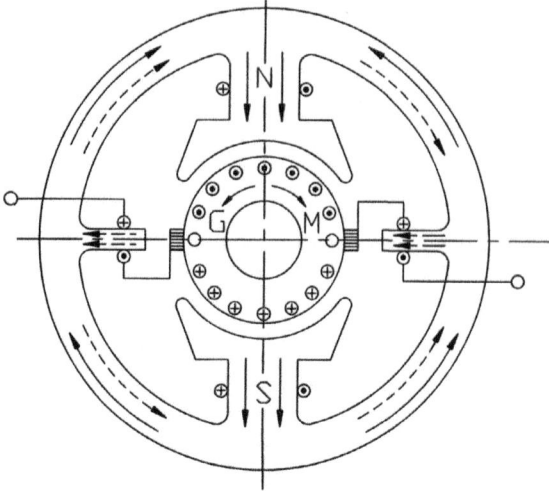

Fig. 4.17. Polarity of interpoles for motor (M) and generator (G) mode.

In machines without interpoles, the commutating magnetic flux necessary to induce the EMF E_c that balances the EMF E_r is produced by shifting the brushes from the geometrical neutral line. To improve commutation when a machine functions as a motor it is necessary to shift the brushes from the neutral line against the direction of armature rotation.

The best and most popular method of improving commutation is to use interpoles. Assuming $E_c = E_r$, the number N_{cp} of commutating turns per pole can be found from the equation

$$2N_{cp}vL_cB_c = \mu_0 \frac{N}{C} vAL_i \left(2\lambda_s + \lambda_e \frac{l_e}{L_i} \right) \qquad (4.57)$$

where L_c is the axial length of the interpole and B_c is its magnetic flux density varying in proportion to the armature current I_a. The polarity of interpoles is shown in Fig. 4.17. If $N_{cp} = N/(2C)$, the magnetic flux density of an interpole is

$$B_c = \mu_0 \frac{L_i}{L_c} A \left(2\lambda_s + \lambda_e \frac{l_e}{L_i} \right) \qquad (4.58)$$

To retain the proportionality between the flux density B_c and the electric loading A for all duties, it is necessary to connect the interpole winding in series with the armature winding and to keep the magnetic circuit of the interpoles unsaturated ($B_c \propto I_a$). Thus, the MMF per interpole is

$$F_{cp} = N_{cp}(2I_a) = F_{aq} + \frac{1}{\mu_o}B_c g'_c = \frac{A_T}{2} + \frac{1}{\mu_o}B_c g'_c \qquad (4.59)$$

where F_{aq} is according to eqn (4.40) and g'_c is the equivalent air gap between the armature core and interpole face including slotting.

4.6 Starting

Combining eqn (4.1) and eqn (4.2), the armature current can be expressed as a function of speed, i.e.,

$$I_a = \frac{V - c_E n \Phi_g - \Delta V_{br}}{\sum R_a} \qquad (4.60)$$

At the first instant of starting the speed $n = 0$ and the EMF $E = 0$. Hence, according to eqn (4.60) the starting current is equal to the locked-rotor (short-circuit) current:

$$I_{ash} = \frac{V - \Delta V_{br}}{\sum R_a} \gg I_{ar} \qquad (4.61)$$

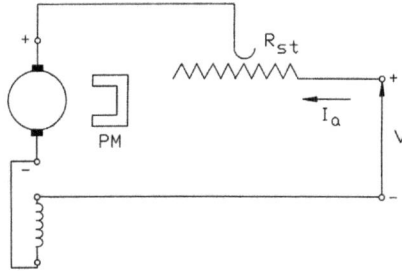

Fig. 4.18. Circuit diagram of a d.c. PM motor with starting rheostat R_{st}.

where I_{ar} is the rated armature current. The torque corresponding to the current I_{ash} is called the *stall torque* or *locked–rotor torque*, i.e.,

$$T_s = k_T I_{ash} = k_T \frac{V - \Delta V_{br}}{\sum R_a} \qquad (4.62)$$

where the torque constant k_T is according to eqns (4.7) and (4.8). The stall torque is the torque developed with voltage applied to the armature winding terminals and rotor locked.

To reduce the starting current I_{ash}, a starting rheostat is connected in series with the armature winding to get the resultant resistance of the armature circuit $\sum R_a + R_{st}$ where R_{st} is the resistance of starting rheostat as shown in Fig. 4.18. The highest resistance must be at the first instant of starting. The locked-rotor armature current I_{ash} drops to the value of

$$I_{amax} = \frac{V - \Delta V_{br}}{\sum R_a + R_{st}} \tag{4.63}$$

As the speed increases from 0 to n', the EMF also increases too, and

$$I_{amin} = \frac{V - E' - \Delta V_{br}}{\sum R_a + R_{st}} \tag{4.64}$$

where $E' = c_E n' \Phi_g < E$. When the speed reaches its rated value $n = n_r$, the starting rheostat can be removed, since

$$I_a = I_{ar} = \frac{V - c_E n_r \Phi_g - \Delta V_{br}}{\sum R_a} \tag{4.65}$$

4.7 Speed control

The following relationship obtained from eqns (4.1) and (4.2)

$$n = \frac{1}{c_E \Phi_g}[V - I_a(\sum R_a + R_{rhe}) - \Delta V_{br}] \tag{4.66}$$

tells that the speed of a d.c. motor can be controlled by changing:

- the supply mains voltage V;
- the armature-circuit resistance $\sum R_a + R_{rhe}$ where R_{rhe} is the resistance of the armature rheostat;
- the air gap (field) flux Φ_g.

The last method is possible only when PMs are furnished with additional field coils for speed control.

From the above eqn (4.66) and eqns (4.4) and (4.8) it is possible to obtain the steady-state speed n of a PM d.c. commutator motor as a function of T_d for a given V, i.e.,

$$n = \frac{1}{k_E}(V - \Delta V_{br}) - \frac{\sum R_a + R_{rhe}}{k_E k_T}T_d \tag{4.67}$$

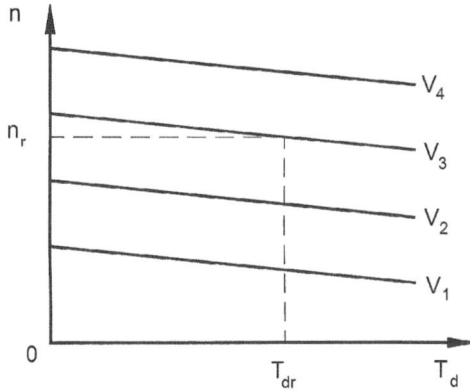

Fig. 4.19. Variable armature terminal-voltage speed control for a d.c. PM motor ($V_1 < V_2 < V_3 < V_4$).

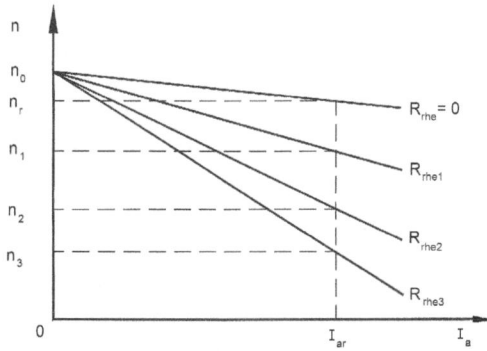

Fig. 4.20. Speed–armature current curves for armature rheostat speed control of a d.c. PM motor.

4.7.1 Armature terminal-voltage speed control

The speed can be easily controlled from zero up to a maximum safe speed. The controlled-voltage source may be a solid-state controlled rectifier, chopper or a d.c. generator. The speed–armature current characteristics are shown in Fig. 4.19.

The speed–torque characteristics in Fig. 4.19 can be shifted vertically by controlling the applied terminal voltage V. As the torque is increased, the speed–torque characteristic at a given V is essentially horizontal, except for the drop due to the voltage $I_a \sum R_a$ across the armature-circuit resistance, brush voltage ΔV_{br} and drop due to the armature reaction.

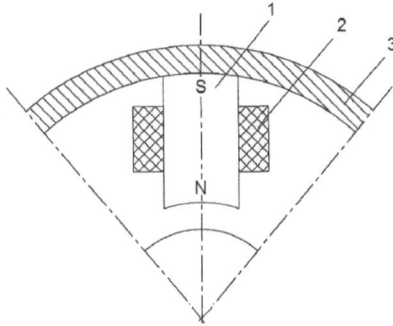

Fig. 4.21. Stator of a PM d.c. commutator motor with additional field winding for speed control: 1 — PM, 2 — additional field winding, 3 — stator yoke.

4.7.2 Armature rheostat speed control

The total resistance $\sum R_a + R_{rhe}$ of the armature circuit may be varied by means of a rheostat in series with the armature winding. This rheostat must be capable of carrying the heavy armature current continuously. It is thus more expensive than a starting rheostat designed for a short time duty. The speed–armature current characteristics are shown in Fig. 4.20.

4.7.3 Shunt-field control

This is the most economical and efficient method of controlling the speed of a d.c. motor, but in the case of PM excitation requires an additional winding around PMs, as shown in Figs 4.4c and 4.21, and an additional solid-state controlled rectifier.

4.7.4 Chopper variable-voltage speed control

In applications in which the power source is a battery and efficiency is an important consideration, various *chopper* drives provide variable armature terminal voltage to d.c. motors as a means of speed control. Examples are battery-driven automobiles, d.c. trams, d.c. underground trains, etc. Choppers may employ thyristors or power transistors. It has been pointed out that the d.c. value of voltage or current is its average value. The chopper is essentially a switch that turns on the battery for short time intervals, in general, a d.c. converter having no a.c. link (Fig. 1.3a). It may vary the average d.c. value of the terminal voltage by varying the pulse width, i.e., pulse width modulation (PWM) or pulse frequency, i.e., pulse frequency modulation (PFM), or both (Fig. 4.22). PM or series motors are often used in these systems.

(a)

(b)   (c)

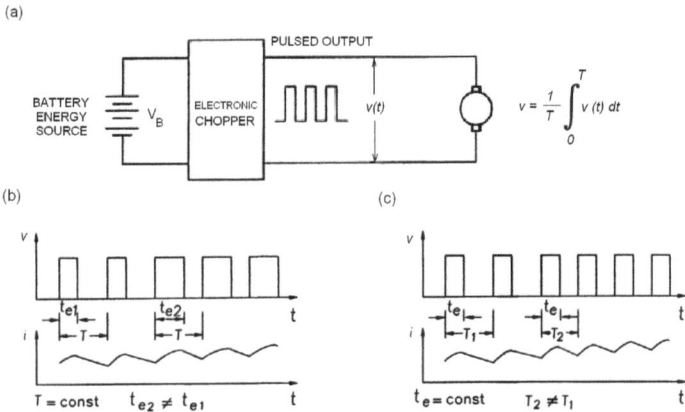

Fig. 4.22. Electronic chopper: (a) general arrangement, (b) PWM, (c) PFM.

4.8 Servo motors

PM d.c. brush *servo motors* can be controlled only by changing the armature terminal voltage (Fig. 4.19). In addition to low inertia of the rotor and very rapid starting, d.c. servo motors must have *linear voltage–speed characteristics* and *linear current–torque characteristics*. These requirements can be met by coreless PM d.c. brush motors (Subsection 4.1.3), i.e., cylindrical inside-field type motors with Faulhaber skew-wound coils (Figs 4.5) or disk type motors with printed rotor winding (Fig. 4.4c). A coreless cylindrical PM d.c. commutator servo motor is shown in Fig. 4.23.

Similar to other types of servo motors used in control systems, a d.c. servo brush motor can be characterized by the following parameters:

- relative torque

$$t = \frac{T}{T_{st}} \tag{4.68}$$

- relative speed

$$\nu = \frac{n}{n_0} \tag{4.69}$$

- control voltage–to–rated armature voltage ratio

$$\alpha = \frac{V_c}{V_r} \tag{4.70}$$

where T_{st} is the starting torque, n_0 is the no-load speed, V_c is the armature terminal control voltage signal and V_r is the armature rated voltage. Mechanical characteristics $t = f(\nu)$ at $\alpha = constant$ and control characteristics

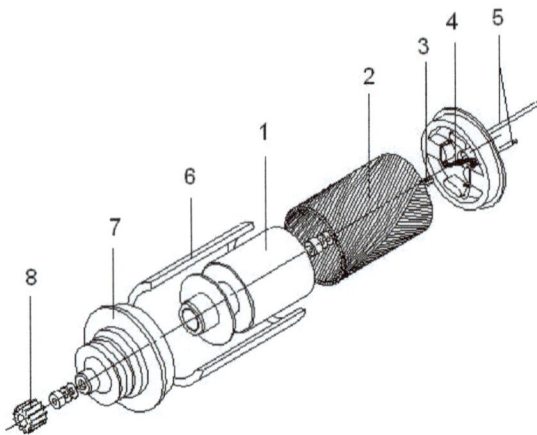

Fig. 4.23. Expanded view of a cylidrical inside-field type coreles PM d.c. commutator servo motor. 1 — PM, 2 — skew-wound armature winding, 3 — commutator, 4 — brushes, 5 — terminal leads, 6 — housing (enclosure), 7 — end cover, 8 — pinion. Source: Faulhaber Micro Drive Systems and Technologies – Technical Library, Croglio, Switzerland.

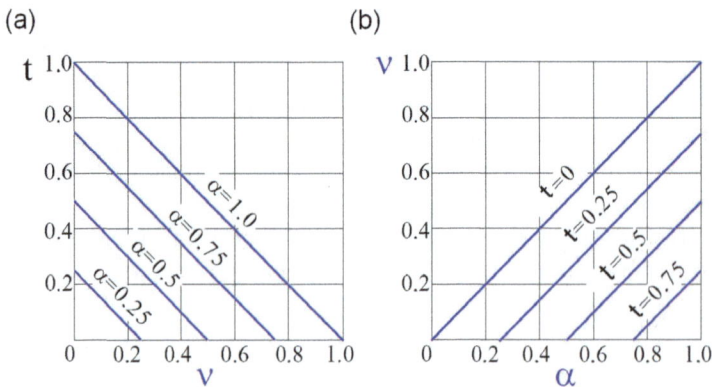

Fig. 4.24. Mechanical characteristics $t = f(\nu)$ at $\alpha = constant$ and control characteristics $\nu = f(\alpha)$ at $t = constant$ of a PM d.c. commutator servo motor.

$\nu = f(\alpha)$ at $t = constant$ are plotted in Fig. 4.24. The control voltage–to–rated armature voltage ratio α is sometimes called the *signal ratio* [15].

4.9 Magnetic circuit

In this section a cylindrical d.c. commutator motor with ferromagnetic rotor core as shown in Figs 4.1, 4.2 or 4.3b will be considered. The equations derived can be easily adjusted to other magnetic circuit configurations, e.g., disk rotor motors.

The air gap magnetic flux per pole

$$\Phi_g = \alpha_i \tau L_i B_g = b_p L_i B_g \tag{4.71}$$

results from eqn (4.29). On the other hand, to obtain an increase in B_g, d.c. PM commutator motors of cylindrical construction have the length L_M of the PM greater than the effective length L_i of the armature core. The air gap magnetic flux Φ_g can also be expressed with the aid of the useful magnetic flux density B_u at the internal surface S_M of the stator PM, i.e.,

$$\Phi_g = b_p L_M B_u = S_M B_u \tag{4.72}$$

where $S_M = b_p L_M$ is the PM cross section area of the pole shoe. For $L_M > L_i$ the air gap magnetic flux density at the surface of the armature core $B_g > B_u$. The magnetic flux density in the magnet $B_M = \sigma_{lM} B_u$ because the total magnetic flux is

$$\Phi_M = B_M S_M = \Phi_g + \Phi_{lM} = B_u S_M + B_{lM} S_M \tag{4.73}$$

where B_{lM} is the leakage magnetic flux density of the PM.

The saturation factor k_{sat} of the magnetic circuit is defined as the total MMF per pole pair to the magnetic voltage drop (MVD) across the air gap taken twice as expressed by eqn (2.48).

4.9.1 MMF per pole

With the stator PMs longer than the armature core, i.e., $L_M > L_i$, the MVD across the air gap is

$$V_g = \frac{\Phi_g}{\mu_0 \alpha_i \tau} \int_0^{k_C g} \frac{dx}{L_i + x(L_M - L_i)/g} \tag{4.74}$$

$$= \frac{\Phi_g}{\mu_0 \alpha_i \tau} \ln\left(\frac{L_M}{L_i}\right) \frac{k_C g}{L_M - L_i} \tag{4.75}$$

where Φ_g is according to eqn (4.71) or (4.72), g is the air gap and k_C is the Carter's coefficient (Appendix A, eqn (A.22)). For $L_M = L_i$ the term $\ln(l_M/L_i)(L_M - L_i)^{-1} \to 1/L_i$ and the air gap MVD is simply

$$V_g = \frac{\Phi_g}{\mu_0 \alpha_i \tau L_i} k_C g = \frac{B_g}{\mu_0} k_C g \tag{4.76}$$

where $B_g = B_u$. From Ampère's circuital law the MMF per pole pair is

$$2F_p = 2\frac{\Phi_g}{\mu_0 \alpha_i \tau} \ln\left(\frac{L_M}{L_i}\right) \frac{k_C g}{L_M - L_i} + 2\frac{B_M}{\mu_0} g_{My} + V_{1y} + 2V_{2t} + V_{2y} \tag{4.77}$$

where $F_p = H_M h_M$, g_{My} is the air gap between the PM and stator yoke, V_{1y} is the MVD in the stator yoke, V_{2t} is the MVD in the rotor tooth, and V_{2y} is the MVD in the rotor yoke — see eqn (2.48). In small d.c. PM commutator motors the air gap $g_{My} = 0.04 \ldots 0.10$ mm.

4.9.2 Air gap permeance

Eqn (4.77) can be brought to the form

$$2F_p = 2\frac{B_u}{\mu_0}\left[\frac{S_M}{\alpha_i \tau} \ln\left(\frac{L_M}{L_i}\right)\frac{k_C g}{L_M - L_i} + \sigma_{lM} g_{My}\right] k_{sat} \tag{4.78}$$

where S_M is the PM cross-sectional area, σ_{lM} is the coefficient of the leakage flux of the PM according to eqns (2.10) and (2.55), and k_{sat} is the saturation factor of the magnetic circuit according to eqn (2.48) in which $V_{1t} = 0$ and

$$2V_g = 2\frac{L_i B_g}{\mu_0} \ln\left(\frac{L_M}{L_i}\right)\frac{k_C g}{L_M - L_i} + 2\frac{B_M}{\mu_0} g_{My}$$

The air gap permeance G_g for Φ_g per pole pair can be found on the basis of eqns (4.72) and (4.78) as

$$G_g = \frac{\Phi_g}{2F_p}$$

$$= \frac{\mu_0}{2\left[1/(\alpha_i \tau) \ln(L_M/L_i)/(L_M - L_i)k_C g + (1/S_M)\sigma_{lM} g_{My}\right] k_{sat}} \tag{4.79}$$

For $L_M = L_i$ the above eqn (4.79) takes the form

$$G_g = \mu_0 \frac{\alpha_i \tau L_i}{2(k_C g + \sigma_{lM} g_{My}) k_{sat}} \tag{4.80}$$

Eqn (4.80) can be used both for motors with segmental and cylindrical PMs. The nonlinearity due to the magnetic saturation of the rotor core and stator yoke has been included since G_g is dependent on k_{sat}.

4.9.3 Leakage permeances

The total leakage permeance of a segmental PM is (Fig. 4.25a)

$$G_{lM} = k_V (G_{ll} + G_{lc}) \tag{4.81}$$

where $k_V = 0.5$ is the coefficient including the MVD along the height h_M of the PM [100].

The leakage permeance for the flux between the yoke and lateral surface of the PM is

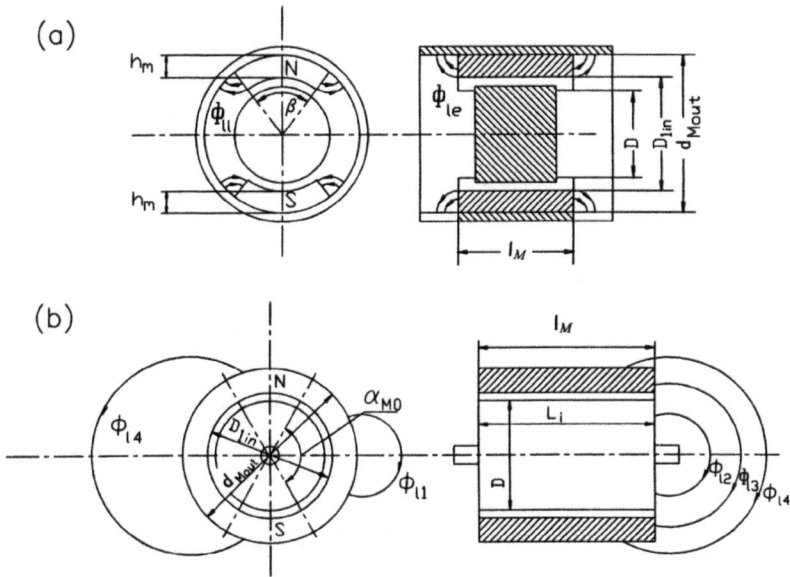

Fig. 4.25. Dimensions and equivalent leakage fluxes of two-pole d.c. motors with: (a) segmental PMs, (b) circular PMs.

$$G_{ll} = \mu_0 \frac{h_M L_M}{\delta_{lav}} \tag{4.82}$$

The leakage permeance for the flux between the yoke and end surface is

$$G_{lc} = \mu_0 \frac{0.5\beta(D_{1in} + h_M)h_M}{\delta_{cav}} \tag{4.83}$$

where $\delta_{lav} = \delta_{cav} = 0.25\pi h_M$ are the average paths for leakage fluxes.

For a circular PM without the stator mild steel yoke the total leakage permeance is (Fig. 4.25b)

$$G_{lM} = G_{l1} + 2G_{l2} + G_{l3} + G_{l4} \tag{4.84}$$

where G_{l1} and G_{l2} are the permeances for leakage fluxes through external cylindrical and end surfaces of the PM within the angle $\alpha_{M0} \approx 2\pi/3$, whilst G_{l3} and G_{l4} are the permeances for leakage fluxes through end and external cylindrical surfaces corresponding to the angle $\pi - \alpha_{M0}$. In practice, these leakage permeances can be calculated using the following formulae [100]

$$G_{l1} \approx 0.3\mu_0(d_{Mout} + l_M) \tag{4.85}$$

$$G_{l2} \approx \mu_0(0.27d_{Mout} - 0.04D_{1in}) \tag{4.86}$$

$$G_{l3} \approx 2\mu_0 d_{Mout} \frac{0.42 d_{Mout} + 0.14 D_{1in}}{7 d_{Mout} + D_{1in}} \tag{4.87}$$

$$G_{l4} \approx \mu_0 (0.14 d_{Mout} + 0.24 l_M) \tag{4.88}$$

where all the dimensions are according to Fig. 4.25b. There are two additional leakage permeances in the case of the rotor (armature) being removed [100]:

- for the flux through the internal cylindrical surface within the angle α_{M0}

$$G_{l5} \approx 0.5 \mu_0 l_M \left(0.75 - 0.1 \frac{d_{Mout}}{D_{1in}} \right) \tag{4.89}$$

- for the flux through the internal cylindrical surface within the angle $\pi - \alpha_{M0}$

$$G_{l6} \approx 0.5 \mu_0 l_M \tag{4.90}$$

Thus, the total permeance per pole pair for leakage magnetic fluxes inside a cylindrical PM with the armature being removed is

$$G'_{lM} = 2 G_{l5} + G_{l6} \tag{4.91}$$

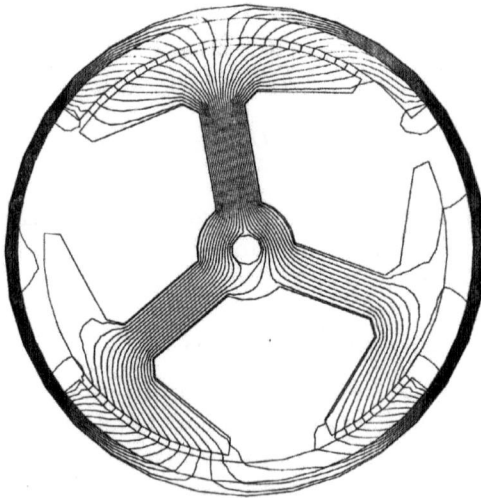

Fig. 4.26. Magnetic field distribution in a three-coil, two-pole PM commutator motor in the 10^0 rotor position.

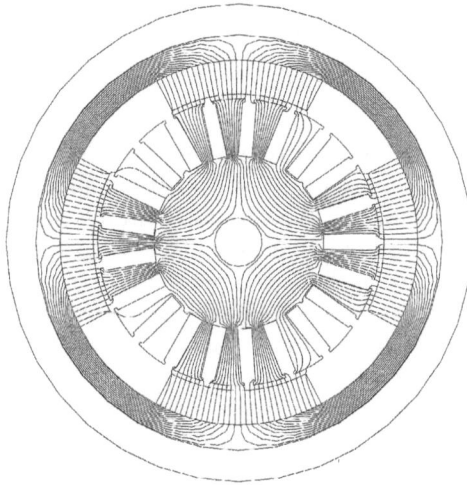

Fig. 4.27. Magnetic field distribution in a four-pole PM commutator motor car starter.

4.10 Applications

4.10.1 Toys

Cylindrical PM commutator micromotors are commonly used in battery-powered toys, e.g., Fig. 4.26. The armature winding is a three-coil lap winding connected to a three-segment commutator. The stator has a two-pole segmental or cylindrical barium ferrite magnet. Three armature salient poles and two-pole excitation system provide starting torque at any rotor position. The magnetic flux distribution is shown in Fig. 4.26.

4.10.2 Car starters

Car starter motors used to be d.c. series commutator motors. Recently, PM excitation systems have also been used in order to reduce the motor volume and increase the efficiency. Most PM car starter motors are rated from 2000 to 3000 W. Starter motors are built both as low-speed and high-speed machines (6000 rpm). In high-speed motors, epicyclic gears reduce the speed.

A large number of armature coils (wave windings) and four-pole or six-pole PM excitation systems are used [228]. Two-pole designs are avoided as the strong armature reaction flux of a two-pole machine could demagnetize ferrite magnets. High armature currents require two brushes in parallel.

The ambient temperature for car motors is from -40 to $+60^0$ C, sometimes to $+100^0$ C (hot car engine). The performance characteristics of starter motors are very sensitive to the temperature as PMs change their magnetic parameters and the armature winding resistance fluctuates with temperature. Fig. 4.27 shows the magnetic field distribution in a 3-kW PM motor starter.

4.10.3 Underwater vehicles

Underwater vehicles (UVs) are of two groups: manned and unmanned, commonly known as underwater robotic vehicles (URVs). URVs are very attractive for operation in unstructured and hazardous environments such as the ocean, hydro power plant reservoirs and at nuclear plants (Fig. 1.30). In the underwater environment, URVs are used for various work assignments. Among them are: pipe-lining, inspection, data collection, drill support, hydrography mapping, construction, maintenance and repairing underwater equipment. High technology developed for on-land systems cannot be directly adapted to UV systems, since such vehicles have different dynamic characteristics from on-land vehicles and because of the high density, nonuniform and unpredictable operating environment [312].

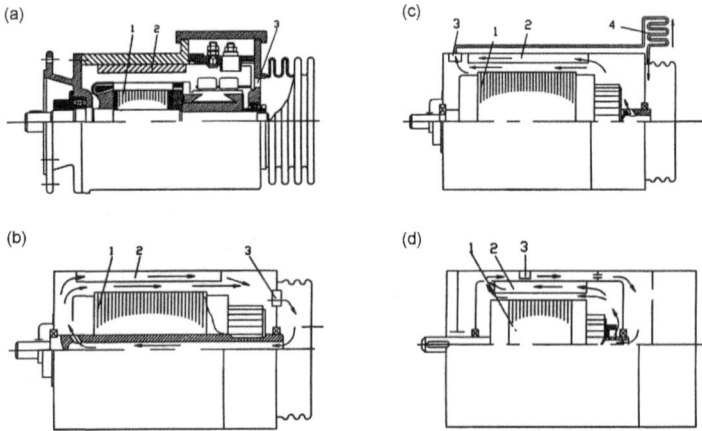

Fig. 4.28. PM d.c. commutator motors with liquid dielectrics for UVs: (a) hermetically sealed motor with natural circulation of the liquid dielectric and frame used as an intermediate heat exchanger, (b) hermetically sealed motor with forced circulation of liquid dielectric and ducts in the rotor, (c) hermetically sealed motor with external heat exchanger, (d) motor in a box with forced circulation of liquid dielectric. 1 — armature, 2 — PM, 3 — filter, 4 — heat exchanger.

The difficult control problem with URVs stems from the following facts [312]:

- the vehicle has nonlinear dynamic behavior,
- hydrodynamics of the vehicle are poorly known and may vary with relative vehicle velocity to fluid motion,

- the vehicle usually has comparable velocities along all three axes,
- the forces and torques generated by high density fluid motion are significant,
- the added mass aspect of the dense ocean medium results in time-delay control response characteristics,
- a variety of unmeasurable disturbances are present due to multi-directional currents,
- the centers of gravity and buoyancy may vary due to payload change during the operation.

Approximately 10% of the total ocean surface has a depth less then 300 m (continental shelves). The ocean bed is up to 8000 m deep for about 98% of the ocean surface. The rest are deep cavities up to 11,000 m, e.g., Marianas Trench, 11,022 m depth. Most UVs operate at a depth of up to 8000 m and pressure up to 5.88×10^7 Pa. As the UV submerges or emerges the pressure and temperature of the seawater changes. The pressure is proportional to the depth, while the temperature is a function of many variables such as depth, region, current, season of year, etc. In the depths, the temperature is approximately 0^0C, and at the surface, in some regions, can reach $+30^0$C. In polar regions the temperature can be as low as -40^0C. Electric motors of UVs are subject to large variation in pressure and temperature.

For UVs without any power cable to the base ship, batteries or other chemical energy sources with rated voltages from 30 to 110 V are used. Depending on the depth of submersion and time of autonomous operation, the mass of payload is only from 0.15 to 0.3 of the mass of vehicle. The major part of an autonomous UV's displacement is taken by the battery. The time of autonomous operation depends on the battery capacity. The motor's efficiency is very important. There are two duty cycles of UVs: continuous duty limited by the capacity of battery (up to a few hours) and short time duty (up to 2 to 3 minutes). The output power of electric motors for propulsion is up to 75 kW for manned UVs (on average 20 kW) and 200 W to 1.1 kW for unmanned URVs.

The following constructions of electric motors can be used for UV propulsion [293]:

- d.c. commutator motors with a high integrity, durable frame which protects against the action of external environment. The major disadvantage is its large mass of frame and high axial forces due to the pressure on the shaft. For example, at the depth of 6000 m and shaft diameter of 30 mm the axial force is 41,500 N. Such a construction is feasible for depths less than 100 m.
- Induction and PM brushless motors operating in seawater at high pressure. These motors are more reliable but require inverters which increase the mass of vehicle.
- a.c. or d.c. motors and additional hydraulic machine.

- d.c. commutator motor filled with a liquid dielectric (LDDCM). This is a simple solution and all the advantages of d.c. motor drives are displayed: high overload capacity, simple speed control and small mass. The pressure of the liquid dielectric is equal to that of the external environment and is secured only by a light waterproof frame and compensator. The motor operates at high pressure. Such LDDCMs were used in the 1950s in J. Picard's bathyscaphe *Trieste*.

UVs with LDDCMs were the most popular until the mid 1980s. The investigated problems in LDDCMs were: hydrodynamic forces in commutator–brush systems, commutation in liquid dielectrics, wear of brushes, losses, heat transfer, choice of electric and magnetic loadings, construction of commutators and brushes and others. LDDCMs are divided into two groups [293]:

- totally enclosed, hermetically sealed motors for operation in liquid aggressive medium in which the frame is utilized as an intermediate heat exchanger (Fig. 4.28a);
- motors of open or protected construction, the stators and rotors of which are immersed in liquid dielectric or placed in special boxes filled with liquid dielectrics (Fig. 4.28d).

With regard to the heat exchange from the frame surface, hermetically sealed LDDCMs are classified into machines with natural and forced cooling systems. Forced circulation of liquid dielectric around the frame is caused by the rotating parts such as blades or screws. With regard to the circulation of inner liquid, the first two groups of machines can be divided into machines with natural circulation of liquid and those with forced circulation of liquid. Directional circulation of liquid is caused by radial or axial ducts in the rotor (Fig. 4.28b). The most intensive cooling of hermetically sealed motors can be achieved with the aid of external heat exchangers (Fig. 4.28c). For larger motors, a double cooling system with built-in frame heat exchanger is recommended [293].

To obtain minimum mass and maximum efficiency from LDDCMs the angular speed is usually from 200 to 600 rad/s. Such high speeds require reduction gears. The rotor and stator surfaces should be smooth to minimize the friction losses. The rotor and connections are subject to mechanical forces, and are cast in epoxy resin. Fuel oil, diesel oil, transformer oil and other synthetic oils can be used as liquid dielectrics.

High-performance rare-earth d.c. PM commutator motors can be used for small URVs [57]. Recently, PM brushless motors have been used almost exclusively in URV thrusters.

4.10.4 Linear actuators

Electromechanical linear actuators convert rotary motion and torque to linear thrust and displacement. Because they are easily interfaced and controlled

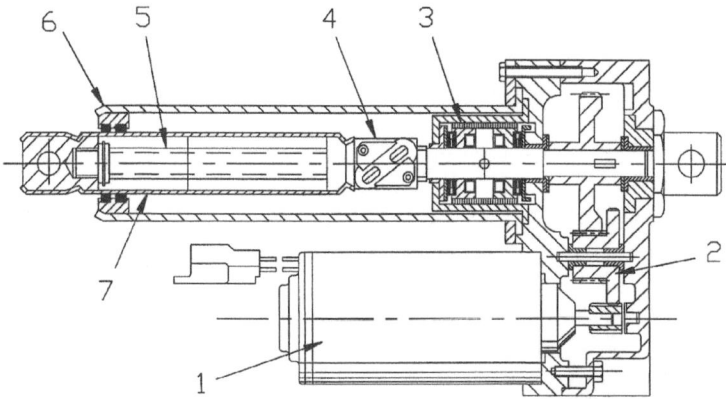

Fig. 4.29. Electromechanical linear actuator: 1 — d.c. PM commutator motor, 2 — reduction gearing, 3 — wrap spring brake, 4 — ball nut, 5 — ball lead screw, 6 — cover tube, 7 — extension tube.

with the aid of programmable logic controllers (PLCs) and microprocessors, they are being widely used in precision motion-control applications. Linear actuators are self-contained ball lead screw or roller screw units driven by an electric motor through a torque multiplier reduction gear. Both d.c. and a.c. PM motors can be used. Fig. 4.29 shows a simple electromechanical linear actuator with a d.c. PM commutator motor.

4.10.5 Wheelchairs

A wheelchair is an armchair with four wheels in which two of them (larger) are driven wheels and the remaining two smaller wheels are steering wheels. A wheelchair drive should meet the following requirements [127]:

- autonomous electrical energy source, adjusted to indoor operation, which should not emit any pollution,
- battery capacity of a minimum of 5 h operation or 30 km distance,
- two or three ranges of speed: 0 to 1.5 m/s for indoor driving, 0 to 3 m/s for sidewalk driving, and 0 to 6 m/s for street and road driving,
- good maneuvering and steerability for driving in small rooms with furniture,
- a simple steering system to allow a handicapped person to use with only one limb or even mouth.

Control with only one limb is possible using one steering lever with two degrees of freedom in the x and y directions provided that the wheelchair is equipped with two independent driving systems for the left and right wheel.

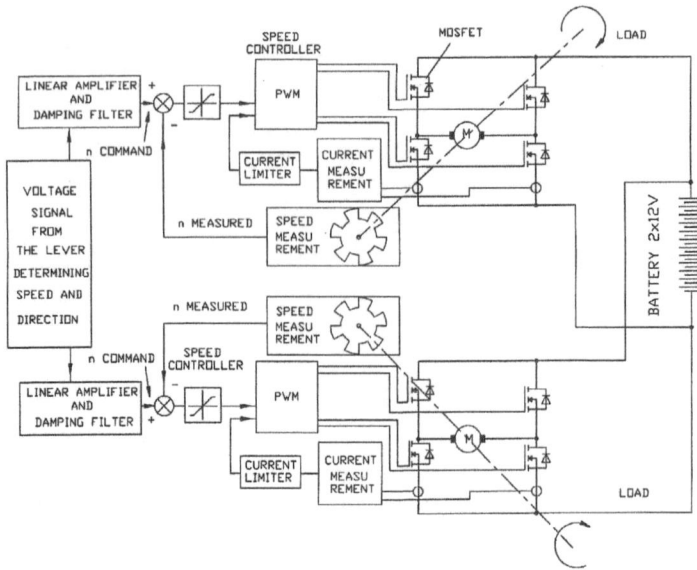

Fig. 4.30. Block diagram of a wheelchair drive system with two d.c. PM commutator motors.

The block diagram of a wheelchair drive system is shown in Fig. 4.30 [127]. Two d.c. PM commutator motors rated at 200 W, 24 V, 3000 rpm have been used. The 24 V battery has a capacity of 60 Ah. One motor drives the left and the second the right wheel. The maximum external diameter of the motor cannot be more than 0.1 m so the rotor stack is about 0.1 m long. The speed of the motor and its armature current are linear functions of torque. Because motors are supplied from the battery, the voltage is not constant and the internal resistance of the battery and voltage drop across transistors and connecting wires must be taken into account in the calculation of characteristics.

Two independent drive systems are controlled with the aid of one voltage signal from the lever proportional to the speed and direction of motion. The control and power systems comprise:

- linear amplifier and damping filter
- speed controller
- current measurement unit
- speed measurement unit
- pulse-width modulator (PWM)

The motion control of the wheelchair is done by the voltage signal from the lever which determines the speed and direction. There are two degrees of freedom on the steering lever and each lever is equipped with a mechanical shutter. One of the shutters has freedom in the x-axis and the second one in the y-axis. The voltage signals (speed and direction) generated by photoelectric sensors are proportional to the position of the shutters. Forward direction of the lever $(+x)$ causes two of the same signals proportional to the speed to be transmitted to the linear amplifier for the right and left wheel drive. The wheelchair moves forward. The more the lever is shifted the higher the control signal and the higher the speed. Backward shift $(-x)$ of the lever causes backward driving. Lateral shift $(+y)$ of the lever at $x = 0$ produces positive speed and direction signals transmitted to the linear amplifier of the left wheel drive and negative signals transmitted to the linear amplifier of the right wheel drive. Wheels rotate with the same speed but in opposite directions and the wheelchair rotates in the same place. Opposite lateral shift $(-y)$ of the lever at $x = 0$ causes a similar rotation of the wheelchair in the opposite direction. Like in the x-axis, the more the lever is shifted in the y direction the faster is the revolution of the wheelchair. Simultaneous shift in the x and y directions of the lever causes the superposition of two movements, and there is a difference in speed of the left and right wheels. The wheelchair can move and turn forward or backward.

During fast shifts of the lever the damping filter prevents sudden switching on the voltage across the armature terminals and limits the wheelchair acceleration. This makes a fluent motion and improves the comfort of riding. The speed controller allows the maximum speed to be set (for a given speed range) and prevents this speed from being exceeded. The speed controller receives a signal proportional to the speed command and a second signal proportional to the measured speed. The speed is measured with the aid of a pulse tachogenerator mounted on the motor shaft. These pulses are converted into voltage with the aid of a D/A converter. The tachogenerator is also equipped with a direction identifier. The voltage signal from the speed controller is transmitted to the PWM. There is also an overcurrent protection system, among others, to protect the ferrite PMs against demagnetization.

The described wheelchair drive is designed for two-directional smooth operation both in motoring and regenerative braking mode [127].

4.10.6 Mars robotic vehicles

A six-wheel robotic vehicle called *Sojourner* driven by Maxon d.c. commutator motors was used in the 1996-97 NASA Mars Pathfinder mission [95]. The vehicle is shown in Fig. 4.31a and its specifications are given in Table 4.6. Specifications of Maxon motors with NdFeB PMs and precious metal brushes are listed in Table 4.4. *Sojourner* was built at NASA Jet Propulsion Laboratory, Los Angeles County, CA, U.S.A. All *Sojourner's* equipment including computers, lasers, motors and radio modem were fed from a lightweight 16.5 W solar array. During the times with not enough sunlight the hardware was powered by batteries.

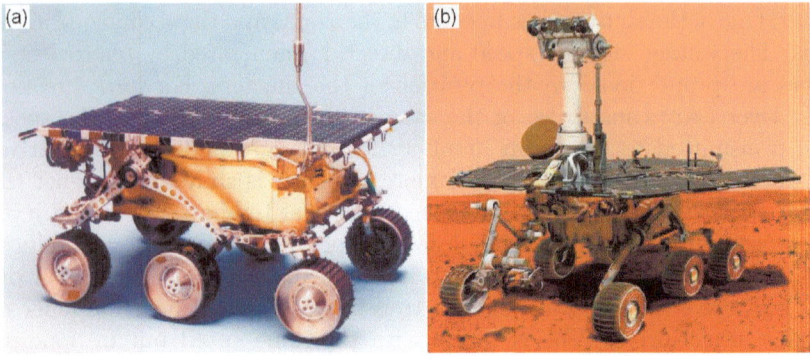

Fig. 4.31. Robotic vehicles for Mars Missions: (a) *Sojourner* (1996-97); (b) exploration rovers *Spirit* and *Opportunity* (2003 ongoing). Photo courtesy of NASA.

Table 4.6. Comparison of robotic vehicles

Height	*Sojourner*	*Spirit* and *Opportunity*
Weight, kg	11	185
Height, m	0.32	1.57
Height above ground, m	0.25	1.54
Communication	8 Bit CPU	32 Bit CPU
Cameras	3 (768 × 484)	9 (1024 × 1024)
Spectrometers	1	3
Speed, m/h	3.6	36 to 100
Maxon motors	11 RE16	17 RE20
		22 RE25

The follow-on roving mission to Mars was the Mars Exploration Rover mission with two rovers called *Spirit* and *Opportunity* (Fig. 4.31b), which landed on Mars in 2004. The two rovers were also equipped with Maxon d.c. commutator motors (Tables 4.4 and 4.6). The most recent Mars lander called *Phoenix* (2007) was supported by Maxon motors too.

Numerical examples

Numerical example 4.1

Find the main dimensions (armature diameter and its effective length and PM length), electric loading, and magnetic loading of a d.c. PM commutator motor of cylindrical construction with a slotted rotor rated at: $P_{out} = 40$ W, $V = 110$ V, and $n = 4000$ rpm. The PMs are made of *Hardferrite* 28/26, the demagnetizion curve of which is shown in Fig. 2.5. The efficiency at rated load should be a minimum of $\eta = 0.6$. The motor has to be designed for continuous duty.

Solution

The electromagnetic power for continuous duty according to eqn (4.24)

$$P_{elm} = \frac{1 + 2\eta}{3\eta} P_{out} = \frac{1 + 2 \times 0.6}{3 \times 0.6} 40.0 \approx 48.9 \text{ W}$$

The armature current

$$I_a = \frac{P_{out}}{\eta V} = \frac{40.0}{0.6 \times 110.0} \approx 0.61 \text{ A}$$

The armature EMF

$$E = \frac{P_{elm}}{I_a} = \frac{48.9}{0.61} = 80.2 \text{ V}$$

Electric and magnetic loadings: according to Fig. 4.9, for $P_{out}/n = 40/4000 = 0.01$ W/rpm $= 10 \times 10^{-3}$ W/rpm and continuous duty the line current density $A = 7500$ A/m and the air gap magnetic flux density $B_g = 0.35$ T.

The output coefficient expressed by eqn (4.36) is

$$\sigma_p = \alpha_i \pi^2 B_g A = 0.67\pi^2 0.35 \times 7500 = 17,358 \frac{\text{VAs}}{\text{m}^3}$$

where the effective arc pole coefficient has been assumed $\alpha_i = 0.67$.

Armature diameter and effective length: assuming $L_i/D \approx 1$, the output coefficent is $\sigma_p = P_{elm}/(D^3 n)$ and

$$D = \sqrt[3]{\frac{P_{elm}}{\sigma_p n}} = \sqrt[3]{\frac{48.9}{17,358 \times (4000/60)}} = 0.0348 \text{ m} \approx 35 \text{ mm}$$

The effective armature length $L_i \approx D = 35$ mm.

The *Hardferrite* 28/26 can produce a useful magnetic flux density B_u not exceeding 0.25 T (see Fig. 2.5). To obtain $B_g = 0.35$ T the length of the PM $L_M > L_i$. On the basis of eqns (4.71) and (4.72)

$$L_M = \frac{L_i B_g}{B_u} = \frac{35 \times 0.35}{0.25} = 49 \text{ mm}$$

Numerical example 4.2

A d.c. PM commutator motor has the following rated parameters: $P_{out} = 10$ kW, $V = 220$ V, $I_a = 50$ A, $n = 1500$ rpm. The armature circuit resistance is $\sum R_a = 0.197$ Ω and brush voltage drop is $\Delta V_{br} = 2$ V. Find the speed and efficiency for the shaft load torque equal to 80% of the rated torque and armature series rheostat $R_{rhe} = 0.591$ Ω.

Solution

The rated shaft torque according to eqn (4.15)

$$T_{sh} = \frac{P_{out}}{2\pi n} = \frac{10,000}{2\pi \times (1500/60)} = 63.66 \text{ Nm}$$

The armature EMF at rated current according to eqn (4.1)

$$E = V - I_a \sum R_a - \Delta V_{br} = 220 - 50 \times 0.197 - 2 = 208.15 \text{ V}$$

The developed torque at rated current according to eqns (4.9) to (4.11)

$$T_d = \frac{EI_a}{2\pi n} = \frac{208.15 \times 50}{2\pi \times (1500/60)} = 66.26 \text{ Nm}$$

The torque covering rotational losses

$$T_{rot} = T_d - T_{sh} = 66.26 - 63.66 = 2.6 \text{ Nm}$$

The torque developed at $0.8T_{sh}$

$$T_d' = 0.8T_{sh} + T_{rot} = 0.8 \times 63.66 + 2.6 = 53.53 \text{ Nm}$$

The armature current at $0.8T_{sh}$

$$I_a' = I_a \frac{T_d'}{T_d} = 50\frac{53.53}{66.26} = 40.4 \text{ A}$$

At no-load $I_a \approx 0$. Thus, the no-load speed

$$n_0 = n\frac{V - \Delta V_{br}}{E} = 1500\frac{220 - 2}{208.15} = 1571 \text{ rpm}$$

The speed at $0.8T$ and with additional armature resistance $R_{rhe} = 0.591$ Ω

$$n' = n_0 \frac{V - I_a'(\sum R_a + R_{rhe}) - \Delta V_{br}}{V - \Delta V_{br}}$$

$$= 1571\frac{220 - 40.4(0.197 + 0.591) - 2}{220 - 2} = 1341.5 \text{ rpm}$$

The output power at $0.8T_{sh}$ and with additional armature resistance

$$P_{out}' = 0.8T_{sh}(2\pi n') = 0.8 \times 63.66 \times (2\pi) \times \frac{1341}{60} = 7154.5 \text{ W}$$

The input power at $0.8T_{sh}$ and with additional armature resistance:

$$P'_{in} = VI'_a = 220 \times 40.4 = 8888 \text{ W}$$

The corresponding efficiency

$$\eta' = \frac{7154.5}{8888.0} = 0.805.$$

The rated efficiency is $\eta = 10,000/(220 \times 50) = 0.909$. There is about a 10% drop of efficiency as compared with rated conditions.

Table 4.7. Design data of a 370-W d.c. PM commutator motor

Quantity	Value
Rated power P_{out}	370 W
Rated terminal voltage V	180 V
Rated speed n	1750 rpm
Number of poles $2p$	2
Number of armature slots	20
Air gap g	2.25 mm
Length of PM l_M	79.6 mm
Length of armature core L_i	63.4 mm
Overlap angle of PMs β	2.6878 rad
Number of commutator segments C	40
Number of armature conductors N	920
Number of coil-sides per slot	2

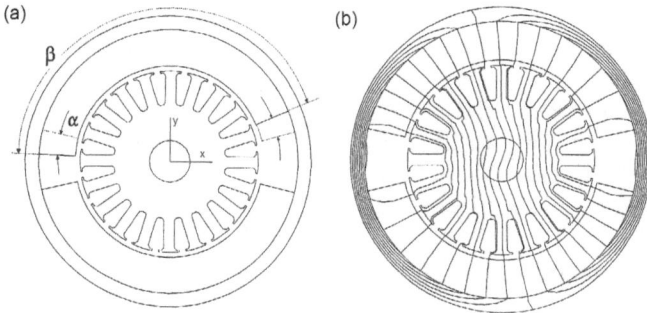

Fig. 4.32. Cross section of the d.c. motor according to example 4.3: (a) shift angle α and overlap angle β, (b) resultant magnetic flux (PM excitation and armature) for $\alpha = 0$.

Numerical example 4.3

In the manufacture of d.c. PM commutator motors, the segmental ferrite magnets are generally glued or clamped into position. This can lead to inaccurate positioning of the magnets and thus to an asymmetrical magnetic circuit. It is thus of interest to manufacturers to obtain tolerances of these inaccuracies with relation to the motors' performance and the forces produced on the shaft due to the imbalanced attraction of the rotor by the magnets. For a 370-W d.c. PM commutator motor find the resultant magnetic flux, MMF, rotor speed, armature current and efficiency for different shifts of one magnet. The cross section of the motor is shown in Fig. 4.32 and the specification data is given in Table 4.7.

Solution

Owing to the asymmetrical nature of this problem, classical analysis techniques are not suitable. The FEM is an ideal method since it can model the entire magnetic circuit [308].

Fig. 4.33 shows the effect on the magnetic flux and MMF (stator and armature reaction MMF) for different directions of rotation in the asymmetrical conditions. The resultant EMF is either increased or decreased. Exaggerated magnet shift of $\alpha = 10.5^0$ has been assumed to simulate the extreme case of magnet misalignment. In the manufacture of d.c. PM commutator motors the error in placing the magnets is usually $\alpha < 2^0$.

The change in performance as the magnet shift increases can be seen from the characteristics in Fig. 4.34.

The following conclusions can be made about this 370-W d.c. PM commutator motor:

- The performance of the motor is not affected in any significant manner due to the asymmetry of the stator magnetic poles. The variation of the calculated speed, armature current, and efficiency has shown a maximum change of 1%.
- The torque produced can either be increased or decreased depending on the direction of rotation. The torque will in general improve if the magnets are moved in the direction of rotation.
- The asymmetry of the magnetic circuit contributes to the unbalanced magnetic pull.
- The force produced by the unbalanced magnetic pull is in the same direction as the magnet movement. This net force can lead to unwanted noise and increased wear of bearings.

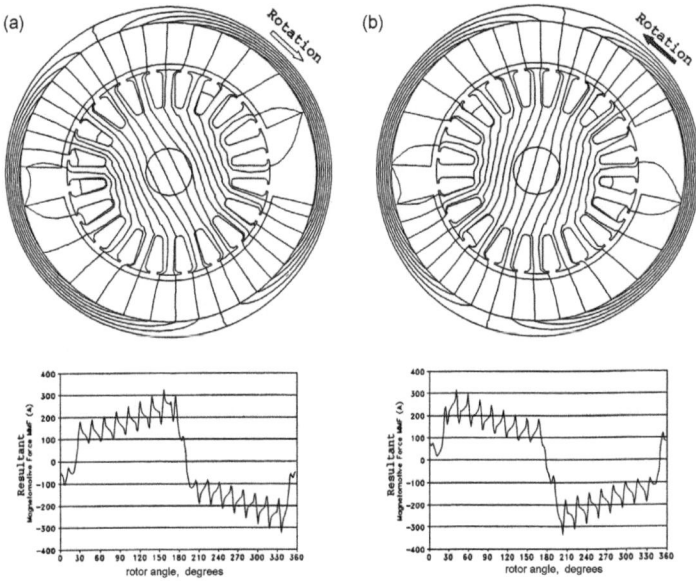

Fig. 4.33. Flux plots and resultant MMF distributions for two different directions of rotor rotation with one magnet shifted 10.5^0. *Numerical example 4.3.*

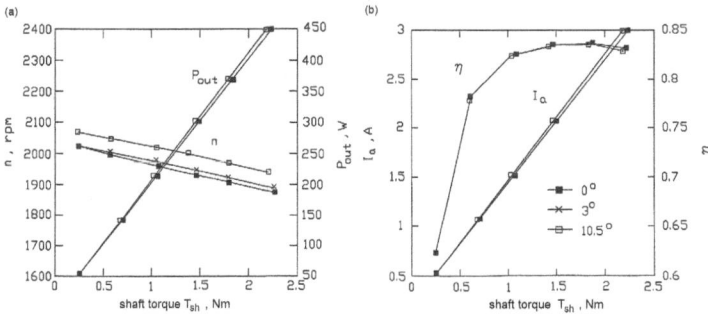

Fig. 4.34. Performance characteristics for $\alpha = 0$, 3, and 10.5^0: (a) speed against torque, (b) armature current and efficiency against torque. The rotor rotates in the direction of the magnet shift. *Numerical example 4.3.*

5

Permanent Magnet Synchronous Motors

5.1 Construction

Synchronous motors operate at a constant speed in absolute synchronism with the line frequency. Synchronous motors are classified according to their rotor's design, construction, materials and operation into the four basic groups:

- electromagnetically-excited motors
- PM motors
- reluctance motors
- hysteresis motors

In electromagnetically excited and PM motors a cage winding is frequently mounted on salient-pole rotors to provide asynchronous starting and to damp oscillations under transient conditions, so-called *damper*.

Recent developments in rare-earth PM materials and power electronics have opened new prospects on the design, construction and application of PM synchronous motors. Servo drives with PM motors fed from static inverters are finding applications on an increasing scale. PM servo motors with continuous output power of up to 15 kW at 1500 rpm are common. Commercially, PM a.c. motor drives are available with ratings up to at least 746 kW. Rare-earth PMs have also been recently used in large power synchronous motors rated at more than 1 MW [23, 24, 276]. Large PM motors can be used both in low-speed drives (ship propulsion) and high-speed drives (pumps and compressors).

PM synchronous motors are usually built with one of the following rotor configurations:

(a) classical (F. Merrill's rotor, US patent 2543639 assigned to General Electric), with salient poles, laminated pole shoes and a cage winding (Fig. 5.1a);
(b) interior-magnet rotor (Fig. 5.1b, 5.1i, 5.1j);
(c) surface-magnet rotor (Fig. 5.1c, 5.1g, 5.1h);
(d) inset-magnet rotor (Fig. 5.1d);

(e) rotor with buried magnets symmetrically distributed (Fig. 5.1e);

(f) rotor with buried magnets asymmetrically distributed according to German patent 1173178 assigned to Siemens, also called *Siemosyn*(Fig. 5.1f).

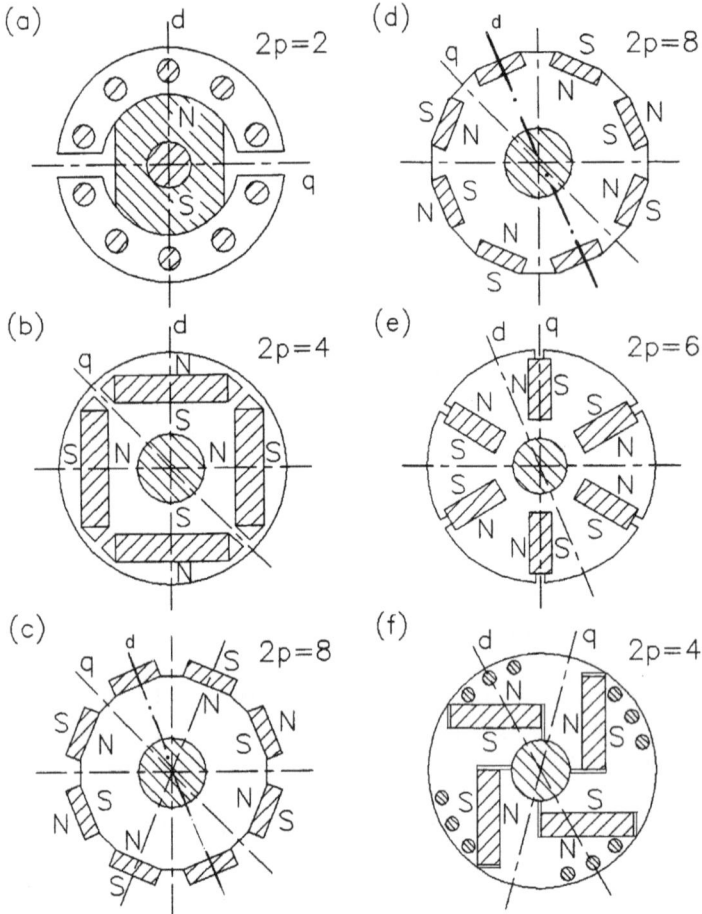

Fig. 5.1. Rotor configurations for PM synchronous motors: (a) classical configuration (US patent 2543639), (b) interior-magnet rotor, (c) surface-magnet rotor, (d) inset-magnet rotor, (e) rotor with buried (spoke) magnets symmetrically distributed, (f) rotor with buried magnets asymmetrically distributed (German patent 1173178).

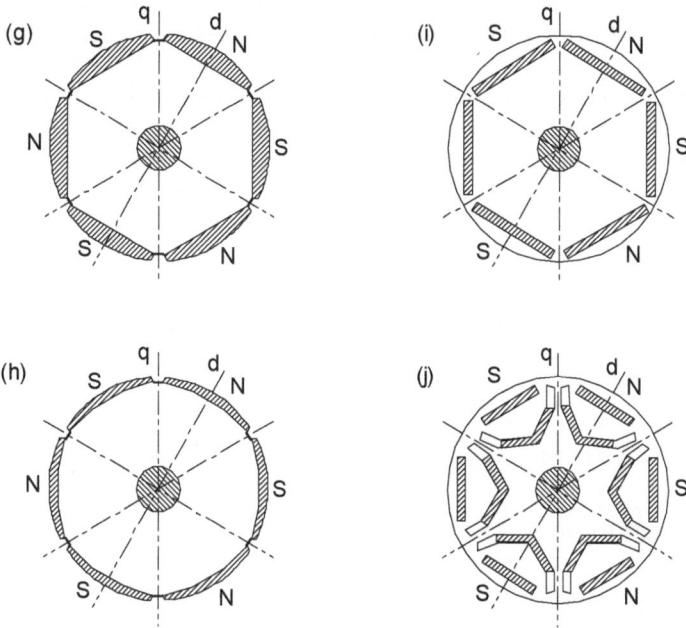

Fig. 5.1. Continued: (g) bread loaf magnets, (h) decentered magnets, (i) interior six-pole rotor, (j) interior double-layer magnets (folded magnets).

5.2 Fundamental relationships

5.2.1 Speed

In the steady-state range, the rotor speed is given by the input frequency–to–number of pole pairs ratio, i.e.,

$$n_s = \frac{f}{p} \tag{5.1}$$

and is equal to the synchronous speed of the rotating magnetic field produced by the stator.

5.2.2 Air gap magnetic flux density

The first harmonic of the air gap magnetic flux density is

$$B_{mg1} = \frac{2}{\pi} \int_{-0.5\alpha_i\pi}^{0.5\alpha_i\pi} B_{mg} \cos\alpha\, d\alpha = \frac{4}{\pi} B_{mg} \sin\frac{\alpha_i\pi}{2} \tag{5.2}$$

where, neglecting the saturation of the magnetic circuit, the magnetic flux density $B_{mg} = \mu_0 F_{exc}/(g'k_C)$ under the pole shoe can be found on the basis of the excitation MMF F_{exc}, equivalent air gap g' which includes the PM height h_M and Carter's coefficient k_C. Carter's coefficient is given by eqns (A.27) and (A.28) in Appendix A. For $\alpha_i = 1$ the fundamental harmonic component B_{mg1} is $4/\pi$ times the B_{mg} peak flat-topped value.

The coefficient α_i is defined as the ratio of the *average–to–maximum value* of the normal component of the air gap magnetic flux density, i.e.,

$$\alpha_i = \frac{B_{avg}}{B_{mg}} \tag{5.3}$$

If the magnetic field distribution in the air gap is sinusoidal, $\alpha_i = 2/\pi$. For zero magnetic voltage drop in the ferromagnetic core and uniform air gap

$$\alpha_i = \frac{b_p}{\tau} \tag{5.4}$$

The coefficient α_i is also called the *pole-shoe arc b_p–to–pole pitch τ ratio.*

5.2.3 Voltage induced (EMF)

The no-load *rms* voltage induced in one phase of the stator winding (EMF) by the d.c. magnetic excitation flux Φ_f of the rotor is

$$E_f = \pi\sqrt{2}fN_1k_{w1}\Phi_f \tag{5.5}$$

where N_1 is the number of the stator turns per phase, k_{w1} is the stator winding coefficient (Appendix A, eqn (A.1), and the fundamental harmonic Φ_{f1} of the excitation magnetic flux density Φ_f without armature reaction is

$$\Phi_{f1} = L_i \int_0^\tau B_{mg1} \sin\left(\frac{\pi}{\tau}x\right) dx = \frac{2}{\pi}\tau L_i B_{mg1} \tag{5.6}$$

Similarly, the voltage E_{ad} induced by the d-axis armature reaction flux Φ_{ad} and the voltage E_{aq} induced by the q-axis flux Φ_{aq} are, respectively

$$E_{ad} = \pi\sqrt{2}fN_1k_{w1}\Phi_{ad} \tag{5.7}$$

$$E_{aq} = \pi\sqrt{2}fN_1k_{w1}\Phi_{aq} \tag{5.8}$$

The first harmonics of armature reaction magnetic fluxes are

$$\Phi_{ad} = \frac{2}{\pi}B_{mad1}\tau L_i \tag{5.9}$$

$$\Phi_{aq} = \frac{2}{\pi}B_{maq1}\tau L_i \tag{5.10}$$

where B_{mad1} and B_{maq1} are the peak values of the first harmonic of armature reaction magnetic flux density in the d and q-axis, respectively.

As shown in Fig 5.1 the *direct* or *d*-axis is the center axis of the magnetic pole while the *quadrature* or *q*-axis is the axis parallel (90^0 electrical) to the *d*-axis. The EMFs E_f, E_{ad}, E_{aq}, and magnetic fluxes Φ_f, Φ_{ad}, and Φ_{aq} are used in construction of phasor diagrams and equivalent circuits. The EMF E_i per phase with the armature reaction taken into account is

$$E_i = \pi\sqrt{2}fN_1 k_{w1}\Phi_g \qquad (5.11)$$

where Φ_g is the air gap magnetic flux under load (excitation flux Φ_f reduced by the armature reaction flux). At no-load (very small armature current) $\Phi_g \approx \Phi_f$. Including the saturation of the magnetic circuit

$$E_i = 4\sigma_f fN_1 k_{w1}\Phi_g \qquad (5.12)$$

The form factor σ_f depends on the magnetic saturation of armature teeth, i.e. the sum of the air gap MVD and the teeth MVD divided by the air gap MVD.

5.2.4 Armature line current density and current density

The peak value of the stator (armature) line current density (A/m) or *specific electric loading* is defined as the number of conductors in all phases $2m_1N_1$ times the peak armature current $\sqrt{2}I_a$ divided by the armature circumference πD_{1in}, i.e.,

$$A_m = \frac{2m_1\sqrt{2}N_1 I_a}{\pi D_{1in}} = \frac{m_1\sqrt{2}N_1 I_a}{p\tau} = \frac{m_1\sqrt{2}N_1 J_a s_a}{p\tau} \qquad (5.13)$$

where J_a is the current density (A/m^2) in the stator (armature) conductors and s_a is the cross section of armature conductors including parallel wires. For air cooling systems $J_a \leq 7.5$ A/mm^2 (sometimes up to 10 A/mm^2) and for liquid coling systems $10 \leq J_a \leq 28$ A/mm^2. The top value is for very intensive oil spray cooling systems.

5.2.5 Electromagnetic power

For an m_1-phase salient pole synchronous motor with negligible stator winding resistance $R_1 = 0$, the electromagnetic power is expressed as

$$P_{elm} = m_1\left[\frac{V_1 E_f}{X_{sd}}\sin\delta + \frac{V_1^2}{2}\left(\frac{1}{X_{sq}} - \frac{1}{X_{sd}}\right)\sin 2\delta\right] \qquad (5.14)$$

where V_1 is the input (terminal) phase voltage, E_f is the EMF induced by the rotor excitation flux (without armature reaction), δ is the power angle, i.e. the angle between V_1 and E_f, X_{sd} is the synchronous reactance in the direct axis (*d*-axis synchronous reactance), and X_{sq} is the synchronous reactance in the quadrature axis (*q*-axis synchronous reactance).

5.2.6 Synchronous reactance

For a salient pole synchronous motor the d-axis and q-axis synchronous reactances are

$$X_{sd} = X_1 + X_{ad} \qquad\qquad X_{sq} = X_1 + X_{aq} \qquad (5.15)$$

where $X_1 = 2\pi f L_1$ is the stator leakage reactance, X_{ad} is the d-axis armature reaction reactance, also called d-axis mutual reactance, and X_{aq} is the q-axis armature reaction reactance, also called q-axis mutual reactance. The reactance X_{ad} is sensitive to the saturation of the magnetic circuit whilst the influence of the magnetic saturation on the reactance X_{aq} depends on the rotor construction. In salient-pole synchronous machines with electromagnetic excitation X_{aq} is practically independent of the magnetic saturation. Usually, $X_{sd} > X_{sq}$ except for some PM synchronous machines.

The leakage reactance X_1 consists of the slot, end-connection differential and tooth–top leakage reactances (Appendix A). Only the slot and differential leakage reactances depend on the magnetic saturation due to leakage fields [231].

5.2.7 Subtransient synchronous reactance

The d-axis subtransient synchronous reactance is defined for generator operation, for the first instant of a sudden short circuit of the stator, when the damper and the excitation winding (if it exists) repel the armature magnetic flux. The same effect appears for motoring, when the rotor is suddenly locked. The subtransient synchronous reactance is the sum of the stator winding leakage reactance X_1 and the parallel connection of the reactances of the damper X_{damp}, field excitation winding X_{exc}, and the d-axis armature reaction reactance, i.e.,

$$X_{sd}'' = X_1 + \frac{X_{damp}X_{exc}X_{ad}}{X_{damp}X_{exc} + X_{exc}X_{ad} + X_{ad}X_{damp}} \qquad (5.16)$$

The resistance R_{damp} of the damper weakens the screening effect of this winding and the short circuit current decays. A similar equation to (5.16) can be written for the q axis, i.e. for the q-axis subtransient reactance X_{sq}''.

5.2.8 Transient synchronous reactance

With the exponential decay of the damping-winding current, the armature MMF is able to force its flux deeper into the pole (despite the opposition of the induced rotor current). The decay of the current in the excitation winding or conductive PMs (see Tables 2.2, 2.3 and 2.4) can be slower than that in the damper since the inductance of the excitation winding is larger. This stage is characterized by the transient reactance

$$X'_{sd} = X_1 + \frac{X_{exc}X_{ad}}{X_{exc} + X_{ad}} \tag{5.17}$$

Eventually, the steady-state short circuit condition is reached, when the stator magnetic flux penetrates freely through the rotor core. The steady-state is characterized by the d-axis and q-axis synchronous reactances (5.15).

5.2.9 Electromagnetic (developed) torque

The electromagnetic torque developed by the synchronous motor is determined by the electromagnetic power P_{elm} and angular synchronous speed $\Omega_s = 2\pi n_s$ which is equal to the mechanical angular speed of the rotor, i.e.,

$$T_d = \frac{P_{elm}}{2\pi n_s} = \frac{m_1}{2\pi n_s}\left[\frac{V_1 E_f}{X_{sd}}\sin\delta + \frac{V_1^2}{2}\left(\frac{1}{X_{sq}} - \frac{1}{X_{sd}}\right)\sin 2\delta\right] \tag{5.18}$$

The above equation neglects the stator winding resistance R_1.

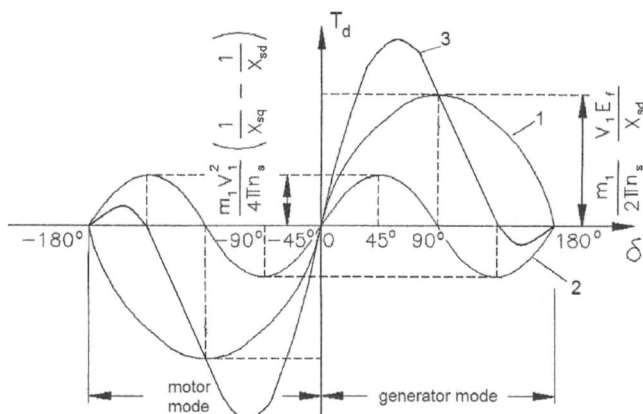

Fig. 5.2. Torque-angle characteristics of a salient-pole synchronous machine with $X_{sd} > X_{sq}$: 1 — synchronous torque T_{dsyn}, 2 — reluctance torque T_{drel}, 3 — resultant torque T_d.

In a salient pole-synchronous motor the electromagnetic torque has two components (Fig. 5.2):

$$T_d = T_{dsyn} + T_{drel} \tag{5.19}$$

where the fundamental torque,

$$T_{dsyn} = \frac{m_1}{2\pi n_s} \frac{V_1 E_f}{X_{sd}} \sin \delta \qquad (5.20)$$

is a function of both the input voltage V_1 and the excitation EMF E_f. The additional torque

$$T_{drel} = \frac{m_1 V_1^2}{4\pi n_s} \left(\frac{1}{X_{sq}} - \frac{1}{X_{sd}} \right) \sin 2\delta \qquad (5.21)$$

depends only on the voltage V_1 and also exists in an unexcited machine ($E_f = 0$) provided that $X_{sd} \neq X_{sq}$. The torque T_{dsyn} is called the *synchronous torque* and the torque T_{drel} is called the *reluctance torque*. For salient-pole synchronous motors with electromagnetic d.c. excitation $X_{sd} > X_{sq}$. For some PM synchronous motors as, for example, according to Figs 5.1b and 5.1d, $X_{sd} < X_{sq}$. The proportion between X_{sd} and X_{sq} strongly affects the shape of curves 2 and 3 in Fig. 5.2. For cylindrical rotor synchronous machines $X_{sd} = X_{sq}$, and

$$T_d = T_{dsyn} = \frac{m_1}{2\pi n_s} \frac{V_1 E_f}{X_{sd}} \sin \delta \qquad (5.22)$$

5.2.10 Form factor of the excitation field

The form factor of the excitation field results from eqn (5.2), i.e.,

$$k_f = \frac{B_{mg1}}{B_{mg}} = \frac{4}{\pi} \sin \frac{\alpha_i \pi}{2} \qquad (5.23)$$

where the *pole-shoe arc–to–pole pitch* ratio $\alpha_i < 1$ is calculated according to eqn (5.4).

5.2.11 Form factors of the armature reaction

The form factors of the armature reaction are defined as the ratios of the *first harmonic amplitudes–to–maximum values of normal components of armature reaction magnetic flux densities* in the d-axis and q-axis, respectively, i.e.,

$$k_{fd} = \frac{B_{ad1}}{B_{ad}} \qquad\qquad k_{fq} = \frac{B_{aq1}}{B_{aq}} \qquad (5.24)$$

The peak values of the first harmonics B_{ad1} and B_{aq1} of the armature magnetic flux density can be calculated as coefficients of Fourier series for $\nu = 1$, i.e.,

$$B_{ad1} = \frac{4}{\pi} \int_0^{0.5\pi} B(x) \cos x \, dx \qquad (5.25)$$

Table 5.1. Factors k_f, k_{fd}, k_{fq}, k_{ad}, and k_{aq} for salient-pole synchronous machines according to eqns (5.23), (5.24), (5.27) and (5.28)

| | $\alpha_i = b_p/\tau$ | | | | | | |
Factor	0.4	0.5	0.6	$2/\pi$	0.7	0.8	1.0
k_f	0.748	0.900	1.030	1.071	1.134	1.211	1.273
k_{fd}	0.703	0.818	0.913	0.943	0.958	0.987	1.00
k_{fq}	0.097	0.182	0.287	0.391	0.442	0.613	1.00
k_{ad}	0.939	0.909	0.886	0.880	0.845	0.815	0.785
k_{aq}	0.129	0.202	0.279	0.365	0.389	0.505	0.785

$$B_{aq1} = \frac{4}{\pi} \int_0^{0.5\pi} B(x) \sin x \, dx \qquad (5.26)$$

For a salient-pole motor with electromagnetic excitation and the air gap $g \approx 0$ (fringing effects neglected), the d- and q-axis form factors of the armature reaction are

$$k_{fd} = \frac{\alpha_i \pi + \sin \alpha_i \pi}{\pi} \qquad k_{fq} = \frac{\alpha_i \pi - \sin \alpha_i \pi}{\pi} \qquad (5.27)$$

5.2.12 Reaction factor

The reaction factors in the d- and q-axis are defined as

$$k_{ad} = \frac{k_{fd}}{k_f} \qquad k_{aq} = \frac{k_{fd}}{k_f} \qquad (5.28)$$

The form factors k_f, k_{fd} and k_{fq} of the excitation field and armature reaction and reaction factors k_{ad} and k_{aq} for salient-pole synchronous machines according to eqns (5.23), (5.24), (5.27) and (5.28) are given in Table 5.1.

5.2.13 Equivalent field MMF

Assuming $g = 0$, the equivalent d-axis field MMF (which produces the same fundamental wave flux as the armature-reaction MMF) is

$$F_{excd} = k_{ad} F_{ad} = \frac{m_1 \sqrt{2} \, N_1 k_{w1}}{\pi} \frac{}{p} k_{ad} I_a \sin \Psi \qquad (5.29)$$

where I_a is the armature current and Ψ is the angle between the resultant armature MMF F_a and its q-axis component $F_{aq} = F_a \cos \Psi$. Similarly, the equivalent q-axis MMF is

$$F_{excq} = k_{aq} F_{aq} = \frac{m_1 \sqrt{2} \, N_1 k_{w1}}{\pi} \frac{}{p} k_{aq} I_a \cos \Psi \qquad (5.30)$$

5.2.14 Armature reaction reactance

The d-axis armature reaction reactance with the magnetic saturation being included is

$$X_{ad} = k_{fd}X_a = 4m_1\mu_0 f \frac{(N_1 k_{w1})^2}{\pi p} \frac{\tau L_i}{g'} k_{fd} \tag{5.31}$$

where μ_0 is the magnetic permeability of free space, L_i is the effective length of the stator core and

$$X_a = 4m_1\mu_0 f \frac{(N_1 k_{w1})^2}{\pi p} \frac{\tau L_i}{g'} \tag{5.32}$$

is the inductive reactance of the armature of a non-salient-pole (cylindrical rotor) synchronous machine. Similarly, for the q-axis

$$X_{aq} \approx k_{fq}X_a = 4m_1\mu_0 f \frac{(N_1 k_{w1})^2}{\pi p} \frac{\tau L_i}{g'_q} k_{fq} \tag{5.33}$$

For most PM configurations the equivalent air gap g' in eqns (5.31) and (5.32) should be replaced by $gk_C k_{sat} + h_M/\mu_{rrec}$ and g'_q in eqn (5.33) by $g_q k_C k_{satq}$ where g_q is the mechanical clearance in the q-axis, k_C is the Carter's coefficient for the air gap according to eqn (A.27) and $k_{sat} \geq 1$ is the saturation factor of the magnetic circuit. For the rotor shown in Fig. 5.1a and salient pole rotors with electromagnetic excitation the saturation factor $k_{satq} \approx 1$, since the q-axis armature reaction fluxes, closing through the large air spaces between the poles, depend only slightly on the saturation.

5.3 Phasor diagram

When drawing phasor diagrams of synchronous machines, two arrow systems are used:

(a) generator arrow system, i.e.,

$$\mathbf{E}_f = \mathbf{V}_1 + \mathbf{I}_a R_1 + j\mathbf{I}_{ad}X_{sd} + j\mathbf{I}_{aq}X_{sq}$$

$$= \mathbf{V}_1 + \mathbf{I}_{ad}(R_1 + jX_{sd}) + \mathbf{I}_{aq}(R_1 + jX_{sq}) \tag{5.34}$$

(b) consumer (motor) arrow system, i.e.,

$$\mathbf{V}_1 = \mathbf{E}_f + \mathbf{I}_a R_1 + j\mathbf{I}_{ad}X_{sd} + j\mathbf{I}_{aq}X_{sq}$$

$$= \mathbf{E}_f + \mathbf{I}_{ad}(R_1 + jX_{sd}) + \mathbf{I}_{aq}(R_1 + jX_{sq}) \tag{5.35}$$

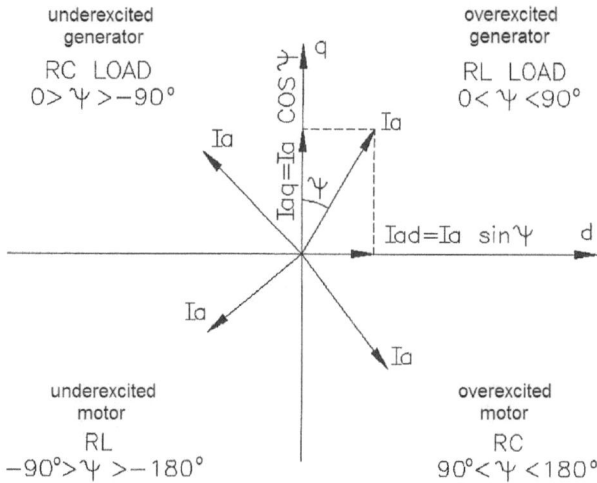

Fig. 5.3. Location of the armature current $\mathbf{I}_a$ in d—q coordinate system.

where

$$\mathbf{I}_a = \mathbf{I}_{ad} + \mathbf{I}_{aq} \tag{5.36}$$

and

$$I_{ad} = I_a \sin \Psi \qquad\qquad I_{aq} = I_a \cos \Psi \tag{5.37}$$

When the current arrows are in the opposite direction the phasors $\mathbf{I}_a$, $\mathbf{I}_{ad}$, and $\mathbf{I}_{aq}$, are reversed by 180^0. The same applies to the voltage drops. The location of the armature current $\mathbf{I}_a$ with respect to the d and q-axis for generator and motor mode is shown in Fig. 5.3.

Phasor diagrams for synchronous generators are constructed using the generator arrow system. The same system can be used for motors; however, the consumer arrow system is more convenient. An underexcited motor (Fig. 5.4a) draws an inductive current and a corresponding reactive power from the line. Fig. 5.4b shows the phasor diagram using the same consumer arrow system for a load current $\mathbf{I}_a$ leading the vector $\mathbf{V}_1$ by the angle ϕ. At this angle the motor is, conversely, overexcited and induces with respect to the input voltage $\mathbf{V}_1$ a capacitive current component $I_a \sin \Psi$. An overexcited motor, consequently, draws a leading current from the circuit and delivers reactive power to it.

In the phasor diagrams according to Fig. 5.4 the stator core losses have been neglected. This assumption is justified only for power frequency synchronous motors with unsaturated armature cores.

(a)

(b)

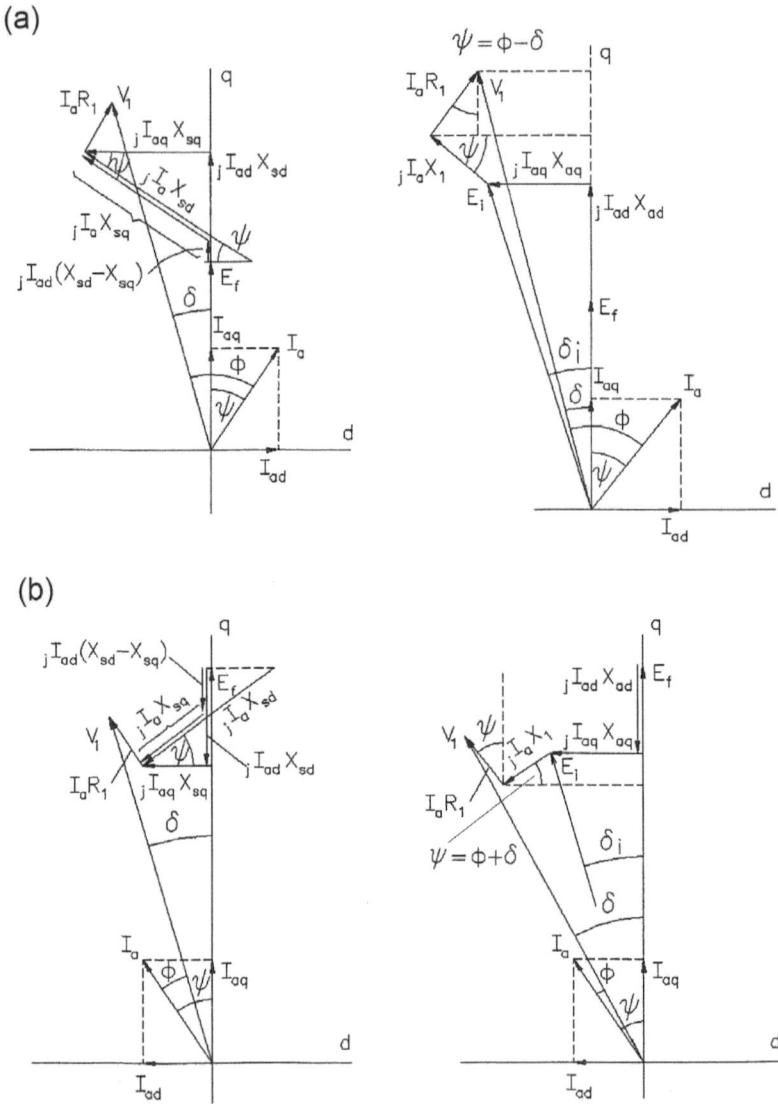

Fig. 5.4. Phasor diagrams of salient-pole synchronous motors for the consumer arrow system: (a) underexcited motor, (b) overexcited motor.

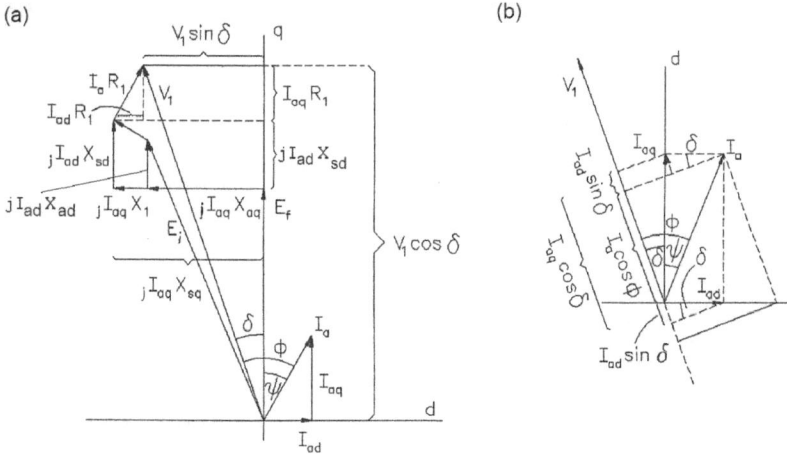

Fig. 5.5. Phasor diagrams for finding: (a) axis currents I_{ad} and I_{aq}; (b) input power P_{in} as a function of I_{ad}, I_{aq}, and δ.

Fig. 5.5 shows the phasor diagram of an underexcited synchronous motor with some necessary details for finding the *rms* axis currents I_{ad} and I_{aq}. The input voltage V_1 projections on the d and q axes are

$$V_1 \sin \delta = I_{aq} X_{sq} - I_{ad} R_1$$

$$V_1 \cos \delta = E_f + I_{ad} X_{sd} + I_{aq} R_1 \tag{5.38}$$

For an overexcited motor

$$V_1 \sin \delta = I_{aq} X_{sq} + I_{ad} R_1$$

$$V_1 \cos \delta = E_f - I_{ad} X_{sd} + I_{aq} R_1 \tag{5.39}$$

The currents of an underexcited motor

$$I_{ad} = \frac{V_1 (X_{sq} \cos \delta - R_1 \sin \delta) - E_f X_{sq}}{X_{sd} X_{sq} + R_1^2} \tag{5.40}$$

$$I_{aq} = \frac{V_1 (R_1 \cos \delta + X_{sd} \sin \delta) - E_f R_1}{X_{sd} X_{sq} + R_1^2} \tag{5.41}$$

are obtained by solving the set of eqns (5.38). Similarly, the currents of an overexcited motor are found by solving the set of eqns (5.39). The d-axis current of an overexcited motor is

$$I_{ad} = \frac{V_1(R_1 \sin\delta - X_{sq}\cos\delta) + E_f X_{sq}}{X_{sd}X_{sq} + R_1^2} \tag{5.42}$$

and the q-axis current is expressed by eqn (5.41). The rms armature current of an underexcited motor as a function of V_1, E_f, X_{sd}, X_{sq}, δ, and R_1 is

$$I_a = \sqrt{I_{ad}^2 + I_{aq}^2} = \frac{V_1}{X_{sd}X_{sq} + R_1^2}$$

$$\times \sqrt{\left[(X_{sq}\cos\delta - R_1\sin\delta) - \frac{E_f X_{sq}}{V_1}\right]^2 + \left[(R_1\cos\delta + X_{sd}\sin\delta) - \frac{E_f R_1}{V_1}\right]^2} \tag{5.43}$$

The angle between the phasor $\mathbf{I}_a$ and q-axis is $\psi = \phi \mp \delta$ where the "$-$" sign is for an underexcited motor and the "$+$" sign is for an overexcited motor.

The phasor diagram (Fig. 5.5b) can also be used to find the input power, i.e.,

$$P_{in} = m_1 V_1 I_a \cos\phi = m_1 V_1 (I_{aq}\cos\delta - I_{ad}\sin\delta) \tag{5.44}$$

Putting eqns (5.38) into eqn (5.44)

$$P_{in} = m_1[I_{aq}E_f + I_{ad}I_{aq}X_{sd} + I_{aq}^2 R_1 - I_{ad}I_{aq}X_{sq} + I_{ad}^2 R_1]$$

$$= m_1[I_{aq}E_f + R_1 I_a^2 + I_{ad}I_{aq}(X_{sd} - X_{sq})]$$

Because the stator core loss has been neglected, the electromagnetic power is the motor input power minus the stator winding loss $\Delta P_{1w} = m_1 I_a^2 R_1 = m_1(I_{ad}^2 + I_{aq}^2)R_1$. Thus

$$P_{elm} = P_{in} - \Delta P_{1w} = m_1[I_{aq}E_f + I_{ad}I_{aq}(X_{sd} - X_{sq})]$$

$$= \frac{m_1[V_1(R_1\cos\delta + X_{sd}\sin\delta) - E_f R_1)]}{(X_{sd}X_{sq} + R_1^2)^2} \tag{5.45}$$

$$\times[V_1(X_{sq}\cos\delta - R_1\sin\delta)(X_{sd} - X_{sq}) + E_f(X_{sd}X_{sq} + R_1^2) - E_f X_{sq}(X_{sd} - X_{sq})]$$

The electromagnetic torque developed by a salient-pole synchronous motor is

$$T_d = \frac{P_{elm}}{2\pi n_s} = \frac{m_1}{2\pi n_s}\frac{1}{(X_{sd}X_{sq} + R_1^2)^2}$$

$$\times\{V_1 E_f(R_1\cos\delta + X_{sd}\sin\delta)[(X_{sd}X_{sq} + R_1^2) - X_{sq}(X_{sd} - X_{sq})]$$

$$-V_1 E_f R_1 (X_{sq} \cos \delta - R_1 \sin \delta)(X_{sd} - X_{sq})$$

$$+V_1^2 (R_1 \cos \delta + X_{sd} \sin \delta)(X_{sq} \cos \delta - R_1 \sin \delta)(X_{sd} - X_{sq})$$

$$-E_f^2 R_1 [(X_{sd} X_{sq} + R_1^2) - X_{sq}(X_{sd} - X_{sq})]\} \qquad (5.46)$$

The last term is the constant component of the electromagnetic torque independent of the load angle δ. Putting $R_1 = 0$, eqn (5.46) becomes the same as eqn (5.18). Small synchronous motors have a rather high stator winding resistance R_1 that is comparable with X_{sd} and X_{sq}. That is why eqn (5.46) is recommended for calculating the performance of small motors.

5.4 Characteristics

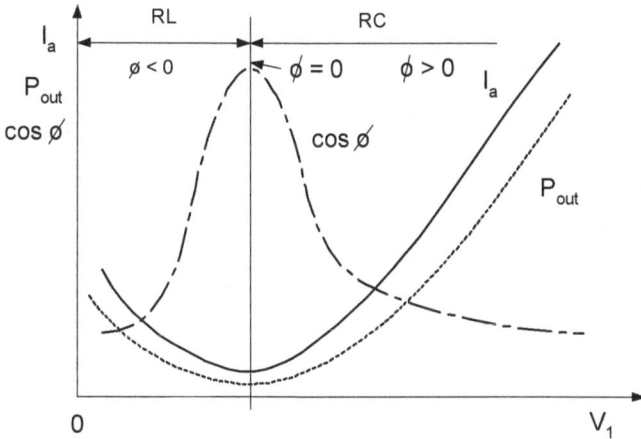

Fig. 5.6. No-load characteristics $I_a = f(V_1)$, $P_{in} = f(V_1)$ and $\cos \phi = f(V_1)$ of a PM synchronous motor.

The most important characteristic of a synchronous motor is the torque T_d—δ angle characteristic (Fig. 5.2). The overload capacity factor is a ratio of the maximum torque or maximum output power to the rated torque or output power. The torque–angle characteristic depends on the input voltage.

Any change in the input voltage at $T_{sh} = 0$ or $T_{sh} = const$ results in a change in the armature current and power factor. The no-load characteristics $I_a = f(V_1)$, $P_{in} = f(V_1)$ and $\cos \phi = f(V_1)$ at $T_{sh} = 0$ are, from the left-hand side, limited by the threshold voltage and from the right-hand side by the maximum current in the armature winding (Fig. 5.6). The minimum armature

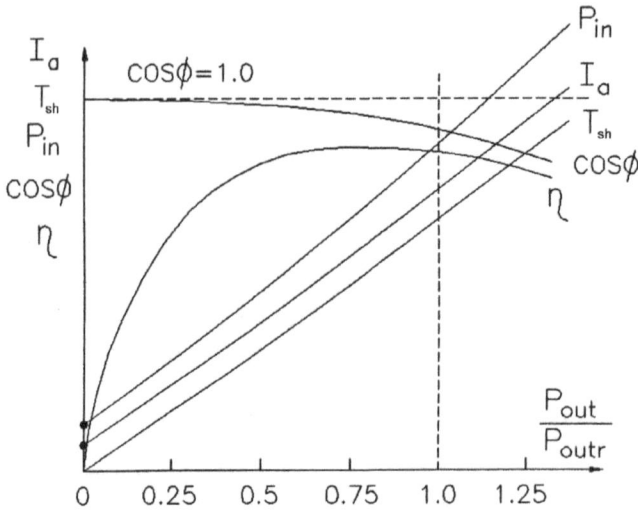

Fig. 5.7. Performance characteristics of a PM synchronous motor: armature current I_a, shaft torque T_{sh}, input power P_{in}, power factor $\cos\phi$, and efficiency η plotted against P_{out}/P_{outr}, where P_{outr} is the rated ouput power.

current is for $\cos\phi = 1$ or $\phi = 0$. The lagging power factor corresponding to an underexcited motor (or RL load) is on the left of the point $\phi = 0$, and the leading power factor corresponding to an overexcited motor (or RC load) is on the right of the point $\phi = 0$. The overexcited motor behaves as an RC load and can compensate the reactive power consumed by, e.g., induction motors and underloaded transformers.

The armature current I_a, shaft torque T_{sh}, input power P_{in}, power factor $\cos\phi$, and efficiency η are plotted against the relative output power (output power–to–rated output power) in Fig. 5.7.

For $I_{ad} = 0$ the angle $\Psi = 0$ (between the armature current $I_a = I_{aq}$ and EMF E_f). Therefore, the angle ϕ between the current and voltage is equal to the load angle δ between the voltage V_1 and EMF E_f, i.e.,

$$\cos\phi = \frac{E_f + I_a R_1}{V_1} \tag{5.47}$$

and

$$V_1^2 = (E_f + I_a R_1)^2 + (I_a X_{sq})^2 \approx E_f^2 + I_a^2 X_{sq}^2 \tag{5.48}$$

Thus

$$\cos\phi \approx \sqrt{1 - \left(\frac{I_a X_{sq}}{V_1}\right)^2} + \frac{I_a R_1}{V_1} \tag{5.49}$$

At constant voltage V_1 and frequency (speed) the power factor $\cos \phi$ decreases with the load torque (proportional to the armature current I_a). The power factor can be kept constant by increasing the voltage in proportion to the current increase, i.e., keeping $I_a X_{sq}/V_1 = const.$

5.5 Starting

5.5.1 Asynchronous starting

A synchronous motor is not self-starting. To produce an *asynchronous starting torque*, its rotor must be furnished with a cage winding or mild steel pole shoes. The starting torque is produced as a result of the interaction between the stator rotating magnetic field and the rotor currents induced in the cage winding or mild steel pole shoes [153].

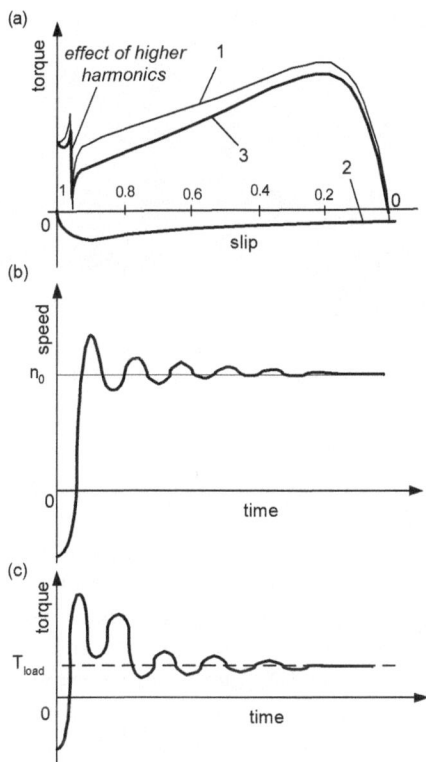

Fig. 5.8. Characteristics of a line start PM brushless motor: (a) steady-state torque–slip characteristic; (b) speed–time characteristic; (c) torque–time characteristic. 1 — asynchronous torque, 2 — braking torque produced by PMs, 3 — resultant torque, n_0 — steady-state speed, T_{load} — load torque.

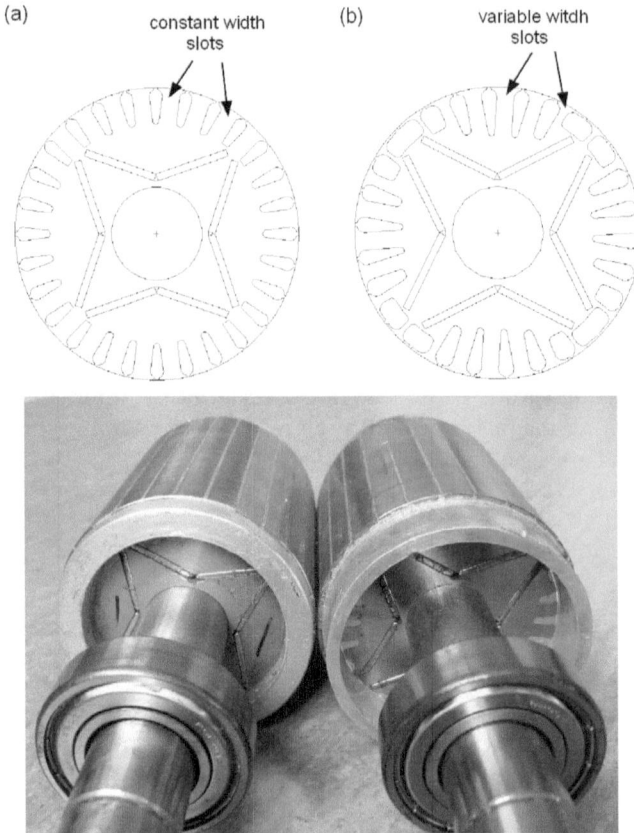

Fig. 5.9. Rotors of line start PM synchronous motors with: (a) constant width slots; (b) variable width slots. Photo courtesy of Technical University of Wroclaw, Poland [313].

PM synchronous motors that can produce asynchronous starting torque are commonly called *line start PM synchronous motors*. These motors can operate without solid state converters. After starting, the rotor is pulled into synchronism and rotates with the speed imposed by the line input frequency. The efficiency of line start PM motors is higher than that of equivalent induction motors and the power factor can be equal to unity.

The rotor bars in line start PM motors are unskewed, because PMs are embedded axially in the rotor core. In comparison with induction motors, line start PM motors produce much higher content of higher space harmonics in the air gap magnetic flux density distribution, current and electromagnetic torque. Further, the line start PM synchronous motor has a major drawback during the starting period as the magnets generate a brake torque which decreases the starting torque and reduces the ability of the rotor to synchronize a load [190]. Starting characteristics of a line start PM brushless motor are plotted in Fig. 5.8.

There are many constructions of line start PM brushless motors, e.g., according to US patent 2543639 (Fig. 5.1a), German patent 1173178 (Fig. 5.1f) or international patent publication WO 2001/06624. Fig. 5.9 shows two rotors for line start PM synchronous motor [313]: rotor with conventioanl cage winding and rotor with slots of different shapes in the d and q-axis. The rotor shown in Fig. 5.9b allows for significant reduction of the 5th, 11th, 13th, 17th and higher odd harmonics [313].

5.5.2 Starting by means of an auxiliary motor

Auxiliary induction motors are frequently used for starting large synchronous motors with electromagnetic excitation. The synchronous motor has an auxiliary starting motor on its shaft, capable of bringing it up to the synchronous speed at which time synchronizing with the power circuit is possible. The unexcited synchronous motor is accelerated to almost synchronous speed using a smaller induction motor. When the speed is close to the synchronous speed, first the armature voltage and then the excitation voltage is switched on, and the synchronous motor is pulled into synchronism.

The disadvantage of this method is it's impossible to start the motor under load. It would be impractical to use an auxiliary motor of the same rating as that of the synchronous motor and expensive installation.

5.5.3 Frequency-change starting

Frequency-change starting is a common method of starting synchronous motors with electromagnetic excitation. The frequency of the voltage applied to the motor is smoothly changed from the value close to zero to the rated value. The motor runs synchronously during the entire starting period being fed from a variable voltage variable frequency (VVVF) solid state inverter.

5.6 Reactances

The accuracy of calculating the steady-state performance of small PM synchronous motors depends largely on the accuracy of calculating the synchronous reactances in the d- and q-axis. For typical medium power and large synchronous motors with electromagnetic excitation analytical methods using form factors of the armature magnetic flux density are good enough. Small PM synchronous motors have sometimes complicated structure and FEM is necessary to obtain an accurate distribution of the magnetic field. This distribution is very helpful to estimate correctly the form factors of the excitation and armature magnetic flux densities. Moreover, the FEM makes it possible to find the d- and q-axis synchronous reactances and armature reaction (mutual) reactances straightforward by computing the corresponding inductances (Chapter 3).

The measurement of the synchronous reactances for small PM synchronous motors with sufficient degree of accuracy is rather a difficult problem. There are several methods of measuring synchronous reactances of medium and large power synchronous machines but the assumptions made do not allow these methods to be applied to small synchronous motors. Reliable results can be obtained using a special laboratory set consisting of the tested motor, additional synchronous motor, prime mover, double-beam oscilloscope, and brake [233]. The measured load angle, input voltage, armature current, armature winding resistance, and power factor allow for finding the synchronous reactances X_{sd} and X_{xq} on the basis of the phasor diagram.

5.6.1 Analytical approach

The analytical approach to calculating the armature reaction reactances is based on the distribution of the armature winding normal component of magnetic flux density. This distribution can be assumed to be a periodical function or can be found using analog or numerical modelling, e.g., a FEM computer software. The d- and q-axis armature reaction reactances are expressed with the aid of so-called form factors of the armature reaction k_{fd} and k_{fq} according to eqns (5.24), (5.27), (5.31) and (5.33), i.e., $X_{ad} = k_{fd}X_a$ and $X_{aq} = k_{fq}X_a$. The armature reaction reactance X_a is the same as that for a cylindrical-rotor synchronous machine and is given by eqn (5.32).

To obtain a saturated synchronous reactance, the equivalent air gap $k_C g$ should be multiplied by the saturation factor $k_{sat} > 1$ of the magnetic circuit, i.e., to obtain $k_C k_{sat} g$. In salient-pole synchronous machines with electromagnetic excitation the magnetic saturation affects only X_{sd} since the q-axis air gap (between neighboring pole shoes) is very large. In some PM synchronous machines the magnetic saturation can affect both X_{sd} and X_{sq} — see eqn (5.33).

For the distributions of d- and q-axis magnetic flux densities according to Fig. 5.8, the first harmonics of the magnetic flux densities are

(a) in the case of inset-type PMs (Fig. 5.10a)

$$B_{ad1} = \frac{4}{\pi} \left[\int_0^{0.5\alpha_i\pi} (B_{ad}\cos x)\cos x dx + \int_{0.5\alpha_i\pi}^{0.5\pi} (c_g B_{ad}\cos x)\cos x dx \right]$$

$$= \frac{1}{\pi} B_{ad}\left[\alpha_i\pi + \sin\alpha_i\pi + c_g(\pi - \alpha_i\pi - \sin\alpha_i\pi)\right] \qquad (5.50)$$

$$B_{aq1} = \frac{4}{\pi} \left[\int_0^{0.5\alpha_i\pi} (B_{aq}\sin x)\sin x dx + \int_{0.5\alpha_i\pi}^{0.5\pi} (c_g B_{aq}\sin x)\sin x dx \right]$$

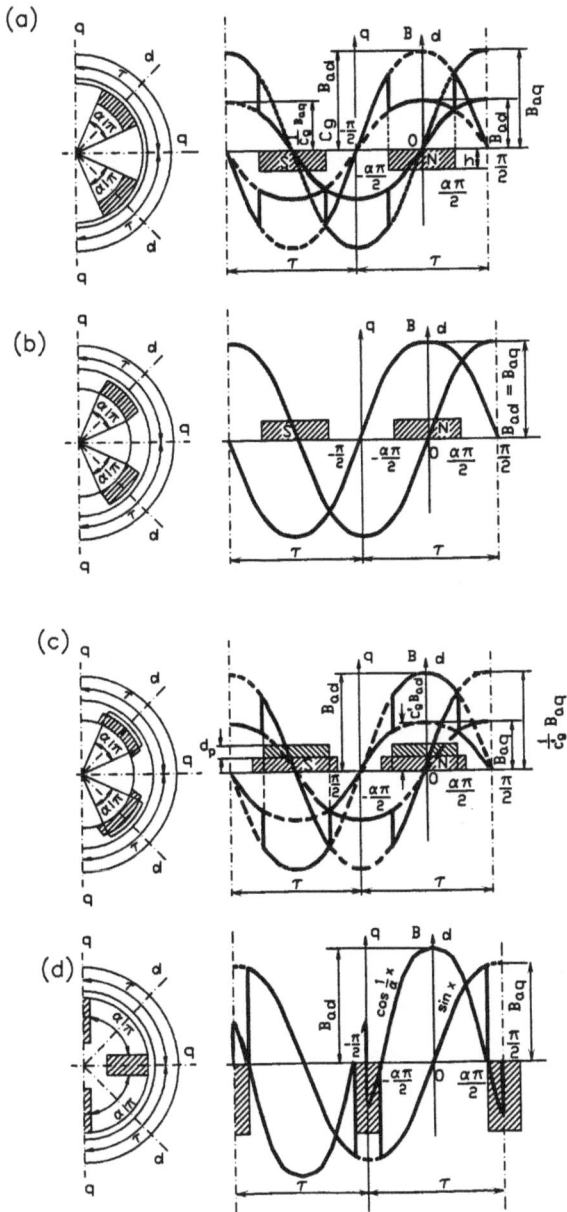

Fig. 5.10. Distribution of the d-axis and q-axis magnetic flux density for rare-earth PM rotors: (a) inset-type PM rotor, (b) surface PMs, (c) surface PMs with mild steel pole shoes, and (d) buried PM rotor.

Table 5.2. Form factors of the armature reaction for PM synchronous machines

Rotor configuration	d-axis	q-axis
Inset type PM rotor	$k_{fd} = \frac{1}{\pi}[\alpha_i\pi + \sin\alpha_i\pi$ $+c_g(\pi - \alpha_i\pi - \sin\alpha_i\pi)]$	$k_{fq} = \frac{1}{\pi}[\frac{1}{c_g}(\alpha_i\pi - \sin\alpha_i\pi)$ $+\pi(1-\alpha_i) + \sin\alpha_i\pi]$
	$c_g \approx 1 + h/g$	
Surface PM rotor	$k_{fd} = k_{fq} = 1$	
Surface PM rotor with mild steel pole shoes	$k_{fd} = \frac{1}{\pi}[\alpha_i\pi + \sin\alpha_i\pi$ $+c_g'(\pi - \alpha_i\pi - \sin\alpha_i\pi)]$	$k_{fq} = \frac{1}{\pi}[\frac{1}{c_g'}(\alpha_i\pi - \sin\alpha_i\pi)$ $+\pi(1-\alpha_i) + \sin\alpha_i\pi]$
	$c_g' \approx 1 - d_p/g_q$	
Buried PMs	$k_{fd} = \frac{4}{\pi}\alpha_i\frac{1}{1-\alpha_i^2}\cos(0.5\alpha_i\pi)$	$k_{fq} = \frac{1}{\pi}(\alpha_i\pi - \sin\alpha_i\pi)$
Salient pole rotor with excitation winding	$k_{fd} = \frac{1}{\pi}(\alpha_i\pi + \sin\alpha_i\pi)$	$k_{fq} = \frac{1}{\pi}(\alpha_i\pi - \sin\alpha_i\pi)$

$$= \frac{1}{\pi}B_{aq}\left[\frac{1}{c_g}(\alpha_i\pi - \sin\alpha_i\pi) + \pi(1-\alpha_i) + \sin\alpha_i\pi\right] \qquad (5.51)$$

(b) in the case of surface PMs (Fig. 5.10b)

$$B_{ad1} = \frac{4}{\pi}\int_0^{0.5\pi}(B_{ad}\cos x)\cos x dx = B_{ad} \qquad (5.52)$$

$$B_{aq1} = \frac{4}{\pi}\int_0^{0.5\pi}(B_{ad}\sin x)\sin x dx = B_{aq} \qquad (5.53)$$

(c) in the case of surface PMs with mild steel pole shoes (Fig. 5.10c)

$$B_{ad1} = \frac{4}{\pi}\left[\int_0^{0.5\alpha_i\pi}(B_{ad}\cos x)\cos x dx + \int_{0.5\alpha_i\pi}^{0.5\pi}(c_g'B_{ad}\cos x)\cos x dx\right]$$

$$= \frac{1}{\pi}B_{ad}\left[\alpha_i\pi + \sin\alpha_i\pi + c_g'(\pi - \alpha_i\pi - \sin\alpha_i\pi)\right] \qquad (5.54)$$

$$B_{aq1} = \frac{4}{\pi} \left[\int_0^{0.5\alpha_i\pi} \frac{1}{c'_g} (B_{aq} \sin x) \sin x dx + \int_{0.5\alpha_i\pi}^{0.5\pi} (B_{aq} \sin x) \sin x dx \right]$$

$$= \frac{1}{\pi} B_{aq} \left[\frac{1}{c'_g} (\alpha_i \pi - \sin \alpha_i \pi) + \pi(1 - \alpha_i) + \sin \alpha_i \pi \right] \qquad (5.55)$$

(d) in the case of buried PMs (Fig. 5.10d)

$$B_{ad1} = \frac{4}{\pi} B_{ad} \int_0^{0.5\alpha_i\pi} \cos(\frac{1}{\alpha_i} x) \cos x dx$$

$$= \frac{2}{\pi} B_{ad} \left[\frac{\alpha_i \sin(1 + \alpha_i) x/\alpha_i}{1 + \alpha_i} + \frac{\alpha_i \sin(1 - \alpha_i) x/\alpha_i}{1 - \alpha_i} \right]_0^{0.5\alpha_i\pi}$$

$$= \frac{4}{\pi} B_{ad} \frac{\alpha_i}{1 - \alpha_i^2} \cos\left(\alpha_i \frac{\pi}{2} \right) \qquad (5.56)$$

$$B_{aq1} = \frac{4}{\pi} \int_0^{0.5\alpha_i\pi} (B_{aq} \sin x) \sin x dx$$

$$= \frac{2}{\pi} B_{aq} \int_0^{0.5\alpha_i\pi} (1 - \cos x) dx = \frac{1}{\pi} B_{aq} (\alpha_i \pi - \sin \alpha_i \pi) \qquad (5.57)$$

For inset-type PMs the coefficient c_g expresses an increase in the d-axis armature magnetic flux density due to a decrease in the air gap (Fig. 5.10a) from $g + h$ to g, where h is the depth of the slot for the PM. Since the MVD across $g + h$ is equal to the sum of the MVDs across the air gap g and ferromagnetic tooth height h, the following equality can be written: $B_{ad}c_g g/\mu_0 + B_{Fe}h/(\mu_0\mu_r) = B_{ad}(g + h)/\mu_0$. Because $B_{Fe}h/(\mu_0\mu_r) << B_{ad}(g + h)/\mu_0$, the coefficient of increase in the magnetic flux density due to a decrease in the air gap is $c_g \approx 1 + h/g$. Of course, when $h = 0$ then $B_{ad1} = B_{ad}$, $B_{aq1} = B_{aq}$, and $k_{fd} = k_{fq} = 1$. It means that the machine behaves as a cylindrical rotor machine. Similarly, for the surface magnet rotor $k_{fd} = k_{fq} = 1$ (Fig. 5.10b) since the relative magnetic permeability of rare-earth PMs $\mu_r \approx 1$. The coefficient $c'_g \approx 1 - d_p/g_q$ for surface PMs with mild steel pole shoes (Fig. 5.10c), where d_p is the thickness of the mild steel pole shoe and g_q is the air gap in the q-axis, is evaluated in the same way as the coefficient c_g for inset-type PMs. For the buried magnet rotor the d-axis armature flux density changes as $\cos(x/\alpha_i)$ and the q-axis armature flux density changes as $\sin x$ (Fig. 5.10d).

The coefficients k_{fd} and k_{fq} for different rotor configurations are given in Table 5.2. The last row shows k_{fd} and k_{fq} for salient pole synchronous motors with electromagnetic excitation [185].

5.6.2 FEM

One of the two methods described in Chapter 3, Section 3.12 can be used in the calculation of synchronous reactances, i.e., (a) the number of flux linkages of the coil divided by the current in the coil and (b) the energy stored in the coil divided by one-half the current squared. The current/energy perturbation method, although more complicated, is recommended for calculating dynamic inductances (converter fed motors).

5.6.3 Experimental method

The experimental methods for measuring the synchronous reactances of large and medium power synchronous machines cannot normally be used in the case of small PM synchronous motors due to unacceptable simplifications and assumptions being made. An accurate method of measurement of synchronous reactances results from the phasor diagram shown in Fig. 5.4a. The following equations can be written:

$$V_1 \cos \delta = E_f + I_{ad} X_{ad} + I_a X_1 \sin(\phi - \delta) + I_a R_1 \cos(\phi - \delta)$$

$$= E_f + I_a X_{sd} \sin(\phi - \delta) + I_a R_1 \cos(\phi - \delta) \tag{5.58}$$

$$V_1 \sin \delta = I_a X_1 \cos(\phi - \delta) + I_{aq} X_{aq} - I_a R_1 \sin(\phi - \delta)$$

$$= I_a X_{sq} \cos(\phi - \delta) - I_a R_1 \sin(\phi - \delta) \tag{5.59}$$

where for an underexcited motor $\Psi = \phi - \delta$. The d-axis synchronous reactance can be found from eqn (5.58) and the q-axis synchronous reactance can be found from equation (5.59), i.e.,

$$X_{sd} = \frac{V_1 \cos \delta - E_f - I_a R_1 \cos(\phi - \delta)}{I_a \sin(\phi - \delta)} \tag{5.60}$$

$$X_{sq} = \frac{V_1 \sin \delta + I_a R_1 \sin(\phi - \delta)}{I_a \cos(\phi - \delta)} \tag{5.61}$$

It is easy to measure the input voltage V_1, phase armature current I_a, armature winding resistance R_1 per phase, and for an underexcited motor the angle $\phi = \Psi + \delta = \arccos[P_{in}/(m_1 V_1 I_a)]$. To measure the load angle δ it is recommended to use the machine set [233] shown in Fig. 5.11. The EMF E_f is assumed to be equal to the no-load EMF E_0, i.e., at $I_a \approx 0$.

The terminals of the two corresponding phases of the TSM and ASM are connected to a double-beam oscilloscope. The no-load EMFs E_{oTSM} of TSM and E_{oASM} of ASM operating as generators should be in phase. In this way

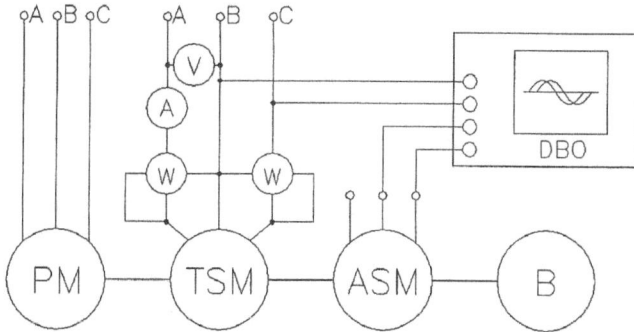

Fig. 5.11. Laboratory set for measuring the load angle δ of PM synchronous motors: TSM — tested synchronous motor, ASM — additional synchronous motor with the same number of poles as TSM, PM — prime mover (synchronous or d.c. motor), B — brake, DBO — double-beam oscilloscope.

the same positions of the TSM and ASM rotors with regard to the same phase windings can be found. When the TSM is connected to a three-phase power supply and the oscilloscope receives the signals V_{TSM} and E_{oASM}, the load angle δ can be measured, i.e., the phase angle between instantaneous values of V_{TSM} and E_{oASM}. Eqns (5.60) and (5.61) allow for the investigation of how the input voltage V_1 and load angle δ affect X_{sd} and X_{sq}. The accuracy of measuring the angle δ depends largely on the higher harmonic contents in V_{TSM} and E_{oASM}.

5.7 Rotor configurations

5.7.1 Merrill's rotor

The first successful construction of a PM rotor for small synchronous motors rated at high frequencies was patented by F.W. Merrill (US patent 2543639) [215]. It was a four-pole motor similar to the two-pole motor shown in Fig. 5.1a. The laminated external ring has deep narrow slots between each of the PM poles (Fig. 5.1a). Owing to cage rotor winding the motor is self-starting. The leakage flux produced by the PM can be adjusted by changing the width of the narrow slots. The Alnico PM is protected against demagnetization because the armature flux at starting and reversal goes through the laminated rings and narrow slots omitting the PM. The PM is mounted on the shaft with the aid of an aluminium or zinc alloy sleeve. The thickness of the laminated rotor ring is chosen such that its magnetic flux density is approximately 1.5 T when the rotor and stator are assembled. Magnetic flux density in the rotor teeth can be up to 2 T.

5.7.2 Interior-type PM motors

The interior-magnet rotor has radially magnetized and alternately poled magnets (Fig. 5.1b). Because the magnet pole area is smaller than the pole area at the rotor surface, the air gap flux density on open circuit is less than the flux density in the magnet [165]. The synchronous reactance in d-axis is smaller than that in q-axis since the q-axis magnetic flux can pass through the steel pole pieces without crossing the PMs. The magnet is very well protected against centrifugal forces. Such a design is recommended for high frequency high speed motors.

5.7.3 Surface PM motors

The surface magnet motor can have magnets magnetized radially (Fig. 5.1c) or sometimes circumferentially. An external high conductivity nonferromagnetic cylinder is sometimes used. It protects the PMs against the demagnetizing action of armature reaction and centrifugal forces, provides an asynchronous starting torque, and acts as a damper. If rare-earth PMs are used, the synchronous reactances in the d- and q-axis are practically the same (Table 5.2).

5.7.4 Inset-type PM rotor

In the inset-type motors (Fig. 5.1d) PMs are magnetized radially and embedded in shallow slots. The rotor magnetic circuit can be laminated or made of solid steel. In the first case a starting cage winding or external nonferromagnetic cylinder is required. The q-axis synchronous reactance is greater than that in the d-axis. In general, the EMF E_f induced by the PMs is lower than that in surface PM rotors.

5.7.5 Buried PM motors

The buried-magnet rotor has circumferentially magnetized PMs embedded in deep slots (Fig. 5.1e). Because of circumferential magnetization, the height h_M of the PM is in tangential direction, i.e., along the pole pitch. The effective pole arc coefficient α_i is dependent on the slot width. The synchronous reactance in q-axis is greater than that in d-axis. A starting asynchronous torque is produced with the aid of both a cage winding incorporated in slots in the rotor pole shoes (laminated core) or solid salient pole shoes made of mild steel. The width of the iron bridge between the inner ends of the neighboring magnets has to be carefully chosen. The application of a nonferromagnetic shaft is essential (Fig. 5.12). With a ferromagnetic shaft, a large portion of useless magnetic flux goes through the shaft [28] (Fig. 5.12a). A buried-magnet rotor should be equipped with a nonferromagnetic shaft (Fig. 5.12b) or a nonferromagnetic sleeve between the ferromagnetic shaft and rotor core should be used.

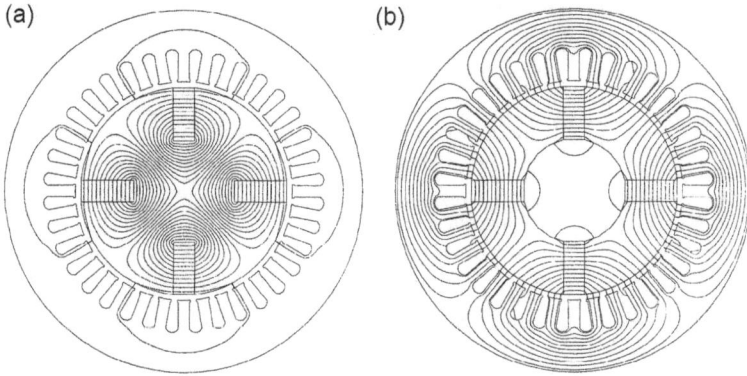

Fig. 5.12. Magnetic flux distribution in the cross section of a buried-magnet synchronous motor: (a) improperly designed rotor with ferromagnetic shaft, (b) rotor with nonferromagnetic shaft.

Table 5.3. Comparison between PM synchronous motors with surface and buried magnets

Surface magnets	Buried magnets
Air gap magnetic flux density is smaller than B_r	Air gap magnetic flux density can be greater than B_r (with more than four poles)
Simple motor construction	Relatively complicated motor construction (a nonferromagnetic shaft is common)
Small armature reaction flux	Higher armature reaction flux, consequently more expensive converter
Permanent magnets not protected against armature fields	Permanent magnets protected against armature fields
Eddy-current losses in permanent magnets (when their conductivity is greater than zero)	No eddy-current losses in permanent magnets
Damper in form of conducting sleeve (if necessary)	Damper in form of cage winding

Table 5.4. Comparison between PM synchronous and induction motors

Quantity	Synchronous motor	Induction motor
Speed	Constant, independent of the load	As the load increases, the speed decreases
Power factor $\cos\phi$	Adjustable pf (controlled by solid state converter). Operation at pf = 1 is possible	Depends on the air gap pf $\approx$ 0.8...0.9 at rated load pf $\approx$ 0.1 at no load
Nonferromagnetic air gap	From a fraction of mm to a few milimeters	As small as possible
Torque-voltage characteristic	Torque directly proportional to the input voltage	Torque directly proportional to the input voltage squared
Cost	More expensive than induction motor	Cost effective motor

A brief comparison between surface and buried magnet synchronous motors is given in Table 5.3. An alternative construction is a rotor with asymmetrically distributed buried magnets (Fig. 5.1e) developed by Siemens (German patent 1173178).

5.8 Comparison between synchronous and induction motors

PM synchronous motors, as compared with their induction counterparts, do not have rotor winding losses and require simple line commutated inverters which are more efficient than forced commutated inverters. Table 5.4 contains a comparison of the speed, power factor $\cos\phi$, air gap, torque-voltage characteristics, and price of synchronous and induction motors. A larger air gap in synchronous motors makes them more reliable than induction motors. The increased air gap is required to minimize the effect of the armature reaction, to reduce the synchronous reactance (if necessary) and to improve the stability.

Table 5.5 compares 50-kW, 6000-rpm, 200-Hz PM synchronous and cage induction motor drives [13]. The total power losses of the PM synchronous motor drive are reduced by 43% as compared with the induction motor drive. Thus the efficiency has been increased from 90.1 to 94.1% by 4% (2 kW power saving) [13].

Table 5.5. Power losses and efficiency of PM synchronous and cage induction motor drives rated at 50 kW, 6000 rpm and 200 Hz

Losses	PM synchronous motor	Cage induction motor
Winding losses		
Stator winding	820 W	
		} 1198 W
Rotor winding	–	
Damper	90 W	–
Losses due to skin effect in the stator winding	30 W	
Losses due to skin effect in the rotor winding	–	} 710 W
Core losses	845 W	773 W
Higher harmonic losses		
Damper	425 W	–
Rotor surface	–	221 W
Flux pulsation	–	301 W
Rotational losses		
Bearing friction	295	
		} 580 W
Windage	70	
Total motor losses	2575 W	3783 W
Total inverter losses	537 W	1700 W
Total drive losses	3112	5483 W
Efficiency		
Motor	95.1%	93.0%
Electromechanical drive system	94.1%	90.1%

5.9 Sizing procedure and main dimensions

The volume of all PMs used in a motor

$$V_M = 2ph_M w_M l_M \tag{5.62}$$

depends on the quality of PM material (maximum energy). In eqn (5.62) $2p$ is the number of poles, and h_M, w_M, and l_M are the height, width and length of the PM, respectively. The output power of a PM synchronous motor is proportional to V_M. Using the operating diagram of a PM (Fig. 2.9), the maximum electromagnetic power developed by a PM synchronous motor can be expressed as follows [20, 233]:

$$P_{max} = \frac{\pi^2}{2} \frac{\xi}{k_f k_{ad}(1+\epsilon)} f B_r H_c V_M \tag{5.63}$$

where $k_f = 0.7 \ldots 1.3$ is the form factor of the rotor excitation flux, k_{ad} is the d-axis armature reaction factor, $\epsilon = E_f/V_1 = 0.60 \ldots 0.95$ for underexcited motors (Fig. 5.11), E_f is the EMF induced by the rotor excitation flux at no-load, V_1 is the input voltage, f is the input frequency, B_r is the remanent magnetic flux density, and H_c is the coercive force. The *coefficient of utilization of the PM* in a synchronous motor is defined as

$$\xi = \frac{E_f I_{aK}}{E_r I_{ac}} = 0.3 \ldots 0.7 \tag{5.64}$$

where I_{aK} is the current corresponding to the MMF F_K for the point K (Fig. 2.9) which determines the beginning of the recoil line. The point K is the intersection point of the demagnetization curve $\Phi = f(F)$ and the straight line $G_{ext} = \Phi_K/F_K$ or $G_t = \Phi_K/(F_K - F'_{admax})$. In general, the current I_{aK} is the highest possible armature current which takes place at reversal, i.e., $I_{aK} = I_{arev}$ [119]. The EMF E_r corresponds to B_r and the armature current I_{ac} corresponds to H_c.

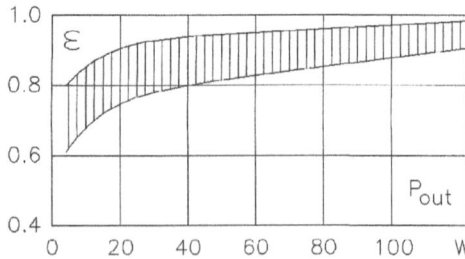

Fig. 5.13. Coefficient $\epsilon = E_f/V_1$ as a function of P_{out} for small a.c. PM motors.

With the aid of the overload capacity factor $k_{ocf} = P_{max}/P_{out}$ where P_{out} is the output rated power ($P_{max} \approx 2P_{out}$), the volume of PMs is

$$V_M = c_V \frac{P_{out}}{f B_r H_c} \tag{5.65}$$

where

$$c_V = \frac{2 k_{ocf} k_{fd}(1+\epsilon)}{\pi^2 \xi} = 0.54 \ldots 3.1 \tag{5.66}$$

The inner stator diameter D_{1in} can be estimated on the basis of the *ouput coefficient* for a.c. machines. The apparent electromagnetic power crossing the air gap is

$$S_{elm} = m_1 E_f I_a = \pi \sqrt{2} f N_1 k_{w1} \Phi_f \frac{m_1 \pi D_{1in} A_m}{2 m_1 \sqrt{2} N_1}$$

$$= \frac{\pi^2}{2}(n_s p) k_{w1} \frac{L_i D_{1in}^2}{p} B_{mg} A_m = 0.5\pi^2 k_{w1} D_{1in}^2 L_i n_s B_{mg} A_m \tag{5.67}$$

where $n_s = f/p$ is the synchronous speed, k_{w1} is the stator winding factor according to eqn (A.15), A_m is the peak value of the stator line current density according to eqn (5.13), and B_{mg} is the peak value of the air gap magnetic flux density approximately equal to the peak value of its first harmonic B_{mg1} according to eqn (5.6). The amplitude of the *stator line current density* A_m ranges from $10,000$ A/m for small motors to $55,000$ A/m for medium-power motors [225].

The electromagnetic torque

$$T_d = \frac{S_{elm} \cos \psi}{2 \pi n_s} = \frac{\pi}{4} k_{w1} D_{1in}^2 L_i B_{mg} A_m \cos \psi \tag{5.68}$$

It has been assumed that $B_{mg} \approx B_{mg1}$ where B_{mg1} is according to eqns (5.2) and (5.23). Since

$$P_{out} = P_{in}\eta = m_1 V_1 I_a \eta \cos \phi = \frac{1}{\epsilon} S_{elm} \eta \cos \phi \tag{5.69}$$

the output coefficient is

$$\sigma_p = \frac{P_{out}\epsilon}{D_{1in}^2 L_i n_s} = 0.5\pi^2 k_{w1} A_m B_{mg1} \eta \cos \phi \tag{5.70}$$

The *air gap magnetic flux density*

$$B_{mg} = \frac{\Phi_f}{\alpha_i \tau L_i} \tag{5.71}$$

for the sizing procedure of NdFeB PM synchronous motors can initially be estimated as $B_{mg} \approx (0.6\ldots0.8)B_r$ and for ferrites as $\approx (0.4\ldots0.7)B_r$. The magnetic flux density can also be approximately estimated on the basis of eqn (2.14).

There is a free choice in the effective length L_i of the armature stack, i.e., the ratio L_i/D_{1in} depends on the motor application.

The *air gap (mechanical clearance)* between the stator core and rotor poles or pole shoes is advised to be 0.3 to 1.0 mm for small PM synchronous motors. The smaller the air gap the lower the starting current. On the other hand, the effect of armature reaction and cogging (detent) torque increases as the air gap decreases.

The MMF F'_{ad} is the d-axis armature reaction MMF acting directly on the PM, i.e.,

$$F'_{ad} = F_{excd}\frac{1}{\sigma_{lM}} = \frac{m_1\sqrt{2}}{\pi}\frac{N_1 k_{w1}}{p}k_{ad}I_{ad}\frac{1}{\sigma_{lM}} \tag{5.72}$$

where F_{excd} is the d-axis armature reaction MMF referred to the field excitation system according to eqn (5.29), σ_{lM} is the coefficient of the PM leakage flux according to eqns (2.10) and (2.55), N_1 is the number of armature turns per phase, and I_{ad} is the d-axis armature current according to eqn (5.37) or (5.40).

5.10 Performance calculation

In the case of induction motors the steady-state characteristics, i.e., output power P_{out}, input power P_{in}, stator current I_1, shaft torque T_{sh}, efficiency η, and power factor $\cos\phi$ are calculated as functions of slip s. The same can be done for synchronous motors replacing the slip s by the load angle δ between the input voltage V and EMF E_f [152] provided that the synchronous reactances X_{sd}, X_{sq} and the armature resistance R_1 are known. Equations for the armature currents (5.40), (5.41), and (5.43), input power (5.44), electromagnetic power (5.45), and developed torque (5.46) are used. Calculations of the rotational, armature core, and stray losses (Appendix B), shaft torque, power factor, and efficiency are similar to those for induction motors. Characteristics plotted against the load angle δ can then be presented in more convenient forms, e.g., as functions of the angle Ψ between the current I_a and EMF E_f.

Another approach is to perform calculations for $I_{ad} = 0$ ($\Psi = 0$) and rated voltage and frequency to obtain rated current, torque, output power, efficiency and power factor. Then, a variable load can be simulated by changing the current I_a, say, from zero to $1.5I_a$ to find load characteristics as functions of current or torque at $n_s = const$ and $\Psi = 0$. Characteristics for $\Psi \neq 0$ can be obtained in a similar way.

5.11 Dynamic model of a PM motor

Control algorithms of sinusoidally excited synchronous motors frequently use the d–q linear model of electrical machines. The d–q dynamic model is expressed in a *rotating reference frame* that moves at synchronous speed ω. The time varying parameters are eliminated and all variables are expressed in *orthogonal* or *mutually decoupled* d and q axes.

A synchronous machine is described by the following set of general equations:

$$v_{1d} = R_1 i_{ad} + \frac{d\psi_d}{dt} - \omega\psi_q \tag{5.73}$$

$$v_{1q} = R_1 i_{aq} + \frac{d\psi_q}{dt} + \omega\psi_d \tag{5.74}$$

$$v_f = R_f I_f + \frac{d\psi_f}{dt} \tag{5.75}$$

$$0 = R_D i_D + \frac{d\psi_D}{dt} \tag{5.76}$$

$$0 = R_Q i_Q + \frac{d\psi_Q}{dt} \tag{5.77}$$

The linkage fluxes in the above equations are defined as

$$\psi_d = (L_{ad} + L_1)i_{ad} + L_{ad}i_D + \psi_f = L_{sd}i_{ad} + L_{ad}i_D + \psi_f \tag{5.78}$$

$$\psi_q = (L_{aq} + L_1)i_{aq} + L_{aq}i_Q = L_{sq}i_{aq} + L_{aq}i_Q \tag{5.79}$$

$$\psi_f = L_{fd}I_f \tag{5.80}$$

$$\psi_D = L_{ad}i_{ad} + (L_{ad} + L_D)i_D + \psi_f \tag{5.81}$$

$$\psi_Q = L_{aq}i_{aq} + (L_{aq} + L_Q)i_Q \tag{5.82}$$

where v_{1d} and v_{1q} are d- and q-axis components of terminal voltage, ψ_f is the maximum flux linkage per phase produced by the excitation system, R_1 is the armature winding resistance, L_{ad}, L_{aq} are d- and q-axis components of the armature self-inductance, $\omega = 2\pi f$ is the angular frequency of the armature current, i_{ad}, i_{aq} are d- and q-axes components of the armature current, i_D, i_Q are d- and q-axes components of the damper current. The field winding resistance which exists only in the case of electromagnetic excitation is R_f, the field excitation current is I_f and the excitation linkage flux is ψ_f. The damper resistance and inductance in the d axis is R_D and L_D, respectively. The damper resistance and inductance in the q axis are R_Q and L_Q, respectively. The resultant armature inductances are

$$L_{sd} = L_{ad} + L_1, \qquad L_{sq} = L_{aq} + L_1 \qquad (5.83)$$

where L_{ad} and L_{aq} are self-inductances in d and q axis, respectively, and L_1 is the leakage inductance of the armature winding per phase. In a three-phase machine $L_{ad} = (3/2)L'_{ad}$ and $L_{aq} = (3/2)L'_{aq}$ where L'_{ad} and L'_{aq} are self-inductances of a single phase machine.

The excitation linkage flux $\psi_f = L_{fd}I_f$ where L_{fd} is the maximum value of the mutual inductance between the armature and field winding. In the case of a PM excitation, the fictitious current is $I_f = H_c h_M$.

For machines with no damper winding $i_D = i_Q = 0$ and the voltage equations in the d and q-axis are

$$v_{1d} = R_1 i_{ad} + \frac{d\psi_d}{dt} - \omega\psi_q = \left(R_1 + \frac{dL_{sd}}{dt} \right) i_{ad} - \omega L_{sq} i_{aq} \qquad (5.84)$$

$$v_{1q} = R_1 i_{aq} + \frac{d\psi_q}{dt} + \omega\psi_d = \left(R_1 + \frac{dL_{sq}}{dt} \right) i_{aq} + \omega L_{sd} i_{ad} + \omega\psi_f \qquad (5.85)$$

The matrix form of voltage equations in terms of inductances L_{sd} and L_{sq} is

$$\begin{bmatrix} v_{1d} \\ v_{1q} \end{bmatrix} = \begin{bmatrix} R_1 + \frac{d}{dt}L_{sd} & -\omega L_{sq} \\ \omega L_{sd} & R_1 + \frac{d}{dt}L_{sq} \end{bmatrix} \begin{bmatrix} i_{ad} \\ i_{aq} \end{bmatrix} + \begin{bmatrix} 0 \\ \omega\psi_f \end{bmatrix} \qquad (5.86)$$

For the steady state operation $(d/dt)L_{sd}i_{ad} = (d/dt)L_{sq}i_{aq} = 0$, $\mathbf{I}_a = I_{ad} + jI_{aq}$, $\mathbf{V}_1 = V_{1d} + jV_{1q}$, $i_{ad} = \sqrt{2}I_{ad}$, $i_{aq} = \sqrt{2}I_{aq}$, $v_{1d} = \sqrt{2}V_{1d}$, $v_{1q} = \sqrt{2}V_{1q}$, $E_f = \omega L_{fd}I_f/\sqrt{2} = \omega\psi_f/\sqrt{2}$ [105]. The quantities ωL_{sd} and ωL_{sq} are known as the $d-$ and q-axis synchronous reactances, respectively. Eqn (5.86) can be brought to the form (5.38).

The instantaneous power input to the three phase armature is

$$p_{in} = v_{1A}i_{aA} + v_{1B}i_{aB} + v_{1C}i_{aC} = \frac{3}{2}(v_{1d}i_{ad} + v_{1q}i_{aq}) \qquad (5.87)$$

The power balance equation is obtained from eqns (5.84) and (5.85), i.e.,

$$v_{1d}i_{ad} + v_{1q}i_{aq} = R_1 i_{ad}^2 + \frac{d\psi_d}{dt}i_{ad} + R_1 i_{aq}^2 + \frac{d\psi_q}{dt}i_{aq} + \omega(\psi_d i_{aq} - \psi_q i_{ad}) \qquad (5.88)$$

The last term $\omega(\psi_d i_{aq} - \psi_q i_{ad})$ accounts for the electromagnetic power of a single phase, two-pole synchronous machine. For a three phase machine

$$p_{elm} = \frac{3}{2}\omega(\psi_d i_{aq} - \psi_q i_{ad}) = \frac{3}{2}\omega[(L_{sd}i_{ad} + \psi_f)i_{aq} - L_{sq}i_{ad}i_{aq}]$$

$$= \frac{3}{2}\omega[\psi_f + (L_{sd} - L_{sq})i_{ad}]i_{aq} \tag{5.89}$$

The electromagnetic torque of a three phase motor with p pole pairs is

$$T_d = p\frac{p_{elm}}{\omega} = \frac{3}{2}p[\psi_f + (L_{sd} - L_{sq})i_{ad}]i_{aq} \quad \text{N} \tag{5.90}$$

Compare eqn (5.90) with eqn (5.18).

The relatioships between i_{ad}, i_{aq} and phase currents i_{aA}, i_{aB} and i_{aC} are

$$i_{ad} = \frac{2}{3}\left[i_{aA}\cos\omega t + i_{aB}\cos\left(\omega t - \frac{2\pi}{3}\right) + i_{aC}\cos\left(\omega t + \frac{2\pi}{3}\right)\right] \tag{5.91}$$

$$i_{aq} = -\frac{2}{3}\left[i_{aA}\sin\omega t + i_{aB}\sin\left(\omega t - \frac{2\pi}{3}\right) + i_{aC}\sin\left(\omega t + \frac{2\pi}{3}\right)\right] \tag{5.92}$$

The reverse relations, obtained by simultaneous solution of eqns (5.91) and (5.92) in conjuction with $i_{aA} + i_{aB} + i_{aC} = 0$, are

$$i_{aA} = i_{ad}\cos\omega t - i_{aq}\sin\omega t$$

$$i_{aB} = i_{ad}\cos\left(\omega t - \frac{2\pi}{3}\right) - i_{aq}\sin\left(\omega t - \frac{2\pi}{3}\right) \tag{5.93}$$

$$i_{aC} = i_{ad}\cos\left(\omega t + \frac{2\pi}{3}\right) - i_{aq}\sin\left(\omega t + \frac{2\pi}{3}\right)$$

5.12 Noise and vibration of electromagnetic origin

The acoustic noise and vibration of electromagnetic origin is caused by parasitic effects due to higher harmonics, eccentricity, phase unbalance and sometimes magnetostriction.

5.12.1 Radial forces

The space and time distribution of the stator and rotor MMFs of a polyphase electrical machine can be expressed by the following equations:

$$\mathcal{F}_1(\alpha, t) = \sum_{\nu=0}^{\infty}\mathcal{F}_{m\nu,n}\cos(\nu p\alpha \pm \omega_n t) \tag{5.94}$$

$$\mathcal{F}_2(\alpha, t) = \sum_{\mu=0}^{\infty} \mathcal{F}_{m\mu,n} \cos(\mu p\alpha \pm \omega_{\mu,n} t + \phi_{\mu,n}) \qquad (5.95)$$

where α is the angular distance from a given axis, p is the number of pole pairs, n is the number of higher time harmonics, e.g., generated by the inverter, ω_n is the angular frequency of the stator current for the nth time harmonic, $\omega_{\mu,n}$ is the angular frequency of the rotor space harmonic μ for given n, $\phi_{\mu,n}$ is the angle between vectors of the stator ν and rotor μ space harmonics for given n, and $\mathcal{F}_{m\nu,n}$ and $\mathcal{F}_{m\mu,n}$ are the peak values of the νth and μth harmonics for given n of the stator and rotor MMFs, respectively. The product $p\alpha = \pi x/\tau$ where τ is the pole pitch and x is the linear distance from a given axis. For symmetrical polyphase stator windings and integral number of slots per pole per phase

$$\nu = 2m_1 l \pm 1 \qquad (5.96)$$

where m_1 is the number of the stator phases and $l = 0, 1, 2, \ldots$. For rotors of synchronous machines

$$\mu = 2l - 1 \qquad (5.97)$$

where $l = 1, 2, 3, \ldots$. The instantaneous value of the normal component of the magnetic flux density in the air gap at a point α can be calculated as

$$b(\alpha, t) = [\mathcal{F}_1(\alpha, t) + \mathcal{F}_2(\alpha, t)]G(\alpha, t)$$

$$= b_1(\alpha, t) + b_2(\alpha, t) \quad \text{T} \qquad (5.98)$$

where

- for the stator

$$b_1(\alpha, t) = \sum_{\nu=0}^{\infty} B_{m\nu,n} \cos(\nu p\alpha \pm \omega_n t) \qquad (5.99)$$

- for the rotor

$$b_2(\alpha, t) = \sum_{\mu=0}^{\infty} B_{m\mu,n} \cos(\mu p\alpha \pm \omega_{\mu,n} t + \phi_{\mu,n}) \qquad (5.100)$$

If $b_2(\alpha, t)$ is estimated on the basis of the known value of the air gap magnetic flux density, the effect of the stator slot openings must be taken into account.

The air gap relative permeance variation can be expressed with the aid of a Fourier series as an even function

$$G(\alpha) = \frac{A_0}{2} + \sum_{k=1,2,3,\ldots}^{\infty} A_k \cos(k\alpha) \qquad \text{H/m}^2 \qquad (5.101)$$

where k is the number of harmonics which replace the air gap variation bounded by the stator and rotor active surfaces.

The first constant term corresponds to the relative permeance G_0 of physical air gap g' increased by Carter's coefficient k_C. For surface magnets $g' = gk_C + h_M/\mu_{rrec}$ and for buried magnets totally enclosed by laminated or solid steel core $g' = gk_C$. Thus,

$$\frac{A_0}{2} = G_0 = \frac{\mu_0}{g'} \qquad \text{H/m}^2 \qquad (5.102)$$

The higher harmonics coefficients are

$$A_k = \frac{2}{\pi} \int_0^\pi G(\alpha)\cos(k\alpha)d\alpha \qquad (5.103)$$

The air gap permeance variation (5.101) is frequently given in the following form [76, 86, 144]

$$G(\alpha) = \frac{\mu_0}{g'} - 2\mu_0 \frac{\gamma_1}{t_1} \sum_{k=1,2,3,...}^\infty k_{ok}^2 \cos(k s_1 \alpha) \qquad (5.104)$$

where the slot opening factor

$$k_{ok} = \frac{\sin\left[k\rho\pi b_{14}/(2t_1)\right]}{k\rho\pi b_{14}/(2t_1)} \qquad (5.105)$$

relative slot opening

$$\kappa = \frac{b_{14}}{g} \qquad (5.106)$$

auxiliary functions [76, 86, 144]

$$\rho = \frac{\kappa}{5+\kappa} \frac{2\sqrt{1+\kappa^2}}{\sqrt{1+\kappa^2}-1} \qquad (5.107)$$

$$\gamma_1 = \frac{4}{\pi}\left[0.5\kappa\arctan(0.5\kappa) - \ln\sqrt{1+(0.5\kappa)^2}\right] \qquad (5.108)$$

and Carter's coefficient k_C is according to eqn (A.27), s_1 is the number of the stator slots, t_1 is the slot pitch and b_{14} is the slot opening. In terms of linear circumferential distance x the angular displacement is

$$\alpha = \frac{2\pi}{s_1 t_1} x \qquad (5.109)$$

Assuming the *relative eccentricity*

$$\epsilon = \frac{e}{g} \qquad (5.110)$$

where e is the rotor (shaft) eccentricity and g is an ideal uniform air gap for $e = 0$, the variation of the air gap around the magnetic circuit periphery is

$$g(\alpha, t) \approx g[1 + \epsilon \cos(\alpha - \omega t)] \qquad (5.111)$$

The *static eccentricity* is when the rotor rotates about its geometric center. The *dynamic eccentricity* is when the rotor rotates about the stator geometric center.

According to Maxwell stress tensor, the magnitude of the *radial force* per unit area at any point of the air gap for $n = 1$ is

$$p_r = \frac{b^2(\alpha, t)}{2\mu_0} = \frac{1}{2\mu_0}[\mathcal{F}_1(\alpha, t) + \mathcal{F}_2(\alpha, t)]^2 G^2(\alpha)$$

$$= \frac{1}{2\mu_0}[\mathcal{F}_1^2(\alpha, t) + 2\mathcal{F}_1(\alpha, t)\mathcal{F}_2(\alpha, t) + \mathcal{F}_2^2(\alpha, t)]^2 G^2(\alpha)$$

$$= \frac{1}{2\mu_0}\{[\sum_{\nu=0}^{\infty}\mathcal{F}_{m\nu}\cos(\nu p\alpha \pm \omega t)]^2$$

$$+ 2\sum_{\nu=0}^{\infty}\mathcal{F}_{m\nu}\cos(\nu p\alpha \pm \omega t)\sum_{\mu=0}^{\infty}\mathcal{F}_{m\mu}\cos(\mu p\alpha \pm \omega_\mu t + \phi_\mu)$$

$$+ [\sum_{\mu=0}^{\infty}\mathcal{F}_{m\mu}\cos(\mu p\alpha \pm \omega_\mu t + \phi_\mu)]^2\}[\frac{A_0}{2} + \sum_{k=1}^{\infty}A_k\cos(k\alpha)]^2 \quad \text{N/m}^2 \quad (5.112)$$

There are three groups of the infinite number of radial force waves [290]. The square terms (both for the stator and rotor) produce constant stresses and radial force waves with double the number of pole pairs and double the pulsation of the source wave magnetic flux density.

The distribution of the radial force density in a medium power PM brushless motor with buried magnets is shown in Fig. 5.14.

Only the interaction of the rotor and stator waves (second term in square brackets) produce low mode and high amplitude of force waves which are important from the acoustic point of view. These forces per unit area can be expressed in the following general form

$$p_r(\alpha, t) = P_r \cos(r\alpha - \omega_r t) \qquad (5.113)$$

where for the mixed products of the stator and rotor space harmonics $r = (\nu \pm \mu)p = 0, 1, 2, 3, \ldots$ is the order (circumferential mode number) of the force wave and $\omega_r = \omega_\nu \pm \omega_\mu$ is the angular frequency of the force of the rth order.

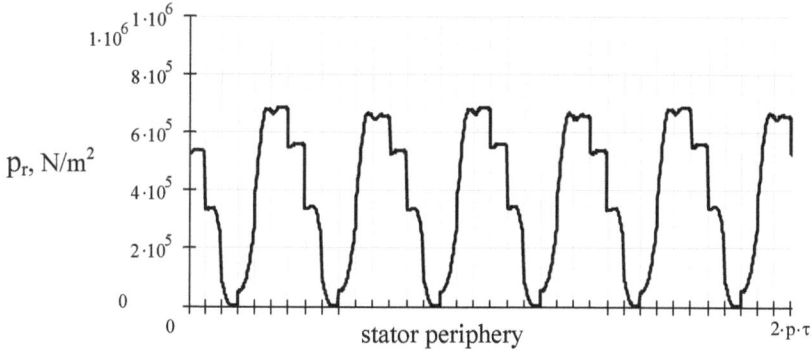

Fig. 5.14. Distribution of the radial force density around the stator periphery for $m_1 = 3$, $p = 4$, $s_1 = 36$ and $\tau = 82$ mm.

The radial forces circulate around the stator bore with the angular speed ω_r/r and frequency $f_r = \omega_r/(2\pi)$. For a small number of the stator pole pairs the radial forces may cause the stator to vibrate.

For d.c. PM brushless motors the full load armature reaction field is normally less than 20% of the open-circuit magnetic field. Thus, the effect of the first and second term in eqn (5.112) on the acoustic noise at no load is minimal [121].

The amplitude of the radial force pressure

$$P_r = \frac{B_{m\nu} B_{m\mu}}{2\mu_0} \quad \text{N/m}^2 \tag{5.114}$$

To obtain the amplitude of the radial force, the force pressure amplitude P_r should be multiplied by $\pi D_{1in} L_i$ where D_{1in} is the stator core inner diameter and L_i is the effective length of the stator core.

The largest deformation of the stator rings is in the case when the frequency f_r is close to the natural mechanical frequency of the stator. The most important from an airborne noise point of view are low mode numbers, i.e., $r = 0, 1, 2, 3$ and 4.

5.12.2 Deformation of the stator core

Vibration mode $r = 0$.

For $r = 0$ (pulsating vibration mode) the radial force density (pressure)

$$p_0 = P_0 \cos \omega_0 t \tag{5.115}$$

is distributed uniformly around the stator periphery and changes periodically with time. It causes a radial vibration of the stator core and can be compared to a cylindrical vessel with a variable internal overpressure [144]. Eqn (5.115) describes an interference of two magnetic flux density waves of equal lengths (the same number of pole pairs) and different velocity (frequency).

Vibration mode $r = 1$.
   For $r = 1$ the radial pressure

$$p_1 = P_1 \cos(\alpha - \omega_1 t) \tag{5.116}$$

produces a single-sided magnetic pull on the rotor. The angular velocity of the pull rotation is ω_1. A heavy vibration of the machine occurs at resonance. Physically, eqn (5.116) describes an interference of two magnetic flux density waves for which the number of pole pairs differ by one.

Vibration modes $r = 2, 3, 4$.
   For $r = 2, 3, 4$ deflections of the stator core will occur. Fig. 5.15 shows space distribution of forces producing vibrations of the order of $r = 0, 1, 2, 3, 4$.

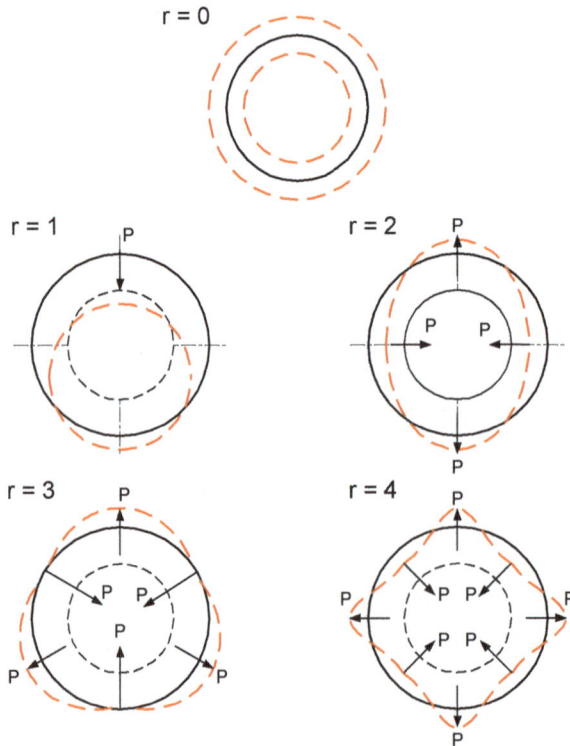

Fig. 5.15. Deformation of the core caused by the space distribution of radial forces.

5.12.3 Natural frequencies of the stator

The natural frequency of the stator core (yoke) of the rth order can be expressed as [290]

$$f_r = \frac{1}{2\pi}\sqrt{\frac{K_r}{M}} \qquad (5.117)$$

where K_r and M are the lumped stiffness and lumped mass of the stator core, respectively. The lumped stiffness is proportional to the Young modulus (elasticity modulus) E_c of the core, radial thickness of the core (yoke) h_c and length L_i of the stator core, and inversely proportional to the core average diameter D_c, i.e.,

$$K_r \propto \frac{E_c h_c L_i}{D_c} \quad \text{N/m} \qquad (5.118)$$

Analytical equations for calculation of K_r and M are given, e.g., in [121, 290, 311]. To take into account the frame and winding, the natural frequency of the stator system can approximately be found as [121]

$$f_r \approx \frac{1}{2\pi}\sqrt{\frac{K_r + K_r^{(f)} + K_r^{(w)}}{M + M_f + M_w}} \qquad (5.119)$$

where $K_r^{(f)}$ is the lumped stiffnes of the frame, $K_r^{(w)}$ is the lumped stiffens of the tooth-slot zone including the winding and insulation, M_f is the mass of the frame and M_w is mass of the winding with insulation and encapsulation.

The modulus of elasticity is 0.21×10^{12} Pa (N/m^2) for steel, $\leq 0.20 \times 10^{12}$ Pa for laminations, 0.11 to 0.13×10^{12} Pa for copper, 0.003×10^{12} Pa for polymer insulation and 0.0094×10^{12} Pa for copper–polymer insulation structure.

5.13 Applications

5.13.1 Open loop control

Motors with laminated rotors and buried PMs can be furnished with an additional cage winding located in the rotor pole shoes. Solid steel pole shoes in surface and inset type PM motors behave in a similar manner as the cage winding. This damper winding also adds a component of asynchronous torque production so that the PM motor can be operated stably from an inverter without position sensors. As a result, a simple *constant voltage–to–frequency* control (Fig. 5.16) using a pre-programmed sinusoidal voltage PWM algorithm can provide speed control for applications such as pumps and fans which do not require fast dynamic response [166]. Thus PM motors can replace induction motors in some variable-speed drive applications to improve the drive efficiency with minimal changes to the control electronics.

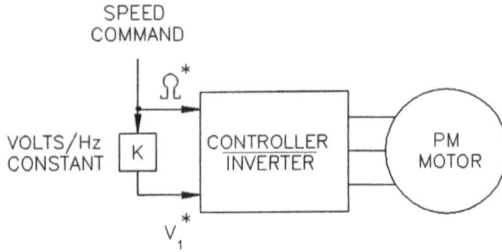

Fig. 5.16. Simplified block diagram of open loop *voltage–to–frequency* control of an interior or buried PM motor.

5.13.2 High-performance closed-loop control

To achieve high performance motion control with a sinusoidal PM motor, a rotor position sensor is typically required. Depending on the specific sinusoidal PM motor drive performance, an absolute encoder or resolver providing an equivalent digital resolution of 6 bits per electrical cycle (5.6° elec.) or higher is typically required. The second condition for achieving high-performance motion control is high-quality phase current control.

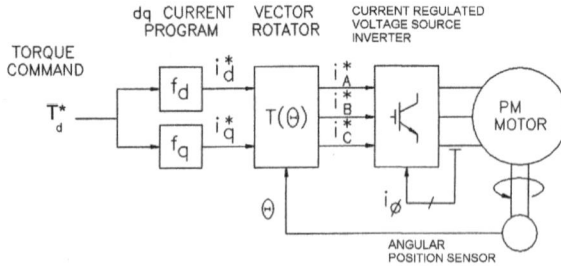

Fig. 5.17. Block diagram of high-performance torque control scheme for sinusoidal PM motor using vector control concept.

One of the possible approaches is the *vector control* shown in Fig. 5.17 [166]. The incoming torque command T_d^* is mapped into commands for i_{ad}^* and i_{aq}^* current components according to eqn (5.90) where $\psi_f = N_1 \Phi_f$ is the PM flux linkage amplitude, L_{sd} and L_{sq} are the synchronous inductances under conditions of alignment with the rotor d and q axes, respectively. Compare eqn (5.90) with eqns (5.45) and (5.46). The current commands in the rotor d—q reference frame (d.c. quantities for a constant torque command) are then transformed into the instantaneous sinusoidal current commands for the individual stator phases i_{aA}^*, i_{aB}^*, and i_{aC}^* using the rotor angle feedback and the basic inverse vector rotation equations [165]. Current regulators for each of the three stator current phases then operate to excite the phase windings with the desired current amplitudes.

The most common means of mapping the torque command T_d^* into values for i_{ad}^* and i_{aq}^* is to set a constraint of maximum *torque–to–current* operation which is nearly equivalent to maximizing operating efficiency [166].

A DSP controlled PM synchronous motor drive with a current PI regulator has been discussed, e.g., in [310]. The space vector modulation method used for generating the PWM voltage is superior to the subharmonic method for the total harmonic distortion.

5.13.3 High-performance adaptive fuzzy control

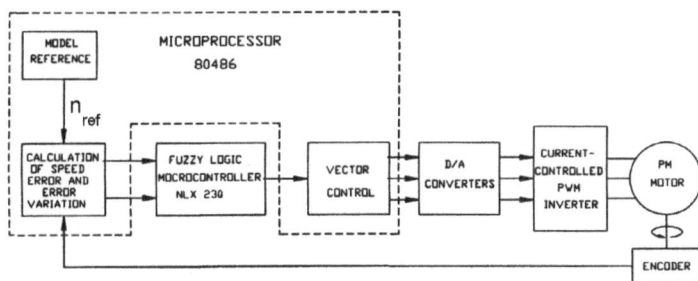

Fig. 5.18. Block diagram of the adaptive fuzzy controller for a PM synchronous motor drive.

Fig. 5.18 shows a block diagram of the experimental speed and position control system based on *fuzzy logic* approach, combined with a simple and effective adaptive algorithm [55]. The PM synchronous motor is fed from a current controlled inverter. The control can be based, e.g., on a general-purpose microprocessor Intel 80486 and a fuzzy logic microcontroller NLX 230 from American Neuralogics. The microprocessor implements a model reference adaptive control scheme in which the control parameters are updated by the fuzzy logic microcontroller. During the drive operation both the load torque and moment of inertia tend to change. In order to compensate for the variations of shaft torque and moment of inertia, fuzzy logic approach has been used.

Numerical examples

Numerical example 5.1

Find the main dimensions and volume of PMs of a three-phase brushless synchronous motor rated at: output power $P_{out} = 1.5$ kW, input frequency $f = 50$ Hz, synchronous speed $n_s = 1500$ rpm, line voltage $V_{1L} = 295$ V. The SmCo PMs with $B_r \approx 1.0$ T and $H_c \approx 700,000$ A/m should be distributed at the rotor surface and furnished with mild steel pole shoes. These pole shoes can replace the cage winding and reduce the starting current. The product $\eta \cos \phi$ at rated load should be minimum 0.75. The motor has to be designed for continuous duty and inverter applications.

Estimate the number of armature turns per phase and cross section of the armature conductor.

<u>Solution</u>

The rated armature current

$$I_a = \frac{P_{out}}{3V_1 \eta \cos \phi} = \frac{1500}{3 \times (295/\sqrt{3}) \times 0.75} = 3.9 \text{ A}$$

The number of pole pairs

$$p = \frac{f}{n_s} = \frac{50}{25} = 2$$

since the synchronous speed in rev/s is 25. This is a four pole motor.

For the SmCo PMs it is recommended to assume the air gap magnetic flux density between 0.65 and 0.85 T, say, $0.75B_r$, i.e., $B_{mg} = 0.75$ T. The stator line current density for a four-pole 1.5-kW motor should be approximately $A_m = 30,500$ A/m (peak value). The winding coefficient for three-phase, four-pole, double layer windings can be assumed as $k_{w1} = 0.96$. The output coefficient according to eqn (5.70) is

$$\sigma_p = 0.5\pi^2 \times 0.96 \times 30,500 \times 0.75 \times 0.75 = 81,276 \ \frac{\text{VAs}}{\text{m}^3}$$

The no-load EMF to phase voltage ratio has been assumed $\epsilon = 0.83$. Thus the product

$$D_{1in}^2 L_i = \frac{P_{out}\epsilon}{\sigma_p n_s} = \frac{1500 \times 0.83}{81,276 \times 25} = 0.000613 \text{ m}^3$$

The product $D_{1in}^2 L_i = 0.000613 \text{ m}^3 = 613 \text{ cm}^3$ corresponds to the inner stator diameter $D_{1in} \approx 78$ mm and the effective stator length $L_i = 100$ mm. Since the motor is designed for variable speed drives (inverter applications), the main dimensions have been increased to $D_{1in} \approx 82.54$ mm and $L_i = 103$ mm. The pole pitch

$$\tau = \frac{\pi D_{1in}}{2p} = \frac{\pi 0.08254}{2p} = 0.0648 \text{ m}$$

Assuming $\alpha_i = 0.5$ the effective pole arc coefficient is $b_p = \alpha_i \tau = 0.5 \times 64.8 = 32.4$ mm. Since the surface magnets are furnished with mild steel pole shoes b_p is the width of the external pole shoe. The thickness of the mild steel pole shoe pole has been assumed as $d_p = 1$ mm. The air gap in the d-axis has been chosen small, i.e., $g = 0.3$ mm, similar as that in cage induction motors. The q-axis air gap can be much larger, say, $g_q = 5.4$ mm.

The form factor of the armature reaction in d-axis according to Table 5.2 is

$$k_{fd} = \frac{1}{\pi}\{0.5\pi + \sin(0.5\pi) + 0.8148[\pi - 0.5\pi - \sin(0.5\pi)]\} = 0.966$$

where $c_g' \approx 1 - d_p/g_q = 1 - 1/5.4 = 0.8148$.

The form factor of the excitation field according to eqn (5.23)

$$k_f = \frac{4}{\pi}\sin\frac{\alpha_i\pi}{2} = \frac{4}{\pi}\sin\frac{0.5\pi}{2} = 0.9$$

The armature reaction factor in d-axis according to eqn (5.28)

$$k_{ad} = \frac{k_{fd}}{k_f} = \frac{0.966}{0.9} = 1.074$$

The overload capacity factor has been assumed $k_{ocf} \approx 2$. The coefficient (5.70) of utilization of the PM can be estimated as $\xi = 0.55$. Thus

$$c_V = \frac{2k_{ocf}k_f k_{ad}(1+\epsilon)}{\pi^2\xi} = \frac{2 \times 2 \times 0.9 \times 1.074 \times (1+0.83)}{\pi^2 0.55} = 1.303$$

The volume of PMs is estimated on the basis of eqn (5.65)

$$V_M = c_V\frac{P_{out}}{fB_r H_c} = 1.303\frac{1500}{50 \times 1.0 \times 700,000} \approx 0.000056 \text{ m}^3 = 56 \text{ cm}^3$$

Three parallel 5-mm thick, 10-mm wide, and 100-mm long surface SmCo PMs per pole have been designed as shown in Fig. 5.19a. The volume of all PMs used in the motor is

$$V_M = 2p(3h_M w_M l_M) = 4(3 \times 0.005 \times 0.01 \times 0.1) = 0.00006 \text{ m}^3 = 60 \text{ cm}^3$$

The rotor has been built from solid carbon steel. Surface SmCo PMs have been fixed with the aid of a special glue. In addition, each PM pole has been furnished with mild-steel pole shoes [119] as shown in Fig. 5.19a. The magnetic flux lines in the cross section of the designed motor, as obtained from the FEM, are plotted in Figs 5.19b and 5.19c.

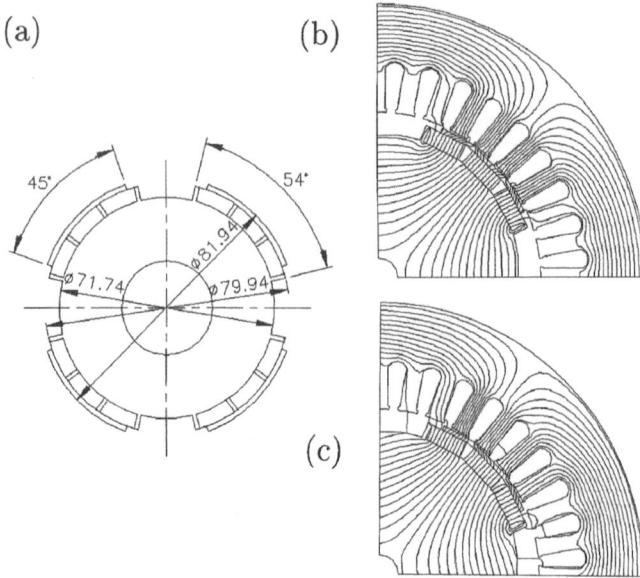

Fig. 5.19. Rotor of the designed motor: (a) dimensions, (b) magnetic flux distribution at no-load, (c) magnetic flux distribution at rated load.

The number of armature turns per phase can be roughly chosen on the basis of the assumed line current density and armature current as given by eqn (5.13), i.e.,

$$N_1 = \frac{A_m p \tau}{m_1 \sqrt{2} I_a} = \frac{30,500 \times 2 \times 0.0648}{3\sqrt{2} \times 3.9} \approx 240$$

A double-layer winding located in $s_1 = 36$ stator (armature) slots has been assumed. For 36 slots and $2p = 4$ the number of armature turns per phase should be $N_1 = 240$. Two parallel conductors with their diameter $d_a = 0.5$ mm have been chosen. The cross section area of the armature conductors is

$$s_a = 2 \times \frac{\pi d_a^2}{4} = 2 \times \frac{\pi 0.5^2}{4} = 0.3927 \text{ mm}^2$$

The current density in the armature winding

$$J_a = \frac{I_a}{s_a} = \frac{3.9}{0.3927} = 9.93 \; \frac{\text{A}}{\text{mm}^2}$$

This value can be accepted for continuous operation of small a.c. motors provided that the motor is equipped with a fan and the insulation class is minimum F.

Other parameters have been found after performing calculations of the electric and magnetic circuit. The motor specifications are given in Table 5.6.

Table 5.6. Data of designed PM synchronous motor

Quantity	Value
Input frequency, f	50 Hz
Input voltage (line-to-line)	295.0 V
Connection	Y
Inner diameter of the stator, D_{1in}	82.5 mm
Outer diameter of the stator, D_{1out}	136.0 mm
Air gap in d axis (clearance), g	0.3 mm
Air gap in q axis (clearance), g_q	5.4 mm
Effective length of the stator core, L_i	103.0 mm
Stacking factor for the stator core, k_i	0.96
Armature winding coil pitch, w_c	64.8 mm
Length of a single overhang, l_e	90.8 mm
Number of turns per phase, N_1	240
Number of parallel conductors	2
Number of stator slots, s_1	36
Stator wire conductivity at 20^0 C, σ	57×10^{-6} S/m
Diameter of stator conductor, d_a	0.5 mm
Width of the stator slot opening, b_{14}	2.2 mm
Height of permanent magnet, h_M	5.0 mm
Width of permanent magnet, w_M	3×10 mm
Length of permanent magnet, l_M	100.0 mm
Remanent magnetic flux density, B_r	1.0 T
Coercive force, H_c	700.0 kA/m
Width of pole shoe, b_p	32.4 mm
Thickness of pole shoe, d_p	1.0 mm

Numerical example 5.2

Find the armature reaction, leakage and synchronous reactances of the motor according to Example 5.1.

Solution

The reactances can be found quickly using the classical approach. The effective pole arc coefficient $\alpha_i = 0.5$ and coefficient $c'_g \approx 1 - d_p/g_q = 0.8148$ have been found in the previous Example 5.1. There have been designed $s_1 = 36$ semi-closed oval slots in the stator with the following dimensions: $h_{11} = 9.9$ mm, $h_{12} = 0.1$ mm, $h_{13} = 0.2$ mm, $h_{14} = 0.7$ mm, $b_{11} = 4.8$ mm, $b_{12} = 3.0$ mm, $b_{14} = 2.2$ mm. The height and the width of the stator slot opening are h_{14} and b_{14}, respectively.

The form factors k_{fd} and k_{fq} can be computed using equations given in Table 5.2, thus

$$k_{fd} = \frac{1}{\pi}[0.5\pi + sin(0.5\pi) + 0.8148(\pi - 0.5\pi - sin(0.5\pi))] = 0.966$$

$$k_{fq} = \frac{1}{\pi}\left[\frac{1}{0.8148}(0.5\pi - \sin(0.5\pi)) + \pi(1 - 0.5) + \sin(0.5\pi)\right] = 1.041$$

The coeffcient k_{fd} is lower than k_{fq}. There is higher magnetic flux in the q-axis than in the d-axis as the PMs and mild steel pole shoes have a low reluctance. The FEM modelling confirms this effect (see Fig. 5.20).

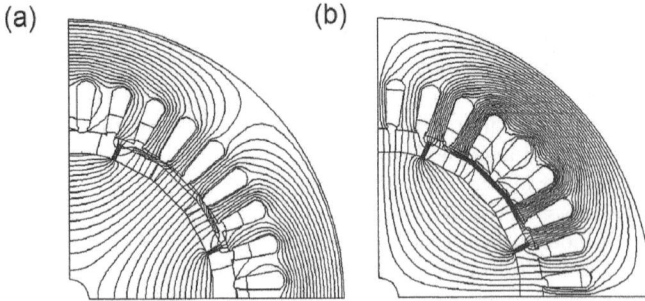

Fig. 5.20. Magnetic flux distribution in the d and q-axis: (a) d-axis flux, (b) q-axis flux.

Carter's coefficient for a resultant air gap in the d-axis $g_t \approx g_q - d_p = 5.4 - 1.0 = 4.4$ mm according to eqn (A.27) is

$$k_C = \frac{t_{s1}}{t_{s1} - \gamma_1 g_t} = \frac{7.2}{7.2 - 0.0394 \times 4.4} = 1.0258$$

where the slot pitch is

$$t_1 = \frac{\pi D_{1in}}{s_1} = \frac{\pi \times 82.54}{36} = 7.2 \text{ mm}$$

and

$$\gamma_1 = \frac{4}{\pi}\left[\frac{2.2}{2 \times 4.4}\arctan\frac{2.2}{2 \times 4.4} - \ln\sqrt{1 + \left(\frac{2.2}{2 \times 4.4}\right)^2}\right] = 0.0394$$

The d-axis armature reaction reactance according to eqn (5.31) is

$$X_{ad} = 4 \times 3 \times 0.4\pi \times 10^{-6}\frac{(240 \times 0.96)^2}{\pi \times 2}\frac{0.0648 \times 0.103}{1.0258 \times 1.0 \times 0.0044} \times 0.966 = 9.10 \ \Omega$$

In a similar way the q-axis synchronous reactance can be found — eqn (5.33):

$$X_{sq} = \frac{k_{fq}}{k_{fd}}X_{ad} = \frac{0.966}{1.041}9.10 = 9.81 \ \Omega$$

It has been assumed that the saturation factor of magnetic circuit $k_{sat} = 1.0$, i.e., the magnetic saturation has been neglected.

The stator armature leakage reactance is given by eqn (A.30), i.e.,

$$X_1 = 4\pi \times 50 \times 0.4\pi \times 10^{-6} \frac{0.103 \times 240^2}{2 \times 3}$$

$$\times (1.9368 + 0.4885 + 0.1665 + 0.1229) = 2.12 \ \Omega$$

where the coefficients of permeances for leakage fluxes are

- slot leakage flux according to eqn (aa-slotoval)

$$\lambda_{1s} = 0.1424 + \frac{9.9}{3 \times 3.0} \times 0.972 + \frac{0.1}{3.0} 8$$

$$+0.5 \arcsin \sqrt{1 - \left(\frac{2.2}{3.0}\right)^2} + \frac{0.7}{0.2} = 1.938$$

$$k_t = 0.972 \qquad \text{for} \qquad t = \frac{b_{11}}{b_{12}} = \frac{4.8}{3.0} = 1.6$$

- end winding connection leakage flux according to eqn (A.19)

$$\lambda_{1e} \approx 0.34 \frac{q_1}{L_i} \left(l_{1e} - \frac{2}{\pi} w_c \right) = 0.34 \frac{3}{0.103} \left(0.0908 - \frac{2}{\pi} 0.0648 \right) = 0.4885$$

$$q_1 = \frac{s_1}{2pm_1} = \frac{36}{4 \times 3} = 3,$$

in which the length of one-sided end connection $l_{1e} = 0.0908$ m (winding layout), and the stator coil pitch $w_c = \tau = 0.0648$ m (full-pitch coils).

- differential leakage flux according to eqn (A.24)

$$\lambda_{1d} = \frac{m_1 q_1 \tau k_{w1}^2}{\pi^2 k_C k_{sat} g_t} \tau_{d1} = \frac{3 \times 3 \times 0.0648 \times 0.96^2}{\pi^2 \times 1.0258 \times 1.0 \times 0.0044} \times 0.0138 = 0.1665$$

in which the stator differential leakage factor for $q_1 = 3$ and $w_c/\tau = 1$ is $\tau_{d1} = 0.0138$ (Fig. A.3).

- tooth-top leakage flux according to eqn (A.29)

$$\lambda_{1t} = \frac{5g/b_{14}}{5 + 4g/b_{14}} = \frac{5 \times 0.3/2.2}{5 + 4 \times 0.3/2.2} = 0.1229$$

Synchronous reactances in the d and q axes according to eqns (5.15)

$$X_{sd} = 2.12 + 9.10 = 11.22 \ \Omega \qquad\qquad X_{sq} = 2.12 + 9.81 = 11.93 \ \Omega$$

Numerical example 5.3

The synchronous motor according to Example 5.1 has been redesigned. The surface magnet rotor with mild steel pole shoes has been replaced by a rotor with buried magnets symmetrically distributed (Figs 5.1e and 5.12). Four Nd-FeB PMs ($h_M = 8.1$ mm, $w_M = 20$ mm, $l_M = 100$ mm) have been embedded in four open slots machined in a rotor body made of mild carbon steel. To minimize the bottom leakage flux, the shaft has been made of nonferromagnetic steel. The motor specifications are listed in Table 5.7.

Table 5.7. Data of tested buried PM synchronous motor

Quantity	Value
Input frequency, f	50 Hz
Input voltage (line-to-line)	380.0 V
Connection	Y
Inner diameter of the stator, D_{1in}	82.5 mm
Outer diameter of the stator, D_{1out}	136.0 mm
Air gap in d axis (clearance), g	0.55 mm
Effective length of the stator core, L_i	100.3 mm
Stacking factor for the stator core, k_i	0.96
Armature winding coil pitch, w_c	64.8 mm
Length of a single overhang, l_e	90.8 mm
Number of turns per phase, N_1	240
Number of parallel wires	2
Number of stator slots, s_1	36
Diameter of stator conductor, d_1	0.5 mm
Width of the stator slot opening, b_{14}	2.2 mm
Height of permanent magnet, h_M	8.1 mm
Width of permanent magnet, w_M	20.0 mm
Length of permanent magnet, l_M	100.0 mm
Remanent magnetic flux density, B_r	1.05 T
Coercive force, H_c	764.0 kA/m

Find: (1) the magnetic flux distributions and normal components of the air gap magnetic flux density distribution excited by PMs and the armature winding; (2) synchronous reactances; and (3) the electromagnetic torque, output power, efficiency and power factor as functions of the load angle δ.

<u>Solution</u>

The magnetic flux distribution and normal components of the air gap magnetic flux density distributions have been obtained on the basis of the FEM (Fig. 5.21). The magnetic fluxes and air gap magnetic flux density waveforms for sinusoidal armature current have been separately plotted for the rotor excitation field (Fig. 5.21a), the *d*-axis field (Fig. 5.21b), and the *q*-axis field (Fig. 5.21c).

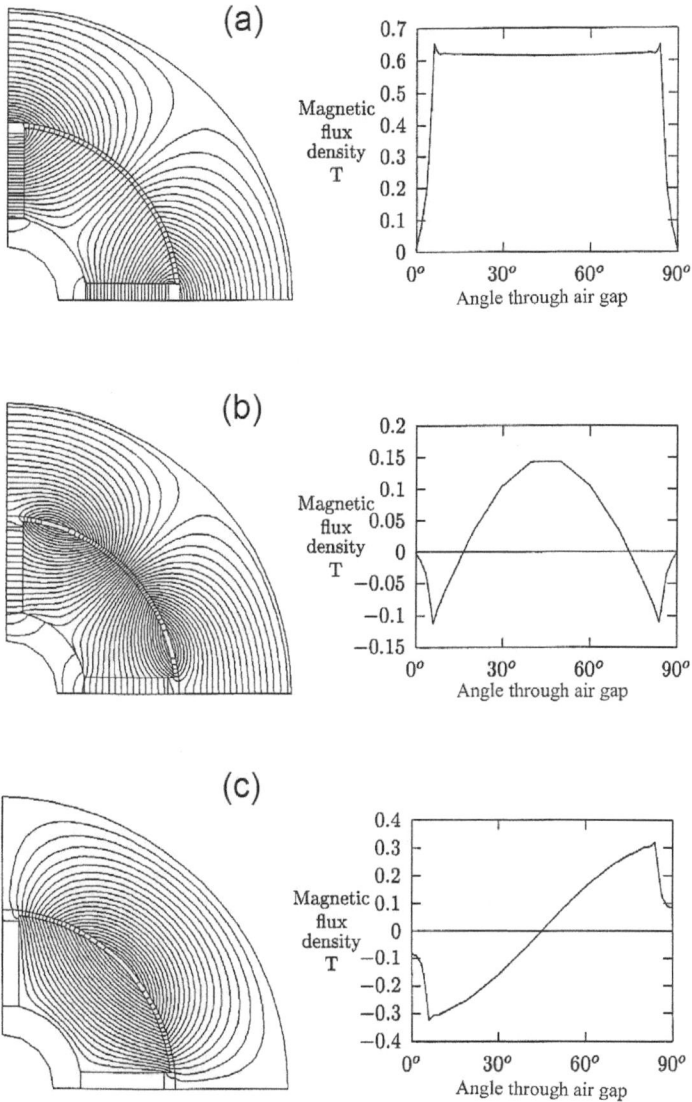

Fig. 5.21. Magnetic flux distribution and normal components of air gap magnetic flux density waveforms for sinusoidal armature current: (a) rotor field, (b) *d*-axis field, (c) *q*-axis field.

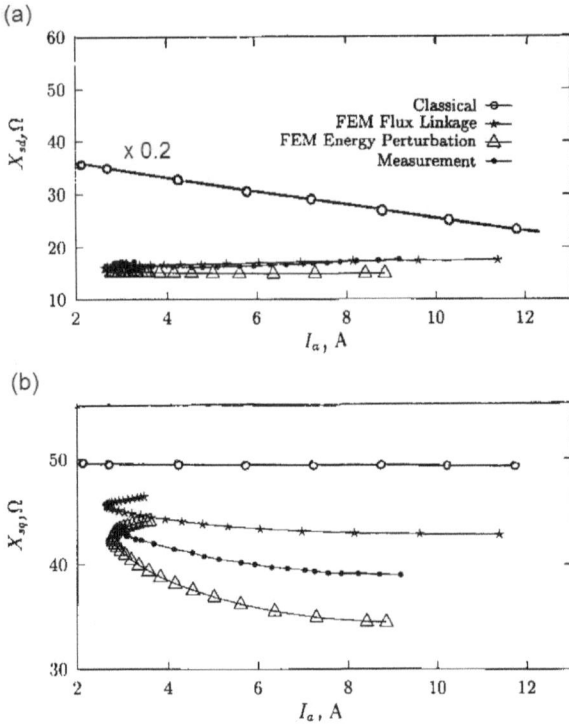

Fig. 5.22. Synchronous reactances for the buried-magnet motor (Table 5.7) obtained from analytical approach, FEM and measurements at 50 Hz and 380 V (line-to-line): (a) X_{sd}, (b) X_{sq}.

Synchronous reactances have been calculated using an analytical approach (classical method), the FEM based on the magnetic vector potential and flux linkage, the FEM based on the current/energy perturbation, and compared with experimental test results.

For $\tau = \pi D_{1in}/(2p) = \pi \times 82.5/4 = 64.8$ mm, the ratio of the pole-shoe arc b_p-to-pole pitch τ is

$$\alpha_i = \frac{b_p}{\tau} = \frac{\tau - h_M}{\tau} = \frac{64.8 - 8.1}{64.8} = 0.875$$

The form factor of the armature reaction in the d-axis according to Table 5.2 is

$$k_{fd} = \frac{4}{\pi} \times 0.875 \frac{1}{1 - 0.875^2} \cos(0.5 \times 0.875 \times \pi) = 0.9273$$

The form factor of the armature reaction in the q-axis according to Table 5.2 is

(a)

(b)

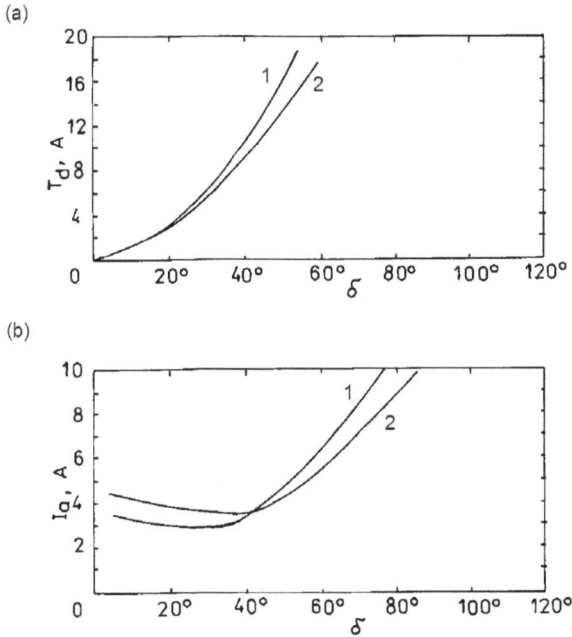

Fig. 5.23. Angle characteristics of the buried PM motor at constant terminal phase voltage of 220 V and $f = 50$ Hz (Table 5.7): (a) electromagnetic torque T_d, (b) stator current I_a. 1 — measurements, 2 — calculation on the basis of analytical approach with X_{sd} and X_{sq} obtained from the FEM.

$$k_{fq} = \frac{1}{\pi}(0.875\pi - \sin 0.875\pi) = 0.7532$$

The d-axis unsaturated armature reaction reactance according to eqn (5.31)

$$X_{ad} = 4 \times 3 \times 0.4\pi \times 10^{-6}$$

$$\times 50 \frac{(240 \times 0.96)^2}{\pi \times 2} \frac{0.0648 \times 0.1008}{1.16553 \times 0.00055 + 0.0081/1.094} \times 0.9273 = 4.796 \ \Omega$$

where the winding factor $k_{w1} = 0.96$ for $m_1 = 3$, $s_1 = 36$ and $2p = 4$, $\mu_{rrec} = B_r/(\mu_0 H_c) = 1.05/(0.4\pi \times 10^{-6} \times 764,000) = 1.094$, and Carter's coefficient $k_C = 1.16553$ for $b_{14} = 2.2$ mm, $g = 0.55$ mm and $D_{1in} = 82.5$ mm.

The q-axis unsaturated armature reaction reactance calculated on the basis of eqn (5.33)

$$X_{aq} = 4 \times 3 \times 0.4\pi \times 10^{-6}$$

$$\times 50 \frac{(240 \times 0.96)^2}{\pi \times 2} \frac{0.0648 \times 0.1008}{1.16553 \times 0.00055} \times 0.7532 = 48.89 \ \Omega$$

The unsaturated armature leakage reactance obtained in the same way as that in Example 5.2 is $X_1 \approx 2.514 \ \Omega$.

Neglecting the magnetic saturation due to main and leakage fields, the synchronous reactances according to eqns (5.15) are

$$X_{sd} = 2.514 + 4.796 \approx 7.31 \ \Omega \qquad\qquad X_{sq} = 2.514 + 48.89 \approx 51.4 \ \Omega$$

The magnetic saturation mainly affects the synchronous reactance in the d-axis.

A computer program has been created for calculating the magnetic circuit, electromagnetic parameters, and performance.

Synchronous reactances obtained using different methods are plotted in Fig. 5.22. The angle characteristics, i.e., T_d and I_a as functions of the load angle δ are plotted in Fig. 5.23. Curves obtained from the analytical method with synchronous reactances according to eqns (5.15), (5.31) and (5.33) are not in agreement with measurements (curve 1). These characteristics are very close to each other if X_{sd} and X_{sq} are calculated on the basis of the FEM and then the values of X_{sd} and X_{sq} are used in the equivalent circuit (curves 2).

There is a contradiction between Fig. 5.2 and Fig. 5.23, i.e., negative load angle for motor mode in Fig. 5.2 and positive angle in Fig. 5.23. However, it is more convenient to assign, in practical considerations, a positive torque for positive load angle in the range $0 \leq \delta \leq 180°$ for both motor and generator mode. At rated conditions ($\delta = 45°$) the output power $P_{out} \approx 1.5$ kW, $T_d = 10.5$ Nm, $\eta \approx 82\%$, $\cos\phi \approx 0.9$, angle $\Psi \approx -19°$ (between I_a and E_f), $I_a = 3.18$ A (I_{aq} is predominant), and $B_{mg} = 0.685$ T.

Numerical example 5.4

Find the losses due to 7th space harmonic in a four-pole surface PM rotor spinning at 10,000 rpm. The air gap $g = 1.5$ mm, the magnitude of the magnetic flux density of the 7th space harmonic is $B_{m7} = 0.007$ T, the electric conductivity of NdFeB PMs is $\sigma_{PM} = 0.5236 \times 10^6$ S/m at 100°C, the relative recoil magnetic permeability $\mu_{rrec} = 1.035$, the magnet pole arc–to–pole pitch ratio $\alpha_i = 0.8$, the rotor outer diameter $D_{2out} = 0.1$ m, the length of PM equal to the length of the stator stack $l_M = L_i = 0.1$ m,

Solution

The pole pitch $\tau = \pi(0.1 + 2 \times 0.0015)/4 = 0.081$ m, the parameter $\beta_7 = 7\pi/0.081 = 271.3$ 1/m, the frequency of the stator current $f = 2 \times 10\,000/60 = 333.3$ Hz and the frequency of magnetic flux in PMs $f_7 = |1-7| \times 333.3 = 2000$ Hz (this harmonic rotates in the same direction as the rotor).

The coefficient of attenuation of the 7th harmonic of the electromagnetic field in PMs according to eqn (B.29)

$$k_7 = \sqrt{\pi 2000 \times 0.4\pi \times 10^{-6} \times 1.035 \times 0.5236 \times 10^6} = 65.4 \; 1/\text{m}$$

The coefficient $a_{R\nu}$ for the 7th harmonic according to eqn (B.28)

$$a_{R7} = \frac{1}{\sqrt{2}} \sqrt{\sqrt{4 + \frac{271.3^4}{65.4^4}} + \frac{271.3^4}{65.4^4}} = 4.156 \; 1/\text{m}$$

The edge effect coefficient for the 7th harmonic according to eqn (B.30)

$$k_{r7} = 1 + \frac{1}{7}\frac{2}{\pi}\frac{0.081}{0.1} = 1.074$$

The surface area of PMs according to eqn (B.26)

$$S_{PM} = 0.8\pi 0.1 \times 0.1 = 0.025 \; \text{m}^2$$

The losses in PMs due to the 7th space harmonic according to eqn (B.27)

$$\Delta P_{PM7} = 4.156 \times 1.074 \frac{65.4^3}{271.3^2} \left(\frac{0.007}{0.4\pi \times 10^{-6} \times 1.035}\right)^2 \frac{1}{0.5236 \times 10^6} \times 0.025 = 23.6 \; \text{W}$$

6

d.c. Brushless Motors

The electric and magnetic circuits of PM synchronous motors and PM d.c. brushless motors are similar, i.e., polyphase (usually three phase) armature windings are located in the stator and moving magnet rotor serves as the excitation system. PM synchronous motors are fed with three phase sinusoidal voltage waveforms and operate on the principle of magnetic rotating field. For *constant voltage–to–frequency* control technique no rotor position sensors are required. PM d.c. brushless motors operate from a d.c. voltage source and use direct feedback of the rotor angular position, so that the input armature current can be switched, among the motor phases, in exact synchronism with the rotor motion. This concept is known as *self-controlled synchronization*, or *electronic commutation*. The solid state inverter and position sensors are equivalent to the mechanical commutator in d.c. brush motors. Variable d.c. bus voltage can be obtained using

- a variable voltage transformer and diode rectifier
- a controlled rectifier (thyristor, GTO or IGBT bridge)

In the second case the d.c. bus voltage is a function of the firing angle of the controlled rectifier.

6.1 Fundamental equations

6.1.1 Terminal voltage

From Kirchhoff's voltage law, the instantaneous value of the motor *terminal phase voltage* is

$$v_1 = e_f + R_1 i_a + L_s \frac{di_a}{dt} \tag{6.1}$$

where e_f is the instantaneous value of the EMF induced in a single phase armature winding by the PM excitation system, i_a is the armature instantaneous current, R_1 is the armature resistance per phase and L_s is the synchronous inductance per phase which includes both the leakage and armature reaction inductances.

Eqn (6.1) corresponds to the half-wave operation of a Y-connected motor with accessible neutral point. For a three-phase bridge inverter with six solid state switches and Y-connected motor, two phase windings are always connected in series during conduction period, e.g., for phases A and B

$$v_1 = (e_{fA} - e_{fB}) + 2R_1 i_a + 2L_s \frac{di_a}{dt} \tag{6.2}$$

where $e_{fA} - e_{fB} = e_{fAB}$ is the line–to–line EMF, in general e_{fL-L}.

6.1.2 Instantaneous current

Assuming zero-impedance solid state switches, $v_1 = V_{dc}$ where V_{dc} is the inverter input d.c. voltage and $L_s \approx 0$, the *instantaneous armature current* is

- for Y-connected windings and half-wave operation

$$i_a(t) = \frac{V_{dc} - e_f}{R_1} \tag{6.3}$$

- for Y-connected windings and full-wave operation

$$i_a(t) = \frac{V_{dc} - e_{fL-L}}{2R_1} \tag{6.4}$$

where e_{fL-L} is the line-to-line EMF induced in two series connected phase windings.

Including the inductance L_s and assuming $e_{fL-L} = E_{fL-L} = const$ (trapezoidal EMF), eqn (6.4) for the conduction period takes the following form [225]

$$i_a(t) = \frac{V_{dc} - E_{fL-L}}{2R_1} \left(1 - e^{(R_1/L_s)t}\right) + I_{amin} e^{(R_1/L_s)t} \tag{6.5}$$

where I_{amin} is the armature current at $t = 0$. The machine is underexcited since $V_{dc} > E_{fL-L}$ for the current rise.

6.1.3 EMF

The EMF can simply be expressed as a function of the rotor speed n, i.e.,

- for a half-wave operation

$$E_f = c_E \Phi_f n = k_E n \tag{6.6}$$

- for a full-wave operation

$$E_{fL-L} = c_E \Phi_f n = k_E n \tag{6.7}$$

where c_E or $k_E = c_E \Phi_f$ is the *EMF constant* also called the *armature constant*. For PM excitation and negligible armature reaction $\Phi_f \approx const$.

6.1.4 Inverter a.c. output voltage

When a PM brushless motor is fed from a solid state converter, and the d.c. bus voltage is known, the fundamental harmonic of the output a.c. line–to–line voltage of the inverter is

- for a six-step three-phase inverter

$$V_{1L} = \frac{\sqrt{6}}{\pi} V_{dc} \approx 0.78 V_{dc} \tag{6.8}$$

- for a sinusoidal PWM three-phase voltage source inverter

$$V_{1L} = \sqrt{\frac{3}{2}} m_a \frac{V_{dc}}{2} \approx 0.612 m_a V_{dc} \tag{6.9}$$

where the *amplitude modulation index*

$$m_a = \frac{V_m}{V_{mcr}} \tag{6.10}$$

is the ratio of peak values V_m and V_{mcr} of the modulating and carrier waves, respectively. For space vector modulation $0 \leq m_a \leq 1$.

6.1.5 d.c. bus voltage of a controlled rectifier

The d.c. output voltage of a controlled rectifier is

- for a three-phase fully controlled rectifier

$$V_{dc} = \frac{3\sqrt{2}V_{1L}}{\pi} \cos \alpha \tag{6.11}$$

- for a three-phase half-controlled rectifier

$$V_{dc} = \frac{3\sqrt{2}V_{1L}}{2\pi}(1 + \cos \alpha) \tag{6.12}$$

where V_{1L} is the line-to-line *rms* voltage and α is the firing angle.

6.1.6 Electromagnetic torque

Assuming that the flux linkage in the stator winding produced by the PM rotor is $\Psi_f = M_{12}i_2$, the *electromagnetic torque* is

$$T_d(i_a, \theta) = i_a \frac{d\Psi_f}{d\theta} \qquad (6.13)$$

where $i_1 = i_a$ is the stator current, i_2 is the current in a fictitious rotor winding which is assumed to have a mutual inductance M_{12} with the stator winding and θ is the rotor angular position.

6.1.7 Electromagnetic torque of a synchronous motor

The electromagnetic torque developed by a synchronous motor is usually expressed as a function of the angle Ψ between the q-axis (EMF E_f axis) and the armature current I_a, i.e.,

$$T_d = c_T \Phi_f I_a \cos \Psi \qquad (6.14)$$

where c_T is the *torque constant*. For a PM motor

$$T_d = k_T I_a \cos \Psi \qquad (6.15)$$

where $k_T = c_T \Phi_f$ is a new torque constant.

The maximum torque is when $\cos \Psi = 1$ or $\Psi = 0^0$. It means that the armature current $I_a = I_{ad}$ is in phase with the EMF E_f.

6.1.8 Electromagnetic torque of a PM brushless d.c. motor

The torque equation is similar to eqn (4.6) for a d.c. commutator motor, i.e.,

$$T_d = c_{Tdc} \Phi_f I_a = k_{Tdc} I_a \qquad (6.16)$$

where c_{Tdc} and $k_{Tdc} = c_{Tdc} \Phi_f$ are torque constants.

6.1.9 Linear and rotational speed of brushless motors

The *linear speed* in m/s is the full angle of rotation or 2τ divided by the period of full rotation $T = 1/(pn)$ i.e.,

$$v = \frac{2\tau}{T} = 2\tau pn \qquad (6.17)$$

where τ is the pole pitch, p is the number of pole pairs and n is the rotational speed in rev/s. Surface linear speed of the rotor cannot exceed the permissible value for a given rotor construction (Chapter 9).

6.1.10 Concentrated-coil armature winding

The armature (stator) winding made of *concentrated non-overlapping coils* is simple to manufafacture and provides short end connections. The concentrated-coil winding is feasible when

$$\frac{N_c}{GCD(N_c, 2p)} = km_1 \tag{6.18}$$

where N_c is the total number of armature coils, GCD is the greatest common divisor of N_c and the number of poles $2p$, m_1 is the number of phases and $k = 1, 2, 3, \ldots$.

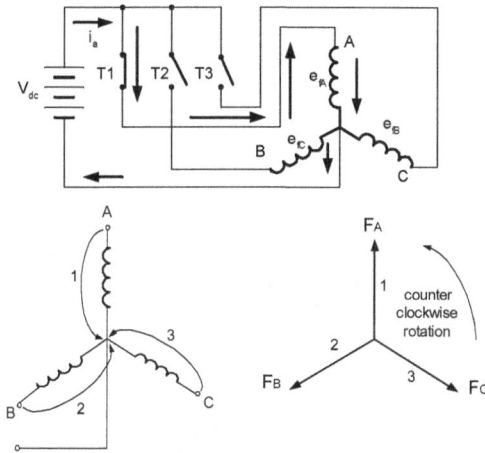

Fig. 6.1. Switching sequence and MMF phasors for three-phase unipolar–driven Y-connected PM d.c. brushless motor.

6.2 Commutation of PM brushless motors

6.2.1 Unipolar-driven PM brushless motor

The unipolar or half-wave operation of PM brushless motors is explained in Fig. 6.1. The three phase winding is Y-connected and neutral point is available. The d.c. voltage V_{dc} is switched across phase–to–neutral terminals with the aid of one solid state switch per phase. Each phase terminal receives positive voltage and the neutral wire is of negative polarity. For the current sequence $i_{aA}, i_{aB}, i_{aC}, \ldots$ the MMF phasors F_A, F_B, F_C rotate counterclockwise. If the switching sequence is reversed, i.e., i_{aA}, i_{aC}, i_{aB}, the direction of rotation of MMF phasors will be clockwise. For a linear magnetic circuit

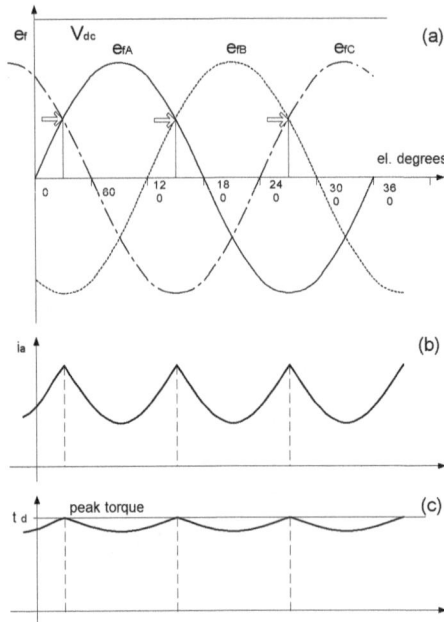

Fig. 6.2. Ideal three-phase unipolar operation of a Y-connected PM d.c. brushless motor: (a) sinusoidal EMF waveforms, (b) current waveforms, (c) electromagnetic torque waveforms. Switching points are marked with arrows.

the phase magnetic fluxes are proportional to MMFs. The EMF, current and torque waveforms are shown in Fig. 6.2. This type of operation (commutation) is called *half-wave operation* (commutation) because conduction occurs only during the positive half of the EMF waveform. The shape of the EMF waveforms depends on the design of PMs and stator windings. Phase EMF waveforms shown in Fig. 6.2 have sinusoidal shapes. In practice, both the shapes of the EMF and current differ from those shown in Fig. 6.2. The d.c. voltage V_{dc} is higher than the peak phase EMF so that the current flows from the phase terminal to the neutral wire during the 120^0 conduction period. Neglecting the winding inductance and assuming zero switching time, the armature instantaneous current is given by eqn (6.3). The electromagnetic power per phase at a given time instant is $p_{elm} = i_a e_f$ and the electromagnetic torque is given by eqns (3.77) or (6.67). Commuting at the zero crossing of the EMF waveform should always be avoided as the torque is zero no matter how much current is injected into the phase winding.

For unipolar operation the torque ripple is high, not acceptable in some PM brushless motor applications.

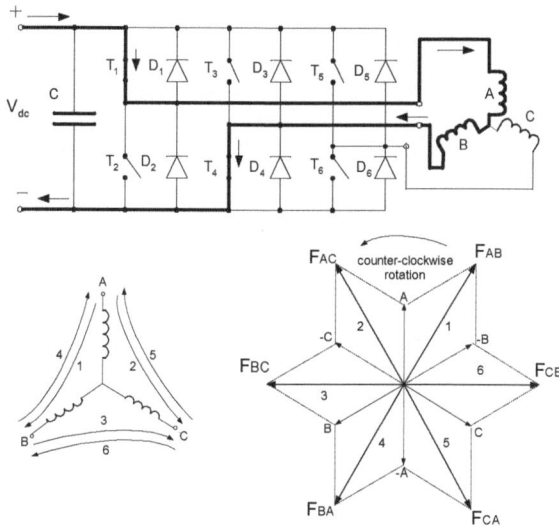

Fig. 6.3. Switching sequence and MMF phasors for six-step commutation of a Y-connected PM d.c. brushless motor. Commutation sequence is AB, AC, BC, BA, CA, CB, etc.

6.2.2 Bipolar-driven PM brushless motor, two phases on

The term "bipolar" indicates the capability of providing the motor phase current of either positive or negative polarity. A PM d.c. brushless motor is driven by a three phase inverter bridge and all six solid state switches are used. In Fig. 6.3 the d.c. voltage V_{dc} is switched between phase terminals and for the Y connection two windings belonging to different phases are series connected during each conduction period. The current sequence is i_{aAB}, i_{aAC}, i_{aBC}, i_{aBA}, i_{aCA}, i_{aCB}, ... For this current sequence, the MMFs F_{AB}, F_{AC}, F_{BC}, F_{BA}, F_{CA}, F_{CB},... rotate counterclockwise (Fig. 6.4). This operation is called *bipolar* or *full-wave operation* because conduction occurs for both the positive and negative half of the EMF waveform. For sinusoidal EMF waveforms the currents can be regulated in such a way as to obtain approximately square waves. The electromagnetic power and torque are always positive because negative EMF times negative current gives a positive product. Each conduction period (one step) for line currents is 60^0. Therefore, this is a *six-step commutation* with only two phases on (120^0 current conduction) at any time leaving the remaining phase floating [164]. As a result, the torque ripple is substantially reduced.

At non-zero speed, the maximum *torque-to-current* ratio is achieved at the peak of EMF waveforms. The current is in phase with the EMF. The commutation timing is determined by the rotor position sensors or estimated on the basis of the motor parameters, e.g., EMF.

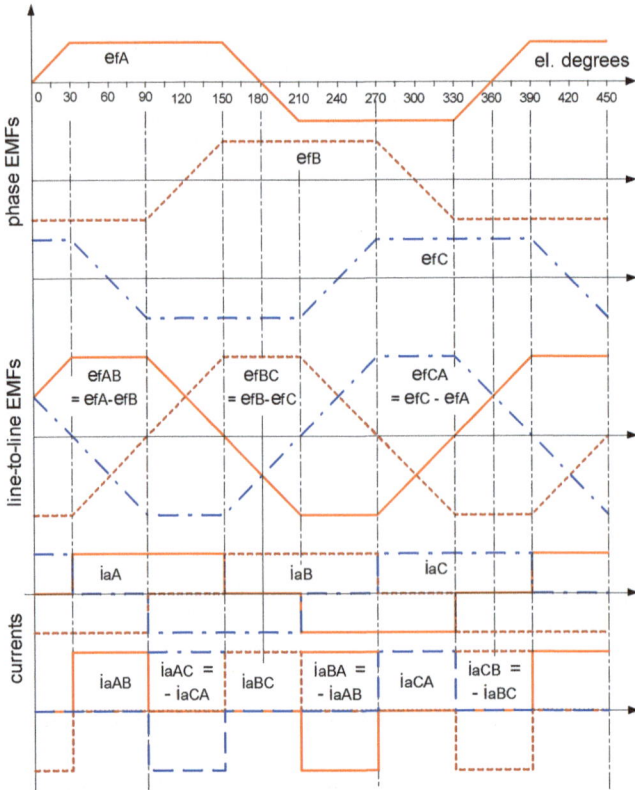

Fig. 6.4. Phase and line-to-line trapezoidal EMF and square current waveforms of a bipolar-driven PM brushless motor with 120^0 current conduction.

The average torque can be maximized and torque ripple can be minimized if the EMF waveform has a trapezoidal shape (Fig. 6.4). For *trapezoidal operation* the peak line-to-line EMF occurs during the whole conduction period, i.e., 60^0 for line current as given in Figs 6.4 and 6.5. The EMF, i.e., $e_{fAC} = e_{fA} - e_{fC} = -e_{fCA} = e_{fC} - e_{fA}$ and the current, e.g., $i_{aAC} = -i_{aCA}$. The trapezoidal shape of the line–to–line EMFs is obtained by proper shaping and magnetizing the PMs and proper designing the stator winding. Theoretically, the flat top EMF waveforms at d.c voltage $V_{dc} = const$ produce square current waveforms and a constant torque independent of the rotor position (Fig. 6.5). Owing to the armature reaction and other parasitic effects, the EMF waveform is never ideally flat. However, the torque ripple below 10% can be achieved. Torque ripple can further be reduced by applying more than three phases.

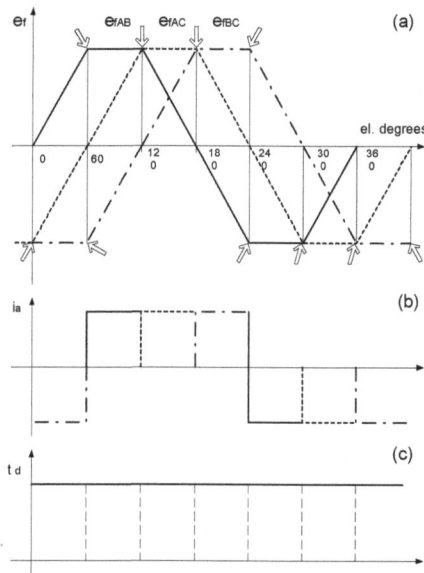

Fig. 6.5. Ideal three-phase six-step operation of a Y-connected PM d.c. brushless motor: (a) trapezoidal line–to–line EMF waveforms, (b) current waveforms, (c) electromagnetic torque waveforms. Switching points are marked with arrows.

6.2.3 Bipolar-driven PM brushless motor, three phases on

In six-step mode operation only one upper and one lower solid state switch are turned on at a time (120^0 conduction). With more than two switches on at a time, a 180^0 current conduction can be achieved, as shown in Fig. 6.6. If the full current flows, say, through one upper leg, two lower legs conduct half of the current.

6.3 EMF and torque of PM brushless motors

6.3.1 Synchronous motor

The three-phase stator winding with distributed parameters produces sinusoidal or quasi-sinusoidal distribution of the MMF. In the case of inverter operation all three solid state switches conduct current at any instant of time.

For a sinusoidal distribution of the air gap magnetic flux density the first harmonic of the excitation flux can be found on the basis of eqn (5.5). The field excitation magnetic flux calculated on the basis of the maximum air gap magnetic flux density B_{mg} is then $\Phi_f \approx \Phi_{f1} = (2/\pi)L_i \tau k_f B_{mg}$ where the *form factor of the excitation field* $k_f = B_{mg1}/B_{mg}$ is according to eqn (5.23). Assuming that the instantaneous value of EMF induced in a single stator conductor by the first harmonic of the magnetic flux density is

(a)   (b)

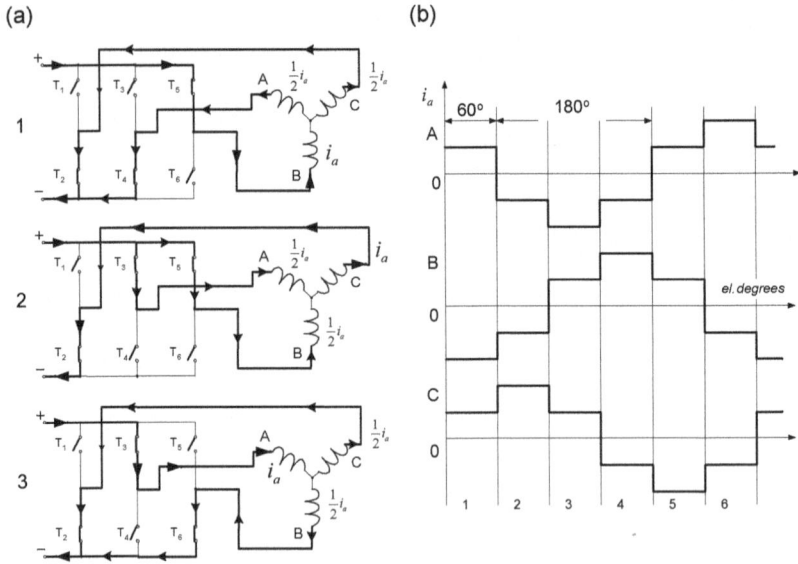

Fig. 6.6. Three-phase bipolar-driven Y-connected PM d.c. brushless motor with three phases on at a time: (a) commutation, (b) current waveforms.

$e_{f1} = E_{mf1} \sin(\omega t) = B_{mg1} L_i v_s \sin(\omega t) = 2 f B_{mg1} L_i \tau \sin(\omega t)$, the *rms* EMF is $E_{mf1}/\sqrt{2} = \sqrt{2} f B_{mg1} L_i \tau = (1/2)\pi\sqrt{2} f(2/\pi) B_{mg1} L_i \tau$. For two conductors or one turn $E_{mf1}/\sqrt{2} = \pi\sqrt{2} f(2/\pi) B_{mg1} L_i \tau$. For $N_1 k_{w1}$ turns, where k_{w1} is the winding factor, the *rms* EMF is

$$E_f \approx E_{f1} = \pi\sqrt{2} N_1 k_{w1} f \alpha_i k_f B_{mg} L_i \tau$$

$$= \pi p \sqrt{2} N_1 k_{w1} \Phi_f n_s = c_E \Phi_f n_s = k_E n_s \qquad (6.19)$$

where $c_E = \pi p \sqrt{2} N_1 k_{w1}$ and $k_E = c_E \Phi_f$ are EMF constants.

The load angle δ of a synchronous machine is usually defined as an angle between the input phase voltage V_1 and the EMF E_f induced in a single stator (armature) winding by the rotor magnetic excitation flux Φ_f. In the phasor diagram for an overexcited salient-pole synchronous motor shown in Fig. 6.7 the angle Ψ between the q-axis (EMF axis) and the armature current I_a is a function of the load angle δ and the angle ϕ between the armature current I_a and the input voltage V_1, i.e., $\Psi = \delta + \phi$. This phasor diagram has been drawn using a consumer arrow system and it corresponds to eqn (5.35).

Assuming a negligible difference between d- and q-axis synchronous reactances, i.e., $X_{sd} - X_{sq} \approx 0$, the electromagnetic (air gap) power $P_{elm} \approx m_1 E_f I_{aq} = m_1 E_f I_a \cos\Psi$. Note that in academic textbooks the electromagnetic power is usually calculated as $P_{elm} \approx P_{in} = m_1 V_1 I_a \cos\phi =$

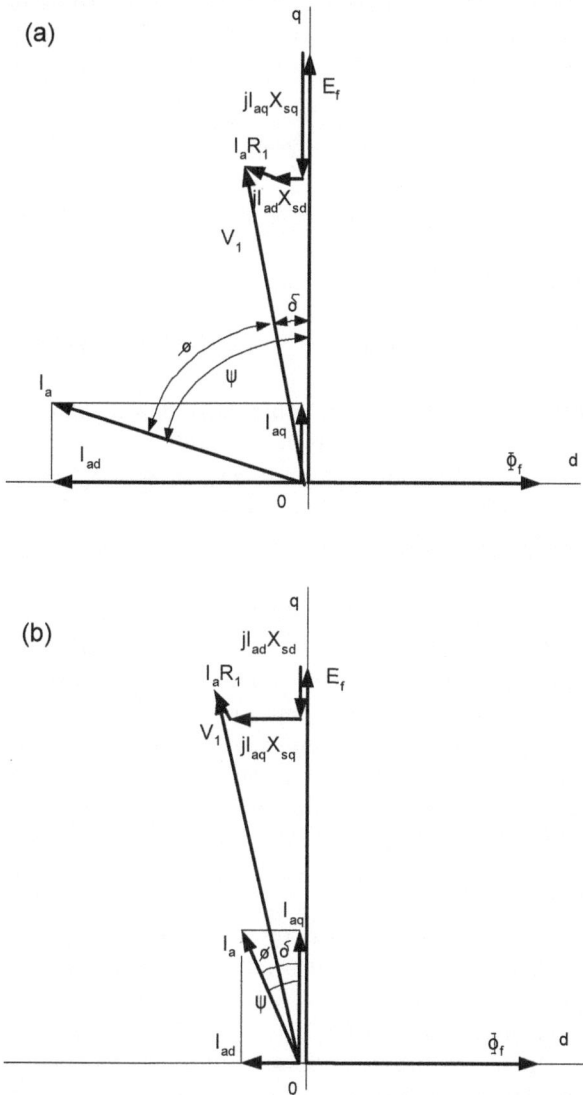

Fig. 6.7. Phasor diagrams for overexcited salient-pole synchronous motors: (a) high d-axis current I_{ad}, (b) low d-axis current I_{ad}.

$m_1 V_1 I_a \cos(\Psi \pm \delta) = m_1 (V_1 E_f / X_{sd}) \sin \delta$. Putting E_f according to eqn (6.19) and $\Omega_s = 2\pi n_s = 2\pi f/p$ the electromagnetic torque developed by the motor is:

$$T_d = \frac{P_{elm}}{2\pi n_s} = \frac{m_1 E_f I_a}{2\pi n_s} \cos \Psi$$

$$= \frac{m_1}{\sqrt{2}} p N_1 k_{w1} \Phi_f I_a \cos \Psi = c_T \Phi_f I_a \cos \Psi = k_T I_a \cos \Psi \qquad (6.20)$$

where the *torque constant* is

$$c_T = m_1 \frac{c_E}{2\pi} = \frac{m_1}{\sqrt{2}} p N_1 k_{w1} \qquad \text{or} \qquad k_T = c_T \Phi_f \qquad (6.21)$$

Similar equations have been derived by other authors [218, 225]. For example, eqn (6.20) has been derived in [218] as an integral of the elementary torque contributions over the whole air gap periphery in which the effective sine distributed turns in series per phase $N_s = (4/\pi) N_1 k_{w1}$.

The maximum torque

$$T_{dmax} = k_T I_a \qquad (6.22)$$

is for $\Psi = 0^0$ which means that $\delta = \phi$ (Fig. 6.8), i.e. the rotor excitation flux Φ_f and the armature current I_a are perpendicular one to the other. There is no demagnetizing component Φ_{ad} of the armature reaction flux and the air gap magnetic flux density takes its maximum value. The EMF E_f is high so it can better balance the input voltage V_1 thus minimizing the armature current I_a. When Ψ approaches 0^0, the low armature current is mainly torque producing (Figs 6.7b and 6.8). An angle $\Psi = 0^0$ results in a decoupling of the rotor flux Φ_f and the armature flux Φ_a which is important in high-performance servo-drives.

If a synchronous motor is supplied by an inverter in which solid state switches are commutated by the synchronous motor voltages, the leading power factor operation is required. For a lagging power factor the inverter must consist of self-controlled switches.

A drop in the input frequency at $\Psi = 0^0$ improves the power factor but due to the lower EMF E_f, according to eqn (6.19), the motor draws higher armature current I_a. In fact, the efficiency of a synchronous motor at $\Psi = 0^0$ and low input frequency f deteriorates [58].

The experimental tests at maximum torque on PM synchronous motors confirm that the ratio *maximum torque–armature current* increases significantly as frequency is reduced [58]. When the input frequency is high, the load angle δ at maximum torque is large and ϕ is small, so Ψ approaches 90^0 (Fig. 6.7a). This phasor diagram also corresponds to flux-weakening regime operation since a negative current I_{ad} has been introduced [225]. Only a fraction of the armature current produces a torque. When the input frequency is

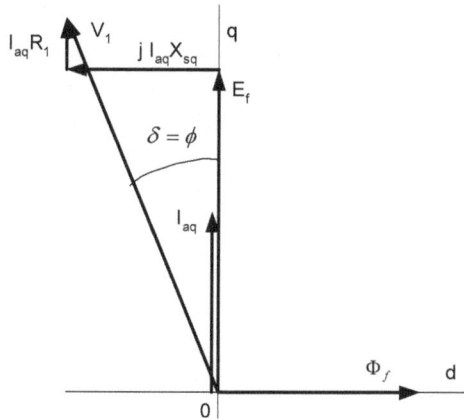

Fig. 6.8. Phasor diagram at $\delta = \phi$ ($\Psi = 0$, $I_{ad} = 0$).

small, the load angle δ at maximum torque is reduced and ϕ is again small, since the stator circuit is then predominantly resistive.

When the d-axis current I_{ad} is high, the angle Ψ and rms stator (armature) current $I_a = \sqrt{I_{ad}^2 + I_{aq}^2}$ are high and input voltage V_1 and power factor $\cos \phi$ are low. When the q-axis current I_{aq} is high, the angle Ψ and rms stator current I_a are low and the input voltage V_1 and power factor $\cos \phi$ are high.

6.3.2 PM d.c. brushless motors

PM d.c. brushless motors predominantly have surface-magnet rotors designed with large effective pole–arc coefficients $\alpha_i^{(sq)} = b_p/\tau$. Assuming Y-connected windings and six-step commutation, as in Fig. 6.3, only two of the three motor phase windings conduct at the same time, i.e., i_{aAB} (T1T4), i_{aAC} (T1T6), i_{aBC} (T3T6), i_{aBA} (T3T2), i_{aCA} (T5T2), i_{aCB} (T5T4), etc. At the on-time interval (120^0) for the phase windings A and B, the solid state switches T1 and T4 conduct. The instantaneous input phase voltage is expressed by eqn (6.2). The solution to eqn (6.2) is given by eqn (6.5). When T1 switches off, the current freewheels through diode D2. For the off-time interval both switches T1 and T4 are turned-off and the diodes D2 and D3 conduct the armature current which charges the capacitor C. The instantaneous voltage has a similar form to eqn (6.2) in which $(e_{fA} - e_{fB})$ is replaced by $-(e_{fA} - e_{fB})$ and fourth term $(1/C) \int i dt$ includes the capacitor C [225]. If the solid state devices are switched at relatively high frequency, the winding inductance keeps the on-off rectangular current waveforms smooth.

For d.c. current excitation $\omega \to 0$; then eqn (5.35) is similar to that describing a steady state condition of a d.c. commutator motor, i.e.,

$$V_{dc} = E_{fL-L} + 2R_1 I_a^{(sq)} \tag{6.23}$$

where $2R_1$ is the sum of two-phase resistances in series (for Y-connected phase windings), E_{fL-L} is the sum of two phase EMFs in series, V_{dc} is the d.c. input voltage supplying the inverter and $I_a^{(sq)}$ is the flat-topped value of the square-wave current equal to the inverter input current. The phasor analysis does not apply to this type of operation since the armature current is nonsinusoidal.

For an ideal rectangular distribution of $B_{mg} = const$ in the interval of $0 \leq x \leq \tau$ or from 0^0 to 180^0

$$\Phi_f = L_i \int_0^\tau B_{mg} dx = \tau L_i B_{mg}$$

Including the pole shoe width $b_p < \tau$ and a fringing flux, the excitation flux is somewhat smaller

$$\Phi_f^{(sq)} = b_p L_i B_{mg} = \alpha_i^{(sq)} \tau L_i B_{mg} \tag{6.24}$$

For a square-wave excitation the EMF induced in a single turn (two conductors) is $2B_{mg}L_i v = 4pn B_{mg} L_i \tau$. Including b_p and fringing flux the EMF for $N_1 k_{w1}$ turns $e_f = 4pn N_1 k_{w1} \alpha_i^{(sq)} B_{mg} L_i \tau = 4pn N_1 k_{w1} \Phi_f$. For the Y-connection of the armature windings, as in Fig. 6.3, two phases are conducting at the same time. The EMF contributing to the electromagnetic power is

$$E_{fL-L} = 2e_f = 8p N_1 k_{w1} \alpha_i^{(sq)} \tau L_i B_{mg} n = c_{Edc} \Phi_f^{(sq)} n = k_{Edc} n \tag{6.25}$$

where $c_{Edc} = 8p N_1 k_{w1}$ and $k_{Edc} = c_{Edc} \Phi_f^{(sq)}$.

The electromagnetic torque developed by the motor is

$$T_d = \frac{P_g}{2\pi n} = \frac{E_{fL-L} I_a^{(sq)}}{2\pi n} = \frac{4}{\pi} p N_1 k_{w1} \alpha_i^{(sq)} \tau L_i B_{mg} I_a^{(sq)}$$

$$= \frac{4}{\pi} p N_1 k_{w1} \Phi_f^{(sq)} I_a^{(sq)} = c_{Tdc} \Phi_f^{(sq)} I_a^{(sq)} = k_{Tdc} I_a^{(sq)} \tag{6.26}$$

where $c_{Tdc} = c_{Edc}/(2\pi) = (4/\pi) p N_1 k_{w1}$, $k_{Tdc} = c_{Tdc} \Phi_f^{(sq)}$ and $I_a^{(sq)}$ is the flat-top value of the phase current.

For $n = n_s$ and $\Psi = 0^0$ the ratio T_d of a square-wave motor-to-T_d of a sine-wave motor is

$$\frac{T_d^{(sq)}}{T_d} = \frac{4}{\pi} \frac{\sqrt{2}}{m_1} \frac{\Phi_f^{(sq)}}{\Phi_f} \frac{I_a^{(sq)}}{I_a} \approx 0.6 \frac{\Phi_f^{(sq)}}{\Phi_f} \frac{I_a^{(sq)}}{I_a} \tag{6.27}$$

Assuming the same motor and the same values of air gap magnetic flux densities, the ratio of the square-wave motor flux to sine-wave motor flux is

$$\frac{\Phi_f^{(sq)}}{\Phi_f} = \frac{1}{k_f} \tag{6.28}$$

Table 6.1. Specifications of three-phase medium power d.c. PM brushless motors manufactured by Powertec Industrial Motors, Inc., Rock Hill, SC, U.S.A.

Parameters	E254E3	E258E3	E259E3
Rated base speed at 640 V d.c., rpm		1000	
Output power at rated speed, kW	34.5 (16.8)	47.3 (24.5)	63.6 (27.2)
Current at rated speed, A	53 (26)	77 (41)	97 (43)
Continuous stall torque, Nm	344 (188)	482 (286)	644 (332)
Continuous stall current, A	57 (31)	82 (49)	102 (54)
Peak torque (theoretical), Nm	626	1050	1280
Current at peak torque, A	93	160	187
Max. torque of static friction, Nm	1.26	1.90	2.41
Torque constant (line-to-line), Nm/A	6.74	6.55	6.86
EMF constant (line-to-line), V/rpm	0.407	0.397	0.414
Resistance (line-to-line) of hot machine, Ω	0.793	0.395	0.337
Resistance (line-to-line) of cold machine, Ω	0.546	0.272	0.232
Inductance (line-to-line), mH	11.6	6.23	5.70
Electrical time constant, ms	19.8	22.9	24.9
Mechanical time constant, ms	1.85	1.53	1.46
Moment of inertia, kgm$^2 \times 10^{-3}$	103	160	199
Coefficient of viscous damping, Nm/rpm	0.00211	0.00339	0.00423
Thermal resistance, ^{0}C/W	0.025 (0.080)	0.023 (0.065)	0.021 (0.064)
Thermal time constant, min	36 (115)	47 (130)	49 (140)
Mass, kg	134 (129)	200 (185)	234 (219)
Power density, kW/kg	0.257 (0.130)	0.236 (0.132)	0.272 (0.124)

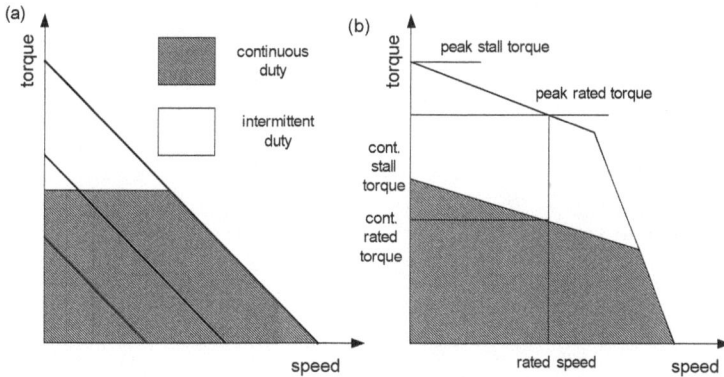

Fig. 6.9. Torque–speed characteristics of a PM brushless motor: (a) theoretical, (b) practical.

Table 6.2. PM d.c. brushless servo motors (R60 series) manufactured by Pacific Scientific, Wilmington, MA, U.S.A.

Parameters	R63	R65	R67
No-load speed, rpm	5400 (10,500)	3200 (6400)	2200 (4000)
Continuous stall torque, Nm	8	13	19
Peak torque, Nm	26	45	63
Continuous stall current, A	13.5 (27.0)	13.1 (26.2)	13.8 (27.6)
Current at peak torque, A		82 (164)	
Torque constant line-to-line, Nm/A	0.66 (0.33)	1.12 (0.56)	1.56 (0.78)
EMF constant line-to-line, V/rpm	0.07 (0.035)	0.117 (0.059)	0.164 (0.082)
Resistance (line-to-line) of hot motor, Ω	1.4 (0.34)	1.81 (0.51)	2.3 (0.55)
Resistance (line-to-line) of cold motor, Ω	0.93 (0.23)	1.2 (0.34)	1.5 (0.37)
Inductance (line-to-line), mH	8.9 (2.2)	13.7 (3.4)	18.2 (4.6)
Rotor moment of inertia, kgm$^2 \times 10^{-3}$	0.79	1.24	1.69
Coefficient of static friction, Nm	0.16	0.26	0.36
Coefficient of viscous damping, Nm/krpm	0.046	0.075	0.104
Thermal resistance, $^\circ$C/W	0.51	0.42	0.30
Thermal time constant, min	19	36	72
Mass (motor only), kg	13	18	22

6.4 Torque–speed characteristics

On the basis of eqns (6.6) and (6.15) or (6.7) and (6.16) the torque–speed characteristic can be expressed in the following simplified form

$$\frac{n}{n_0} = 1 - \frac{I_a}{I_{ash}} = 1 - \frac{T_d}{T_{dst}} \qquad (6.29)$$

where the no-load speed, locked rotor armature current and stall torque are, respectively,

$$n_0 = \frac{V_{dc}}{k_E} \qquad T_{dst} = k_{Tdc}I_{ash} \qquad I_{ash} = \frac{V_{dc}}{R} \qquad (6.30)$$

where $R = R_1$ for half-wave operation and $R = 2R_1$ for full-wave operation. Eqn (6.29) neglects the armature reaction, rotational and switching losses.

The torque-speed characteristics are shown in Fig. 6.9. Eqns (6.29) and (6.30) are very approximate and cannot be used in calculation of performance characteristisc of commercial PM d.c. brushless motors. Theoretical torque-speed characteristics (Fig. 6.29a) differ from practical characteristics (Fig.

6.29b). The *continuous torque* line is set by the maximum rated temperature of the motor. The intermittent duty operation zone is bounded by the *peak torque* line and the maximum input voltage.

Table 6.1 shows specifications of medium power PM d.c. brushless motors manufactured by Powertec Industrial Motors, Inc., Rock Hill, SC, U.S.A. Class H (180⁰C) insulation has been used but rated conservatively for class F (155⁰C). The stator windings are Y-connected. The rotor is with surface NdFeB PMs. The standard drip-proof blower ventilated design significantly increases the rated power and torque. Values in brackets are for totally enclosed non-ventilated motors.

Table 6.2 shows specifications of small PM d.c. brushless servo motors manufactured by Pacific Scientific, Wilmington, MA, U.S.A. Rated speed is for operation at 240 V a.c. three-phase line. Peak torque ratings are for 5 s. All values are for 25⁰C ambient temperature, motor mounted to aluminum sink and class F winding insulation.

6.5 Winding losses

The *rms* armature current of the d.c. brushless motor is ($T = 2\pi/\omega$):

- for a 120⁰ square wave

$$I_a = \sqrt{\frac{2}{T} \int_0^{T/2} i_a^2(t)dt} = \sqrt{\frac{\omega}{\pi} \int_{\pi/(6\omega)}^{5\pi/(6\omega)} [I_a^{(sq)}]^2 dt}$$

$$= I_a^{(sq)} \sqrt{\frac{\omega}{\pi} \left(\frac{5}{6}\frac{\pi}{\omega} - \frac{1}{6}\frac{\pi}{\omega}\right)} = I_a^{(sq)} \sqrt{\frac{2}{3}} \tag{6.31}$$

- for a 180⁰ square wave

$$I_a = \sqrt{\frac{\omega}{\pi} \int_0^{\pi/\omega} [I_a^{(sq)}]^2 dt} = I_a^{(sq)} \sqrt{\frac{\omega}{\pi} \left(\frac{\pi}{\omega} - 0\right)} = I_a^{(sq)} \tag{6.32}$$

where $I_a^{(sq)}$ is the flat-topped value of the phase current.

The losses in the three-phase armature winding due to a 120⁰ square wave current are:

- for a Y-connected winding

$$\Delta P_a = 2R_{1dc}[I_a^{(sq)}]^2 \tag{6.33}$$

- for a Δ-connected winding

$$\Delta P_a = \frac{2}{3}R_{1dc}[I_a^{(sq)}]^2 \tag{6.34}$$

where R_{1dc} is the armature winding resistance per phase for d.c. current. The current $I_a^{(sq)}$ is equal to the inverter d.c. input voltage.

According to harmonic analysis

$$\Delta P_a = m_1 \sum_{n=1,5,7}^{\infty} R_{1dc} k_{1Rn} \left(\frac{I_{amn}}{\sqrt{2}} \right)^2 = m_1 \frac{1}{2} R_{1dc} \sum_{n=1,5,7}^{\infty} k_{1Rn} I_{amn}^2 \quad (6.35)$$

where k_{1Rn} is the skin effect coefficient (Appendix B) for the nth time harmonic and I_{amn} is the amplitude of the phase harmonic current. Neglecting the skin effect

$$\Delta P_a = m_1 \frac{1}{2} R_{1dc} \sum_{n=1,5,7}^{\infty} I_{amn}^2 = m_1 R_{1dc} I_a^2 \quad (6.36)$$

The *rms* current I_a in eqn (6.36) is composed of sinusoidal harmonic currents of different frequencies

$$I_a = \sqrt{\sum_{n=1}^{\infty} I_{an}^2} \quad (6.37)$$

This *rms* current is equivalent to that according to eqn (6.31). Putting $I_a^{(sq)} = I_a \sqrt{3/2}$ into eqn (6.33) or $I_a^{(sq)} = I_a \sqrt{3} \sqrt{3/2}$ into eqn (6.34) these equations become the same as eqn (6.36).

6.6 Torque ripple

The instantaneous torque of an electrical motor

$$T(\alpha) = T_0 + T_r(\alpha) \quad (6.38)$$

has two components (Fig. 6.10), i.e.:

- constant or average component T_0;
- periodic component $T_r(\alpha)$, which is a function of time or angle α, superimposed on the constant component.

The periodic component causes the *torque pulsation* called also *torque ripple*. There are many definitions of the torque ripple. Torque ripple can be defined in any of the following ways:

$$t_r = \frac{T_{max} - T_{min}}{T_{max} + T_{min}} \quad (6.39)$$

$$t_r = \frac{T_{max} - T_{min}}{T_{av}} \tag{6.40}$$

$$t_r = \frac{T_{max} - T_{min}}{T_{rms}} \tag{6.41}$$

$$t_r = \frac{[\text{torque ripple}]_{rms}}{T_{av}} = \frac{T_{rrms}}{T_{av}} \tag{6.42}$$

where the average torque in eqn (6.40)

$$T_{av} = \frac{1}{T_p} \int_{\alpha}^{\alpha+T_p} T(\alpha)d\alpha = \frac{1}{T_p} \int_0^{T_p} T(\alpha)d\alpha \tag{6.43}$$

and the *rms* or effective torque in eqn (6.41)

$$T_{rms} = \sqrt{\frac{1}{T_p} \int_0^{T_p} T^2(\alpha)d\alpha} \tag{6.44}$$

The torque ripple *rms* T_{rrms} in eqn (6.42) is calculated according to eqn (6.44) in which $T(\alpha)$ is replaced by $T_r(\alpha)$.

In eqns (6.43) and (6.44) T_p is the period of the torque waveform. For a sinusoidal waveform the half-cycle average value is $(2/\pi)T_m$, where T_m is the peak torque and the *rms* value is $T_m/\sqrt{2}$.

For the waveform containing higher harmonics, the *rms* value of the torque ripple is

$$T_{rrms} = \sqrt{T_{rrms1}^2 + T_{rrms2}^2 + \ldots + T_{rrms\nu}^2} \tag{6.45}$$

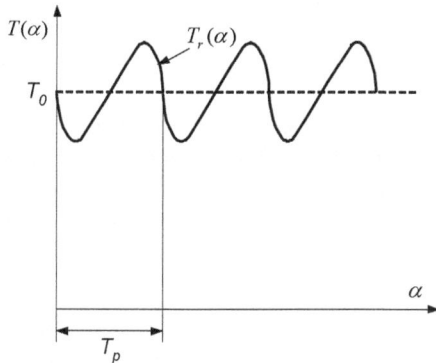

Fig. 6.10. Constant and periodic components of the torque.

6.6.1 Sources of torque pulsations

There are three sources of torque ripple coming from the machine:

(a) cogging effect (detent effect), i.e., interaction between the rotor magnetic flux and variable permeance of the air gap due to the stator slot geometry;
(b) distortion of sinusoidal or trapezoidal distribution of the magnetic flux density in the air gap;
(c) the difference between permeances of the air gap in the d and q axis.

The cogging effect produces the so-called *cogging torque*, higher harmonics of the magnetic flux density in the air gap produce the *field harmonic electromagnetic torque* and the unequal permeance in the d and q axis produces the *reluctance torque*.

The causes of torque pulsation coming from the supply are:

(d) current ripple resulting, e.g., from PWM;
(e) phase current commutation.

6.6.2 Numerical methods of instantaneous torque calculation

The FEM methods of the instantaneous torque calculation are given in Chapter 3, i.e., virtual work method — eqn (3.76), Maxwell stress tensor — eqn (3.72) and Lorentz force theorem — eqn (3.77). FEM offers more precise calculation of the magnetic field distribution than the analytical approaches. On the other hand, the FEM cogging torque calculations are subject to errors introduced by mesh generation or sometimes steel and PM characteristics [85].

6.6.3 Analytical methods of instantaneous torque calculation

Cogging torque

Neglecting the armature reaction and magnetic saturation, the *cogging torque* is independent of the stator current. The fundamental frequency of the cogging torque is a function of the number of slots s_1, number of pole pairs p and input frequency f. One of the cogging frequencies (usually fundamental) can be estimated as

$$f_c = s_1 n_s = s_1 \frac{f}{p} \qquad \text{if} \qquad N_{cog} = \frac{2p}{GCD(s_1, 2p)} = 1 \qquad (6.46)$$

$$f_c = 2n_{cog}f; \qquad n_{cog} = \frac{LCM(s_1, 2p)}{2p} \qquad \text{if} \qquad N_{cog} \geq 1 \quad (6.47)$$

where $LCM(s_1, 2p)$ is the least common multiple of the number of slots s_1 and number of poles $2p$, $GCD(s_1, 2p)$ is the greatest common divisor and n_{cog} is sometimes called the fundamental cogging torque index [142]. For example, for $s_1 = 36$ and $2p = 2$ the fundamental cogging torque index $n_{cog} = 18$ ($LCM = 36$, $GCD = 2$, $N_{cog} = 1$), for $s_1 = 36$ and $2p = 6$ the index $n_{cog} = 6$ ($LCM = 36$, $GCD = 6$, $N_{cog} = 1$), for $s_1 = 36$ and $2p = 8$ the index $n_{cog} = 9$ ($LCM = 72$, $GCD = 4$, $N_{cog} = 2$), for $s_1 = 36$ and $2p = 10$ the index $n_{cog} = 18$ ($LCM = 180$, $GCD = 2$, $N_{cog} = 5$), for $s_1 = 36$ and $2p = 12$ the index $n_{cog} = 3$ ($LCM = 36$, $GCD = 12$, $N_{cog} = 1$), etc. The larger the $LCM(s_1, 2p)$, the smaller the amplitude of the cogging torque.

Analytical methods usually ignore the magnetic flux in the stator slots and magnetic saturation of the stator teeth [42, 80, 316]. Cogging torque is derived from the magnetic flux density distribution either by calculating the rate of change of total stored energy in the air gap with respect to the rotor angular position [1, 42, 80, 141] or by summing the lateral magnetic forces along the sides of the stator teeth [316]. Neglecting the energy stored in the ferromagnetic core, the cogging torque is expressed as

$$T_c = \frac{dW}{d\theta} = \frac{D_{2out}}{2}\frac{dW}{dx} \tag{6.48}$$

where $D_{2out} \approx D_{1in}$ is the rotor outer diameter and $\theta = 2x/D_{2out}$ is the mechanical angle. Thus, the rate of change of the air gap energy is

$$W = gL_i \int \frac{b_g^2(x)}{2\mu_0} dx \tag{6.49}$$

where $b_g(x)$ is the air gap magnetix flux density expressed as a function of the x coordinate. In general,

$$b_g(x) = \frac{b_{PM}(x)}{k_C} + b_{sl}(x) \tag{6.50}$$

The first term $b_{PM}(x)$ is the magnetic flux density waveform excited by PM system with the stator slots being neglected and the second term $b_{sl}(x)$ is the magnetic flux density component due to slot openings. The quotient B_{mg}/k_C is the average value of the magnetic flux density in the air gap [76]. For trapezoidal shape of the magnetic flux density waveform

$$b_{PM}(x) = \frac{4}{\pi}B_{mg}\frac{1}{S}\sum_{\mu=1,3,5,...}^{\infty}\frac{1}{\mu^2}\sin(\mu S)\sin\left(\mu\frac{\pi}{\tau}x\right)k_{sk\mu} \tag{6.51}$$

where B_{mg} is the flat-topped value of the trapezoidal magnetic flux density waveform according to eqn (5.2) and $S = 0.5(\tau - b_p)\pi/\tau$. The PM skew factor for the cogging torque calculation is

$$k_{fsk\mu} = \frac{\sin[\mu b_{fsk}\pi/(2\tau)]}{\mu b_{fsk}\pi/(2\tau)} \tag{6.52}$$

where the skew of PMs $b_{fsk} \approx t_1$. It is equivalent to one slot pitch skew of the stator slots. If $b_{fsk} = 0$ and the stator slot skew $b_{sk} > 0$, the equivalent PM skew factor is

$$k_{sk\mu} = \frac{\sin(\mu b_{sk}\pi/\tau)}{\mu b_{sk}\pi/\tau} \qquad (6.53)$$

The magnetic flux density component due to openings of the stator slots with magnetic field in slots taken into account is [86]

$$b_{sl}(x) = -2\gamma_1 \frac{g'}{t_1} b_{PM}(x) \sum_{k=1,2,3,...}^{\infty} k_{ok}^2 k_{skk} \cos\left(k\frac{2\pi}{t_1}x\right) \qquad (6.54)$$

where the stator slot skew factor for the cogging torque calculation

$$k_{skk} = \frac{\sin[kb_{sk}\pi/(2\tau)]}{kb_{sk}\pi/(2\tau)} \qquad (6.55)$$

and the coefficient k_{ok} of the stator slot opening is according to eqn (5.105). The coefficient k_{ok} contains auxiliary function ρ according to eqn (5.107) and auxiliary function γ_1 according to eqn (5.108). Eqn (6.54) originally derived by Dreyfus [86] does not contain the stator slot skew factor k_{skk}. However, it can easily be generalized for stators with skewed slots. For most PM brushless motors the equivalent air gap in eqn (6.54) is $g' \approx g + h_M/\mu_{rrec}$. For buried magnet rotors according to Figs 5.1b, 5.1i and 5.1j the equivalent air gap is $g' = g$.

The cogging torque then can be calculated as follows

$$T_c(X) = \frac{gL_i}{2\mu_0}\frac{D_{1in}}{2}\frac{d}{dX}\int_{X+a}^{X+b} b_g^2(x)dx \qquad (6.56)$$

With a stationary stator, only the magnetic flux density excited by the rotor PMs depends on the rotor position with respect to the coordinate system fixed to the stator [144]. Thus, it is justified to assume $b_{PM}(x) = B_{mg}/k_C$ and, after performing integration with respect to x, the cogging torque has the following form

$$T_c(X) = \frac{gL_i}{2\mu_0}\frac{D_{1in}}{2}A_T \sum_{k=1,2,3,...}^{\infty} \left\{0.5A_T\zeta_k^2 \left[\cos[\frac{4k\pi}{t_1}(X+b)] - \cos[\frac{4k\pi}{t_1}(X+a)]\right]\right.$$

$$\left. +\zeta_k\frac{B_{mg}}{k_C}\left[\cos[\frac{2k\pi}{t_1}(X+b)] - \cos[\frac{2k\pi}{t_1}(X+a)]\right]\right\} \qquad (6.57)$$

where

$$A_T = -2\gamma\frac{g}{t_1}B_{mg} \qquad \text{and} \qquad \zeta_k = k_{ok}^2 k_{skk} \qquad (6.58)$$

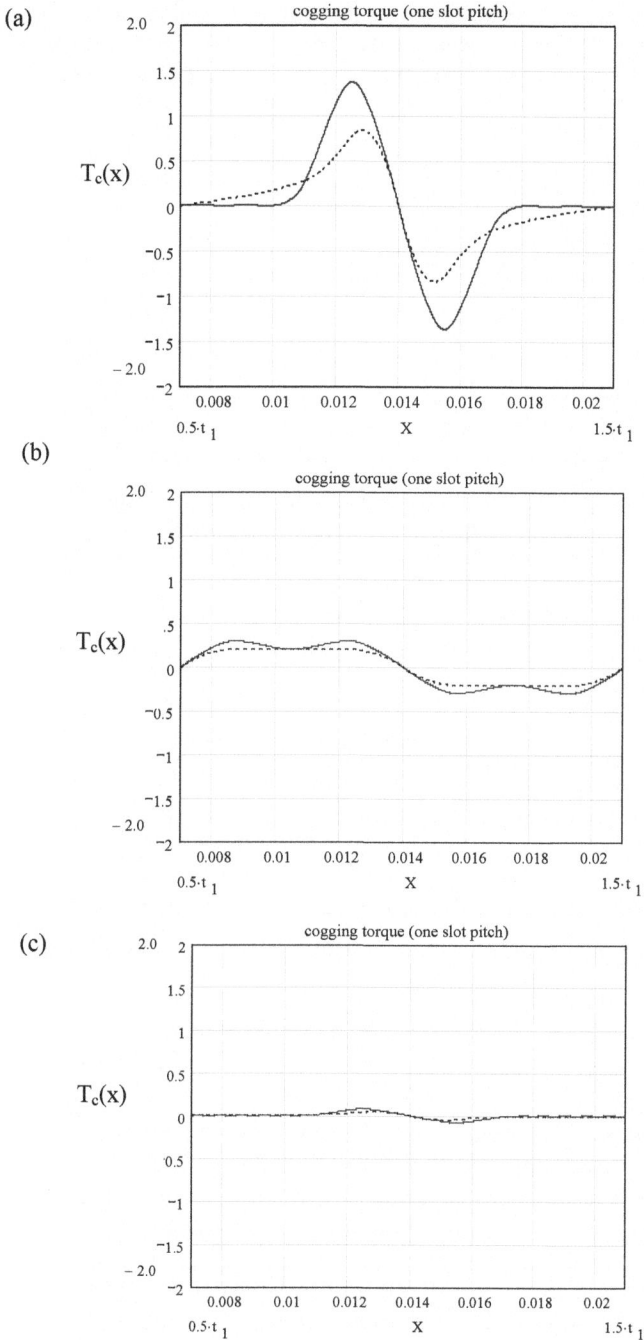

Fig. 6.11. Cogging torque as a function of the rotor position ($0.5t_1 \leq x \leq 1.5t_1$) for a medium power PM brushless motor ($m_1 = 3$, $2p = 10$, $s_1 = 36$, $t_1 = 13.8$ mm, $b_{14} = 3$ mm, $g = 1$ mm, $B_{mg} = 0.83$ T): (a) $b_{sk}/t_1 = 0.1$, (b) $b_{sk}/t_1 = 0.5$, (c) $b_{sk}/t_1 = 0.95$. Solid line: calculations according to eqn (6.60), dash line: calculations according to eqn (6.64).

The coefficent k_{ok} is acording to eqn (5.105) and coefficient k_{skk} is according to eqn (6.55). The first two terms $\cos[4k\pi(X+b)/t_1]$ and $\cos[4k\pi(X+a)/t_1]$ are negligible and eqn (6.57) can be brought to the following simpler form

$$T_c(X) = \frac{gL_i}{2\mu_0}\frac{D_{1in}}{2}A_T\frac{B_{mg}}{k_C}\sum_{k=1,2,3,...}^{\infty}\zeta_k\left[\cos[\frac{2k\pi}{t_1}(X+b)] - \cos[\frac{2k\pi}{t_1}(X+a)]\right]$$

(6.59)

Putting $a = 0.5b_{14}$ and $b = 0.5b_{14} + c_t$ [80] where $c_t = t_1 - b_{14}$ is the stator tooth width, the cogging torque equation becomes

$$T_c(X) = -\frac{gL_i}{\mu_0}\frac{D_{1in}}{2}A_T\frac{B_{mg}}{k_C}\sum_{k=1,2,3,...}^{\infty}(-1)^k\zeta_k\sin\left(k\frac{\pi}{t_1}c_t\right)\sin\left(\frac{2k\pi}{t_1}X\right)$$

(6.60)

In the literature, e.g. [1, 42, 141], the variation of T_c with the mechanical angle $\theta = 2\pi X/(s_1 t_1)$ is expressed by the following empirical equation

$$T_c(\theta) = \sum_{k=1,2,3,...}^{\infty}T_{mk}\chi_{skk}\sin(kN_{cm}\theta)$$

(6.61)

where the skew factor

$$\chi_{skk} = \frac{\sin[kb_{sk}N_{cm}\pi/(t_1 s_1)]}{kb_{sk}N_{cm}\pi/(t_1 s_1)}$$

(6.62)

and b_{sk} is the circumferential skew of stator slots. Putting

$$T_m = \frac{gL_i}{2\mu_0}\frac{D_{1in}}{2}A_T\frac{B_{mg}}{k_C}$$

(6.63)

the following approximate cogging torque equation can also be used

$$T_c(X) = T_m\sum_{k=1}^{\infty}\zeta_k\sin\left(k\frac{2\pi}{t_1}X\right)$$

(6.64)

Eqn (6.64) gives a similar distribution of the cogging torque waveform within one slot pitch interval as eqn (6.60). However, the shape of the cogging torque according to eqns (6.60) and (6.64) is different.

Calculation results of the cogging torque for a medium power PM brushless motor with $m_1 = 3$, $2p = 10$ (embedded magnets) and $s_1 = 36$ are shown in Figs 6.11 and 6.12. Assuming two components $b_{PM}(x)$ and $b_{sl}(x)$ of the magnetic flux density in eqn (6.50) and integrating and differentiating $b_g^2(x)$ numerically, the peak value of the cogging torque waveform depends on the number of slots s_1, slot opening b_{14}, slot (tooth) pitch t_1, air gap g, coefficient ζ_k, number of poles $2p$ and shape of $b_{PM}(x)$ (Fig. 6.12). The fundamental cogging torque index $n_{cog} = 18$ and according to eqn (6.47) the predominant

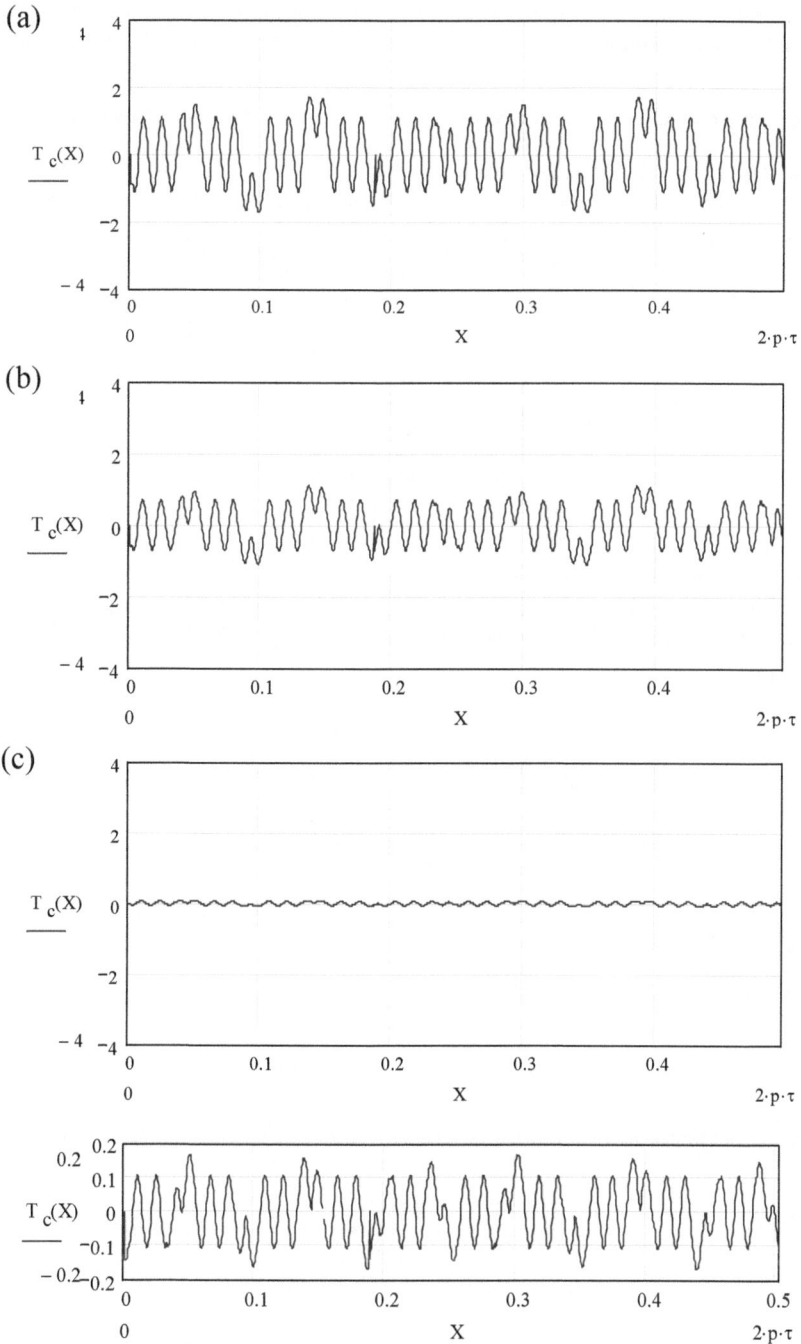

Fig. 6.12. Cogging torque as a function of the rotor position ($0 \leq x \leq 2p\tau$) for a medium power PM brushless motor ($m_1 = 3$, $2p = 10$, $s_1 = 36$, $t_1 = 13.8$ mm, $b_{14} = 3$ mm, $g = 1$ mm, $B_{mg} = 0.83$ T): (a) $b_{sk}/t_1 = 0.1$, (b) $b_{sk}/t_1 = 0.5$, (c) $b_{sk}/t_1 = 0.95$. Calculations according to eqns (6.48) and (6.49).

cogging torque frequency $f_c = 36f$. The same frequency can be obtained from eqn (6.46), if mutiplied by $N_{cog} = 5$.

Torque ripple due to distortion of EMF and current waveforms

Torque ripple due to the shape and distortion of the EMF and current waveforms is called the *commutation torque*. Assuming no rotor currents (no damper, very high resistivity of magnets and pole faces) and the same stator phase resistances, the Kirchhoff's voltage equation for a three phase machine can be expressed in the following matrix form [265]

$$
\begin{bmatrix} v_{1A} \\ v_{1B} \\ v_{1C} \end{bmatrix} = \begin{bmatrix} R_1 & 0 & 0 \\ 0 & R_1 & 0 \\ 0 & 0 & R_1 \end{bmatrix} \begin{bmatrix} i_{aA} \\ i_{aB} \\ i_{aC} \end{bmatrix}
$$

$$
+ \frac{d}{dt} \begin{bmatrix} L_A & L_{BA} & L_{CA} \\ L_{BA} & L_B & L_{CB} \\ L_{CA} & L_{CB} & L_C \end{bmatrix} \begin{bmatrix} i_{aA} \\ i_{aB} \\ i_{aC} \end{bmatrix} + \begin{bmatrix} e_{fA} \\ e_{fB} \\ e_{fC} \end{bmatrix} \tag{6.65}
$$

For inductances independent of the rotor angular position the self-inductances $L_A = L_B = L_C = L$ and mutual inductances between phases $L_{AB} = L_{CA} = L_{CB} = M$ are equal. For no neutral wire $i_{aA} + i_{aB} + i_{aC} = 0$ and $Mi_{aA} = -Mi_{aB} - Mi_{aC}$. Hence

$$
\begin{bmatrix} v_{1A} \\ v_{1B} \\ v_{1C} \end{bmatrix} = \begin{bmatrix} R_1 & 0 & 0 \\ 0 & R_1 & 0 \\ 0 & 0 & R_1 \end{bmatrix} \begin{bmatrix} i_{aA} \\ i_{aB} \\ i_{aC} \end{bmatrix}
$$

$$
+ \begin{bmatrix} L_1 - M & 0 & 0 \\ 0 & L_1 - M & 0 \\ 0 & 0 & L_1 - M \end{bmatrix} \frac{d}{dt} \begin{bmatrix} i_{aA} \\ i_{aB} \\ i_{aC} \end{bmatrix} + \begin{bmatrix} e_{fA} \\ e_{fB} \\ e_{fC} \end{bmatrix} \tag{6.66}
$$

The instantaneous electromagnetic torque is the same as that obtained from the Lorentz equation (3.77), i.e.,

$$
T_d = \frac{1}{2\pi n} [e_{fA} i_{aA} + e_{fB} i_{aB} + e_{fC} i_{aC}] \tag{6.67}
$$

For a bipolar commutation and 120^0 conduction only two phases conduct at any time instant. For example, if $e_{fA} = E_f^{(tr)}$, $e_{fB} = -E_f^{(tr)}$, $e_{fC} = 0$, $i_{aA} = I_a^{(sq)}$, $i_{aB} = -I_a^{(sq)}$ and $i_{aC} = 0$, the instantaneous electromagnetic torque according to eqn (6.67) is [52, 245]

$$
T_d = \frac{2E_f^{(tr)} I_a^{(sq)}}{2\pi n} \tag{6.68}
$$

where $E_f^{(tr)}$ and $I_a^{(sq)}$ are flat topped values of trapezoidal EMF and square wave current (Fig. 6.13). For constant values of EMF and currents, the torque (6.68) does not contain any pulsation [245].

Since $e_f = \omega\psi_f = (2\pi n/p)\psi_f$ where ψ_f is the flux linkage per phase produced by the excitation system, the instantaneous torque (6.67) becomes

$$T_d = p(\psi_{fA}i_{aA} + \psi_{fB}i_{aB} + \psi_{fC}i_{aAC}) \qquad (6.69)$$

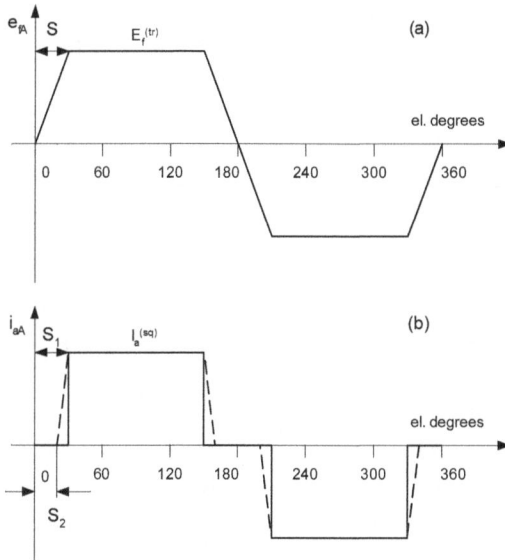

Fig. 6.13. Idealized (solid lines) and practical (dash lines) waveforms of (a) phase EMF, (b) phase current.

To obtain torque for periodic waves of the EMF and current, it is necessary to express those waveforms in form of a Fourier series. The magnetic flux on the basis of eqns (5.6) and (6.51)

$$\Phi_f(t) = L_i \int_0^{w_c} b_{PM}(x)\cos(\mu\omega t)dx$$

$$= \frac{4}{\pi^2}B_{mg}\frac{\tau L_i}{S}\sum_{\mu=1,3,5,\ldots}^{\infty}\frac{1}{\mu^3}\sin(\mu S)k_{sk\mu}[1-\cos(\mu\frac{\pi}{\tau}w_c)]\cos(\mu\omega t) \qquad (6.70)$$

where B_{mg} is the flat-topped value of the trapezoidal magnetic flux density waveform (5.2), w_c is the coil pitch, S is according to Fig. 6.13 and $k_{sk\mu}$ is according to eqn (6.53).

The EMF induced in the phase A is calculated on the basis of electromagnetic induction law according to eqn (3.12)

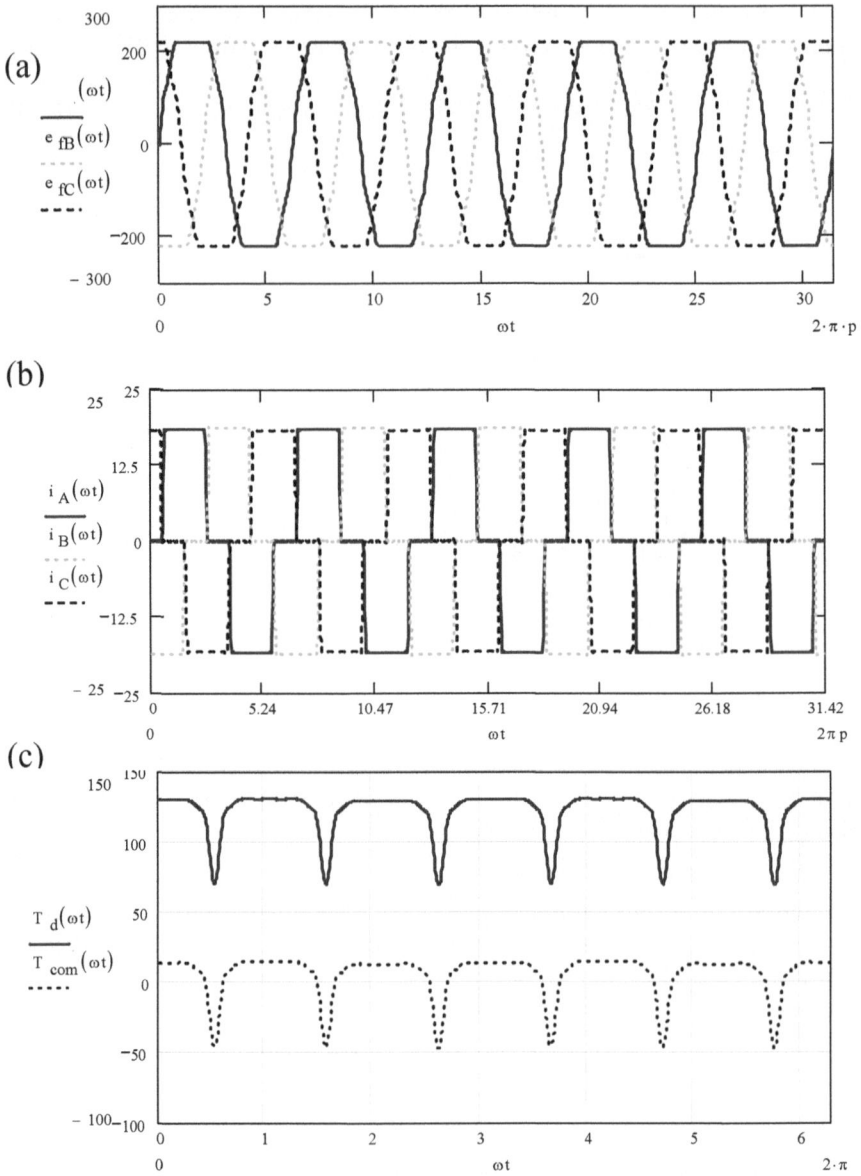

Fig. 6.14. Electromagnetic torque produced by a medium power PM brushless motor ($m_1 = 3$, $2p = 10$, $s_1 = 36$): (a) trapezoidal EMF waveforms, (b) rectangular current waveforms, (c) resultant electromagnetic torque T_d and its commuation component T_{com}. The current commutation angle is 5^0.

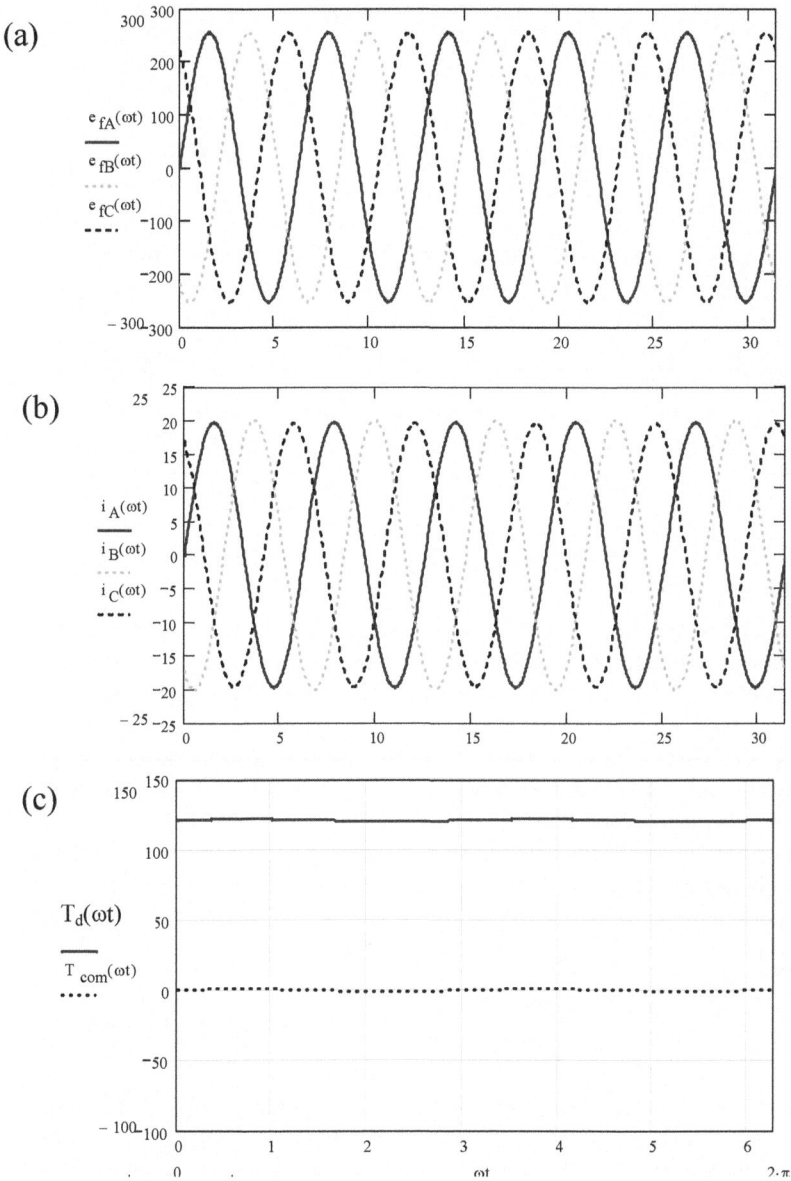

Fig. 6.15. Electromagnetic torque produced by a medium power PM brushless motor ($m_1 = 3$, $2p = 10$, $s_1 = 36$): (a) sinusoidal EMF waveforms, (b) sinusoidal current waveforms, (c) resultant electromagnetic torque T_d and its commuation component T_{com}.

$$e_{fA}(t) = -N_1 \left[k_{d1} \frac{d\Phi_{f1}(t)}{dt} + k_{d3} \frac{d\Phi_{f3}(t)}{dt} + \ldots + k_{d\mu} \frac{d\Phi_{f\mu}(t)}{dt} \right]$$

$$= \frac{8}{\pi} B_{mg} \frac{\tau L_i}{S} f N_1 \sum_{\mu=1,3,5,\ldots}^{\infty} \frac{1}{\mu^2} \sin(\mu S) k_{d\mu} k_{sk\mu} \left[1 - \cos\left(\mu \frac{\pi}{\tau} w_c\right) \right] \sin(\mu\omega t)$$

$$(6.71)$$

where $k_{d\mu} = \sin[\mu\pi/(2m_1)]/\{q_1 \sin[\mu\pi/(2m_1 q_1)]\}$ is the distribution factor for the μth harmonic. The EMFs $e_{fB}(t)$ and $e_{fC}(t)$ induced in the remaining phase windings are shifted by $2\pi/3$ and $-2\pi/3$, respectively. If $b_{sk} > 0$, the EMF is practically sinusoidal, independent of the number of higher harmonics μ taken into account.

For phase A, the current rectangular waveform contains higher time harmonics $n = 1, 3, 5\ldots$ and can be expressed with the aid of the following Fourier series

$$i_{aA}(t) = \frac{4}{\pi} I_a^{(sq)} \sum_{n=1,3,5,\ldots}^{\infty} \frac{1}{n} \cos(nS_1) \sin(\omega t) \qquad (6.72)$$

Harmonics of the magnetic flux and current of the same order produce constant torque. In other words, the constant torque is produced by all harmonics $2m_1 l \pm 1$ where $l = 0, 1, 2, 3, \ldots$ of the magnetic flux and armature currents. Harmonics of the magnetic flux and current of different order produce pulsating torque. However, when the 120^0 trapezoidal flux density waveform interacts with 120^0 rectangular current, only a steady torque is produced with no torque pulsations [245, 246].

In practice, the armature current waveform is distorted and differs from the rectangular shape. It can be approximated by the following trapezoidal function

$$i_{aA}(t) = \frac{4 I_a^{(sq)}}{\pi(S_1 - S_2)} \sum_{n=1,3,5,\ldots}^{\infty} \frac{1}{n^2} [\sin(nS_1) - \sin(nS_2)] \sin(n\omega t) \qquad (6.73)$$

where $S_1 - S_2$ is the commutation angle in radians. The EMF wave also differs from 120^0 trapezoidal functions. In practical motors, the conduction angle is from 100^0 to 150^0, depending on the construction. Deviations of both current and EMF waveforms from ideal functions results in producing torque pulsations [52, 245]. Peak values of individual harmonics can be calculated on the basis of eqns (6.71), (6.72) and (6.73).

Using the realistic shapes of waveforms of EMFs and currents as those in practical motors, eqn (6.67) takes into account all components of the torque ripple except the cogging (detent) component.

Fig. 6.14 shows trapezoidal EMF, trapezoidal current and electromagnetic torque waveforms while Fig. 6.15 shows the same waveforms in the case of sinusoidal EMFs and currents. The torque component due to phase commutation currents $T_{com} = T_d - T_{av}$ where T_{av} is the average electromagnetic torque. Predominant frequencies of torque ripple due to phase commutation are $f_{com} = 2lm_1 f$ where $l = 1, 2, 3, \ldots$. For sinusoidal EMF and current waveforms the electromagnetic torque is constant and does not contain any ripple. Torques shown in Fig. 6.15 are not ideally smooth due to slight current unbalance.

6.6.4 Minimization of torque ripple

The torque ripple can be minimized both by the proper motor design and motor control. Measures taken to minimize the torque ripple by motor design include elimination of slots, skewed slots, special shape slots and stator laminations, selection of the number of stator slots with respect to the number of poles, decentered magnets, skewed magnets, shifted magnet segments, selection of magnet width, direction-dependent magnetization of PMs. Control techniques use modulation of the stator current or EMF waveforms [45, 102].

Slotless windings

Since the cogging torque is produced by the PM field and stator teeth, a slotless winding can totally eliminate the cogging torque. A slotless winding requires increased air gap which in turn reduces the PM excitation field. To keep the same air gap magnetic flux density, the height h_M of PMs must be increased. Slotless PM brushless motors use more PM material than slotted motors.

Skewing stator slots

Normally, the stator slot skew b_{sk} equal to one slot pitch t_1 can reduce the cogging torque practically to zero value — eqns (6.61) and (6.62). Sometimes, the optimal slot skew is less than one slot pitch [197]. On the other hand, the stator slot skew reduces the EMF which results in deterioration of the motor performance. Skewed slots are less effective in the case of rotor eccentricity.

Shaping stator slots

Fig. 6.16 shows methods of reducing the cogging torque by shaping the stator slots, i.e.,

(a) bifurcated slots (Fig. 6.16a),
(b) empty (dummy) slots (Fig. 6.16b),
(c) closed slots (Fig. 6.16c),
(d) teeth with different width of the active surface (Fig. 6.16d).

Bifurcated slots (Fig. 6.16a) can be split in more than two segments. The cogging torque increases with the increase of the slot opening. When designing closed slots (Fig. 6.16c), the bridge between the neighboring teeth must be properly designed. Too thick a bridge (thickness measured in radial direction) increases the stator slot leakage to an unacceptable level. Too narrow a bridge can be ineffective due to high saturation. Because closed slots can only accept "sewed" coils, it is better to close the slots by inserting an internal sintered powder cylinder or make separately the stator yoke and tooth-slot section. In the second case stator slots are open externally from the frame side.

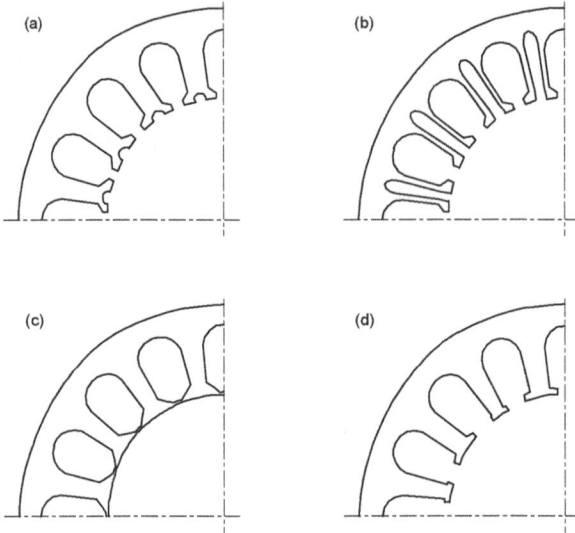

Fig. 6.16. Minimization of the cogging torque by shaping the stator slots: (a) bifurcated slots, (b) empty slots, (c) closed slots, (d) teeth with different width of the active surface.

Selection of the number of stator slots

The least common multiple $LCM(s_1, 2p)$ of the the slot number s_1 and pole number $2p$ has a significant effect on the cogging torque. As this number increases the cogging torque decreases [141]. Similarly, the cogging torque increases as the greatest common divisor $GCD(s_1, 2p)$ of the slot number and pole number increases [61, 141].

Shaping PMs

PMs thinner at the edges than in the center (Fig. 6.17) can reduce both the cogging and commutation torque ripple. Magnet shapes according to Fig.

6.17b require a polygonal cross section of the rotor core. Decentered PMs together with bifurcated stator slots can supress the cogging torque as effectively as skewed slots with much less reduction of the EMF [270].

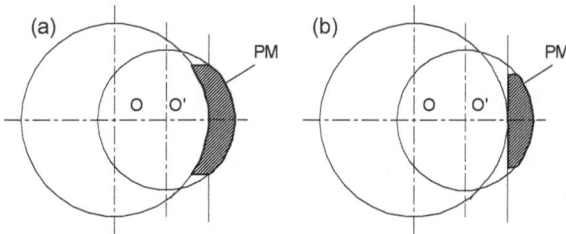

Fig. 6.17. Decentered PMs: (a) arc shaped, (b) bread loaf type.

Skewing PMs

The effect of skewed PMs on the cogging torque suppression is similar to that of skewing the stator slots. Fabrication of twisted magnets for small rotor diameters and small number of poles is rather difficult. Bread loaf-shaped PMs with edges cut aslant are equivalent to skewed PMs.

Shifting PM segments

Instead of designing one long magnet per pole, it is sometimes more convenient to divide the magnet axially into $K_s = 3$ to 6 shorter segments. Those segments are then shifted one from each other by equal distances t_1/K_s or unequal distances the sum of which is t_1. Fabrication of short, straight PM segments is much easier than long, twisted magnets.

Selection of PM width

Properly selected PM width with respect to the stator slot pitch t_1 is also a good method to minimize the cogging torque. The magnet width (pole shoe width) is to be $b_p = (k + 0.14)t_1$ where k is an integer [197] or including the pole curvature $b_p = (k + 0.17)t_1$ [159]. Cogging torque reduction require wider pole shoe than the multiple of slot pitch.

Magnetization of PMs

There is a choice between parallel, radial and direction-dependent magnetization. For example, if a ring-shaped magnet of a small motor is placed around the magnetizer poles without external magnetic circuit, the magnetization vectors will be arranged similar to Mallinson–Halbach array. This method also minimizes the torque ripple.

Creating magnetic circuit asymmetry

Magnetic circuit asymmetry can be created by shifting each pole by a fraction of the pole pitch with respect to the symmetrical position or designing different sizes of North and South magnets of the same pole pair [42].

6.7 Rotor position sensing of d.c. brushless motors

Rotor position sensing in PM d.c. brushless motors is done by *position sensors*, i.e., Hall elements, encoders or resolvers. In rotary machines position sensors provide feedback signals proportional to the rotor angular position.

6.7.1 Hall sensors

The *Hall element* is a magnetic field sensor. When placed in a stationary magnetic field and fed with a d.c. current it generates an output voltage

$$V_H = k_H \frac{1}{\delta} I_c B \sin \beta \qquad (6.74)$$

where k_H is Hall constant in m^3/C, δ is the semiconductor thickness, I_c is the applied current, B is the magnetic flux density and β is the angle between the vector of B and Hall element surface. The polarity depends on whether the pellet is passing a North or a South pole. Thus it can be used as a magnetic flux detector (Fig. 6.18).

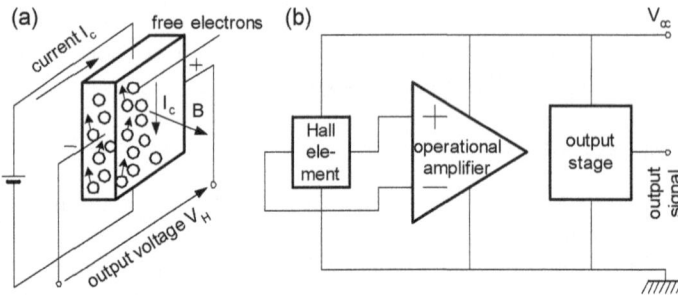

Fig. 6.18. Hall element: (a) principle of operation, (b) block diagram of a Hall integrated circuit.

Rotor position sensing of three phase d.c. brushless motors requires three Hall elements (Fig. 6.19). All the necessary components are often fabricated in an integrated chip (IC) in the arrangement shown in Fig. 6.18b. In most cases satisfactory operation requires the mechanical separation of the Hall elements to be given by

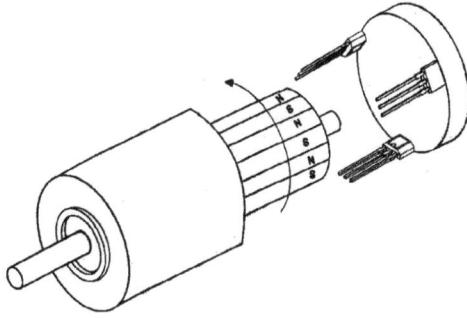

Fig. 6.19. Arrangement of Hall elements in a three-phase PM brushless motor.

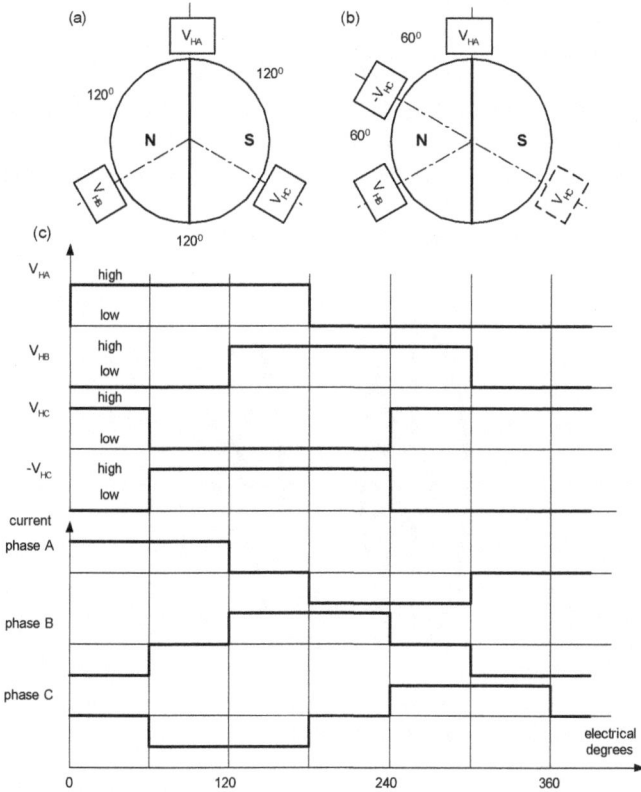

Fig. 6.20. Hall sensor-based three-phase commutation: (a) sensor spacing for 120 electrical degrees, (b) sensor spacing for 60 electrical degrees, (c) sensor signals and phase currents.

$$\alpha_H = \frac{360°}{m_1 p} \qquad (6.75)$$

For example, in the case of a two-pole ($p = 1$), three-phase ($m_1 = 3$) d.c. brushless motor, a mechanical displacement of 120° between individual Hall-effect devices is required. The sensors should be placed 120° apart as in Fig. 6.20a. However, they can also be placed at 60° intervals as shown in Fig. 6.20b. Hall sensors generate a square wave with 120^0 phase difference, over one electrical cycle of the motor. The inverter or servo amplifier drives two of the three motor phases with d.c. current during each specific Hall sensor state (Fig. 6.20c).

6.7.2 Encoders

There are two types of *optical encoders*: absolute and incremental encoders.

In optical encoders a light passes through the transparent areas of a grating and is sensed by a photodetector. To increase the resolution, a collimated light source is used and a mask is placed between the grating and detector. The light is allowed to pass to the detector only when the transparent sections of the grating and mask are in alignment (Fig. 6.21).

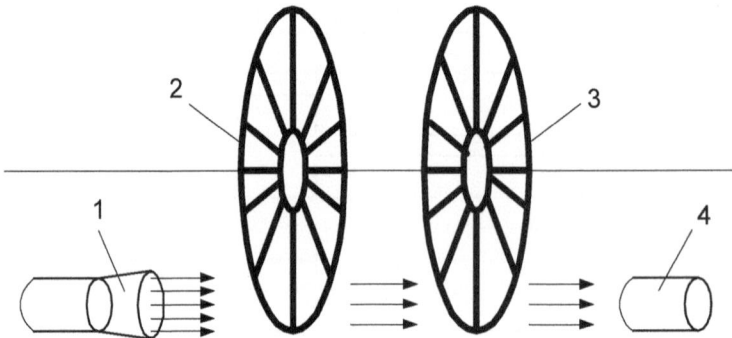

Fig. 6.21. Principle of operation of an optical encoder: 1 — collimated light source, 2 — grating, 3 — mask, 4 — detector of light.

In an *incremental encoder* a pulse is generated for a given increment of shaft angular position which is determined by counting the encoder output pulses from a reference. The rotating disk (grating) has a single track (Fig. 6.22a). In the case of power failure an incremental encoder loses position information and must be reset to known zero point.

To provide information on direction of rotation a two-channel encoder is used. The output square signals of two channels are shifted by 90^0; i.e.,

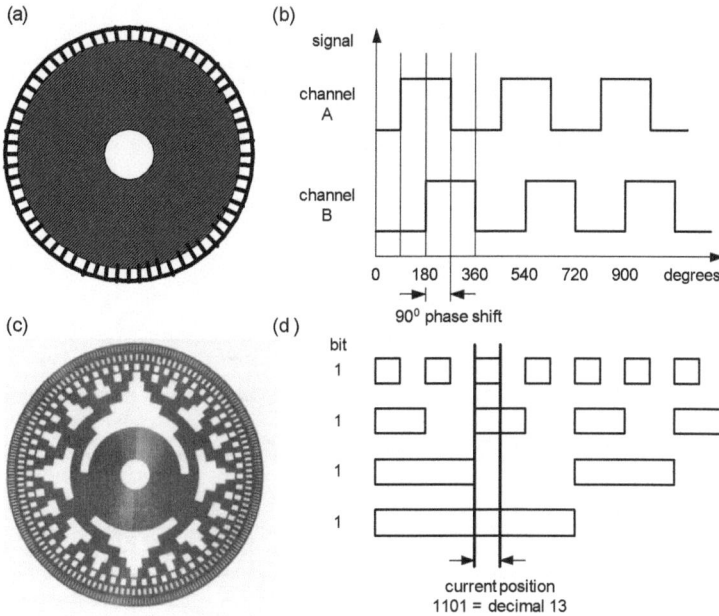

Fig. 6.22. Incremental and absolute encoders: (a) disk of an incremental encoder, (b) quadrature output signals, (c) disk of an absolute encoder, (d) absolute encoder output signals. Courtesy of Parker Hannifin Corporation, Rohnert Park, CA, U.S.A.

channels are arranged to be in *quadrature* (Fig. 6.22b). Four pulse edges from channels A and B may be seen during the same period. By processing the two channel outputs to produce a separate pulse for each square wave edge the resolution of the encoder is quadrupled. The range of resolution of square wave output encoders is wide, up to a few thousand lines/rev.

The *slew rate* of an incremental encoder is the maximum speed determined by the maximum frequency at which it operates. If this speed is exceeded, the accuracy will significantly deteriorate making the output signal unreliable.

An *absolute encoder* is a position verification device that provides unique position information for each shaft angular location. Owing to a certain number of output channels, every shaft angular position is described by its own unique code. The number of channels increases as the required resolution increases. An absolute encoder is not a counting device like an incremental encoder and does not lose position information in the case of loss of power.

The disk (made of glass or metal) of an absolute encoder has several concentric tracks to which independent light sources are assigned (Fig. 6.22c). The tracks vary in the slot size (smaller slots at the outer edge enlarging towards the center) and pattern of slots. A high state or "1" is created when the light passes through a slot. A low state or "0" is created when the light does

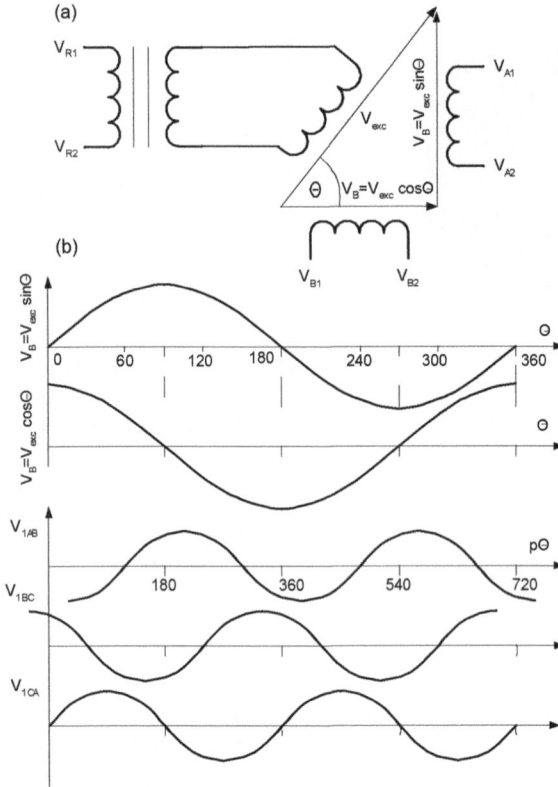

Fig. 6.23. Principle of operation of a rotary resolver: (a) winding configuration, (b) resolver output waves and 4-pole motor voltages. The interval $0 \leq p\theta \leq 720^0$ corresponds to one mechanical revolution.

not pass through the disk. Patterns "1" and "0" provide information about the shaft position. The resolution or the amount of position information that can be obtained from the absolute encoder disk is determined by the number of tracks; i.e., for 10 tracks the resolution is usually $2^{10} = 1,024$ positions per one revolution. The disk pattern is in machine readable code, i.e., binary or gray codes. Fig. 6.22d shows a simple binary output with 4 bits of information. The indicated position corresponds to the decimal number 13 (1101 binary). The next position moving to the right corresponds to 5 (0101 binary) and the previous position moving to the left corresponds to 3 (0011 binary).

Multi-turn absolute encoders have additional disks geared to the main high resolution disk with step-up gear ratio. For example, adding a second disk with 3 tracks and 8:1 gear ratio to the main disk with 1024 positions per revolution, the absolute encoder will have 8 complete turns of the shaft equivalent to 8192 discrete positions.

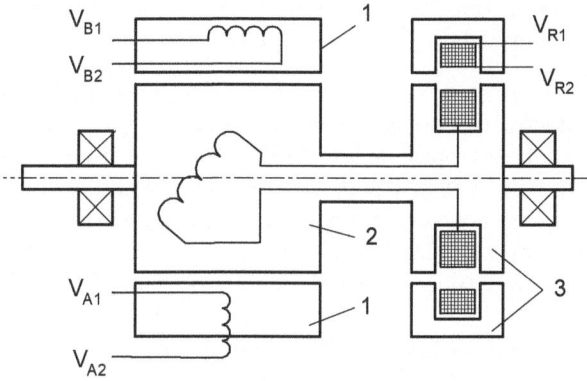

Fig. 6.24. Brushless resolver. 1 — stator, 2 — rotor, 3 — rotary transformer.

6.7.3 Resolvers

A *resolver* is a rotary electromechanical transformer that provides outputs in forms of trigonometric functions of its inputs. For detecting the rotor position of brushless motors, the excitation or primary winding is mounted on the resolver rotor and the output or secondary windings are wound at right angles to each other on the stator core. As a result the output signals are sinusoidal waves in quadrature; i.e., one wave is a sinusoidal function of the angular displacent θ and the second wave is a cosinusoidal function of θ (Fig. 6.23a).

There is one electrical cycle for each signal for each revolution of the motor (Fig. 6.23b). The analog output signals are converted to digital form to be used in a digital positioning system. The difference between the two waves reveals the position of the rotor. The speed of the motor is determined by the period of the waveforms and the direction of rotation is determined by the leading waveform.

Instead of delivering the excitation voltage to the rotor winding by brushes and slip rings, an inductive coupling system is frequently used. A *brushless resolver* with rotary transformer is shown in Fig. 6.24.

6.8 Sensorless motors

There are several reasons to eliminate electromechanical position sensors:

- cost reduction of electromechanical drives
- reliability improvement of the system
- temperature limits on Hall sensors
- in motors rated below 1 W the power consumption by position sensors can substantially reduce the motor efficiency

- in compact applications, e.g., computer hard disk drives, it may not be possible to accommodate position sensors.

For PM motors rated up to 10 kW the cost of an encoder is below 10% of the motor manufacturing cost and depends on the motor rating and encoder type. Elimination of electromechanical sensors and associated cabling not only improves the reliability but also simplifies the installation of the system.

By using sensorless control, computer hard disk drive (HDD) manufacturers can push the limits of size, cost, and efficiency. Owing to the sensorless PM motor, a 2.5-inch hard disk drive requires less than 1 W power to operate [25].

Temperature limits on Hall elements may prevent d.c. brushless motors from being used in freon-cooled compressors. Sensorless motors are applicable in refrigerators and air conditioners.

Sensorless control strategies are different for PM d.c. brushless motors with trapezoidal EMF waveforms where only two out of three phases are simultaneuosly excited and PM synchronous motors or motors with sinusoidal EMF waveforms where all three phases are excited at any instant of time. The simplest methods for PM d.c. brushless motors are based on *back EMF detection* in an unexcited phase winding. Sensorless controllers measure back EMF signals from the unenergized winding to determine the commutation point.

In general, the position information of the shaft of PM brushless motors can be obtained using one of the following techniques [254]:

(a) detection of back EMF (zero crossing approach, phase-locked loop technique, EMF integration approach);

(b) detection of the stator third harmonic voltage;

(c) detection of the conducting interval of free-wheeling diodes connected in antiparallel with the solid state switches;

(d) sensing the inductance variation (in the d and q axis [6]), terminal voltages and currents.

Techniques (a), (b) and (c) are usually used for PM d.c. brushless motors with trapezoidal EMF waveforms. Methods (b) and (d) are used for PM synchronous motors and brushless motors with sinusoidal EMF waveforms.

From the capability at zero speed operation point of view the above methods can be classified into two categories:

- not suitable for detection of signals at standstill or very low speed
- suitable for zero and very low speed.

Normally, methods (a), (b) and (c) cannot be used at zero and very low speed. The principle of inductance variation (d) is well suited for variable speed drives where speed changes in a wide range including standstill. On the other hand, at reversals and overloadings the high armature current increases the saturation of magnetic circuit and the difference between d and q-axis inductance may not be sufficient to detect the rotor position.

Details of different methods of sensorless control are given, e.g., in [6, 25, 73, 101, 254].

6.9 Motion control of PM brushless motors

6.9.1 Converter-fed motors

Most PM d.c. brushless motors are fed from voltage-source, PWM solid state converters. A power electronics converter consists of a rectifier, intermediate circuit (filter) and inverter (d.c. to a.c. conversion). IGBT inverter power circuits are shown in Fig. 6.25. To avoid circulating currents in Δ-connected windings, the resistances and inductances of all phases should be the same and the winding distribution around the armature core periphery should be symmetrical.

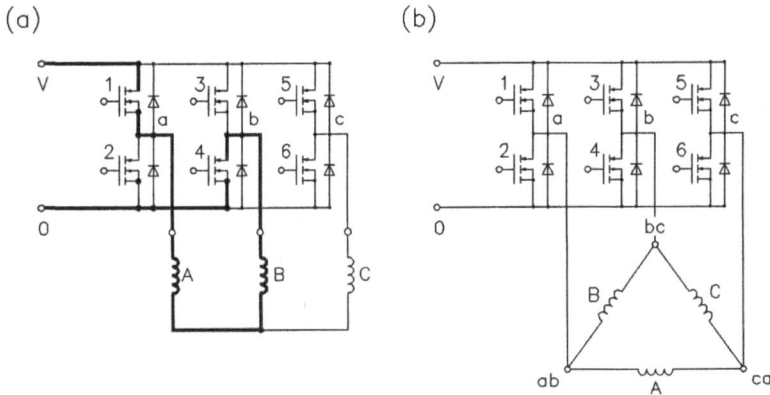

Fig. 6.25. IGBT inverter-fed armature circuits of d.c. brushless motors: (a) Y-connected phase windings, (b) Δ-connected armature windings.

A wiring diagram for a converter-fed motor is shown in Fig. 6.26. To obtain proper operation, minimize radiated noise and prevent shock hazard, proper interconnection wiring, grounding and shielding are important. Many solid state converters require minimum 1 to 3% line impedance calculated as

$$z\% = \frac{V_{10L-L} - V_{1rL-L}}{V_{1rL-L}} \times 100\% \tag{6.76}$$

where V_{10L-L} is the line-to-line voltage measured at no load and V_{1rL-L} is the line-to-line voltage measured at full rated load. The minimum required inductance of the line reactor is

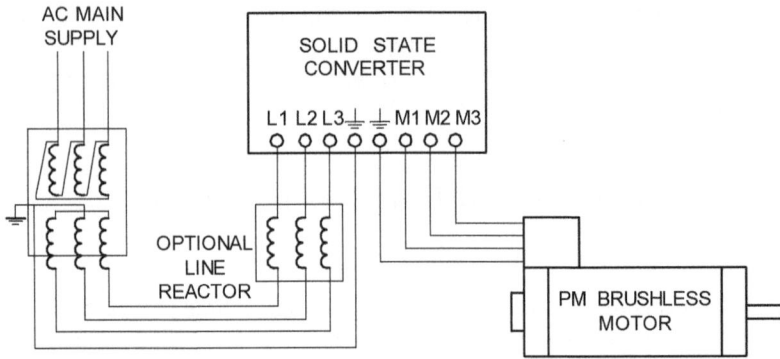

Fig. 6.26. Wiring diagram for a solid state converter-fed PM brushless motor.

$$L = \frac{1}{2\pi f} \frac{V_{1L-L}}{I_a} \frac{z\%}{100} \quad \text{H} \qquad (6.77)$$

where f is the power supply frequency (50 or 60 Hz), V_{1L-L} is the input voltage measured line to line and I_a is the input current rating of control.

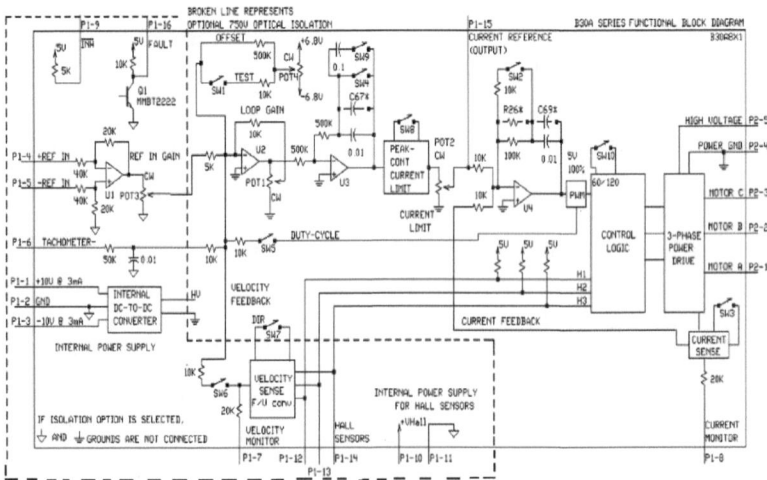

Fig. 6.27. PWM servo amplifier for a PM d.c brushless motor. Courtesy of Advanced Motion Control, Camarillo, CA, U.S.A.

6.9.2 Servo amplifiers

Servo amplifiers are used in motion control systems where precise control of position and/or velocity is required. The amplifier translates the low energy reference signals from the controller into high energy signals (input voltage and armature current). The rotor position information can be supplied either by Hall sensors or encoders. Usually, a servo amplifier has full protection against overvoltage, undervoltage, overcurrent, short circuit and overheating.

PWM brushless amplifiers manufactured by Advanced Motion Control have been designed to drive brushless d.c. motors at high switching frequency (Fig. 6.27). They interface with digital controllers or can be used as stand-alone solid state converters. These models require only a single unregulated d.c. power supply. Specifications are given in Table 6.3.

Table 6.3. PWM brushless amplifiers manufactured by Advanced Motion Control, Camarillo, CA, U.S.A.

Power stage specifications	B30A8	B25A20	B40A8	B40A20
d.c. supply voltage, V	20 to 30	40 to 190	20 to 80	40 to 190
Peak current				
2 s maximum, A	±30	±25	±40	±40
Maximum continuous current, A	±15	±12.5	±20	±20
Minimum load inductance				
per motor phase, mH	200	250	200	250
Switching frequency, kHz	22 ±15%			
Heatsink temp. range, ^{0}C	-25^0 to $+65^0$, disables if $> 65^0$			
Power dissipation				
at continuous current, W	60	125	80	200
Overvoltage shut-down, V	86	195	86	195
Bandwidth, kHz	2.5			
Mass, kg	0.68			

6.9.3 Microcontrollers

The state of the art for brushless PM motors calls for a single chip controller that could be used for low-cost applications. There are a number of commercially available integrated circuits (IC) such as LM621 [249], UC3620 [255], L6230A, [267], MC33035 [198] that can be used to perform simple speed control.

The MC33035 (Fig. 6.28) is a high performance second generation *monolithic brushless d.c. motor integrated controller* [198] containing all of the active functions required to implement a full featured open-loop, three or four phase motor control system (Fig. 6.28). This device consists of a rotor position

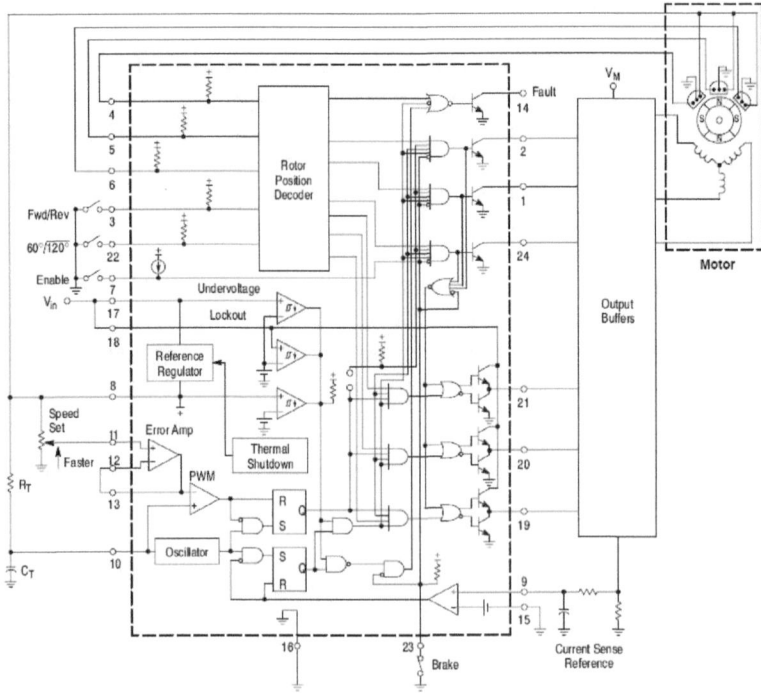

Fig. 6.28. The MC33035 open-loop, three-phase, six-step, full wave brushless motor controller from Motorola.

decoder for proper commutation sequencing, temperature compensated reference capable of supplying sensor power, frequency programmable sawtooth oscillator, fully accessible error amplifier, pulse width modulator comparator, three open collector top drivers, and three high current totem pole bottom drivers ideally suited for driving power MOSFETs. Also included are protective features consisting of undervoltage lockout, cycle-by-cycle current limiting with a selectable time-delayed latched shutdown mode, internal thermal shutdown, and a unique fault output that can be interfaced into microprocessor controlled systems.

Typical motor functions include open-loop speed control, forward or reverse direction, run enable and dynamic braking. In a closed-loop control the MC33035 must be supported by an additional chip MC33039 to generate the required feedback voltage without the need of a costly tachometer.

The three-phase application shown in Fig. 6.28 is a full featured *open-loop motor controller* with full wave, six-step drive. The upper switch transistors are Darlingtons while the lower devices are power MOSFETs. Each of these devices contains an internal parasitic catch diode that is used to return the stator inductive energy back to the power supply. The outputs are capable of driving a Y or Δ connected stator and a grounded neutral Y if split supplies

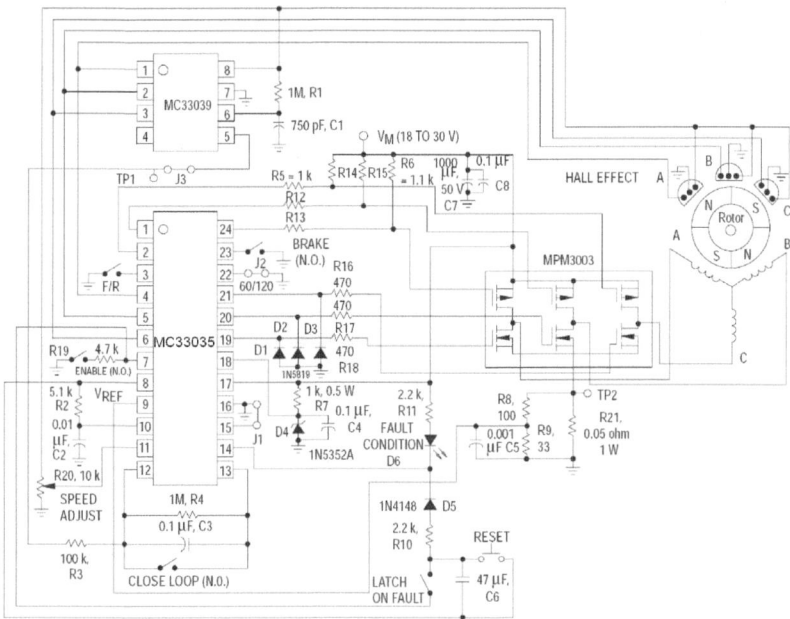

Fig. 6.29. Closed-loop brushless d.c. motor control using the MC33035, MC33039, and MPM3003 ICs from Motorola.

are used. At any given rotor position, only one top and one bottom power switch (of different totem poles) is enabled. This configuration switches both terminals of the stator winding from supply to ground which causes the current flow to be bidirectional or full wave. A leading edge spike is usually present on the current waveform and can cause a current-limit instability. The spike can be eliminated by adding an RC filter in series with the current sense input.

For *closed-loop speed control* the MC33035 requires an input voltage proportional to the motor speed. Traditionally this has been accomplished by means of a tachometer to generate the motor speed feedback voltage. Fig. 6.29 shows an application whereby an MC33039 powered from the 6.25 V reference (pin 8) of the MC33035 is used to generate the required feedback voltage without the need of a costly speed sensor. The same Hall sensor signals used by the MC33035 for rotor position decoding are utilized by the MC33039. Every positive- or negative-going transition of the Hall sensor signals on any of the sensor lines causes the MC33039 to produce an output pulse of defined amplitude and time duration, as determined by the external resistor $R1$ and capacitor $C1$. The output train of pulses at pin 5 of the MC33039 are integrated by the error amplifier of the MC33035 configured as an integrator to produce a d.c. voltage level that is proportional to the motor speed. This speed proportional voltage establishes the PWM reference level at pin 13 of the MC33035 motor controller and closes the feedback loop.

The MC33035 outputs drive an MPM3003 TMOS power MOSFET 3-phase bridge circuit capable of delivering up to 25 A of surge current. High currents can be expected during conditions of start-up, breaking, and change of direction of the motor. The system shown in Fig. 6.29 is designed for a motor having $120^0/240^0$ Hall sensor electrical phasing.

6.9.4 DSP control

A *digital signal processor* (DSP) provides high speed, high resolution and sensorless control algorithms at lower system costs [89, 90, 205, 291]. A more precise control often means performing more calculations implemented in a DSP.

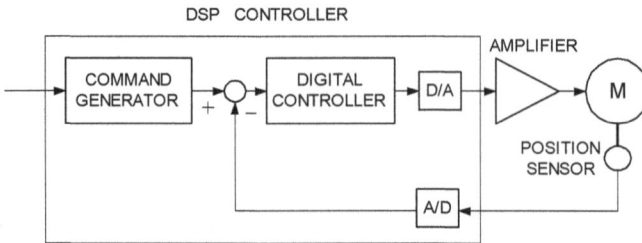

Fig. 6.30. DSP control scheme of brushless servo systems.

For motor control, in general, *fixed-point* DSPs are sufficient because they cost less than *floating point* DSPs and for most applications a dynamic range of 16 bits is enough [89]. When necessary, the dynamic range in a fixed-point processor can simply be increased by performing floating-point calculation in software.

A DSP controller (Fig. 6.30) can be described as a powerful processor that [89]:

- reduces the cost of the system by an efficient control in wide speed range implying a proper rating of solid state devices
- enables reduction of higher harmonics and torque ripple using enhanced algorithms
- implements sensorless algorithms of control
- reduces the amount of memory required by decreasing the number of look-up tables
- generates smooth near-optimal reference profiles and move trajectories in real time
- controls solid state switches of the inverter and generates high resolution PWM outputs
- enables single chip control system.

DSPs can also provide control of multi-variable and complex systems, use neural networks and fuzzy logic, perform adaptive control, provide diagnostic monitoring, e.g., vibration with FFT of spectrum analysis, implement notch filters that eliminate narrow-band mechanical resonance, and more.

6.10 Universal brushless motor electromechanical drives

The universal motor has two modes of operation, a d.c. brushless mode and an a.c. synchronous mode. It combines the ease of speed control of the d.c. motor with the high efficiency and smooth torque production (with minimum torque ripple) of the a.c. synchronous motor. As discussed in previous sections, an a.c. synchronous PM motor and a d.c. brushless PM motor have, in principle, the same magnetic and electric circuits. The difference between the two types of motors is essentially in the shape and control of the input phase currents.

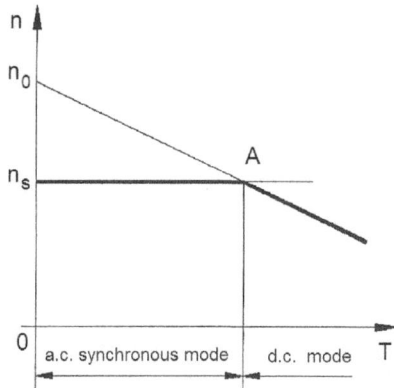

Fig. 6.31. Speed—torque characteristics of a brushless PM motor in a.c. and d.c. modes.

The speed–torque characteristic for these two motors is shown in Fig. 6.31. The speed-torque characteristic for the a.c. synchronous motor is constant and equal to the synchronous speed $n_s = f/p$. The d.c. brushless motor has a so-called *shunt characteristic* and its speed is described by the equation:

$$n = n_o - \frac{R}{c_{Edc}\Phi_f^{(sq)}}I_a^{(sq)} \tag{6.78}$$

where the no-load speed $n_o = V/(c_{Edc}\Phi_f^{(sq)})$.

The universal motor starts in d.c. mode and accelerates until synchronous speed is reached, at which time it switches to synchronous operation. The motor remains in synchronous mode until the stability limit is reached, at point A

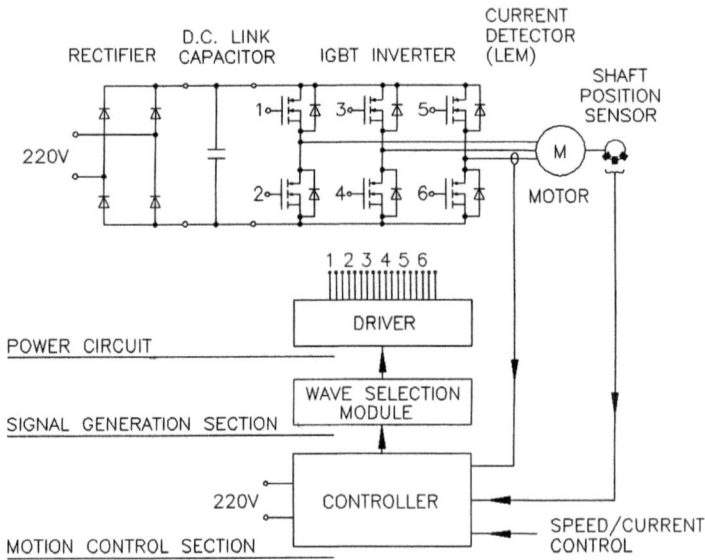

Fig. 6.32. A three-phase a.c.–d.c. universal brushless PM motor.

(Fig. 6.31). As the torque increases above point A the motor switches back to d.c. mode. The intersection point A is selected to obtain a synchronous torque close to either the stability limit or the maximum permissible input armature current of the motor. A torque greater than the stability limit torque will cause the PM rotor to fall out of step. The armature current must be limited to ensure that the stator winding does not overheat or that the PMs are not demagnetized. Once the motor has switched to d.c. mode the speed decreases with an increase in shaft torque, thus delivering approximately constant output power with $V = const$.

The basic elements of a universal PM motor drive (Fig. 6.32) are PM motor, full bridge diode rectifier, a 3-phase IGBT inverter, shaft position sensors, current sensor, controller and a mode selector. A simplified schematic representation of the inverter circuit is shown in Fig. 6.33.

Starting of a PM synchronous motor without position sensors is possible only if the rotor is furnished with a cage winding.

The synchronous mode of operation is obtained using a sinusoidal PWM switching topology. A synchronous regular sampled PWM with a frequency modulation ratio m_f set as an odd integer and a multiple of three is used to ensure the maximum reduction in harmonic content, as discussed in [36, 222].

The sinusoidal PWM switching consists of the constant frequency clock, a set of binary counters and an erasable programmable read-only memory (EPROM). A simplified schematic is shown in Fig. 6.33. The clock frequency determines the synchronous speed which is fixed for 50 Hz. Six bits of each

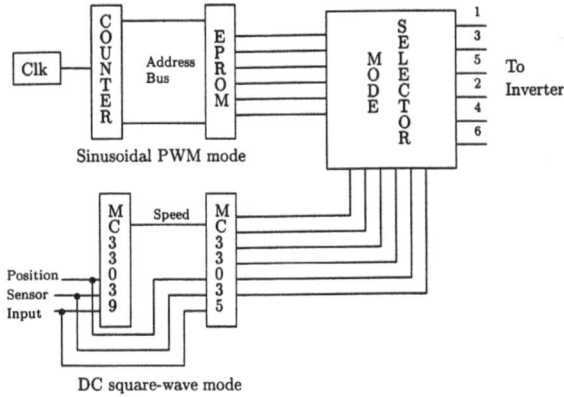

Fig. 6.33. Block diagram showing the link between sinusoidal PWM and d.c. modes.

byte of data stored in the EPROM drive the six transistors of the inverter. Each of the individual data lines of the EPROM thus represents a transistor gate signal.

The d.c. square wave controller is designed around the monolithic brushless d.c. motor controller integrated circuit MC33035 from Motorola [198].

Table 6.4. Comparison of rms voltages for different switching modes (d.c. bus voltage $V_{dc} = const$)

Switching mode	Amplitude modulation index m_a	rms voltage–to–d.c. bus voltage ratio
Square wave		0.8159
Sinusoidal PWM	1.0	0.7071
	1.2	0.7831
	1.4	0.8234
	1.6	0.8502

The speed control function allows for a steady rise in speed from start-up with the added benefit of having current limiting at low speeds. This ensures that the motor's stator windings do not overheat or that the magnets get demagnetized due to high starting currents. The speed is controlled using PWM switching, which is built into the MC33035 chip. The PWM reduces the average voltage across the windings and is thus able to control the motor's

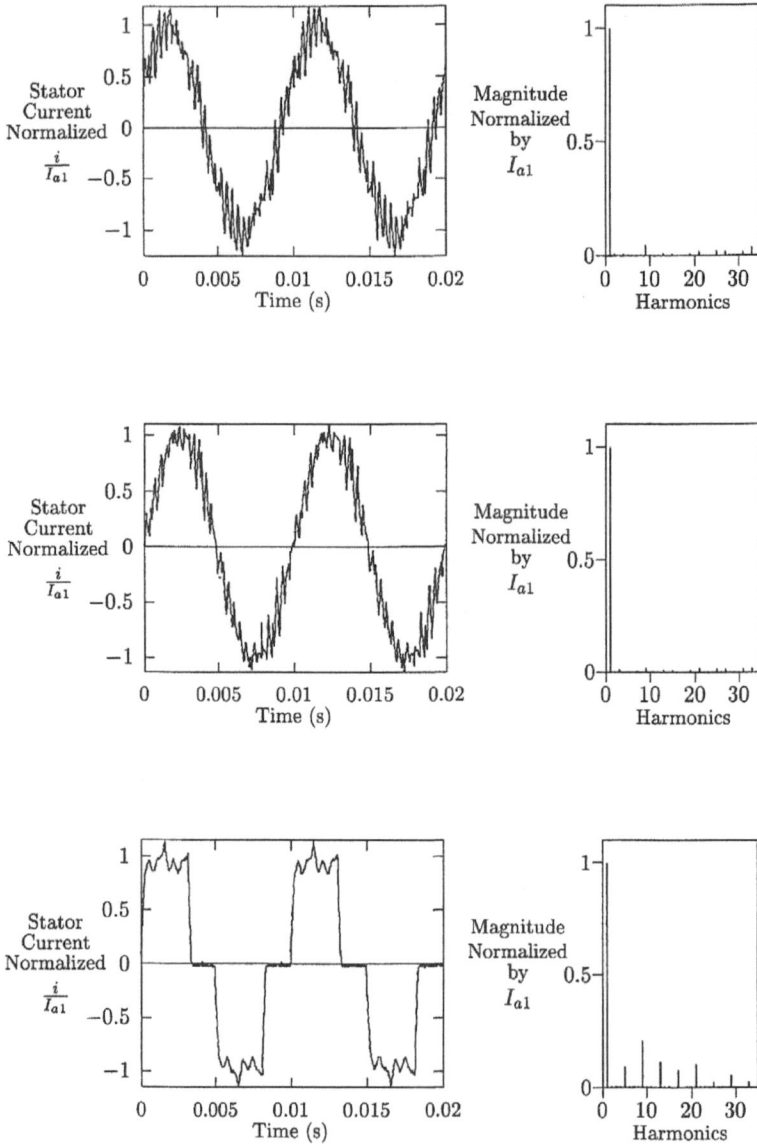

Fig. 6.34. Stator current waveforms and frequency spectrum magnitudes: (a) sinusoidal PWM, $m_a = 1.0$ and $I_a = 2.66$ A, (b) sinusoidal PWM, $m_a = 1.4$ and $I_a = 2.73$ A, (c) d.c. square-wave, $I_a = 2.64$ A. I_{a1} is the *rms* fundamental harmonic armature current.

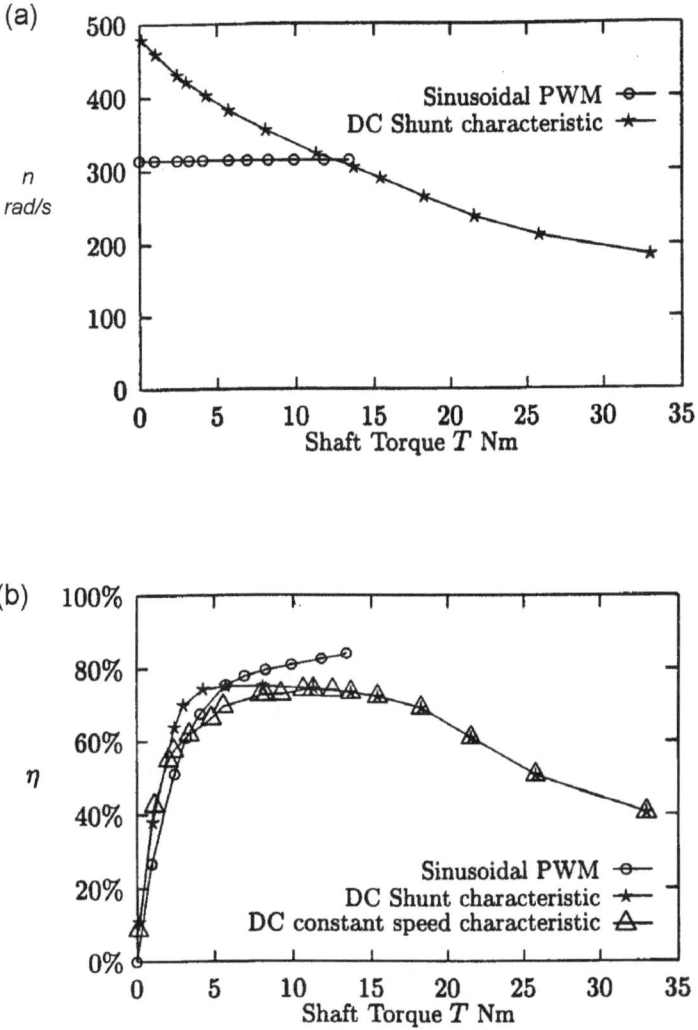

Fig. 6.35. Test results of steady-state characteristics for an integrated universal brushless motor drive: (a) speed against shaft torque, (b) efficiency against shaft torque.

speed from standstill up to full speed. The current is regulated using a single current sensor placed in the d.c. link, since the bus current flows through the two active phases connected in series at each instant in time. The bottom drive of each phase is switched to control the current flow.

The d.c. controller is used in a three-phase closed-loop control using the rotor position and speed feedback. The speed is obtained from the speed control adapter integrated circuit MC33039 specifically designed for use with the MC33035 Motorola IC.

For a smooth transition from d.c. to synchronous mode it has been found that the *rms* current levels at the point of switching should be similar for the two modes of operation. To ensure that this is possible, the *rms* voltage of the two modes should be equal. The torque–to–current ratio is approximately equal for the two operating modes.

The amplitude modulation index m_a of the synchronous mode is increased to match the d.c. mode's *rms* voltage. This increased voltage, due to over-modulation ($m_a > 1$), is to ensure that the synchronous motor can instantaneously produce enough torque to match the d.c. mode's torque at the time of switching. Table 6.4 shows the *rms* voltage ratio for a 120° square-wave and sinusoidal PWMs with different m_a values. The bus voltage V_{dc} is kept constant. An amplitude modulation index of $m_a = 1.4$ is thus used to ensure a stable transition from d.c. to a.c. mode.

The effect of increasing m_a on the harmonic content of the stator current has to be considered. A significant increase in the number of harmonics would lead to greater core losses and thus reduced efficiency. Fig. 6.34 shows the current waveforms for sinusoidal PWM with $m_a = 1$ and $m_a = 1.4$, and d.c. square wave operation. The harmonic content of the current waveform for d.c. square wave operation shows significantly more harmonics than the two sinusoidal PWM waves. The sinusoidal PWM mode with $m_a = 1.4$ is thus expected to be more efficient than the d.c. mode.

The motor used in this universal drive is the buried PM motor shown in Example 5.3. The steady-state characteristics of the complete drive for different modes are shown in Fig. 6.35. The d.c. mode has been tested for constant speed operation and for maximum speed operation. The constant speed d.c. mode uses PWM switching to reduce the average armature voltage and thus maintains a constant speed, but with increased switching losses as compared to the shunt torque-speed characteristic d.c. mode. Two operating modes do intersect close to the synchronous stability limit (about 14 Nm) and that mode switching is possible.

6.11 Smart motors

The *integrated electromechanical drive* also called a *smart motor* combines the electromechanical, electrical and electronic components, i.e., motor, power

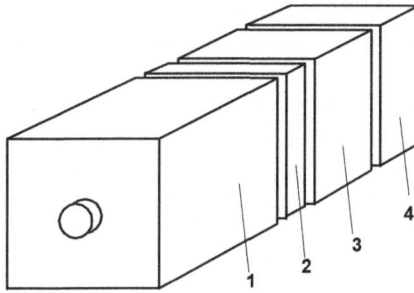

Fig. 6.36. Basic components of a smart motor: 1 — brushless motor, 2 — speed and position sensors, 4 — amplifier or power electronics converter,4 — control circuitry box.

Table 6.5. Smart PM d.c. brushless motors (3400 Series) manufactured by Animatics, Santa Clara, CA, U.S.A.

Specifications	3410	3420	3430	3440	3450
Rated continuous power, W	120	180	220	260	270
Continuous torque, Nm	0.32	0.706	1.09	1.48	1.77
Peak torque, Nm	1.27	3.81	4.06	4.41	5.30
No load speed, rpm	5060	4310	3850	3609	3398
Number of poles			4		
Number of slots			24		
EMF constant, V/krpm	9.2	10.8	12.1	12.9	13.7
Torque constant, Nm/A	0.0883	0.103	0.116	0.123	0.131
Rotor moment of inertia, $\text{kgm}^2 \times 10^{-5}$	4.2	9.2	13.0	18.0	21.0
Length, mm	88.6	105	122	138	155
Width, mm			82.6		
Mass, kg	1.1	1.6	2.0	2.5	2.9

electronics, position, speed and current sensors, controller and protection circuit together in one package (Fig. 6.36).

The traditional concept of an electrical drive is to separate the mechanical functions from the electronic functions which in turn requires a network of cables. In the smart motor or integrated drive the electronic control, position sensors and power electronics are mounted inside the motor against the casing, thus reducing the number of input wires to the motor and forming a structurally sound design. The cables connected to a smart motor are generally the power supply and a single speed signal. In addition, traditional compatibility problems are solved, standing voltage wave between the motor and converter (increase in the voltage at the motor terminals) is reduced and installation of a smart motor is simple. To obtain an even more compacted design, sensorless microprocessor control is used. Careful attention must be given to thermal compatibility of components, i.e., excessive heat generated by the motor winding or power electronics module can damage other components.

Table 6.5 shows specifications of smart PM d.c. brushless servo motors (3400 Series) manufactured by Animatics, Santa Clara, CA, U.S.A. These compact units consist of high power density PM d.c. brushless servo motor, encoder, PWM amplifier, controller and removable 8 kB memory module which holds the application program for stand-alone operation, PC or PLC control.

6.12 Applications

6.12.1 Electric and hybrid electric vehicles

Combustion engines of automobiles are one of the major oil consumers and sources of air pollutions. Oil conservation and road traffic congestion call for new energy sources for propulsion of motor vehicles and protection of the natural environment.

An *electric vehicle* (EV) is driven solely by an electric motor fed from an onboard *rechargeable energy storage system* (RESS), e.g., battery.

A *hybrid electric vehicle* (HEV) has a conventional combustion engine (gasoline or diesel), electric motor and RESS, so that the wheels of the vehicle are driven by both a combustion engine and electric motor. All the energy wasted during braking and idling in conventional vehicles is collected, stored in RESS and utilized in HEVs. The electric motor assists in acceleration (energy saved by the RESS), which allows for a smaller and more efficient combustion engine.

In most contemporary HEVs, called "charge-sustaining," the energy for the battery charging is produced by the internal combustion engine. Some HEVs, called "plug-in" or "charge-depleting," can charge the battery from the utility grid.

HEVs have many advantages over classical vehicles with gasoline or diesel engines: most important are:

(a) smaller size of combustion engine, lower fuel consumption since part of the energy is derived from the RESS and improved efficiency (about 40% better fuel efficiency than that for conventional vehicles of similar ratings);
(b) high torque of electric motor at low speed with high torque of combustion engine in higher speed range make the torque–speed characteristic suitable for traction requirements;
(c) utilization of wasted energy at braking (regenerative braking), idling and low speeds;
(d) the use of electric motor reduces air poultion and acoustic noise;
(e) wear and tear on the combustion engine components decrease, so they can work for a longer period of time;

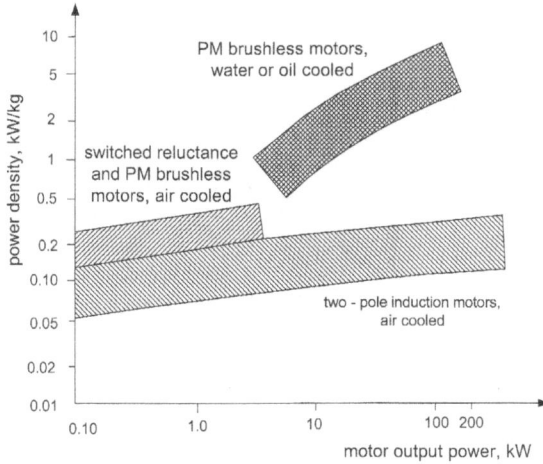

Fig. 6.37. Continuous power density of electric motor for EVs and HEVs [195].

(f) lower maintenance costs due to reduced fuel consumption;
(g) although the initial cost of HEVs is higher than conventional cars, their operating costs are lower over time.

EVs and HEVs use brushless electric motors, i.e., induction motors (IMs), switched reluctance motors (SRMs) and PM brushless motors. The power density as a function of the shaft power of different types of electric motors is shown in Fig. 6.37 [195]. Simulations indicate that a 15% longer driving range is possible for an EV with PM brushless motor drive systems compared with induction types [99]. PM brushless motor drives show the best efficiency, output power to mass, output power to volume (compactness) and overload capacity factor.

Figs 1.23 and 6.38 show the principle of operation of HEVs. In *series* HEVs (Fig. 6.38a) an electric motor drives wheels, while the combustion engine drives the electric generator to produce electricity. In *parallel* HEVs (Fig. 6.38b) the combustion engine is the main way of driving wheels and the electric motor assists only for acceleration. A *series/parallel* HEV (similar to Toyota Prius) is equipped with the so called *power split device* (PSD), which delivers a continuously variable ratio of combustion engine–to–electric motor power to wheels. It can run in "stealth mode" on its stored electrical energy alone.

The PSD is a planetary gear set that removes the need for a traditional stepped gearbox and transmission components in an ordinary gas powered car. It acts as a continuously variable transmission (CVT) but with a fixed gear ratio.

Toyota Prius NHW20 is equipped with a 1.5 l, 57 kW (5000 rpm), four-cylinder gasoline engine, 50 kW (1200 to 1540 rpm), 500 V (maximum) PM

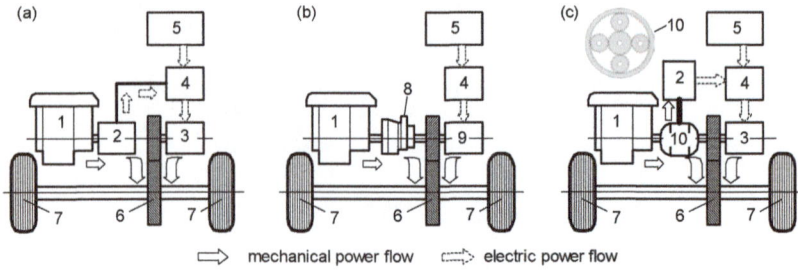

Fig. 6.38. Power trains of HEVs: (a) series hybrid; (b) parallel hybrid; (c) series/parallel hybrid. 1 — gasoline engine, 2 — electric generator, 3 — electric motor, 4 — solid state converter, 5 — battery, 6 — reduction gear, 7 — wheels, 8 — transmission, 9 — motor/generator, 10 — power split device (PSD).

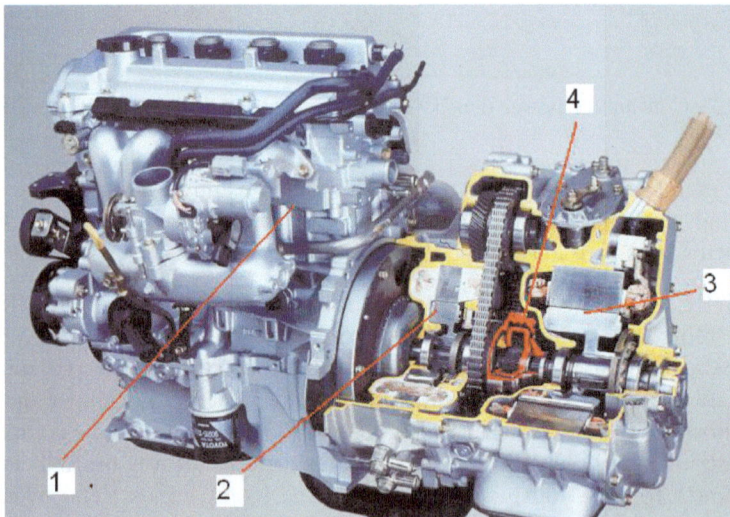

Fig. 6.39. Toyota Prius engine cutaway: 1 — four-cylinder combustion engine, 2 — generator/starter, 3 — electric motor, 4 — PSD.

brushless motor and nickel-metal hydride (NiMh) battery pack as a RESS. To simplify construction, improve transmission and achieve smoother acceleration, the gearbox is replaced by a single reduction gear (Fig. 6.38). This is because the engine and electric motor have different torque-speed characteristics, so that they can act jointly each other to meet the driving performance requirements [168]. Fig. 6.39 shows integration of combustion engine with generator/starter, electric motor and PSD of Toyota Prius. Fig. 6.40 shows a single rotor lamination of Toyota Prius PM brushless motor [168, 269, 271]. The rotor with interior PMs has been selected because it provides wider torque-speed range under the size and weight restrictions than other rotor configu-

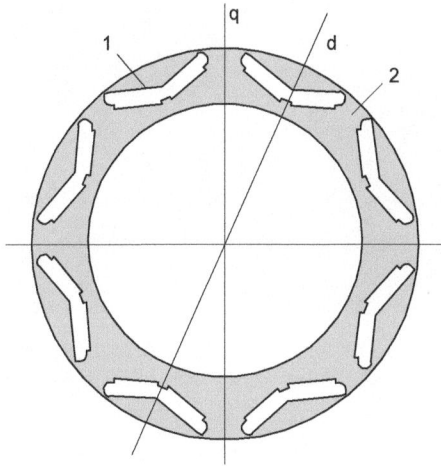

Fig. 6.40. Rotor punching of Toyota Prius electric motor: 1 — groove for PM, 2 — ferromagnetic bridge between PMs.

rations (Figs 5.1b and 5.1i). To utilize the reluctance torque in addition to synchronous torque the q–axis permeance is maximized while keeping low d–axis permeance [168]. A double layer PM arrangement (Fig. 5.1j) seems to be impractical in mass production due to the high cost of manufacturing [168].

Electric motors for passenger hybrid cars are typically rated from 30 to 75 kW. Water cooling offers superior cooling performance, compactness and lightweight design over forced-air motor cooling. The water cooling permits weight reductions of 20% and size reductions of 30% as compared to forced-air designs, while the power consumption for cooling system drops by 75% [169]. The use of a single water cooling system for the motor and solid state converter permits further size reductions.

6.12.2 Variable-speed cooling fans

Computers, instruments, and office equipment use *cooling fans* with simple PM d.c. brushless motors that have inner salient pole stator and outer rotor ring magnet as ahown in Fig. 6.41. The rotor is integrated with fan blades. Running the fan from a low voltage d.c. source, usually 12, 24, 42, or 115 V, makes it easy to control the fan speed electronically, in response to the actual temperature inside the instrument. The simplest method of controlling the d.c. brushless motor is on/off switching with a single transistor. More sophisticated methods for fan control are digital interface integrated circuits (ICs) equipped with remote temperature sensors, e.g., MAX1669 (Dallas Semiconductor).

The rated power of fan brushless motors is in the range 0.6 to 180 W and maximum speed can vary from 2000 to 6000 rpm. A variable speed brushless

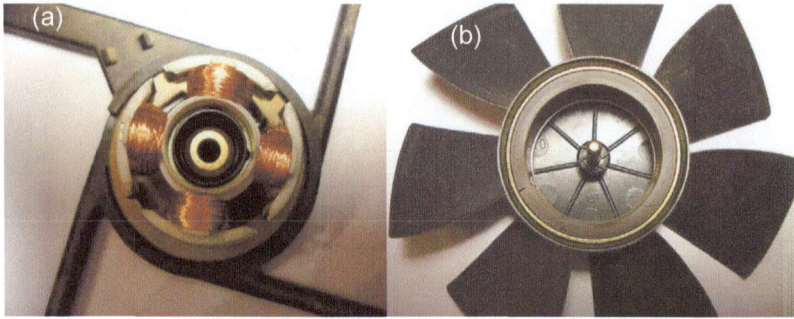

Fig. 6.41. PM brushless motor for a 12 V, 0.15 A fan: (a) two-phase inner stator with four concentrated coils; (b) outer rotor with ring-shaped PM. Control electronics is behind the stator.

motor with temperature sensor can suppress the noise since under most conditions the fan speed and consequently the noise are far less than that at full speed condition (25 to 40 dB).

6.12.3 Computer hard disk drives

The data storage capacity of a *hard disk drive* (HDD) is determined by the *aerial recording density* and number of disks. The aerial density is now 51 Gbit/cm^2 = 329 Gbit/in^2 (2009). Mass of the rotor, moment of inertia and vibration increase with the number of disks. Circumferential vibration of mode $r = 0$ and $r = 1$ (Section 5.12) causes deviations of the rotor from the geometric axis of rotation. Disk drive spindle motors are brushless d.c. motors with outer rotor designs. Drives with a large number of disks have the upper end of the spindle fixed with a screw to the top cover (Fig. 6.42a). This "tied" construction reduces vibration and deviations of the rotor from the center axis of rotation. For a smaller number of disks the so called "untied" construction with fixed shaft (Fig. 6.42b) or rotary shaft (Fig. 6.42c) has been adopted.

Special design features of spindle motors are their high starting torque, limited current supply, reduced vibration and noise, physical constraints on volume and shape, contamination and scaling problems [161, 162, 163]. High starting torque, 10 to 20 times the running torque is required, since the read/write head tends to stick to the disk when not moving. The starting current is limited by the computer power supply which severely limits the starting torque. For a 2.5-inch, 20,000-rpm, 12-V HDD the starting current is less that 2 A at starting torque of 6.2 mNm. The acoustic noise is usually below 30 dB(A) and nonrepeatable run out maximum 2.5×10^{-5} μmm.

The choice of the number of poles determines the frequency of torque ripple and switching frequency. Although larger numbers of poles reduce the torque ripple, it increases switching and hysteresis losses and complicates commutation tuning and installation of rotor position sensors. Most commonly used

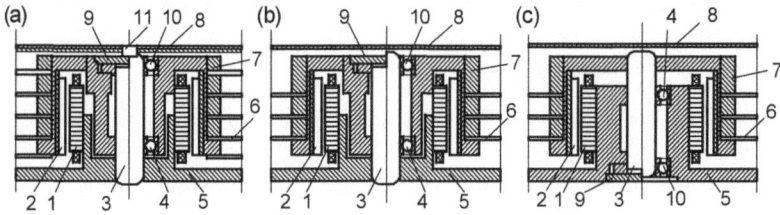

Fig. 6.42. Construction of spindle motors for HDDs: (a) tied type, (b) untied type with fixed shaft, (c) untied type with rotary shaft. 1 — stator, 2 — PM, 3 — shaft, 4 — ball bearing, 5 — base plate, 6 — disk, 7 — disk clamp, 8 — top cover, 9 — thrust bearing, 10 — radial bearing, 11 — screw.

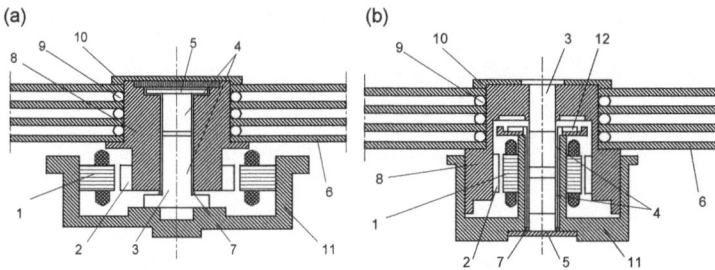

Fig. 6.43. Construction of FDB spindle motors for HDDs: (a) fixed-shaft spindle motor, (b) rotating-shaft spindle motor. 1 — stator, 2 — PM, 3 — shaft, 4 — radial bearing, 5 — thrust bearing, 6 — disk, 7 — stopper/seal, 8 — hub, 9 — spacer, 10 — clamp, 11 — base plate, 12 — attractive magnet.

are four and eight pole motors. The pole–slot combination is important in reducing the torque ripple. Pole–to–slot ratios with high least common multiple $LCM(s_1, 2p)$ such as 8-pole/9-slot ($LCM(9,8) = 72$) and 8-pole/15-slot ($LCM(15,8) = 120$) produce very small cogging torque.

Drawbacks of ball bearings include noise, low damping, limited bearing life and nonrepeatable run out. The HDD spindle motor is now changing from ball bearing to a *fluid dynamic bearing* (FDB) motor. Contact-free FDBs (Fig. 6.43) produce less noise and are serviceable for an extended period of time.

6.12.4 CD players

Information on a compact disk (CD) is recorded at a constant linear velocity. Miniature PM motors are used for:

- spindle rotation
- drawer/tray opening/closing

- pickup position (coarse tracking) unless the unit uses a linear motor or rotary positioner drive
- disk changing (changers only).

A CD mechanism is shown in Fig. 6.44. Electric motors run on a d.c. voltage from a fraction of a volt up to 10 or 12 V, e.g., drawer. Most CD players or CD–ROM drives have electric motor-driven *loading drawers*.

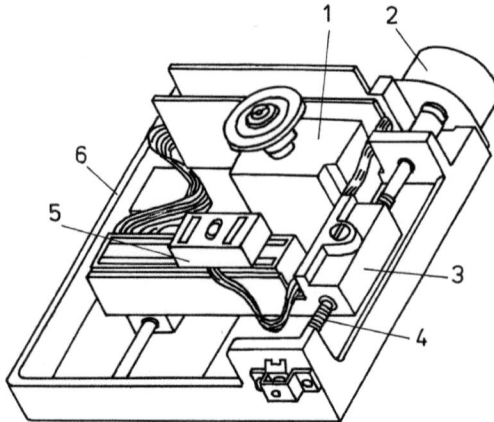

Fig. 6.44. CD mechanism: 1 — spindle motor, 2 — traverse stepping motor (pickup position), 3 — sled, 4 — precision drive screw, 5 — optical laser pickup, 6 — die-cast base.

Spindle speed maintains constant linear velocity (CLV) of disk rotation based on a phase locked loop locking to the clock signal recovered from the disk. A "rough servo control" establishes an initial relationship between the spindle speed, i.e., 200 to 500 rpm and the diameter of the spiral being scanned. Spindle drive is most often done with a PM brushless motor connected to the disk platform (Fig. 6.45). In most cases CD spindle electromechanical drives use pancake d.c. brushless motors with many stator poles, so that motors can run at low and very stable speed. PM brushless motors shown in Fig. 6.45 can be found in CD players, video cassette, and other consumer electronics drive systems. In old CD players, a d.c. commutator motor very similar to the common motors in toys and other battery operated devices can be found.

Pickup motor is either a conventional miniature PM d.c. motor with a belt or gear with a worm, ball, or rack–and–pinion mechanism, a stepping motor, or a direct drive linear motor or rotary positioner with no gears or belts. The mechanism on which the optical laser pickup is mounted is called the *sled*. The entire pickup moves on the sled during normal play or for rapid access

Fig. 6.45. Pancake PM brushless motor with 18 concentrated coil stator and outer rotor with ring-shaped PM . The stator is integrated with control electronics.

to musical selections or CD–ROM data. The sled is supported on guide rails and is moved by a worm or ball gear as shown in Fig. 6.44.

The *optical pickup* is the "stylus" that reads the optical information encoded on the disk. It includes the laser diode, associated optics, focus and tracking actuators, and photodiode array. The optical pickup is mounted on the sled and connects to the servo and readback electronics using flexible printed wiring cables (Fig. 6.44). Since focus must be accurate to 1 μm a focus servo is used. The *focus actuator* is actually a coil of wire in a PM field like the voice coil actuator in a loudspeaker. The focus actuator can move the objective lens up and down – closer or farther from the disk based on focus information taken from the photodiode array.

Fine tracking centers the laser beam on the disk track (to within a fraction of a μm) and compensates for side-to-side runout of the disk and player movement. This also uses a voice coil positioner and optical feedback from the disk surface. *Coarse tracking* moves the entire pickup assembly as a function of fine tracking error exceeding a threshold or based on user or microcontroller requests (like search or skip).

The most common causes of CD motor failures are

- open or shorted windings
- dry/worn bearings
- partial short caused by dirt or carbon buildup on commutator (in the case of d.c. commutator motors).

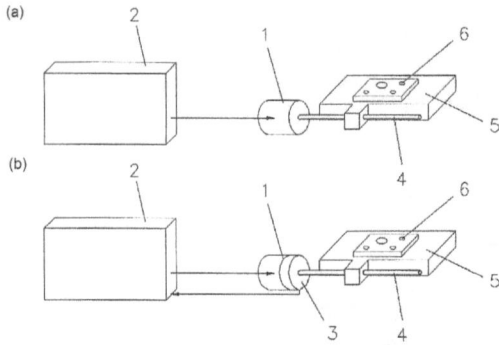

Fig. 6.46. Axis position control schemes for numerically controlled (NC) machine tools using PM brushless motors: (a) open-loop control, (b) closed-loop vector control. 1 — PM brushless motor, 2 — controller, 3 — encoder, 4 — ballscrew, 5 — table, 6 — part.

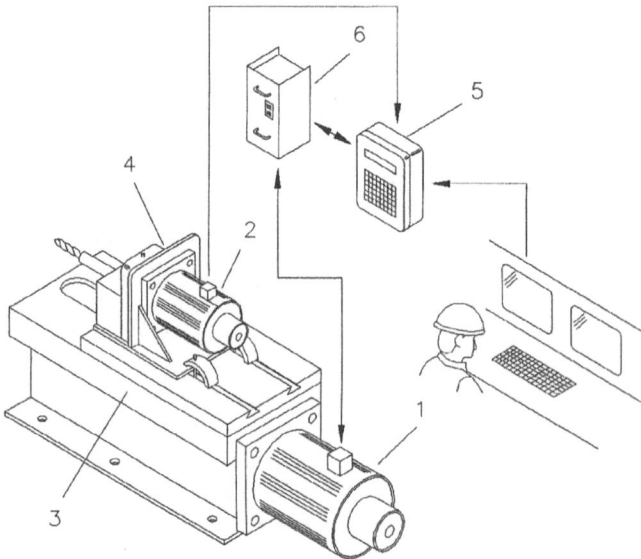

Fig. 6.47. Numerically controlled (NC) machine tool with two brushless motors. 1 — a.c. servo motor, 2 — spindle motor, 3 — slide unit, 4 — transducer for measuring the torque and thrust at the tool, 5 — programmable controller, 6 — digital servo controller.

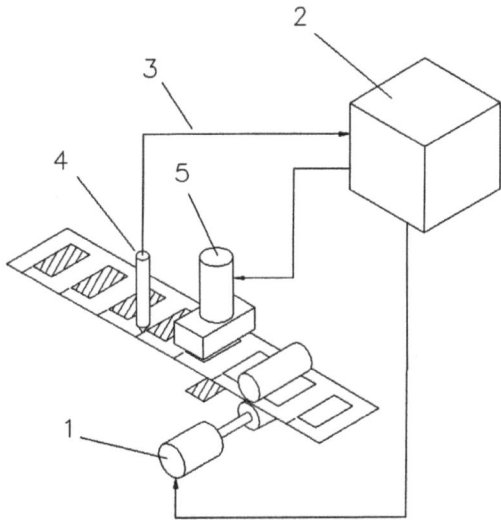

Fig. 6.48. Registration facility: 1 — PM brushless motor, 2 — programmable motion controller, 3 — registration signal, 4 — mark sensor, 5 — punch head. Courtesy of Parker Hannifin Corporation, CA, U.S.A.

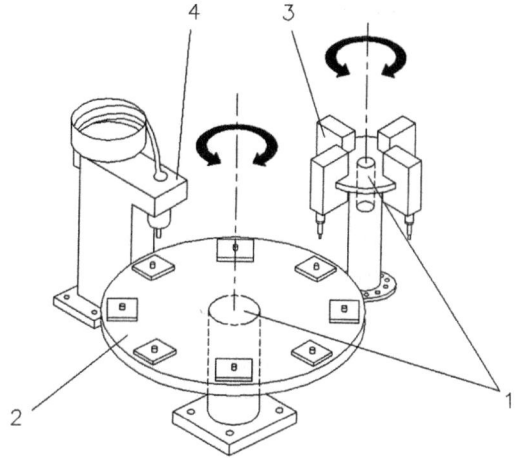

Fig. 6.49. Indexing table for assembly. 1 — PM brushless motor, 2 — rotary table, 3 — tool turret, 4 — tool. Courtesy of Parker Hannifin Corporation, CA, U.S.A.

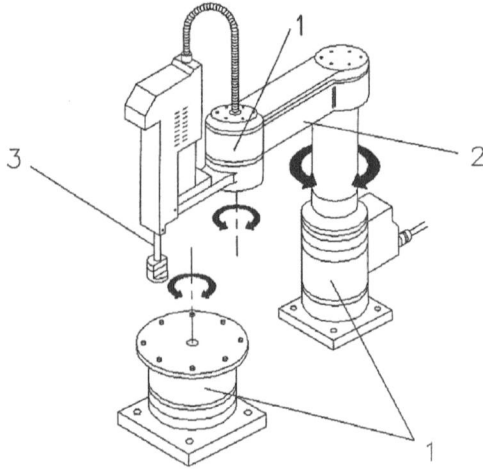

Fig. 6.50. SCARA robot for assembly. 1 — PM brushless motor, 2 — arm, 3 — tool. Courtesy of Parker Hannifin Corporation, CA, U.S.A.

6.12.5 Factory automation

Fig. 6.46 shows the open- and closed-loop control for a numerically controlled (NC) machine tool [79]. The closed-loop vector control scheme provides a very high positioning accuracy.

An NC machine tool with two a.c. motors is presented in Fig. 6.47 [79]. Either induction or PM brushless motors can be used. Diagnostic feedback information is easily accomplished. Sensors and transducers monitor torque and thrust at the tool. Tool breakage is therefore automatically prevented by reducing feed rate. Programmable drives like those in Fig. 6.47 replace feed-boxes, limit switches and hydraulic cylinders, and eliminate changing belts, pulleys, or gears to adjust feed rates and depths [79]. This makes the system very flexible.

Figs 6.48 to 6.50 show applications of PM synchronous motors to industrial processes automation   according to Parker Hannifin Corporation, Rohnert Park, CA, U.S.A. [236].

The registration facility (Fig. 6.48) allows a move to be programmed to end a specified distance after a registration pulse appears at one of the inputs of the programmable motion controller [70]. The indexing table (Fig. 6.49) uses two PM brushless motors for positioning the table and tool turret [70]. A selective compliance assembly robot arm (SCARA), shown in Fig. 6.50, has been developed for applications requiring repetitive and physically taxing operations [70].

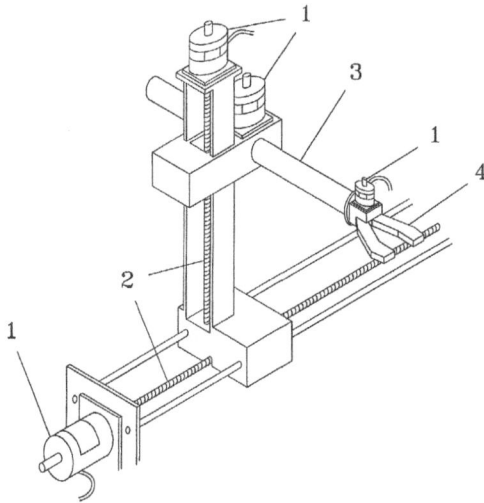

Fig. 6.51. Three-axis manipulator. 1 — PM brushless servomotor, 2 — lead screw, 3 — arm, 4 — gripper. Courtesy of Parker Hannifin Corporation, CA, U.S.A.

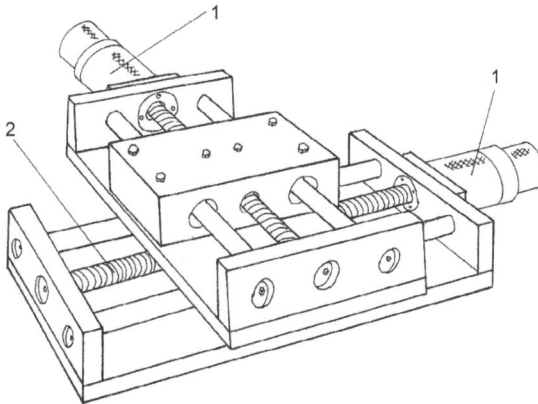

Fig. 6.52. Precise X–Y table. 1 — PM brushless motor, 2 — lead screw.

Fig. 6.51 shows a three-axis pick and place application that requires the precise movement of a manipulator arm in three dimensions. The arm has specific linear paths it must follow to avoid other pieces of machinery [70].

6.12.6 X–Y tables

Fig. 6.52 shows an example of a precise X–Y table, which must meet considerable control rigidity, wide speed control range, smooth rotation, high acceleration, and very precise reproducibility [90]. All these factors affect directly the quality of the process.

6.12.7 Space mission tools

The power provided by *space stations* and *space shuttles* is very limited, so that rechargable hand tools require very efficienct compact motors. PM brushless motors are the best motors for space mission tools. The high-technology *power ratchet tool* and *pistol grip tool* driven by PM brushless motors were used, e.g., in a servicing mission in 1993 to conduct planned maintenance on the Hubble Space Telescope (deployed approximately 600 km above the Earth in 1990).

Numerical examples

Numerical example 6.1

Find the armature current, torque, electromagnetic power and winding losses of the R65 PM d.c. brushless servo motor (Table 6.2) at $n = 2000$ rpm and d.c. bus voltage of a six-step inverter is $V_{dc} = 240$ V.

Solution
The line-to-line EMF is

$$E_{fL-L} = 0.117 \times 2000 = 234 \text{ V}$$

because the EMF constant $k_{Edc} = 0.117$ V/rpm is for line–to–line voltage.
    Assuming that the d.c. bus voltage is approximately equal to the input voltage, the armature current at 2000 rpm

$$I_a \approx \frac{240 - 234}{1.81} = 3.31 \text{ A}$$

because the line-to-line resistance of the hot motor is 1.81 Ω.
    The a.c. line–to–line output voltage of a six-step inverter according to eqn (6.8) is

$$V_{1L} \approx 0.78 \times 240 = 187.2 \text{ V}$$

The shaft torque at 2000 rpm

$$T = 1.12 \times 3.31 = 3.71 \text{ Nm}$$

because the torque constant is $k_T = 1.12$ Nm/A.

Assuming that only two phases conduct current at the same time, the electromagnetic power is

$$P_{elm} = E_{fL-L}I_a = 234 \times 3.31 \approx 776 \text{ W}$$

The electromagnetic torque developed by the motor

$$T_d = \frac{776}{2\pi 2000/60} = 3.70 \text{ Nm}$$

EMF and torque constants k_{Edc} and k_T given in Table 6.2 are not accurate, because the electromagnetic torque T_d should be slightly greater than the shaft torque T. The winding losses for 1.81 Ω line-to-line resistance

$$\Delta P_w = 1.81 \times 2.98^2 = 16 \text{ W}$$

Numerical example 6.2

A three phase, Y-connected, $2p = 8$ pole, surface configuration PM brushless motor has the stator internal diameter $D_{1in} = 0.132$ m, effective length of the stator stack $L_i = 0.153$ m and the PM pole shoe width $b_p = 0.0435$ m. The number of turns per phase $N_1 = 192$, the winding factor is $k_{w1} = 0.926$ and the peak value of the air gap magnetic flux density $B_{mg} = 0.923$ T. Neglecting the armature reaction, find approximate values of the EMF, electromagnetic torque developed by the motor and electromagnetic power at $n = 600$ rpm and rms current $I_a = 14$ A for:

(a) sinewave operation ($\Psi = 0°$)

(b) 120° square wave operation

Solution

(a) **sinewave operation at $\Psi = 0°$**

Pole pitch

$$\tau = \frac{\pi \times 0.132}{8} = 0.0518 \text{ m}$$

Pole shoe–to–pole pitch ratio

$$\alpha_i = \frac{b_p}{\tau} = \frac{43.5}{51.8} = 0.84$$

Form factor of the excitation field according to eqn (5.23)

$$k_f = \frac{4}{\pi} \sin \frac{0.84 \times \pi}{2} = 1.233$$

Excitation flux according to eqn (5.6)

$$\Phi_f = \Phi_{f1} = \frac{2}{\pi} 0.0518 \times 0.153 \times 1.233 \times 0.923 = 0.00574 \text{ Wb}$$

where $B_{mg1} = k_f B_{mg}$. EMF constants according to eqn (6.19)

$$c_E = \pi p \sqrt{2} N_1 k_{w1} = \pi \times 4\sqrt{2} \, 192 \times 0.926 = 3159.6$$

$$k_E = c_E \Phi_f = 3159.6 \times 0.00574 = 18.136 \text{ Vs} = 0.3023 \text{ Vmin}$$

EMF per phase according to eqn (6.19)

$$E_f = k_E n = 0.3023 \times 600 = 181.36 \text{ V}$$

Torque constants according to eqn (6.20)

$$c_T = m_1 \frac{c_E}{2\pi} = 3 \frac{3159.6}{2\pi} = 1508.6$$

$$k_T = c_T \Phi_f = 1508.6 \times 0.00574 = 8.66 \text{ Nm/A}$$

Electromagnetic torque developed at 14 A according to eqn (6.22)

$$T_d = k_T I_a = 8.66 \times 14 = 121.2 \text{ Nm}$$

Electromagnetic power

$$P_{elm} = m_1 E_f I_a \cos \Psi = 3 \times 181.36 \times 14 \times 1 = 7617 \text{ W}$$

(b) 120° square-wave operation

Flat-topped value of the phase current according to eqn (6.31)

$$I_a^{(sq)} = \sqrt{\frac{3}{2}} I_a = \sqrt{\frac{3}{2}} 14 = 17.15 \text{ A}$$

Excitation flux according to eqn (6.24)

$$\Phi_f^{(sq)} = 0.0435 \times 0.153 \times 0.923 = 0.006143 \text{ Wb}$$

The ratio of the square-wave flux to sinewave flux

$$\frac{\Phi_f^{(sq)}}{\Phi_f} = \frac{0.006143}{0.00574} = 1.07$$

Note that it has been assumed $\alpha_i = 2/\pi = 0.6366$ for sinewave mode and $\alpha_i = b_p/\tau = 0.84$ for square wave mode. For the same α_i the flux ratio is equal to $1/k_f$.

EMF constants according to eqn (6.25)

$$c_{Edc} = 8pN_1k_{w1} = 8 \times 4 \times 192 \times 0.926 = 5689.3$$

$$k_{Edc} = 5689.3 \times 0.006143 = 34.95 \text{ Vs} = 0.582 \text{ Vmin}$$

EMF (two phases in series) according to eqn (6.25)

$$E_{fL-L} = 0.582 \times 600 = 349.5 \text{ V}$$

Torque constant according to eqn (6.26)

$$k_{Tdc} = \frac{k_{Edc}}{2\pi} = \frac{34.95}{2\pi} = 5.562 \text{ Nm/A}$$

Electromagnetic torque at $I_a^{(sq)} = 17.15$ A according to eqn (6.26)

$$T_d = k_{Tdc}I_a^{(sq)} = 5.562 \times 17.15 = 95.4 \text{ Nm}$$

Electromagnetic power

$$P_{elm} = E_{fL-L}I_a^{(sq)} = 349.5 \times 17.15 = 5994 \text{ W}$$

Numerical example 6.3

Simulate the armature current waveform and developed torque of a d.c. brushless motor fed from an IGBT VSI shown in Fig. 6.25a. The motor armature winding is Y-connected, the inverter is switched for 120° square wave and the d.c. bus voltage is $V_{dc} = 380$ V. The motor is assumed to be operating at a constant speed of 1500 rpm and is a four-pole motor (50 Hz fundamental input frequency). The induced EMF at 1500 rpm $E_f = 165$ V, the armature winding resistance $R_1 = 4.95$ Ω, the armature winding self inductance per phase is $L_1 = 0.0007$ H and the armature winding mutual inductance per phase is $M = 0.0002$ H.

Show the armature current waveform for one phase and the developed torque using a state space time stepping simulation method.

Solution

Assume switches 4 and 5 (Fig. 6.25a) of the inverter are closed at the start of the simulation. Switch 5 is then opened and at the same time switch 1 is closed. The current in phase C cannot decay to zero instantaneously due to the winding inductance. This decaying current continues to flow in phase C through phase winding B, switch 4 and the diode 6. The current in phase A begins to flow after switch 1 is closed. This increasing current flows through switch 1, into windings of phase A and B, and out of switch 4. There are thus two components of current flowing in phase winding B. The voltage equations for each phase are

$$V_{1A} = E_{fA} + R_1 i_{aA} + L_1 \frac{di_{aA}}{dt} + M \frac{di_{aB}}{dt} + M \frac{di_{aC}}{dt}$$

$$V_{1B} = E_{fB} + R_1 i_{aB} + M \frac{di_{aA}}{dt} + L_1 \frac{di_{aB}}{dt} + M \frac{di_{aC}}{dt}$$

$$V_{1C} = E_{fC} + R_1 i_{aC} + M \frac{di_{aA}}{dt} + M \frac{di_{aB}}{dt} + L_1 \frac{di_{aC}}{dt}$$

The voltage drop between phase A and B, after switch 1 is closed, is

$$V_{1AB} = V_{1A} - V_{1B} = V_{dc} - 2\Delta V_s$$

or can be written in terms of the phase voltages as

$$V_{1AB} = E_{fA} + R_1 i_{aA} + L' \frac{di_{aA}}{dt} - E_{fB} - R_1 i_{aB} - L' \frac{di_{aB}}{dt}$$

where ΔV_s is the voltage drop across each closed IGBT switch and $L' = L_1 - M$. The voltage drop between phase B and C, after switch 5 is opened and while i_{aC} is flowing, is

$$V_{1BC} = V_{1B} - V_{1C} = \Delta V_d + \Delta V_s$$

or can be written in terms of the phase voltages as

$$V_{1BC} = E_{fB} + R_1 i_{aB} + L' \frac{di_{aB}}{dt} - E_{fC} - R_1 i_{aC} - L' \frac{di_{aC}}{dt}$$

where ΔV_d is the voltage drop across a diode. Using the above equations the change in current in each phase, while phase C has a current flow, is:

$$\frac{di_{aA}}{dt} = \frac{1}{3L'} (2V_{1AB} - 2E_{fA} - 2R_1 i_{aA} + E_{fB} + R_1 i_{aB} + E_{fC} + R_1 i_{aC} + V_{1BC})$$

$$\frac{di_{aB}}{dt} = \frac{1}{L'} (E_{fA} + R_1 i_{aA} + L' \frac{di_{aA}}{dt} - E_{fB} - R_1 i_{aB} + V_{1AB})$$

$$\frac{di_{aC}}{dt} = -\frac{di_{aA}}{dt} - \frac{di_{aB}}{dt}$$

since $i_{aA} + i_{aB} + i_{aC} = 0$ and $di_{aA}/dt + di_{aB}/dt + di_{aC}/dt = 0$.

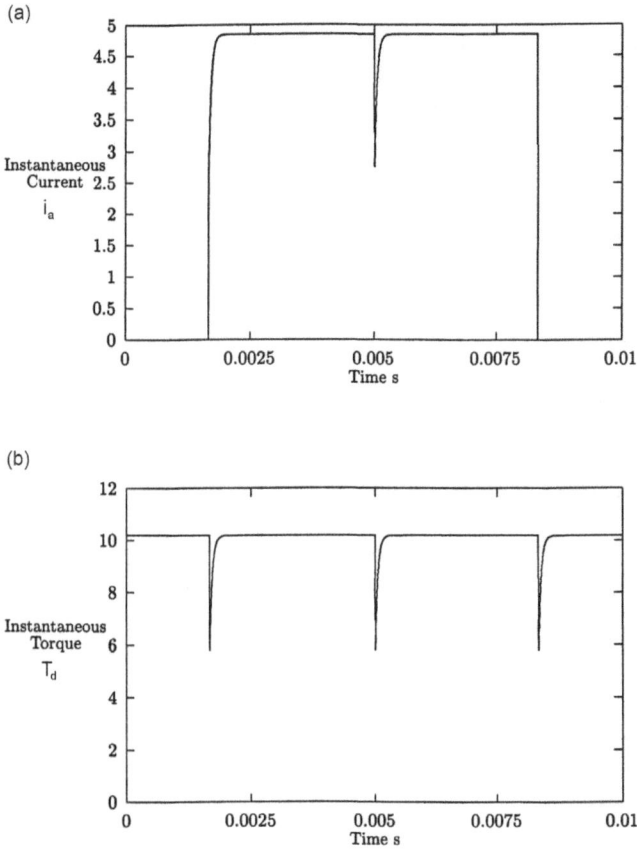

(a)

(b)

Fig. 6.53. Simulation results: (a) instantaneous current i_A versus time, (b) instantaneous electromagnetic torque T_d developed by the motor versus time.

After current in phase C has decayed to zero the derivatives of the phase currents are

$$\frac{di_{aA}}{dt} = \frac{1}{L'}(V_{1AB} - E_{fA} - R_1 i_{aA} + E_{fB} + R_1 i_{aB})$$

$$\frac{di_{aB}}{dt} = -\frac{di_{aA}}{dt} \qquad \frac{di_{aC}}{dt} = 0$$

The initial current flowing through phase B and C is assumed to be a steady state current. The simulation steps over 180^0 visualized in Fig. 6.53a show the current waveform for phase A where at time 0.001667 s switch 1 is closed and switch 5 is opened, at time 0.005 s switch 6 is closed and switch 4 is opened, and at time 0.008333 s switch 3 is closed and switch 1 is opened (refer to Fig.

6.25 for switch numbers). From Fig. 6.53a it is clear that the current decays more quickly than it increases. This results in a dip in the phase A current i_{aA} as the current is switched from phase B to phase C, at time 0.005 s.

The instantaneous electromagnetic torque is calculated from eqn (6.67). Fig. 6.53b shows that the changes in phase current are translated into pulsations in the torque. This example neglects the effects on torque due to a variation in the induced EMF, which in most brushless motors increases the length of these pulsation torques as well as the torque pulsations due to stator slots.

Numerical example 6.4

The following measurements have been taken on a 1500-rpm, four-pole, 120^0 square-wave, three-phase brushless d.c. PM motor: $P_{in} = 1883$ W, $I_a = 3.18$ A, $V_1 = 220$ V (Y-connection), $B_{mg} = 0.6848$ T. The dimensions of the magnetic circuit are: $D_{1in} = 0.0825$ m, $D_{1out} = 0.1360$ m, $L_i = 0.1008$ m, $l_{1e} = 0.0908$ m, $h_{11} = 9.5$ mm, $h_{12} = 0.4$ mm, $h_{13} = 0.2$ mm, $h_{14} = 0.7$ mm, $b_{11} = 4.8$ mm, $b_{12} = 3.0$ mm, $b_{14} = 2.2$ mm, $b_p = 56.7$ mm. The stator laminated core is made of cold rolled electrotechnical steel sheets with the specific core losses $\Delta p_{1/50} = 2.4$ W/kg and stacking coefficient $k_i = 0.96$. The copper wire stator winding is distributed in $s_1 = 36$ slots and has the following parameters: number of turns per phase $N_1 = 240$, number of parallel wires $a = 2$ and wire diameter $d_a = 0.5$ mm without insulation.

Find the losses and motor efficiency.

Solution
The average length of the armature turn according to eqn (B.2) is

$$l_{1av} = 2(0.1008 + 0.0908) = 0.3832 \text{ m}$$

The cross section of the armature conductor

$$s_a = \frac{\pi \times 0.5^2}{4} = 0.1963 \text{ mm}$$

The skin effect in the stator winding wound with a round conductor of small diameter can be neglected. Thus, the armature winding resistance at 75^0C according to eqn (B.3) is

$$R_1 = R_{1dc} = \frac{240 \times 0.3832}{2 \times 47 \times 10^6 \times 0.1963 \times 10^{-6}} = 4.984 \ \Omega$$

The flat-topped value of the current in the armature phase winding according to eqn (6.31) is

$$I_a^{(sq)} = 3.18\sqrt{\frac{3}{2}} = 3.895 \text{ A}$$

The armature winding losses according to eqn (6.33) are

$$\Delta P_a = 2 \times 4.984 \times 3.895^2 = 151.2 \text{ W}$$

The armature winding losses calculated on the basis of eqn (6.36) are the same, i.e.,

$$\Delta P_a = 3 \times 4.984 \times 3.18^2 = 151.2 \text{ W}$$

The stator slot pitch

$$t_1 = \frac{\pi D_{1in}}{s_1} = \frac{\pi \times 82.5}{36} = 7.2 \text{ mm}$$

The width of the stator tooth

$$c_{1t} = \frac{\pi(D_{1in} + 2h_{14} + b_{12})}{s_1} - b_{12} = \frac{\pi(82.5 + 2 \times 0.7 + 3.0)}{36} - 3.0 = 4.58 \text{ mm}$$

The magnetic flux density in the stator teeth

$$B_{1t} = \frac{B_{mg}t_1}{c_{1t}k_i} = \frac{0.6848 \times 7.2}{4.58 \times 0.96} = 1.12 \text{ T}$$

The height of the stator tooth

$$h_{1t} = 0.5b_{11} + h_{11} + h_{12} + 0.5b_{12} + h_{14}$$

$$= 0.5 \times 4.8 + 9.5 + 0.4 + 0.5 \times 3.0 + 0.7 = 14.5 \text{ mm}$$

The height of the stator yoke

$$h_{1y} = 0.5(D_{1out} - D_{1in}) - h_{1t} = 0.5(136.0 - 82.5) - 14.5 = 12.25 \text{ mm}$$

The pole pitch

$$\tau = \frac{\pi D_{1in}}{2p} = \frac{\pi \times 82.5}{4} = 64.8 \text{ mm}$$

The ratio of the pole shoe arc–to–pole pitch

$$\alpha_i^{(sq)} = \frac{b_p}{\tau} = \frac{56.7}{64.8} = 0.875$$

The stator magnetic flux according to eqn (6.24)

$$\Phi_f^{(sq)} = \alpha_i^{(sq)} \tau L_i B_{mg} = 0.875 \times 0.0648 \times 0.1008 \times 0.6848 = 39.14 \times 10^{-4} \text{ Wb}$$

The magnetic flux density in the stator yoke with leakage flux being neglected

$$B_{1y} = \frac{\Phi_f^{(sq)}}{2h_{1y}L_ik_i} = \frac{39.14 \times 10^{-4}}{2 \times 0.01225 \times 0.1008 \times 0.96} = 1.65 \text{ T}$$

The mass of the stator teeth

$$m_{1t} = 7700c_{1t}h_{1t}L_ik_is_1$$

$$= 7700 \times 0.00458 \times 0.0145 \times 0.1008 \times 0.96 \times 36 = 1.781 \text{ kg}$$

The mass of the stator yoke

$$m_{1y} = 7700\pi(D_{1out} - h_{1y})h_{1y}L_ik_i$$

$$= 7700\pi(0.136 - 0.01225) \times 0.01225 \times 0.1008 \times 0.96 = 3.548 \text{ kg}$$

The stator core losses due to fundamental harmonic according to eqn (B.19) are

$$[\Delta P_{Fe}]_{n=1} = 2.4 \left(\frac{50}{50}\right)^{4/3} (1.7 \times 1.12^2 \times 1.781 + 2.4 \times 1.65^2 \times 3.548) = 64.75 \text{ W}$$

It has been assumed $k_{adt} = 1.7$ and $k_{ady} = 2.4$. The higher harmonic stator core losses can approximately be evaluated on the assumption that the *rms* value of the fundamental harmonic voltage $V_{1,1} \approx V_1 = 220$ V and higher harmonics *rms* voltages are $V_{1,5} \approx 220/5 = 44$ V, $V_{1,7} \approx 220/7 = 31.42$ V, $V_{1,11} \approx 220/11 = 20$ V and $V_{1,13} \approx 220/13 = 16.92$ V. Only harmonics $n = 5.7, 11$ and 13 will be included. Through the use of eqn (B.49)

$$\Delta P_{Fe} = 64.75 \times [\left(\frac{220}{220}\right)^2 1^{4/3} + \left(\frac{44}{220}\right)^2 5^{4/3} + \left(\frac{31.42}{220}\right)^2 7^{4/3} + \left(\frac{16.92}{220}\right)^2 13^{4/3}$$

$$+ \left(\frac{20}{220}\right)^2 11^{4/3}] = 64.75 \times 1.998 = 129.4 \text{ W}$$

The rotational losses are calculated on the basis of eqn (B.34), i.e.,

$$\Delta P_{rot} = \frac{1}{30}(0.0825 + 0.15)^4\sqrt{0.1008}\left(\frac{1500}{100}\right)^{2.5} = 0.0269 \text{ kW} \approx 27 \text{ W}$$

The additional losses are assumed to be 3% of the output power. Thus, the output power is

$$P_{out} = \frac{1}{1.03}(P_{in} - \Delta P_a - \Delta P_{Fe} - \Delta P_{rot}) = \frac{1}{1.03}(1883 - 151.2 - 129.4 - 27) \approx 1490 \text{ W}$$

The efficiency

$$\eta = \frac{1490}{1883} = 0.813 \qquad \text{or} \qquad 81.3\%$$

Numerical example 6.5

Find the minimum, average and *rms* torque and values of the relative torque ripple t_r according to eqns (6.39) to (6.42) of the instantaneous torque expressed by the following functions:

(a) sum of constant value T_0 and sinusoidal waveform $T_{rm} \cos \alpha$, i.e.,

$$T(\alpha) = T_0 + T_{rm} \cos \alpha$$

where T_{rm} is the maximum value of the torque ripple. The period of the torque pulsation is $T_p = 10$ electrical degrees.

(b) function created by tips of sinusoids shifted by the electrical angle $2\pi/P$ one from each other, i.e.,

$$T(\alpha) = T_m \cos \alpha \quad \text{for} \quad -\frac{\pi}{P} \leq \alpha \leq \frac{\pi}{P}$$

$$T(\alpha) = T_m \cos(\alpha - \frac{\pi}{P}) \quad \text{for} \quad \frac{\pi}{P} \leq \alpha \leq 3\frac{\pi}{P}$$

$$T(\alpha) = T_m \cos(\alpha - 3\frac{\pi}{P}) \quad \text{for} \quad 3\frac{\pi}{P} \leq \alpha \leq 5\frac{\pi}{P}$$

....

where T_m is the maximum value of the torque and P is the number of pulses per one full period $T_p = 360$ electr. degrees (integer). Such a function is sometimes called a "rectifier function." After resolving into Fourier series

$$T(\alpha) = \frac{2}{\pi} T_m P \sin\left(\frac{\pi}{P}\right) \left[\frac{1}{2} - \sum_{\nu=1}^{\infty} \frac{(-1)^{\nu}}{(\nu P)^2 - 1} \cos(\nu P \alpha) \right]$$

Solution

(a) Instantaneous torque is expressed as $T(\alpha) = T_0 + T_{rm} \cos \alpha$

For $T_p = 10^0 = \pi/18$ the trigonometric function which expresses the torque ripple is $\cos[(2\pi/T_p)\alpha] = \cos(36\alpha)$.

The maximum torque is for $\alpha = 0$, i.e., $T_{max} = T_0 + T_{rm}$. The minimum torque is for $\alpha = 5^0 = \pi/72$, i.e., $T_{min} = T_0 - T_{rm}$.

The average value of the instantaneous torque according to eqn (6.43)

$$T_{av} = \frac{36}{\pi} \int_{-\pi/72}^{\pi/72} [T_0 + T_{rm} \cos(36\alpha)] d\alpha$$

$$= \frac{36}{\pi}\left[T_0\alpha + T_{rm}\frac{1}{36}\sin(36\alpha)\right]_{-\pi/72}^{\pi/72} = T_0 + \frac{2}{\pi}T_{rm}$$

The *rms* value of the torque ripple according to eqn (6.44)

$$T_{rrms} = \sqrt{\frac{36}{\pi}\int_{-\pi/72}^{\pi/72}T_{rm}^2\cos^2(36\alpha)d\alpha} = \sqrt{\frac{36}{\pi}T_{rm}^2\int_{-\pi/72}^{\pi/72}\frac{1}{2}[1+\cos(72\alpha)]d\alpha}$$

$$= \sqrt{\frac{36}{\pi}T_{rm}^2\left[\frac{1}{2}\alpha + \frac{1}{144}\sin(72\alpha)\right]_{-\pi/72}^{\pi/72}} = \frac{1}{\sqrt{2}}T_{rm}$$

For example, for $T_0 = 100$ Nm and $T_{rm} = 10$ Nm, the torque ripple according to eqn (6.39)

$$t_r = \frac{T_{rm}}{T_0} = \frac{10}{100} = 0.1 \quad \text{or} \quad 10\%$$

the torque ripple according to eqn (6.40)

$$t_r = \frac{T_{rm}}{T_0 + (2/\pi)T_{rm}} = \frac{10}{100 + (2/\pi)10} = 0.094 \quad \text{or} \quad 9.4\%$$

the torque ripple according to eqn (6.42)

$$t_r = \frac{T_{rm}}{\sqrt{2}(T_0 + (2/\pi)T_{rm})} = \frac{10}{\sqrt{2}(100 + (2/\pi)10)} = 0.665 \quad \text{or} \quad 6.65\%$$

(b) Torque variation is described by tips of sinusoids shifted by the electrical angle $2\pi/P$

The maximum torque is for $\alpha = 0$, i.e., $T_{max} = T_m$. The minimum torque is for $\alpha = \pi/P$, i.e.,

$$T_{min} = T_m\cos\frac{\pi}{P}$$

The average torque according to eqn (6.43)

$$T_{av} = \frac{P}{2\pi}\int_{-\pi/P}^{\pi/P}T_m\cos\alpha d\alpha = \frac{P}{2\pi}T_{max}[\sin\alpha]_{-\pi/P}^{\pi/P} = T_m\frac{P}{\pi}\sin\frac{\pi}{P}$$

The *rms* torque according to eqn (6.44) in which $T(\alpha) = T_m\cos\alpha$

$$T_{rms} = \sqrt{\frac{P}{2\pi}\int_{-\pi/P}^{\pi/P}T_m^2\cos^2\alpha d\alpha} = \sqrt{\frac{P}{2\pi}T_m^2\int_{-\pi/P}^{\pi/P}\frac{1}{2}(1+\cos 2\alpha)d\alpha}$$

$$= \sqrt{\frac{P}{2\pi}T_m^2 \left[\frac{1}{2}(\alpha + \frac{1}{2}\sin 2\alpha)\right]_{-\pi/P}^{\pi/P}} = T_m\sqrt{\frac{1}{2} + \frac{P}{4\pi}\sin\left(\frac{2\pi}{P}\right)}$$

The torque ripple according to eqn (6.39)

$$t_r = \frac{T_m - T_m\cos(\pi/P)}{T_m + T_m\cos(\pi/P)} = \frac{1 - \cos(\pi/P)}{1 + \cos(\pi/P)}$$

The torque ripple according to eqn (6.40)

$$t_r = \frac{T_m - T_m\cos(\pi/P)}{T_m(P/\pi)\sin(\pi/P)} = \frac{1 - \cos(\pi/P)}{(P/\pi)\sin(\pi/P)}$$

The torque ripple according to eqn (6.41)

$$t_r = \frac{T_m - T_m\cos(\pi/P)}{T_m\sqrt{\frac{1}{2} + \frac{P}{4\pi}\sin\left(\frac{2\pi}{P}\right)}} = \frac{1 - \cos(\pi/P)}{\sqrt{\frac{1}{2} + \frac{P}{4\pi}\sin\left(\frac{2\pi}{P}\right)}}$$

For example, for $P = 6$, the minimum torque $T_{min} = 0.866T_m$, the average torque $T_{av} = 0.9549T_m$, the rms torque $T_{rms} = 0.9558T_m$, the torque ripple according to eqn (6.39) is 0.0718 or 7.18%, the torque ripple according to eqn (6.40) is 0.14 or 14%, and the torque ripple according to eqn (6.41) is also approximately 14%.

Thus, when comparing the torque ripple of different electrical motors, the same definition of torque ripple must be used.

7

Axial Flux Motors

The *axial flux PM motor* is an attractive alternative to the cylindrical radial flux motor due to its pancake shape, compact construction and high power density. These motors are particularly suitable for electrical vehicles, pumps, valve control, centrifuges, fans, machine tools, robots and industrial equipment. They have become widely used for low-torque servo and speed control applications [172]. Axial flux PM motors also called *disk-type motors* can be designed as double-sided or single-sided machines, with or without armature slots, with internal or external PM rotors and with surface mounted or interior type PMs. Low power axial flux PM machines are usually machines with slotless windings and surface PMs.

As the output power of the axial flux motor increases, the contact surface between the rotor and shaft becomes smaller. Careful attention must be given to the design of the rotor-shaft mechanical joint as this is the principal cause of failures of disk type motors.

In some cases, rotors are embedded in power-transmission components to optimize the volume, mass, power transfer and assembly time. For EVs with built-in wheel motors the payoff is a simpler electromechanical drive system, higher efficiency and lower cost. Dual-function rotors may also appear in pumps, blowers, elevators and other types of machinery, bringing new levels of performance to these products.

7.1 Force and torque

In the design and analysis of axial flux motors the topology is complicated by the presence of two air gaps, high axial attractive forces, changing dimensions with radius and the fact that torque is produced over a continuum of radii, not just at a constant radius as in cylindrical motors.

The tangential force acting on the disk can be calculated on the basis of Ampere's circuital law

$$\mathbf{dF}_x = I_a(\mathbf{dr} \times \mathbf{B}_g) = A(r)(\mathbf{dS} \times \mathbf{B}_g) \qquad (7.1)$$

where $I_a\mathbf{dr} = A(r)\mathbf{dS}$, $A(r) = A_m(r)/\sqrt{2}$ according to eqn (5.13) for $D_{1in} = 2r$, $\mathbf{dr}$ is the radius element, $\mathbf{dS}$ is the surface element and $\mathbf{B_g}$ is the vector of the normal component (perpendicular to the disk surface) of the magnetic flux density in the air gap at given radius r.

Assuming the magnetic flux density in the air gap B_{mg} is independent of the radius r, the electromagnetic torque on the basis of eqn (7.1) is

$$dT_d = rdF_x = r[k_{w1}A(r)B_{avg}dS] = 2\pi\alpha_i k_{w1}A(r)B_{mg}r^2 dr \qquad (7.2)$$

where $B_{avg} = \alpha_i B_{mg}$ according to eqn (5.3) and $dS = 2\pi rdr$. The line current density $A(r)$ is the electric loading per one stator active surface in the case of a typical stator winding with distributed parameters (double sided stator and inner rotor) or electric loading of the whole stator in the case of an internal toroidal type or coreless stator.

Table 7.1. Specifications of PM disk brushless servo motors manufactured by E. Bautz GmbH, Weiterstadt, Germany

Quantity	S632D	S634D	S712F	S714F	S802F	S804F
Rated power, W	680	940	910	1260	1850	2670
Rated torque, Nm	1.3	1.8	2.9	4.0	5.9	8.5
Maximum torque, Nm	7	9	14	18	28	40
Standstill torque, Nm	1.7	2.3	3.5	4.7	7.0	10.0
Rated current, A	4.0	4.9	4.9	6.6	9.9	11.9
Maximum current, A	21	25	24	30	47	56
Standstill current, A	5.3	6.3	5.9	7.8	11.7	14.0
Rated speed, rpm	5000	5000	3000	3000	3000	3000
Maximum speed, rpm	6000	6000	6000	6000	6000	6000
Armature constant, V/1000 rpm	23	25	42	42	42	50
Torque constant, Nm/A	0.35	0.39	0.64	0.64	0.64	0.77
Resistance, Ω	2.5	1.8	2.4	1.5	0.76	0.62
Inductance, mH	3.2	2.8	5.4	4.2	3.0	3.0
Moment of inertia, $kgm^2 \times 10^{-3}$	0.08	0.12	0.21	0.3	0.6	1.0
Mass, kg	4.5	5.0	6.2	6.6	9.7	10.5
Diameter of frame, mm	150	150	174	174	210	210
Length of frame, mm	82	82	89	89	103	103
Power density, W/kg	151.1	188.0	146.8	190.9	190.7	254.3
Torque density, Nm/kg	0.289	0.36	0.468	0.606	0.608	0.809

Table 7.2. Specifications of PM disk brushless motors for medium duty electrical vehicles according to Premag, Cohoes, NY, U.S.A.

Quantity	HV2002	HV3202	HV4020	HV5020
Continuous output power, kW	20	32	40	50
Short duration				
Output power, kW	30	48	60	75
Input voltage, V	200	182	350	350
Torque, Nm	93.8	150.0	191.0	238.7
"Base" speed, rpm	2037	2037	2000	2000
Maximum speed, rpm	6725	6725	6600	6600
Efficiency	0.902	0.868	0.906	0.901
Diameter of frame, mm	238.0	286.0	329.2	284.2
Length of frame, m	71.4	85.6	68.1	70.1
Mass, kg	9	12	14	14
Power density, kW/kg	2.22	2.67	2.86	3.57
Torque density, Nm/kg	10.42	12.5	13.64	17.05

7.2 Performance

A three-dimensional FEM analysis is required to calculate the magnetic field, winding inductances, induced EMF and torque. The model can be simplified to a two-dimensional model by introducing a cylindrical cutting plane at the mean radius of the magnets [112]. This axial section is unfolded into a two-dimensional surface on which the FEM analysis can be done, as discussed for cylindrical PM motors in Chapters 3 and 5.

The performance characteristics can also be calculated analytically, using simplifications and adjusting the equations derived for cylindrical motors to disk type motors.

Table 7.1 shows specifications of axial flux PM brushless servo motors rated up to 2.7 kW, manufactured by E. Bautz GmbH, Weiterstadt, Germany.

Table 7.2 shows specifications of axial flux PM brushless motors rated from 20 to 50 kW for medium capacity (1300 to 4500 kg) electrical vehicles. Their pancake shapes make them ideal for direct wheel attachment.

7.3 Double-sided motor with internal PM disk rotor

In the *double-sided motor with internal PM disk rotor*, the armature winding is located on two stator cores. The disk with PMs rotates between two stators.

An eight-pole configuration is shown in Fig. 7.1. PMs are embedded or glued in a nonferromagnetic rotor skeleton. The nonferromagnetic air gap is large, i.e., the total air gap is equal to two mechanical clearances plus the thickness of a PM with its relative magnetic permeability close to unity. A double-sided motor with parallel connected stators can operate even if one stator winding is broken. On the other hand, a series connection can provide equal but opposing axial attractive forces.

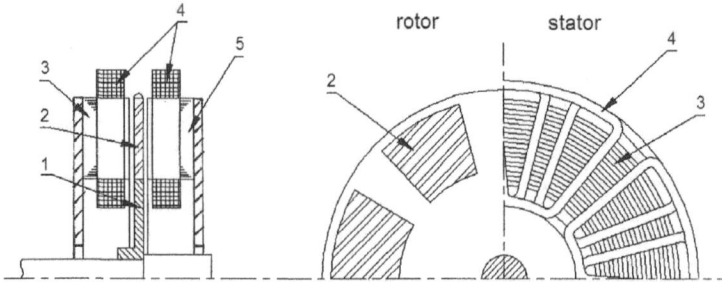

Fig. 7.1. Configuration of axial flux double-sided PM brushless motor with internal disk rotor: 1 — rotor, 2 — PM, 3 — stator core, 4 — stator winding.

A practical three-phase, 200 Hz, 3000 rpm, double-sided axial flux PM brushless motor with built-in brake is shown in Fig. 7.2 [182]. The three-phase winding is Y-connected, two stator windings in series. This motor is used as a flange-mounted servo motor. The ratio $X_{sd}/X_{sq} \approx 1.0$ so the motor can be analyzed as a cylindrical non-salient rotor synchronous machine [155, 180, 182].

7.3.1 Stator core

Normally, the stator cores are wound from electrotechnical steel strips and the slots are machined by shaping or planing. An alternative method is first to punch the slots with variable distances between them and then to wind the steel strip into the form of the slotted toroidal core (Research and Development Institute of Electrical Machines VÚES in Brno, Republic of Czech). In addition, this manufacturing process allows for making skewed slots to minimize the cogging torque and effect of slot harmonics. Each stator core has skewed slots in opposite directions. It is recommended to make a wave stator winding to obtain shorter end connections and more space for the shaft. An odd number of slots, e.g., 25 instead of 24 can also reduce the cogging torque (VÚES Brno).

Another technique is to form the stator core segments [296]. Each segment corresponds to one slot pitch (Fig. 7.3). The lamination strip of constant width is folded at distances proportional to the radius. To make folding easy, the strip has transverse grooves on opposite sides of the alternative steps. The zigzag laminated segment is finally compressed and fixed using a tape or thermosetting, as shown in Fig. 7.3 [296].

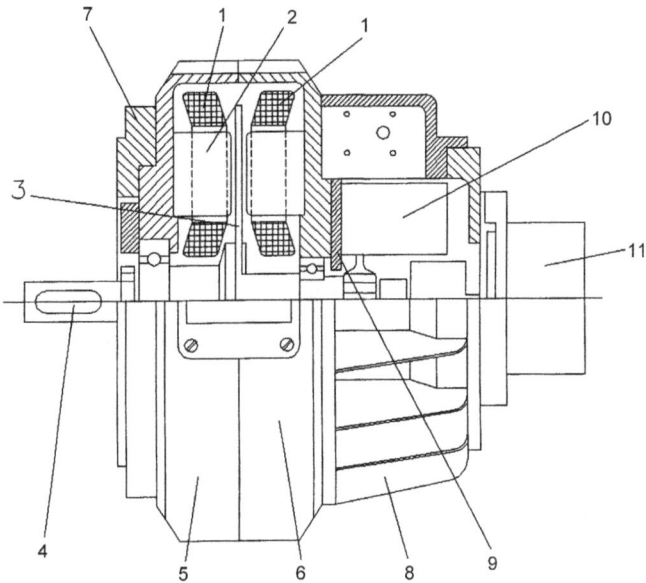

Fig. 7.2. Double-sided axial flux PM brushless motor with built-in brake: 1 — stator winding, 2 — stator core, 3 — disk rotor with PMs, 4 — shaft, 5 — left frame, 6 — right frame, 7 — flange, 8 — brake shield, 9 — brake flange, 10 — electromagnetic brake, 11 — encoder or resolver. Courtesy of Slovak University of Technology STU, Bratislava and Electrical Research and Testing Institute, Nová Dubnica, Slovakia.

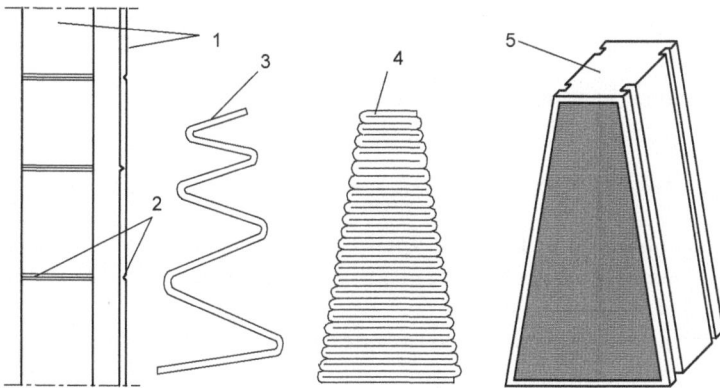

Fig. 7.3. Stator core segment formed from lamination strip: 1 — lamination strip, 2 — groove, 3 — folding, 4 — compressed segment, 5 — finished segment.

7.3.2 Main dimensions

The main dimensions of a double-sided PM brushless motor with internal disk rotor can be determined using the following assumptions: (a) the electric and magnetic loadings are calculated on an average diameter of the stator core; (b) the number of turns per phase per one stator is N_1; (c) the phase armature current in one stator winding is I_a; (d) the back EMF per phase per one stator winding is E_f.

The line current density per one stator is expressed by eqn (5.13) in which the inner stator diameter should be replaced by an average diameter

$$D_{av} = 0.5(D_{ext} + D_{in}) \tag{7.3}$$

where D_{ext} is the outer diameter and D_{in} is the inner diameter of the stator core. The pole pitch and the effective length of the stator core in a radial direction are

$$\tau = \frac{\pi D_{av}}{2p} \qquad\qquad L_i = 0.5(D_{ext} - D_{in}) \tag{7.4}$$

The EMF induced in the stator winding by the rotor excitation system, according to eqn (5.5), for the disk rotor synchronous motor has the following form:

$$E_f = \pi\sqrt{2}n_s p N_1 k_{w1} \Phi_f = \pi\sqrt{2}n_s N_1 k_{w1} D_{av} L_i B_{mg} \tag{7.5}$$

where the magnetic flux can approximately be expressed as

$$\Phi_f \approx \frac{2}{\pi}\tau L_i B_{mg} = \frac{D_{av}}{p} L_i B_{mg} \tag{7.6}$$

The electromagnetic apparent power in two stators

$$S_{elm} = m_1(2E_f)I_a = m_1 E_f(2I_a) = \pi^2 k_{w1} D_{av}^2 L_i n_s B_{mg} A_m \tag{7.7}$$

For series connection the EMF is equal to $2E_f$ and for parallel connection the current is equal to $2I_a$. For a multidisk motor the number "2" should be replaced by the number of stators.

It is convenient to use the ratio of *inner–to–outer stator diameter*

$$k_d = \frac{D_{in}}{D_{ext}} \tag{7.8}$$

Theoretically, a PM axial flux motor develops maximum electromagnetic torque when $k_d = 1/\sqrt{3}$ [4]. The product $D_{av}^2 L_i$ proportional to the volume of one stator is

$$D_{av}^2 L_i = \frac{1}{8}(1 + k_d)(1 - k_d^2)D_{ext}^3$$

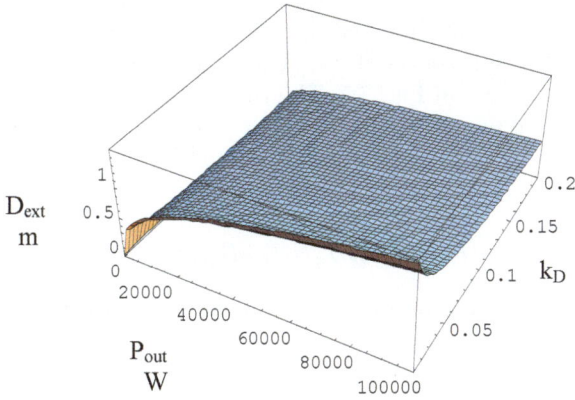

Fig. 7.4. Outer diameter D_{ext} as a function of the output power P_{out} and parameter k_D for $\epsilon = 0.9$, $k_{w1}\eta\cos\phi = 0.84$, $n_s = 1000$ rpm $= 16.67$ rev/s and $B_{mg}A_m = 26,000$ TA/m.

Putting

$$k_D = \frac{1}{8}(1 + k_d)(1 - k_d^2) \tag{7.9}$$

the volume of one stator is proportional to $D_{av}^2 L_i = k_D D_{ext}^3$. In connection with eqns (5.68) and (5.69) the stator outer diameter is

$$D_{ext} = \sqrt[3]{\frac{\epsilon P_{out}}{\pi^2 k_{w1} k_D n_s B_{mg} A_m \eta \cos\phi}} \tag{7.10}$$

The outer diameter of the stator is the most important dimension of disk rotor PM motors. Since $D_{ext} \propto \sqrt[3]{P_{out}}$ the outer diameter increases rather slowly with the increase of the output power (Fig. 7.4). This is why small power disk motors have relatively large diameter. The disk rotor is preferred for medium and large power motors. Motors with output power over 10 kW have reasonable diameters. Also, disk construction is recommended for a.c. servo motors fed with high frequency voltage.

7.4 Double-sided motor with one stator

A *double-sided motor with internal stator* is more compact than the previous construction with internal PM rotor [113, 203, 277, 314]. In this machine the *toroidal stator core* is also formed from a continuous steel tape, as in the motor with internal PM disk. The polyphase slotless armature winding (toroidal type) is located on the surface of the stator core. The total air gap is equal to the thickness of the armature winding, mechanical clearance and the thickness of the PM in the axial direction. The double-sided rotor with PMs is located at two sides of the stator. The configurations with internal and external rotors are shown in Fig. 7.5. The three phase winding arrangement, magnet polarities and flux paths in the magnetic circuit are shown in Fig. 7.6.

The average electromagnetic torque developed by the motor according to eqn (7.2) is

$$dT_d = 2\alpha_i m_1 I_a N_1 k_{w1} B_{mg} r dr$$

Integrating the above equation from $D_{in}/2$ to $D_{ext}/2$ with respect to x

$$T_d = \frac{1}{4}\alpha_i m_1 I_a N_1 k_{w1} B_{mg}(D_{ext}^2 - D_{in}^2)$$

$$= \frac{1}{4}\alpha_i m_1 N_1 k_{w1} B_{mg} D_{ext}^2 (1 - k_d^2) I_a \qquad (7.11)$$

where k_d is according to eqn (7.8). The magnetic flux per pole pitch is

$$\Phi_f = \alpha_i B_{mg} \frac{2\pi}{2p} \int_{0.5 D_{in}}^{0.5 D_{ext}} r dr = \frac{1}{8}\alpha_i \frac{\pi}{p} B_{mg} D_{ext}^2 (1 - k_d^2) \qquad (7.12)$$

The above eqn (7.12) is more accurate than eqn (7.6). Putting eqn (7.12) into eqn (7.11) the average torque is

$$T_d = 2\frac{p}{\pi} m_1 N_1 k_{w1} \Phi_f I_a \qquad (7.13)$$

To obtain the *rms* torque for sinusoidal current and sinusoidal magnetic flux density eqn (7.13) should be multiplied by the coefficient $\pi\sqrt{2}/4 \approx 1.11$, i.e.,

$$T_d = \frac{m_1}{\sqrt{2}} p N_1 k_{w1} \Phi_f I_a = k_T I_a \qquad (7.14)$$

where the torque constant

$$k_T = \frac{m_1}{\sqrt{2}} p N_1 k_{w1} \Phi_f \qquad (7.15)$$

The EMF at no-load can be found by differentiating the first harmonic of the magnetic flux waveform $\phi_{f1} = \Phi_f \sin \omega t$ and multiplying by $N_1 k_{w1}$, i.e.

$$e_f = N_1 k_{w1} \frac{d\phi_{f1}}{dt} = 2\pi f N_1 k_{w1} \Phi_f \cos \omega t$$

The *rms* value is obtained by dividing the peak value $2\pi f N_1 k_{w1} \Phi_f$ of the EMF by $\sqrt{2}$, i.e.,

$$E_f = \pi\sqrt{2} f N_1 k_{w1} \Phi_f = \pi\sqrt{2} p N_1 k_{w1} \Phi_f n_s = k_E n_s \qquad (7.16)$$

where the EMF constant (armature constant)

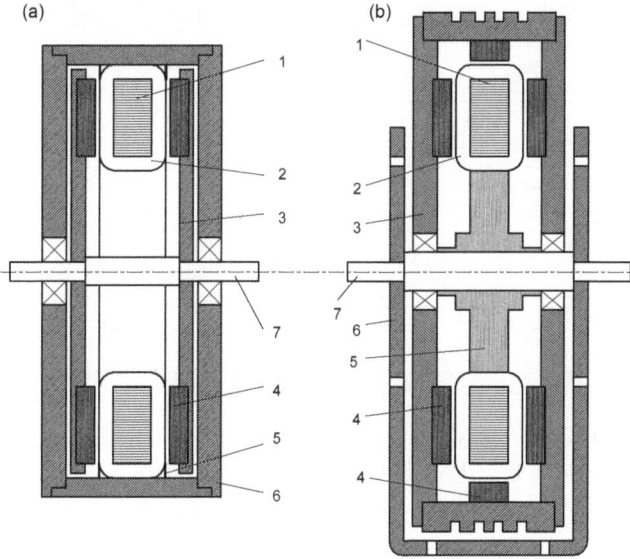

Fig. 7.5. Double-sided motors with one slotless stator: (a) internal rotor, (b) external rotor. 1 — stator core, 2 — stator winding, 3 — steel rotor, 4 — PMs, 5 — resin, 6 — frame, 7 — shaft.

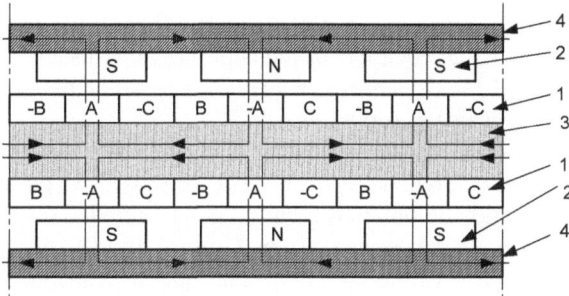

Fig. 7.6. Three-phase winding, PM polarities and magnetic flux paths of a double-sided disk motor with one internal slotless stator. 1 — winding, 2 — PM, 3 — stator yoke, 4 — rotor yoke.

$$k_E = \pi\sqrt{2}pN_1k_{w1}\Phi_f \tag{7.17}$$

The same form of eqn (7.16) can be obtained on the basis of the developed torque $T_d = m_1 E_f I_a/(2\pi n_s)$ in which T_d is according to eqn (7.14). For the toroidal type winding the winding factor $k_{w1} = 1$.

A motor with external rotor, according to Fig. 7.5b, has been designed for hoist applications. A similar motor can be used as an electric car wheel propulsion machine. Additional magnets on cylindrical parts of the rotor are

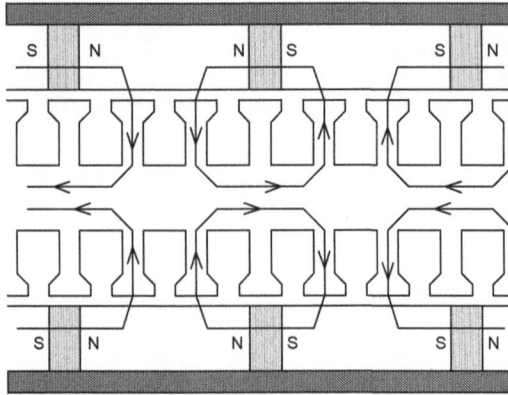

Fig. 7.7. Double-sided motor with one internal slotted stator and buried PMs. 1 — stator core with slots, 2 — PM, 3 — mild steel core (pole), 4 — nonferromagnetic rotor disk.

sometimes added [203] or U-shaped magnets can be designed. Such magnets embrace the armature winding from three sides and only the internal portion of the winding does not produce any electromagnetic torque.

Owing to the large air gap the maximum magnetic flux density does not exceed 0.65 T. To produce this flux density sometimes a large volume of PMs is required. As the permeance component of the flux ripple associated with the slots is eliminated, the cogging torque is practically absent. The magnetic circuit is unsaturated (slotless stator core). On the other hand, the machine structure lacks the necessary robustness [277].

The stator can also be made with slots (Fig. 7.7). For this type of motor, slots are progressively notched into the steel tape as it is passed from one mandrel to another and the polyphase winding is inserted [314]. In the case of the slotted stator the air gap is small ($g \approx 0.5$ mm) and the air gap magnetic flux density can increase to 0.85 T [113]. The magnet thickness is less than 50% that of the previous design, shown in Figs 7.5 and 7.6.

There are a number of applications for medium and large power axial flux motors with external PM rotors, especially in electrical vehicles [113, 314]. Disk-type motors with external rotors have a particular advantage in low speed high torque applications, such as buses and shuttles, due to their large radius for torque production. For small electric cars, the possibility of mounting the electric motor directly into the wheel has many advantages; it simplifies the drive system and the constant velocity joints are no longer needed [113].

Table 7.3. Specifications of single-sided PM disk brushless motors for gearless elevators manufactured by Kone, Hyvinkää, Finland

Quantity	MX05	MX06	MX10	MX18
Rated output power, kW	2.8	3.7	6.7	46.0
Rated torque, Nm	240	360	800	1800
Rated speed, rpm	113	96	80	235
Rated current, A	7.7	10	18	138
Efficiency	0.83	0.85	0.86	0.92
Power factor	0.9	0.9	0.91	0.92
Cooling	natural	natural	natural	forced
Diameter of sheave, m	0.34	0.40	0.48	0.65
Elevator load, kg	480	630	1000	1800
Elevator speed, m/s	1	1	1	4
Location	hoistway	hoistway	hoistway	machine room

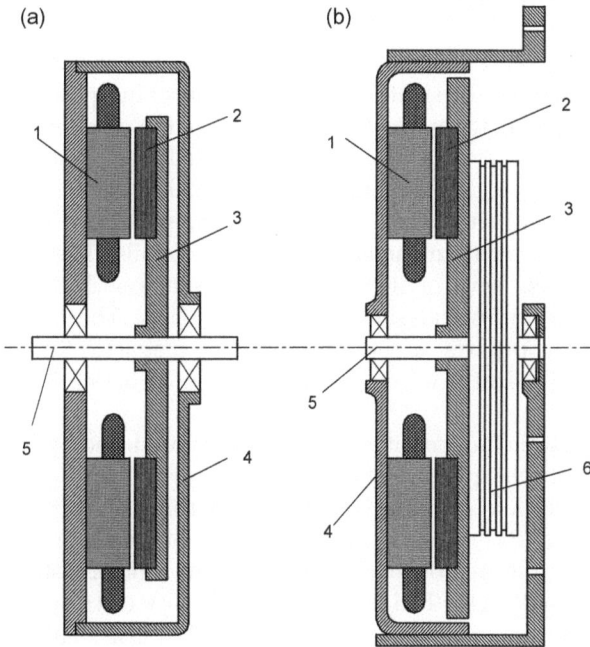

Fig. 7.8. Single sided disk motors: (a) for industrial and traction electromechanical drives, (b) for hoist applications. 1 — stator, 2 — PM, 3 — rotor, 4 — frame, 5 — shaft, 6 — sheave.

7.5 Single-sided motors

Single-sided construction of an axial flux motor is simpler than double-sided, but the torque produced is lower. Fig. 7.8 shows typical constructions with surface PM rotors and laminated stators wound from electromechanical steel strips. A single-sided motor according to Fig. 7.8a has a standard frame and shaft. It can be used in industrial, traction and servo electromechanical drives. The motor for hoist applications shown in Fig. 7.8b is integrated with a sheave (drum for ropes) and brakes (not shown). It is used in gearless elevators [134].

Specifications of single-sided disk type PM motors for gearless passenger elevators are given in Table 7.3 [134]. Stators have from 96 to 120 slots with three-phase short-pitch winding, insulation class F. For example, the MX05 motor rated at 2.8 kW, 280 V, 18.7 Hz has the stator winding resistance $R_1 = 3.5$ Ω, stator winding reactance $X_1 = 10$ Ω, $2p = 20$, sheave diameter 340 mm and weighs 180 kg.

7.6 Ironless double-sided motors

The *ironless* disk type PM brushless motor has neither armature nor excitation ferromagnetic core. The stator winding consists of full-pitch or short-pitch coils wound from insulated wires. Coils are arranged in overlapping layers like petals around the center of a flower and embedded in a plastic of very high mechanical integrity, e.g., U.S. Patent No. 5744896. The winding is fixed to the cylindrical part of the frame. To minimize the winding diameter the end connections are thicker than the active portions of coils. The twin nonferromagnetic rotor disks have cavities of the same shape as PMs. Magnets are inserted in these cavities and glued to the rotor disks. The PMs of opposite polarity fixed to two parts of the rotor produce magnetic flux, the lines of which criss-cross the stator winding. The motor construction is shown in Fig. 7.9.

A strong magnetic flux density in the air gap is produced by PMs arranged in Mallinson–Halbach array. Mallinson–Halbach array does not require any ferromagnetic cores and excites magnetic flux density closer to the sinusoids than a conventional PM array. The key concept of Mallinson–Halbach array is that the magnetization vector should rotate as a function of distance along the array (Figs 7.10 and 7.11). The magnetic flux density distribution plotted in Fig. 7.10 has been produced with the aid of a two dimensional FEM analysis of an ironless motor with magnet-to-magnet air gap of 10 mm (8 mm winding thickness, two 1 mm air gaps). The thickness of each PM is $h_M = 6$ mm. The remanent magnetic flux density is $B_r = 1.23$ T and the coercivity is $H_c = 979$ kA/m. The peak value of the magnetic flux density in the air gap exceeds 0.6 T. Three Mallinson–Halbach arrays have been simulated, i.e., 90^0, 60^0 and 45^0. As the angle between magnetic flux density vectors of neighboring

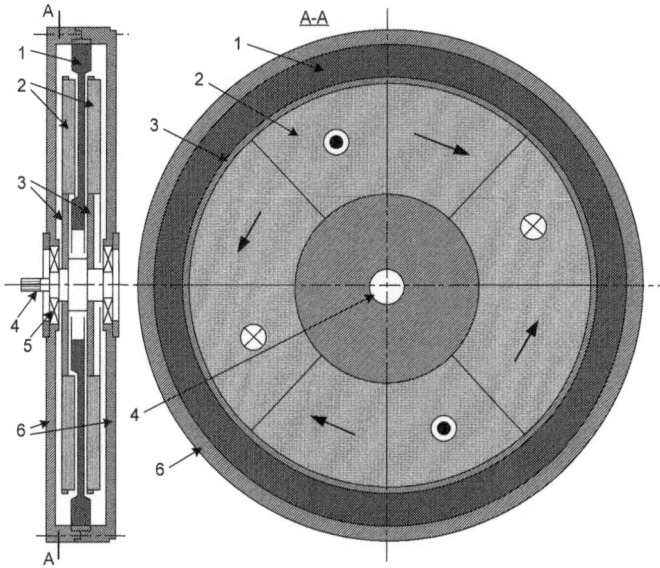

Fig. 7.9. Ironless double-sided PM brushless motor of disk type: 1 — stator winding, 2 — PMs, 3 — rotor, 4 — shaft, 5 — bearing, 6 — frame.

magnets decreases, the peak value of the normal component of the magnetic flux density increases slightly.

Ironless motors do not produce any torque pulsations at zero current state and can reach very high efficiency impossible for standard motors with ferromagnetic cores. Elimination of core losses is extremely important for high speed motors operating at high frequencies. Another advantage is very small mass of the ironless motor and consequently high power density and torque density. These motors are excellent for propulsion of solar powered electric cars [257]. The drawbacks include mechanical integrity problems, high axial forces between PMs on the opposite disks, heat transfer from the stator winding and its low inductance.

Small ironless motors may have *printed circuit stator windings* or *film coil windings*. The film coil stator winding has many coil layers while the printed circuit winding has one or two coil layers. Fig. 7.12 shows an ironless brushless motor with film coil stator winding. Small film coil motors are used in computer peripherals, pagers, mobile phones, flight recorders, card readers, copiers, printers, plotters, micrometers, labeling machines, video recorders and medical equipment.

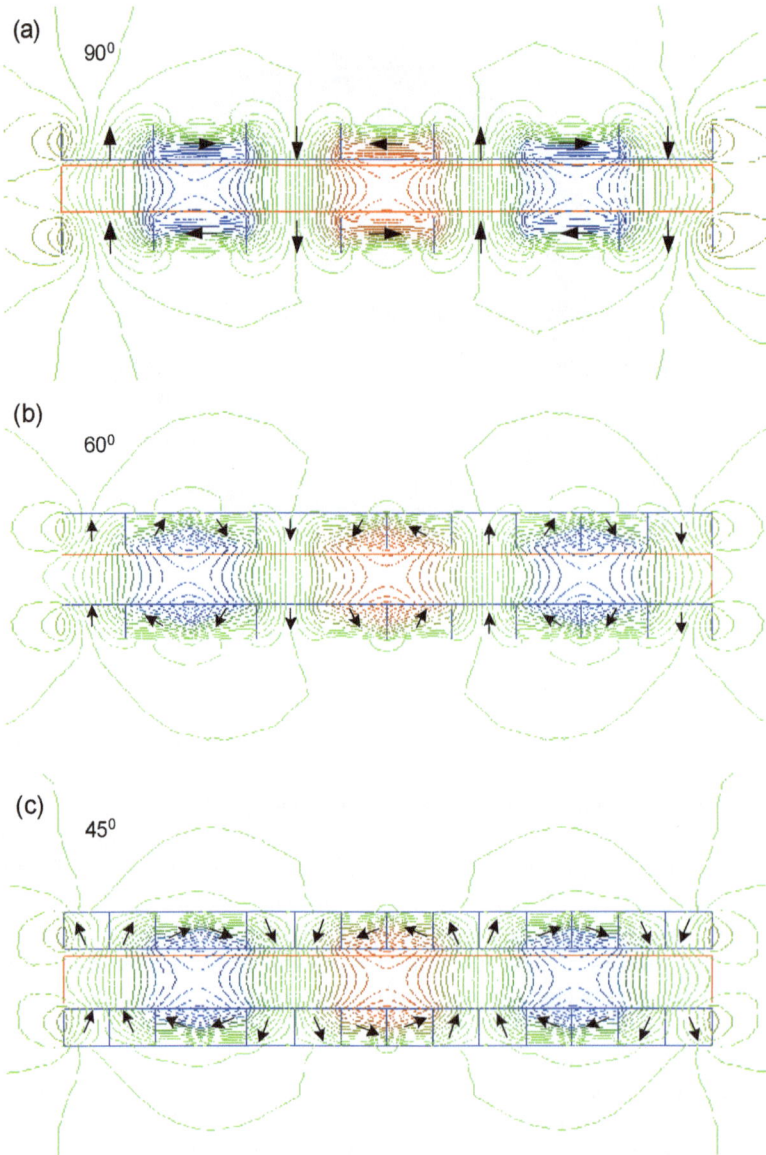

Fig. 7.10. Magnetic flux distribution in an ironless double sided brushless motor excited by Mallinson–Halbach arrays of PMs: (a) 90^0, (b) 60^0, and (c) 45^0 PM array.

(a)

(b)

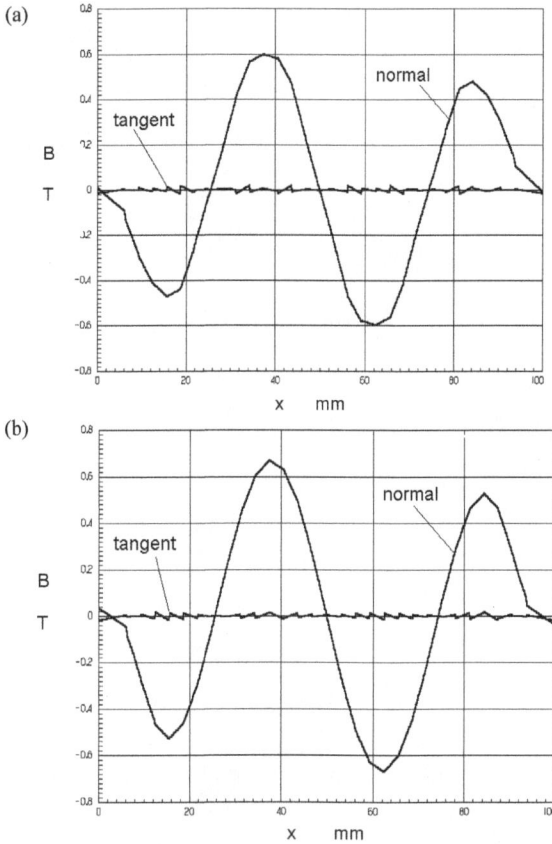

Fig. 7.11. Waveforms of the normal and tangent components of the magnetic flux density in the center of an ironless double-sided brushless motor excited by Mallinson–Halbach arrays of PMs: (a) 90^0, (b) 45^0. The magnetic flux density waveforms are functions of the circumferential distance at the mean radius of the magnets.

7.7 Multidisk motors

There is a limit on the increase of motor torque that can be achieved by enlarging the motor diameter. Factors limiting the single disk design are (a) axial force taken by bearings, (b) integrity of mechanical joint between the disk and shaft and (c) disk stiffness. A more reasonable solution for large torques are double or triple disk motors.

There are several constructions of multidisk motors [2, 3, 5, 63]. Large multidisk motors rated at least 300-kW have a water cooling system with radiators around the winding end connections [63]. To minimize the winding losses the cross section of conductors is bigger in the slot area (skin effect) than in the end connection region. Using a variable cross section means a gain

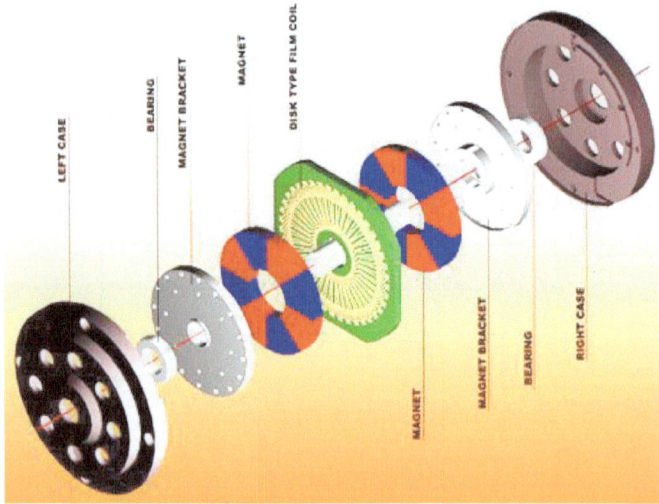

Fig. 7.12. Exploded view of the axial flux PM brushless motor with film coil ironless stator winding. Courtesy of Embest, Soeul, South Korea.

Fig. 7.13. Double disk PM brushless motor for gearless elevators. Courtesy of Kone, Hyvinkää, Finland.

Table 7.4. Specifications of double disk PM brushless motors manufactured by Kone, Hyvinkää, Finland

Quantity	MX32	MX40	MX100
Rated output power, kW	58	92	315
Rated torque, Nm	3600	5700	14,000
Rated speed, rpm	153	153	214
Rated current, A	122	262	1060
Efficiency	0.92	0.93	0.95
Power factor	0.93	0.93	0.96
Elevator load, kg	1600	2000	4500
Elevator speed, m/s	6	8	13.5

of 40% in the rated power [63]. Owing to high mechanical stresses titanium alloy is recommended for disk rotors.

A double disk motor for gearless elevators is shown in Fig. 7.13 [134]. Table 7.4 lists specification data of double-disk PM brushless motors rated from 58 to 315 kW [134].

Ironless disk motors provide a high level of flexibility to manufacture multidisk motors composed of the same segments (modules). Fractional horsepower motors can be "on-site" assembled from modules (Fig. 7.14) by simply removing one of the bearing covers and connecting terminal leads to the common terminal board. The number of modules depends on the requested shaft power or torque. The disadvantage of this type of multidisk motor is that a large number of bearings equal to double the number of modules are required.

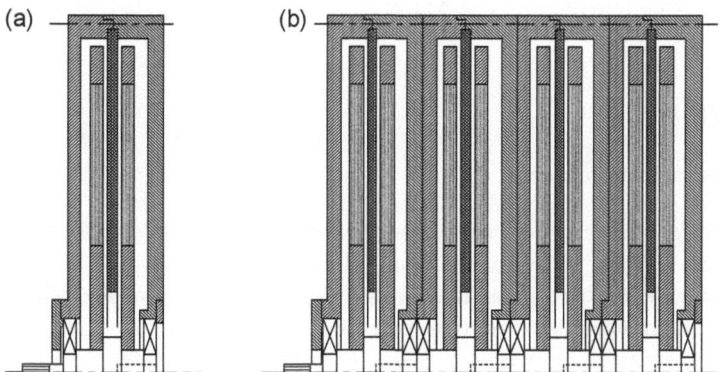

Fig. 7.14. Fractional horsepower ironless multidisk PM brushless motor: (a) single module, (b) four module motor.

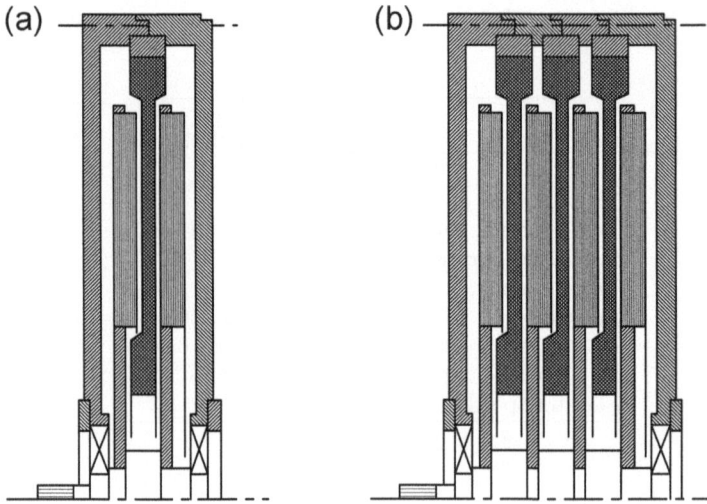

Fig. 7.15. Ironless multidisk PM brushless motors assembled using the same stator and rotor units: (a) single stator motor, (b) three stator motor.

Motors rated at kWs or tens of kWs must be assembled using separate stator and rotor units (Fig. 7.15). Multidisk motors have the same end bells with cylindrical frames inserted between them. The number of rotors is $K_2 = K_1 + 1$ where K_1 is the number of stators, while the number of cylindrical frames is $K_1 - 1$. The shaft must be tailored to the number of modules. Like a standard motor, this kind of motor has only two bearings.

Table 7.5 shows the specifications of single disk and multidisk PM brushless motors manufactured by Lynx Motion Technology Corporation, New Albany, IN, U.S.A. The mutidisk motor M468 consists of T468 single disk motors.

7.8 Applications

7.8.1 Electric vehicles

In a new *electromechanical drive system for electric vehicles* the differential mechanism is replaced by an electronic differential system [113]. The configuration shown in Fig. 7.16a illustrates the use of a pair of electric motors mounted on the chassis to drive a pair of wheels through drive shafts which incorporate constant velocity joints. In the configuration shown in Fig. 7.16b, the motors forming the electronic differential are mounted directly in the wheels of the vehicle. The drive system is considerably simplified, when the motor is mounted in the wheel, because the drive shafts and constant velocity joints are now no longer needed. However, the resultant 'unsprung' wheel

Table 7.5. Specifications of ironless single disk and multidisk PM brushless motors manufactured by Lynx Motion Technology Corporation, New Albany, IN, U.S.A.

Quantity	T468 single disk motor	M468 multidisk motor
Output power, kW	32.5	156
Speed, rpm	230	1100
Torque, Nm	1355	1355
Efficiency	0.94	0.94
Voltage line-to-line, V	432 (216)	400
Current, A	80 (160)	243
Armature constant line-to-line, V/rpm	1.43	0.8
Torque constant, Nm/A	17.1 (8.55)	5.58
Resistance d.c., phase-to-phase, Ω	7.2 (1.8)	0.00375
Inductance line-to-line, mH	4.5 (1.125)	-
Rotor inertia, kgm^2	0.48	1.3
Outer diameter, m	0.468	0.468
Mass, kg	58.1	131.0
Power density, kW/kg	0.56	1.19
Torque density, Nm/kg	23.3	10.34

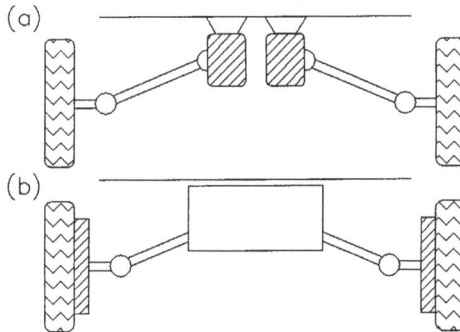

Fig. 7.16. Alternative forms of 'electronic differential' drive schemes: (a) onboard motor, (b) wheel mounted motor [113].

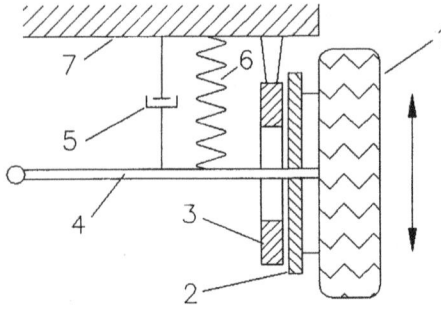

Fig. 7.17. Schematic representation of the disk machine with lower unsprung mass: 1 — wheel, 2 — disk rotor, 3 — stator, 4 — shaft, 5 — damper, 6 — spring, 7 — chassis [113].

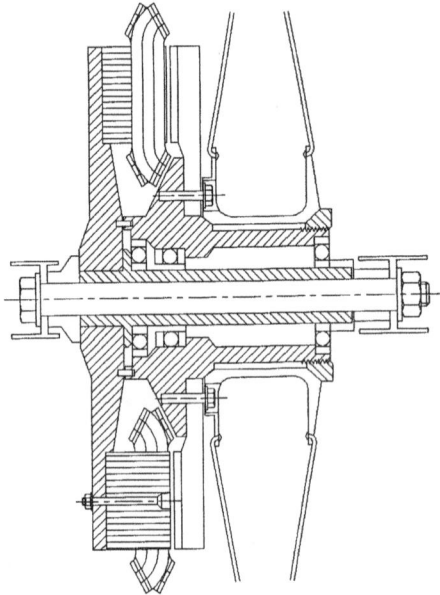

Fig. 7.18. Disk-rotor motor fitted to spoked wheel in rear forks of a solar powered vehicle [237].

mass of the vehicle is increased by the mass of the motor. Wheel motors of this direct drive configuration also suffer because the speed of the rotor is lower than would be the case with a geared arrangement. This leads to an increase of active materials volume required in the motor.

The disadvantages of conventional wheel motors can be overcome by using the arrangement shown in Fig. 7.17. The two stators are directly attached to the vehicle body while the PM rotor is free to move in radial directions. It will be observed that the wheel and disk rotor form the unsprung mass, while the stators of the motor become sprung mass supported on the chassis [113].

The PM disk rotor has to be carefully constructed so that it has adequate mechanical integrity. This is achieved by "canning" the rotor magnet assembly within a nonferromagnetic steel cover. The increase in the air gap due to the nonferromagnetic steel cover (1.2 mm) must be compensated for by increasing the PM thickness (an increase of about 2 mm) [113].

Modern axial flux PM brushless motors can also meet all requirements of high-performance in power limited vehicles, for example solar powered vehicles [237, 257]. Fig. 7.18 shows an axial flux motor with outer diameter of 0.26 m fitted to a spoked wheel [237].

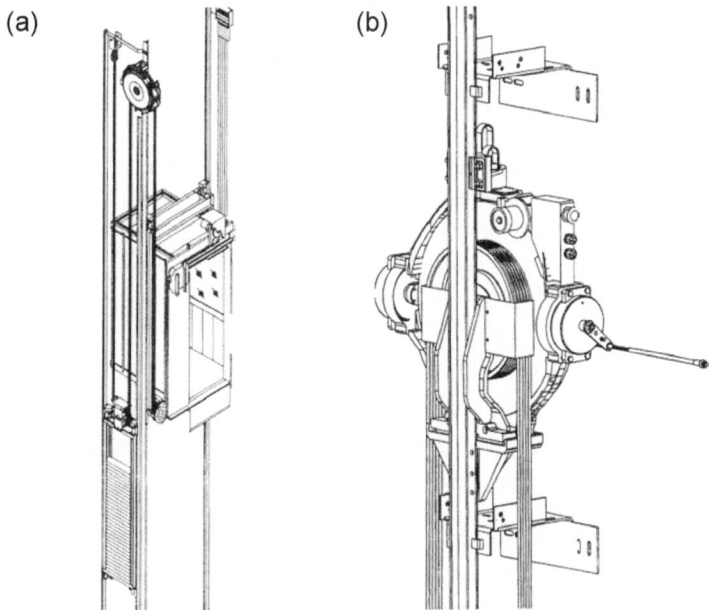

Fig. 7.19. MonoSpace^{TM} elevator: (a) elevator propulsion system; (b) Ecodisk^{TM} motor. Courtesy of Kone, Hyvinkää, Finland [134].

Table 7.6. Comparison of hoisting technologies for 630 kg elevators

Quantity	Hydraulic elevator	Warm gear elevator	Direct PM brushless motor elevator
Elevator speed, m/s	0.63	1.0	1.0
Motor shaft power, kW	11.0	5.5	3.7
Speed of motor, rpm	1500	1500	95
Motor fuses, A	50	35	16
Annual energy consumption, kWh	7200	6000	3000
Hoisting efficiency	0.3	0.4	0.6
Oil requirements, l	200	3.5	0
Mass, kg	350	430	170
Noise level, dB(A)	60...65	70...75	50...55

7.8.2 Gearless elevator propulsion system

The concept of *gearless electromechanical drive for elevators* was introduced in 1992 by Kone Corporation in Hyvinkää, Finland [134]. With the aid of a disk type low speed compact PM brushless motor (Table 7.3) the penthouse machinery room can be replaced by a space-saving direct electromechanical drive. In comparison with a low speed axial flux cage induction motor of similar diameter, the PM brushless motor has double the efficiency and three times higher power factor.

Fig. 7.19a shows the propulsion system of the elevator while Fig. 7.19b shows how the disk PM brushless motor is installed between the guide rails of the car and the hoistway wall.

Table 7.6 contains key parameters for comparison of different hoisting technologies [134]. The direct drive PM brushless motor is a clear winner.

7.8.3 Propulsion of unmanned submarines

An electric *propulsion system for submarines* requires high output power, high efficiency, a very low level of noise and compact motors [69, 223]. Disk-type brushless motors can meet these requirements and run for over 100,000 h without a failure, cooled only by ambient sea water. These motors are virtually silent and operate with a minimum vibration level. The output power at rated operating conditions can exceed 2.2 kW/kg and torque density 5.5 Nm/kg. Typical rotor linear speed of large scale marine propulsion motors is 20 to 30 m/s [223].

7.8.4 Counterrotating rotor ship propulsion system

An axial flux PM motor can be designed with the *counterrotation of two rotors* [50]. This machine topology can find applications in ship propulsion systems

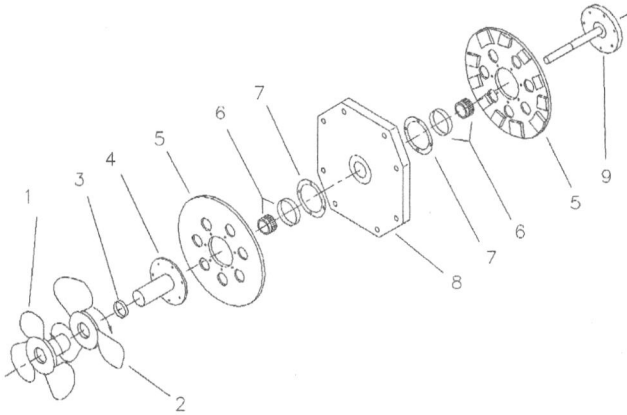

Fig. 7.20. Exploded view of the axial flux PM motor with counter-rotating rotor: 1 — main propeller, 2 — counter-rotating propeller, 3 — radial bearing, 4 — outer shaft, 5 — PM rotor, 6 — motor bearing, 7 — assembly ring, 8 — stator, 9 — inner shaft [50].

which use an additional counter-rotating propeller in order to recover energy from the rotational flow of the main propeller stream. In this case the use of an axial flux motor having counter-rotating rotors allows the elimination of the motion reversal epicyclical gear.

The stator winding coils have a rectangular shape which depends on the cross section of the toroidal core [50]; also see Fig. 7.5. Each coil has two active surfaces and each coil surface interacts with the facing PM rotor. In order to achieve the opposite motion of the two rotors, the stator winding coils have to be arranged in such a manner that counter-rotating magnetic fields are produced in the machine's annular air gaps. The stator is positioned between two rotors which consist of mild steel disks and axially magnetized NdFeB PMs. The magnets are mounted on the disk's surface from the stator sides. Each rotor has its own shaft that drives a propeller; i.e., the motor has two coaxial shafts that are separated by a radial bearing. The arrangement is shown in Fig. 7.20 [50].

Numerical examples

Numerical example 7.1

Find the main dimensions, approximate number of turns per phase and the approximate cross section of the stator slot for a three-phase, double-sided, double-stator disk rotor PM brushless motor rated at: $P_{out} = 75$ kW, $V_{1L} =$

460 V (Y connection), $f = 100$ Hz, $n_s = 1550$ rpm. The stator windings are connected in series.

Solution

For $f = 100$ Hz and $n_s = 1500$ rpm $= 25$ rev/s the number of poles is $2p = 8$. Assuming $D_{ext}/D_{in} = \sqrt{3}$ [4], the parameter k_D according to eqn (7.9) is

$$k_D = \frac{1}{8}\left(1 + \frac{1}{\sqrt{3}}\right)\left[1 - \left(\frac{1}{\sqrt{3}}\right)^2\right] = 0.131$$

For a 75 kW motor the product $\eta \cos\phi \approx 0.9$. The phase current for series connected stator windings is

$$I_a = \frac{P_{out}}{m_1(2V_1)\eta \cos\phi} = \frac{75,000}{3 \times 265.6 \times 0.9} = 104.6 \ \ \text{A}$$

where $2V_1 = 460/\sqrt{3} = 265.6$ V. The electromagnetic loading can be assumed as $B_{mg} = 0.65$ T and $A_m = 40,000$ A/m. The ratio $\epsilon = E_f/V_1 \approx 0.9$ and the stator winding factor has been assumed $k_{w1} = 0.96$. Thus, the stator outer diameter according to eqn (7.10) is

$$D_{ext} = \sqrt[3]{\frac{0.9 \times 75\ 000}{\pi^2 \times 0.96 \times 0.131 \times 25 \times 0.65 \times 40\ 000 \times 0.9}} = 0.453 \ \text{m}$$

The inner diameter, average diameter, pole pitch and effective stator length according to eqns (7.3), (7.4) and (7.8) are, respectively,

$$D_{in} = \frac{D_{ext}}{\sqrt{3}} = \frac{0.453}{\sqrt{3}} = 0.262 \ \text{m}, \qquad D_{av} = 0.5(0.453 + 0.262) = 0.3575 \ \text{m}$$

$$\tau = \frac{\pi \times 0.3575}{8} = 0.14 \ \text{m}, \qquad L_i = 0.5(0.453 - 0.262) = 0.0955 \ \text{m}$$

The number of stator turns per phase per stator calculated on the basis of line current density according to eqn (5.13) is

$$N_1 = \frac{A_m p \tau}{m_1 \sqrt{2} I_a} = \frac{40,000 \times 4 \times 0.14}{3\sqrt{2} \times 104.6} \approx 50$$

The number of stator turns per phase per stator calculated on the basis of eqn (7.5) for EMF and eqn (7.6) for magnetic flux is

$$N_1 = \frac{\epsilon V_1}{2\sqrt{2}fk_{w1}\tau L_i B_{mg}} = \frac{0.9 \times 265.6/2}{2\sqrt{2} \times 100 \times 0.96 \times 0.14 \times 0.0955 \times 0.65} \approx 49$$

A double layer winding can be located, say, in 16 slots per phase, i.e., $s_1 = 48$ slots for a three phase machine. The number of turns should be rounded to

48. This is an approximate number of turns which can be calculated exactly only after performing detailed electromagnetic and thermal calculations of the machine.

The number of slots per pole per phase according to eqn (A.5) is

$$q_1 = \frac{48}{8 \times 3} = 2$$

The number of stator coils (double-layer winding) is the same as the number of slots, i.e., $2pq_1m_1 = 8 \times 2 \times 3 = 48$. If the stator winding is made of four parallel conductors $a_w = 4$, the number of conductors in a single coil is

$$N_{1c} = \frac{a_w N_1}{(s_1/m_1)} = \frac{4 \times 48}{(48/3)} = 12$$

The current density in the stator conductor can be assumed $J_a \approx 4.5 \times 10^6$ A/m^2 (totally enclosed a.c. machines rated up to 100 kW). The cross section area of the stator conductor is

$$s_a = \frac{I_a}{a_w J_a} = \frac{104.6}{4 \times 4.5} = 5.811 \text{mm}^2$$

The stator winding of a 75 kW machine is made of a copper conductor of rectangular cross section. The slot fill factor for rectangular conductors and low voltage machines can be assumed to be 0.6. The cross section of the stator slot should, approximately, be

$$\frac{5.811 \times 12 \times 2}{0.6} \approx 233 \text{ mm}^2$$

where the number of conductors in a single slot is $12 \times 2 = 24$. The minimum stator slot pitch is

$$t_{1min} = \frac{\pi D_{in}}{s_1} = \frac{\pi \times 0.262}{48} = 0.0171 \text{ m} = 17.1 \text{ mm}$$

The stator slot width can be chosen to be 11.9 mm; this means that the stator slot depth is $233/11.9 \approx 20$ mm, and the stator narrowest tooth width is $c_{1min} = 17.1 - 11.9 = 5.2$ mm. Magnetic flux density in the narrowest part of the stator tooth is

$$B_{1tmax} \approx \frac{B_{mg}t_{1min}}{c_{1min}} = \frac{0.65 \times 17.1}{5.2} = 2.14 \text{ T}$$

This is a permissible value for the narrowest part of the tooth. The maximum stator slot pitch is

$$t_{1max} = \frac{\pi D_{ext}}{s_1} = \frac{\pi \times 0.453}{48} = 0.0296 \text{ m} = 29.6 \text{ mm}$$

Magnetic flux density in the widest part of the stator tooth

$$B_{1tmin} \approx \frac{B_{mg}t_{1max}}{c_{1max}} = \frac{0.65 \times 29.6}{29.6 - 11.9} = 1.09 \text{ T}$$

Numerical example 7.2

A three-phase, 2.2-kW, 50-Hz, 380-V (line-to-line), Y-connected, 750-rpm, $\eta = 78\%$, $\cos \phi = 0.83$, double-sided disk PM synchronous motor has the following dimensions of its magnetic circuit: rotor external diameter $D_{ext} = 0.28$ m, rotor internal diameter $D_{in} = 0.16$ m, thickness of the rotor (PMs) $2h_M = 8$ mm, single-sided mechanical clearance $g = 1.5$ mm. The PMs are distributed uniformly and create a surface configuration. The rotor does not have any soft ferromagnetic material. The rotor outer and inner diameters correspond to the outer and inner outline of PMs and the stator stack. The dimensions of semi-closed rectangular slots (Fig. A.2b) are: $h_{11} = 11$ mm, $h_{12} = 0.5$ mm, $h_{13} = 1$ mm, $h_{14} = 1$ mm, $b_{12} = 13$ mm and $b_{14} = 3$ mm. The number of stator slots (one unit) is $s_1 = 24$, the number of armature turns of a single stator per phase is $N_1 = 456$, the diameter of stator copper conductor is 0.5 mm (without insulation), the number of stator parallel wires is $a = 2$ and the air gap magnetic flux density is $B_{mg} = 0.65$ T. The rotational losses are $\Delta P_{rot} = 80$ W. The core losses and additional losses are $\Delta P_{Fe} + \Delta P_{str} = 0.05P_{out}$. There are two winding layers in each stator slot. The two twin stator Y-connected windings are fed in parallel.

Find the motor performance at the load angle $\delta = 11^0$. Compare the calculations obtained from the circuital approach with the finite element results.

Solution

The phase voltage is $V_1 = 380/\sqrt{3} = 220$ V. The number of pole pairs is $p = f/n_s = 50 \times 60/750 = 4$ and $2p = 8$. The minimum slot pitch is

$$t_{1min} = \frac{\pi D_{in}}{s_1} = \frac{\pi \times 0.16}{24} = 0.0209 \text{ m} \approx 21 \text{ mm}$$

The width of the slot is $b_{12} = 13$ mm; that means the narrowest tooth width $c_{1min} = t_{1min} - b_{12} = 21 - 13 = 8$ mm. The magnetic flux density in the narrowest part of the stator tooth

$$B_{1tmax} \approx \frac{B_{mg}t_{1min}}{c_{1min}} = \frac{0.65 \times 21}{8} = 1.7 \text{ T}$$

is rather low.

The average diameter, average pole pitch and effective length of the stator stack according to eqns (7.3) and (7.4) are

$$D_{av} = 0.5(0.28 + 0.16) = 0.22 \text{ m}$$

$$\tau = \frac{\pi \times 0.22}{8} = 0.0864 \text{ m}, \qquad L_i = 0.5(0.28 - 0.16) = 0.06 \text{ m}$$

Because the number of slots per pole per phase (A.5)

$$q_1 = \frac{24}{8 \times 3} = 1$$

the winding factor as expressed by eqns (A.1), (A.6) and (A.3) $k_{w1} = k_{d1}k_{p1} = 1 \times 1 = 1$.

The magnetic flux according to eqn (5.6) and EMF induced by the rotor excitation system according to eqns (5.5) are

$$\Phi_f = \frac{2}{\pi}0.0864 \times 0.06 \times 0.65 = 0.002145 \text{ Wb}$$

$$E_f = \pi\sqrt{2}\, 50 \times 456 \times 1 \times 0.002145 = 217.3 \text{ V}$$

It has been assumed that $B_{mg1} \approx B_{mg}$.

Now, it is necessary to check the electric loading, current density and space factor of the stator slot. For two parallel wires $a_w = 2$, the number of conductors per coil of a double layer phase winding is

$$N_{1c} = \frac{a_w N_1}{(s_1/m_1)} = \frac{2 \times 456}{(24/3)} = 114$$

Thus, the number of conductors in a single slot is equal to the (number of layers) × (number of conductors per coil N_{1c}) = $2 \times 114 = 228$.

The rated input current in a single stator

$$I_a = \frac{P_{out}}{2m_1 V_1 \eta \cos\phi} = \frac{2200}{2 \times 3 \times 220 \times 0.78 \times 0.83} = 2.57 \text{ A}$$

The stator line current density (peak value) on the basis of eqn (5.13)

$$A_m = \frac{3\sqrt{2}\, 456 \times 2.57}{0.0864 \times 4} = 14,386 \text{ A/m}$$

which is rather a low value even for small PM a.c. motors. The cross section of the stator (armature) conductor

$$s_a = \frac{\pi d_a^2}{4} = \frac{\pi 0.5^2}{4} = 0.197 \text{ mm}^2$$

gives the following current density under rated conditions

$$J_a = \frac{2.57}{2 \times 0.197} = 6.54 \text{ A/mm}^2$$

This is an acceptable value of the current density for disk rotor a.c. machines rated from 1 to 10 kW.

For the class F enamel insulation of the armature conductors, the diameter of the wire with insulation is 0.548 mm. Hence, the total cross sectional area of all conductors in the stator slot is

$$228\frac{\pi 0.548^2}{4} \approx 54 \text{ mm}^2$$

The cross section area of a single slot is approximately $h_{11}b_{12} = 11 \times 13 = 143$ mm^2. The space factor $54/143 = 0.38$ shows that the stator can be easily wound, since the average fill factor for low voltage machines with round stator conductors is about 0.4.

The average length of the stator end connection for a disk rotor a.c. machine (compare eqn (A.20)) is

$$l_{1e} \approx (0.083p + 1.217)\tau + 0.02 = (0.083 \times 4 + 1.217)0.0864 + 0.02 = 0.154 \text{ m}$$

The average length of the stator turn according to eqn (B.2) is

$$l_{1av} = 2(L_i + l_{1e}) = 2(0.06 + 0.154) = 0.428 \text{ m}$$

The stator winding resistance per phase at a temperature of 75°C (hot motor) according to eqns (B.1) and (B.11) is

$$R_1 = \frac{N_1 l_{1av}}{a\sigma_1 s_a} = \frac{456 \times 0.428}{47 \times 10^6 \times 2 \times 0.1965} = 10.57 \ \Omega$$

Carter's coefficient is calculated on the basis of eqns (A.27) and (A.28), i.e.,

$$k_C = \left(\frac{28.8}{28.8 - 0.00526 \times 11}\right)^2 = 1.004$$

$$\gamma = \frac{4}{\pi}\left[\frac{3}{2 \times 11}\arctan\left(\frac{3}{2 \times 11}\right) - \ln\sqrt{1 + \left(\frac{3}{2 \times 11}\right)}\right] = 0.00526$$

where $t_1 = \pi D_{av}/s_1 = \pi \times 0.22/24 = 0.0288$ m $= 28.8$ mm. The nonferromagnetic air gap in calculation of Carter's coefficient is $g_t = 2g + 2h_M = 2 \times 1.5 + 8 = 11$ mm. Since there are two slotted surfaces of twin stator cores, Carter's coefficient must be squared.

The stator (one unit) leakage reactance has been calculated according to eqn (A.30), i.e.,

$$X_1 = 4 \times 0.4\pi \times 10^{-6}\pi \times 50\frac{456^2 \times 0.06}{4 \times 1}(0.779 + \frac{0.154}{0.06}0.218$$

$$+0.2297 + 0.9322) = 6.158 \ \Omega$$

in which

- the coefficient of slot leakage reactance — eqn (A.12)

$$\lambda_{1s} = \frac{11}{3 \times 13} + \frac{0.5}{13} + \frac{2 \times 1}{13 + 3} + \frac{1}{3} = 0.779$$

- the coefficient of end connection leakage permeance — eqn (A.19) in which $(w_c = \tau)$

$$\lambda_{1e} \approx 0.34 \times 1 \left(1 - \frac{2}{\pi} \frac{0.0864}{0.154}\right) = 0.218$$

- the coefficient of differential leakage permeance — eqns (A.24) and (A.26)

$$\lambda_{1d} = \frac{3 \times 1 \times 0.0864 \times 1^2}{\pi^2 \times 0.011 \times 1.004} 0.0966 = 0.2297$$

$$\tau_{d1} = \frac{\pi^2 (10 \times 1^2 + 2)}{27} \sin \frac{30^0}{1} - 1 = 0.0966$$

- the coefficient of tooth-top leakage permeance — eqn (A.29)

$$\lambda_{1t} = \frac{5 \times 11/3}{5 + 4 \times 11/3} = 0.9322$$

According to eqns (5.31) and (5.33) in which $k_{fd} = k_{fq}$, the armature reaction reactances for surface type PM rotors and unsaturated machines are

$$X_{ad} = X_{aq} = 4 \times 3 \times 0.4 \times \pi \times 10^{-6} \frac{0.0864 \times 0.06}{1.004 \times 0.011} = 5.856 \ \Omega$$

where the air gap, in the denominator, for the armature flux should be equal to $g_t \approx 2 \times 1.5 + 8 = 11$ mm ($\mu_{rrec} \approx 1$). The synchronous reactances are

$$X_{sd} = X_{sq} = 6.158 + 5.856 = 12.01 \ \Omega$$

The armature currents are calculated on the basis of eqns (5.40), (5.41) and (5.43). For $\delta = 11^0$ ($\cos \delta = 0.982$, $\sin \delta = 0.191$) the current components are $I_{ad} = -1.82$ A, $I_{aq} = 1.88$ A and $I_a = 2.62$ A.

The input power absorbed by one stator is expressed by eqn (5.44). The input power absorbed by two stators in parallel is twice as much, i.e.,

$$P_{in} = 2 \times 3 \times 220(1.88 \times 0.982 - (-1.82)0.191) = 2892.4 \ \text{W}$$

The input apparent power absorbed by two stators

$$S_{in} = 2 \times 3 \times 220 \times 2.62 = 3458.4 \ \text{VA}$$

The power factor is

$$\cos \phi = \frac{2892.4}{3458.4} = 0.836$$

The losses in two stator windings according to eqn (B.12) in which the skin effect coefficient $k_{1R} = 1$ are

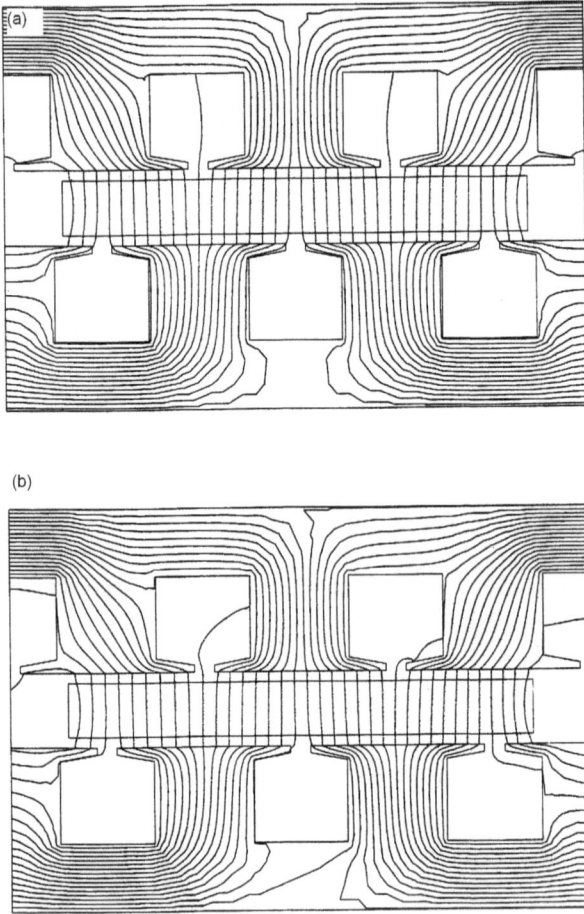

Fig. 7.21. Flux plots in the disk rotor PM rotor: (a) zero armature current, (b) rated armature current.

$$\Delta P_a = 2 \times 3 \times 2.62^2 \times 10.57 = 435.2 \text{ W}$$

The output power assuming that $\Delta P_{Fe} + \Delta P_{str} = 0.05 P_{out}$ is

$$P_{out} = \frac{1}{1.05}(P_{in} - \Delta P_{1w} - \Delta P_{rot}) = \frac{1}{1.05}(2892.4 - 435.2 - 80.0) = 2264 \text{ W}$$

The motor efficiency is

$$\eta = \frac{2264.0}{2892.4} = 0.783 \qquad \text{or} \qquad \eta = 78.3\%$$

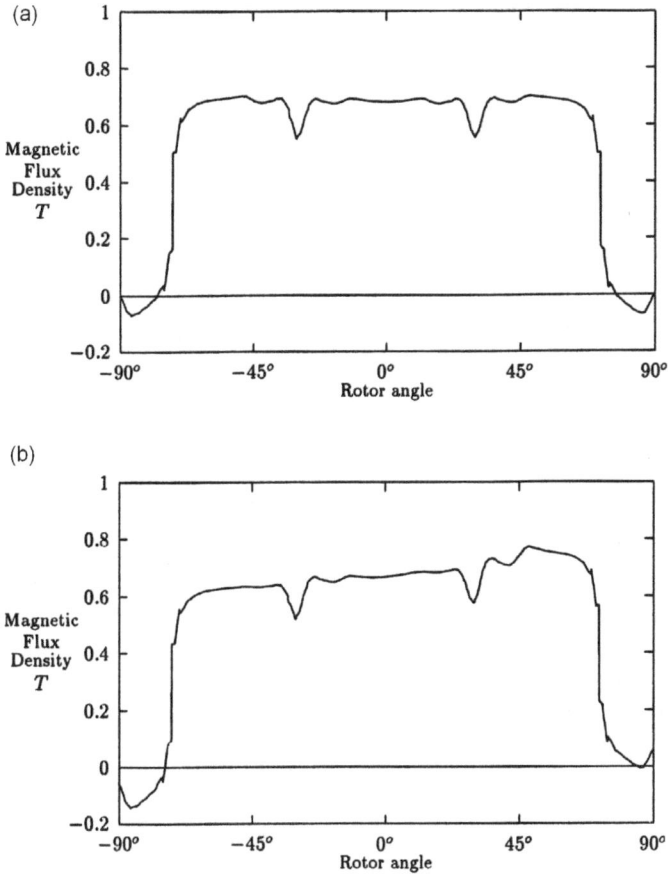

Fig. 7.22. Magnetic flux density distribution along the pole pitch: (a) zero armature current, (b) rated armature current.

The shaft torque

$$T_{sh} = \frac{2264}{2\pi(750/60)} = 28.83 \text{ Nm}$$

The electromagnetic power according to eqn (5.45) is

$$P_{elm} = 2892.4 - 435.2 = 2457.2 \text{ W}$$

The electromagnetic torque developed by the motor

$$T_d = \frac{2457.2}{2\pi \times 750/60} = 31.2 \text{ Nm}$$

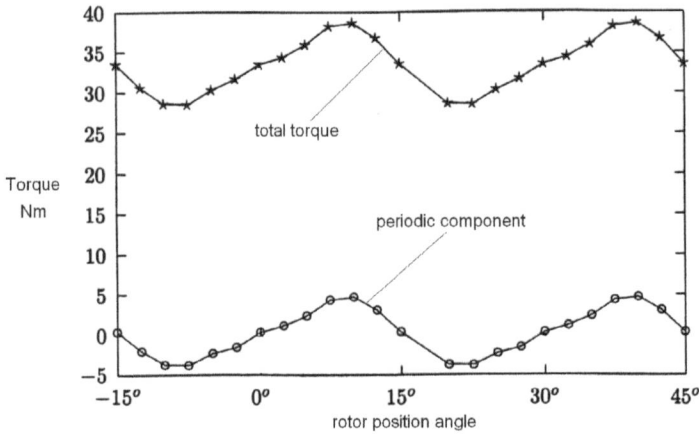

Fig. 7.23. Torque as a function of the rotor position for the disk rotor PM motor.

The results of the FEM analysis are shown in Figs 7.21, 7.22 and 7.23. The FEM gives slightly higher values of average developed torque than the analytical approach. Eqns (5.15), (5.31), (5.33) and (A.25) do not give accurate values of X_{sd} and X_{sq} for disk rotor motors. The electromagnetic torque plotted in Fig. 7.23 against the rotor position has significant periodic component (period equal to double slot pitch, i.e., 30^0).

Numerical example 7.3

A three-phase, 2400 rpm PM disk motor with center mounted ironless stator has a twin PM external rotor. The NdFeB PMs with $B_r = 1.3$ T and $H_c = 1000$ kA/m are fixed to two rotating mild steel disks. The masss of the twin rotor is $m_r = 3.8$ kg and the mass of the shaft is $m_{sh} = 0.64$ kg. The nonferromagnetic distance between opposite PMs is $d = 12$ mm, the winding thickness is $t_w = 9$ mm and the height of PMs (in axial direction) is $h_M = 6$ mm. The outer diameter of PMs equal to the external diameter of the stator conductors is $D_{ext} = 0.24$ m and the parameter $k_d = 1/\sqrt{3}$. The number of poles is $2p = 6$, the number of single layer winding bars (equivalent to the number of slots) is $s_1 = 54$, the number of turns per phase is $N_1 = 270$, the number of parallel conductors $a_w = 2$, the diameter of wire $d_w = 0.912$ mm (AWG 19) and the coil pitch is $w_c = 7$ coil sides. The electric conductivity of copper wire is $\sigma = 47 \times 10^6$ S/m.

Find the motor steady state performance, i.e., output power, torque, efficiency and power factor assuming that the total armature current $I_a = 8.2$ A is torque producing ($I_{ad} = 0$). The magnetic circuit is unsaturated ($k_{sat} \approx 1.0$), the motor is fed with sinusoidal voltage, eddy current losses in conductors are $\Delta P_e = 60$ W and the coefficient of bearing friction in eqn (B.31) $k_{fb} = 2.5$ W/(kg rpm). Windage losses can be neglected.

Solution

The number of coils per phase for a single layer winding is $0.5s_1/m_1 = 0.5 \times 54/3 = 9$. The number of turns per coil is $N_c = 270/9 = 30$. The number of coils sides per pole per phase (equivalent to the number of slots per pole per phase) $q_1 = s_1/(2pm_1) = 54/(6 \times 3) = 3$.

The air gap (mechanical clearance) is $g = 0.5(d - t_w) = 0.5(12 - 9) = 1.5$ mm and the pole pitch measured in number of coil sides is $\tau = s_1/(2p) = 54/6 = 9$.

Input frequency at 2400 rpm

$$f = n_s p = \frac{2400}{60} \times 3 = 120 \text{ Hz}$$

The relative recoil magnetic permeability according to eqn (2.5)

$$\mu_{rrec} = \frac{1}{\mu_0} \frac{\Delta B}{\Delta H} = \frac{1}{0.4\pi \times 10^{-6}} \frac{1.3 - 0}{1\,000\,000 - 0} = 1.035$$

The magnetic voltage drop equation per pole pair

$$4\frac{B_r}{\mu_0 \mu_{rrec}} h_M = 4\frac{B_{mg}}{\mu_0}\left[\frac{h_M}{\mu_{rrec}} + \left(g + \frac{1}{2}t_w\right)k_{sat}\right]$$

Hence

$$B_{mg} = \frac{B_r}{1 + [\mu_{rrec}(g + 0.5t_w)/h_M]k_{sat}}$$

$$= \frac{1.3}{1 + [1.035(1.5 + 0.5 \times 9)/6] \times 1.0} = 0.639 \text{ T}$$

Magnetic flux according to eqn (7.12)

$$\Phi_f = \frac{1}{8}\frac{2}{\pi}\frac{\pi}{3}0.639 \times 0.24^2 \left[1 - \left(\frac{1}{\sqrt{3}}\right)^2\right] = 0.00204 \text{ Wb}$$

Winding factor according to eqns (A.1), (A.6) and (A.3)

$$k_{d1} = \frac{\sin \pi/(2 \times 3)}{3\sin \pi/(2 \times 3 \times 3)} = 0.9598; \qquad k_{p1} = \sin\left(\frac{7}{9}\frac{\pi}{2}\right) = 0.9397$$

$$k_{w1} = 0.9598 \times 0.9397 = 0.9019$$

The EMF constant according to eqn (7.17) and torque constant according to eqn (7.15)

$$k_E = \pi\sqrt{2} \times 3 \times 270 \times 0.9019 \times 0.00204 = 6.637 \text{ V/rev/s} = 0.111 \text{ V/rev/min}$$

$$k_T = k_E \frac{m_1}{2\pi} = 6.637\frac{3}{2\pi} = 3.169 \text{ Nm/A}$$

EMF at 2400 rpm

$$E_f = k_E n_s = 0.111 \times 2400 = 265.5 \text{ V}$$

Electromagnetic torque at $I_a = I_{aq} = 8.2$ A

$$T_d = k_T I_a = 3.169 \times 8.2 = 25.98 \text{ Nm}$$

Electromagnetic power

$$P_{elm} = 2\pi n_s T_d = 2\pi \frac{2400}{60} \times 25.98 = 6530.5 \text{ W}$$

The inner diameter $D_{in} = D_{ext}/\sqrt{3} = 0.24/\sqrt{3} = 0.139$ m [4], average diameter $D_{av} = 0.5(D_{ext} + D_{in}) = 0.5(0.24 + 0.138) = 0.1893$ m, average pole pitch $\tau_{av} = \pi 0.189/6 = 0.099$ m, length of conductor (equal to the radial length of the PM) $L_i = 0.5(D_{ext} - D_{in}) = 0.5(0.24 - 0.139) = 0.051$ m, length of shorter end connection $l_{1emin} = (7/9)\pi D_{in}/(2p) = (7/9)\pi \times 0.139/6 = 0.0564$ m and the length of longer end connection $l_{1emax} = 0.0564 \times 0.24/0.139 = 0.0977$ m.

The average length of the stator turn

$$l_{1av} \approx 2L_i + l_{1emin} + l_{1emax} + 4 \times 0.015$$

$$= 2 \times 0.051 + 0.0564 + 0.0977 + 0.06 = 0.3156 \text{ m}$$

Stator winding resistance at 75^0C according to eqn (B.1)

$$R_1 = \frac{270 \times 0.3156}{47 \times 10^6 \times 2 \times \pi \times (0.912 \times 10^{-3})^2/4} = 1.3877 \text{ }\Omega$$

The maximum width of the coil at the diameter D_{in} is $w_w = \pi 0.138/54 = 0.0081$ m $= 8.1$ mm. The thickness of the coil is $t_w = 8$ mm. The number of conductors per coil is $a_w \times N_c = 2 \times 30 = 60$. The maximum value of the coil packing factor (fill factor) is at D_{in}, i.e.,

$$\frac{d_w^2 \times N_c}{t_w w_w} = \frac{0.912^2 \times 60}{9 \times 8.1} = 0.688$$

Stator current density

$$j_a = \frac{8.2}{2 \times \pi 0.912^2/4} = 6.28 \text{ A/mm}^2$$

Armature winding losses at 75^0C according to eqn (B.12)

$$\Delta P_a = 3 \times 8.2^2 \times 1.3877 = 279.9 \text{ W}$$

Bearing friction losses according to eqn (B.31)

$$\Delta P_{fr} = 2.5(3.8 + 0.64) \times 2400 \times 10^{-3} = 26.6 \text{ W}$$

Output power

$$P_{out} = P_{elm} - \Delta P_{fr} = 6530.5 - 26.6 == 6503.8 \text{ W}$$

Shaft torque

$$T_{sh} = \frac{6503.8}{2\pi \times 2400/60} = 25.88 \text{ Nm}$$

Input power

$$P_{in} = P_{elm} + \Delta P_a + \Delta P_e = 6530.5 + 279.9 + 60 = 6870.4 \text{ W}$$

Efficiency

$$\eta = \frac{6503.8}{6870.4} = 0.947$$

The specific permeance of the end winding connection (overhang) is approximately estimated as

$$\lambda_{1e} \approx 0.3q_1 = 0.3 \times 3 = 0.9$$

The specific permeance for leakage flux about radial parts of conductors

$$\lambda_{1s} \approx \lambda_{1e}$$

The specific permeance of the differential leakage flux

$$\lambda_{1d} = \frac{3 \times 3 \times 0.099 \times 0.9019^2}{\pi^2(2 \times 0.0015 + 0.009) \times 1.0} \times 0.011 = 0.068$$

The leakage reactance of the stator (armature) winding

$$X_1 = 4\pi \times 0.4\pi \times 10^{-6} \times 120 \frac{270^2 \times 0.051}{3 \times 3} \left(0.9 + \frac{0.0564}{0.051}\frac{0.9}{2} + \frac{0.0977}{0.051}\frac{0.9}{2} + 0.068 \right)$$

$$= 1.818 \; \Omega$$

where $k_C = 1$ and the differential leakage factor $\tau_{d1} = 0.011$ (Fig. A.3).

Armature reaction reactances according to [122]

- in the d-axis

$$X_{ad} = 2m_1\mu_0 f \left(\frac{N_1 k_{w1}}{p}\right)^2 \frac{(0.5D_{ext})^2 - (0.5D_{in})^2}{g_{eq}} k_{fd}$$

$$= 2 \times 3 \times 0.4\pi \times 10^{-6} \times 120 \left(\frac{270 \times 0.9019}{3}\right)^2 \frac{(0.5 \times 0.24)^2 - (0.5 \times 0.139)^2}{0.0236} \times 1.0$$

$$= 2.425 \ \Omega$$

where $k_{fd} = 1$ and the equivalent air gap

$$g_{eq} = 2\left[(g + 0.5t_w)k_{sat} + \frac{h_M}{\mu_{rrec}}\right] = 2\left[(1.5 + 0.5 \times 9) \times 1.0 + \frac{6}{1.035}\right] = 23.6 \ \text{mm}$$

- in the q-axis

$$X_{aq} = 2m_1\mu_0 f \left(\frac{N_1 k_{w1}}{p}\right)^2 \frac{(0.5D_{ext})^2 - (0.5D_{in})^2}{g_{eqq}} k_{fq}$$

$$= 2 \times 3 \times 0.4\pi \times 10^{-6} \times 120 \left(\frac{270 \times 0.9019}{3}\right)^2 \frac{(0.5 \times 0.24)^2 - (0.5 \times 0.139)^2}{0.024} \times 1.0$$

$$= 2.385 \ \Omega$$

where $k_{fq} = 1$ and the equivalent air gap

$$g_{eqq} = 2(g + 0.5t_w + h_M) = 2(1.5 + 0.5 \times 9 + 6) = 24.25 \ \text{mm}$$

Synchronous reactances according to eqn (5.15)

$$X_{sd} = 1.818 + 2.425 = 4.243 \ \Omega$$

$$X_{sq} = 1.818 + 2.385 = 4.203 \ \Omega$$

Input phase voltage according to eqn (5.48)

$$V_1 = \sqrt{(E_f + I_a R_1)^2 + (I_a X_{sq})^2} = \sqrt{(265.5 + 8.2 \times 1.3877)^2 + (8.2 \times 4.203)^2} = 279 \ \text{V}$$

Line voltage $V_{1L-L} = \sqrt{3} \times 279 = 483.2 \ \text{V}$.

Power factor according to eqn (5.47)

$$\cos\phi = \frac{E_f + I_a R_1}{V_1} = \frac{265.5 + 8.2 \times 1.3877}{279} = 0.992$$

8

High Power Density Brushless Motors

8.1 Design considerations

A utilization of active materials of an electric motor can be characterized by:

- *power density*, i.e., output (shaft) *power–to–mass* or output *power–to–volume* ratio;
- *torque density*, i.e., shaft *torque–to–mass* or shaft *torque–to–volume* ratio.

Torque density is a preferred parameter to power density when comparing low speed motors, e.g., gearless electromechanical drives, hoisting machinery, rotary actuators, etc. The utilization of active materials increases with the intensity of the cooling system, increase of the service temperature of insulation and PMs, increase of the rated power, rated speed and electromagnetic loading, i.e., $P_{out}/(D_{1in}^2 L_i) \propto S_{elm}/(D_{1in}^2 L_i) = 0.5\pi^2 k_{w1} B_{mg} A_m n_s$ as expressed by eqns (5.67) and (5.70).

All types of electrical machines and electromagnetic devices show a lower ratio of *energy losses–to–output power* with an increase in output power. This means that the efficiency of electric motors increases with an increase in the rated power. Large PM brushless motors can achieve a higher possible efficiency than any other electric motor (except those with superconducting excitation windings). The limitation, however, is the high price of PM materials. NdFeB magnets offer the highest energy density at reasonable costs. Their major drawback, compared to SmCo, is temperature sensitivity. Performance deteriorates with increased temperature which has to be taken into account when the motor is designed. Above a certain temperature, the PM is irreversibly demagnetized. Therefore, the motor's temperature must be kept below the service temperature (180^0 for most NdFEB PMs, maximum 200^0C) when using NdFeB magnets. A natural air cooling system is sometimes not efficient and the stator must be cooled by water or oil circulating through the stator housing. The rotor losses in PM synchronous motors are small so most PM machines employ passive cooling of their rotors.

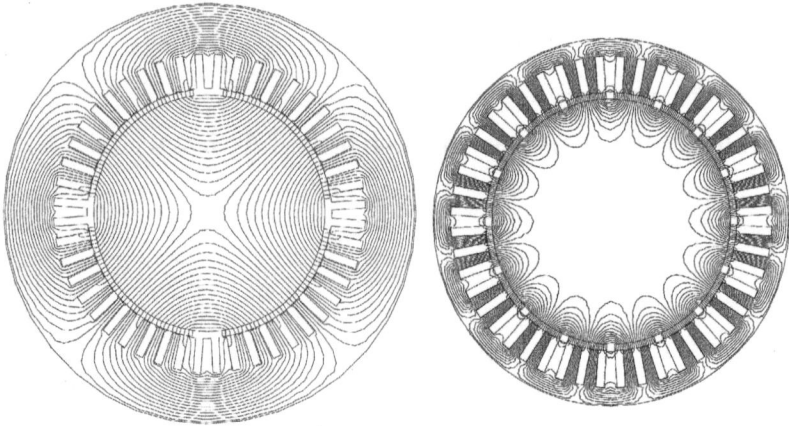

Fig. 8.1. Magnetic flux distribution in a 4- and 16-pole motor.

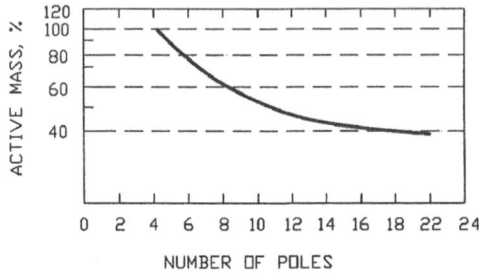

Fig. 8.2. PM motor mass versus number of poles for constant stator inner diameter [99].

The main dimensions (inner stator diameter D_{1in} and effective length L_i of core) of an electric motor are determined by its rated power $P_{out} \propto S_{elm}$, speed n_s, air gap magnetic flux density B_{mg}, and armature line current density A_m – see eqn (5.70). Magnetic flux density in the air gap is limited by the remanent magnetic flux density of PMs and saturation magnetic flux density of ferromagnetic core. The line current density can be increased if the cooling is intensified.

For a given stator inner diameter the mass of the motor can be reduced by using more poles. Fig. 8.1 illustrates this effect in which cross sections of 4-pole and 16-pole motors are compared [99]. The magnetic flux per pole is decreased in proportion to the inverse of the number of poles. Therefore, the outer diameter of the stator core is smaller for a motor with a large number of poles at the same magnetic flux density maintained in the air gap. PM motor mass as a function of the number of poles for the same stator inner diameter is shown in Fig. 8.2 [99]. However, according to eqn (5.67) the electromagnetic power decreases as the number of poles increases, i.e., $S_{elm} = 0.5\pi^2 k_{w1} D_{1in}^2 L_i B_{mg} A_m f/p$.

The absence of the exciter in large PM brushless motors reduces the motor drive volume significantly. For example, in a 3.8 MW PM synchronous motor, about 15% of its volume can be saved [132].

8.2 Requirements

PM brushless motors in the megawatt range tend to replace the conventional d.c. motors in those drives in which a commutator is not acceptable. This can be both in high speed (compressors, pumps, blowers) and low speed applications (mills, winders, electrical vehicles, marine electromechanical drives).

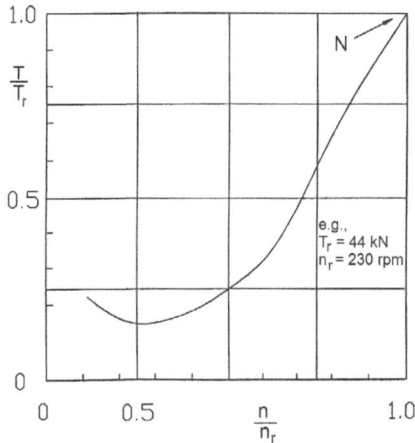

Fig. 8.3. Typical torque–speed characteristic for ship propulsion.

Ships have been propelled and maneuvered by electrical motors since the late 1970s. Recently, rare earth PMs allow the design of brushless motors with very high efficiency over a wide speed range. This is the most important factor in ship and road vehicle propulsion technology. For ship propulsion the typical torque-speed characteristic is shown in Fig. 8.3 [23, 24, 223]. The motor and converter has to be designed with the highest efficiency to meet the rated point N ("corner power" point). It is advantageous to use constant flux motors here in contrast to the hyperbolic characteristics of road vehicles. Point N represents the "worst case conditions" for the efficiency because the core losses as well as the winding losses reach their maximum values. In addition, the efficiency should not decrease significantly at a partial load down to 20% of rated speed ($0.2\,n_r$) as this is the speed of long distance journeys [23, 24].

The solid state converter should operate in such a way as to obtain the lowest possible winding and switching losses. The last feature requires a speed-dependent rearrangement of the winding and inverter components. A subdivision of the stator winding and converter into modules is necessary due to reliability reasons. In case of failure of the drive system the best solution is a modular concept where the damaged module can be quickly replaced by a new one. Armature windings with more than three phases are a promising option [23, 24].

Like most PM brushless motors, a large motor should be controlled by the shaft position angle to obtain the electromagnetic torque directly proportional to the armature currents. The current waveform coincides with the induced voltage and the stator losses reach their minimum value.

8.3 Multiphase motors

For some large PM brushless motors the number of armature phases $m_1 > 3$ is recommended. The armature phase current is inversely proportional to the number of phases, i.e., $I_a = P_{out}/(m_1 V_1 \eta \cos \phi)$. For the constant output power $P_{out} = const$, constant input phase voltage $V_1 = const$ and approximately the same power factor $\cos \phi$ and efficiency η, the armature current is lower for the greater number of phases. This means that a multiphase a.c. motor of the same dimensions as a three-phase motor has similar mechanical characteristics but draws lower phase currents.

Another distinguishing feature is the possibility of a step change in the speed by changing the supply phase voltage sequence. The synchronous speed is inversely proportional to the number of phase voltage sequences k, i.e.,

$$n_s = \frac{f}{kp} \tag{8.1}$$

where f is the input frequency. For the three-phase motor, only two voltage sequences are possible and the switching only causes the reversal of the motor speed. For the m_1-phase motor with an odd number of phases m_1, the first $k = (m_1 - 1)/2$ sequences change the synchronous speed. For a nine-phase motor it will be $k = 1, 2, 3, 4$ [91, 92]. The remaining frequencies excluding the zero sequence produce rotation in the opposite direction (for a nine-phase motor $k = 5, 6, 7, 8$). For an even number of armature phases, the number of sequences changing the speed is $k = m_1/2$ [91, 92]. This is due to harmonic fields. In addition, a synchronous motor must meet the requirement that the number of rotor poles must be adjusted to match each speed.

A multiphase system (m_1 phases) of voltages can be expressed by the following equation [91]:

$$v_l = \sqrt{2} cos[2\pi ft - (l-1)k2\pi/m_1] \tag{8.2}$$

where $l = 1, 2, 3, ...m_1$ or $l = A, B, C, ...m_1$. The voltages v_l create a star-connected m_1 phase voltage source. Changing the sequence of phase voltages is done by selection of the appropriate value of k. The sources v_l or $v_A, v_B, v_C, ...$ can be replaced by a VSI or voltage sources modeling the inverter output voltages.

As in three-phase systems, multiphase systems can be *star* or *polygon* connected. There are the following relationships between line and phase voltages and armature currents for the star connection:

$$V_{1L} = 2V_1 \sin\left(\frac{360^0}{2m_1}\right), \qquad\qquad I_{aL} = I_a \qquad (8.3)$$

For the polygon connection

$$I_{aL} = 2I_a \sin\left(\frac{360^0}{2m_1}\right), \qquad\qquad V_{1L} = V_1 \qquad (8.4)$$

The operation of the motor can be described by the following voltage equation in the stator co-ordinate system:

$$[v] = [R_a][i_a] + \frac{d}{dt}\{[\psi_a] + [\psi_f]\} = [R_a][i_a] + [L_s]\frac{d[i_a]}{dt} + [e_f] \qquad (8.5)$$

where $[R_a]$ is the matrix of armature resistances, $[\psi_a] = [L_s][i_a]$ is the matrix of the armature reaction fluxes caused by the armature currents $[i_a]$, and $[\psi_f]$ is the matrix of PM flux linkages which create the no-load EMFs $[e_f] = d/dt[\psi_f]$. The synchronous inductance matrix $[L_s]$ contains the leakage and mutual inductances.

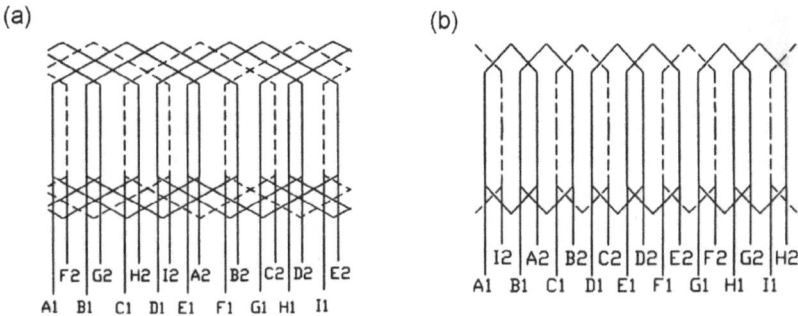

Fig. 8.4. Nine-phase windings located in 18 stator slots: (a) symmetrical, (b) asymmetrical.

The instantaneous electromagnetic torque developed by a multiphase synchronous motor is

$$T_d = \frac{p_{elm}}{2\pi n_s} = \frac{1}{2\pi n_s} \sum_{l=1}^{l=m_1} e_{fl} i_{al} \qquad (8.6)$$

where the instantaneous electromagnetic power is

$$p_{elm} = \sum_{l=1}^{l=m_1} e_{fl} i_{al} \qquad (8.7)$$

Nine-phase motors are easy to design since their windings can be placed in stator cores of most standard three-phase motors. As has been mentioned, a nine-phase motor has four different synchronous speeds for four phase voltage sequences. Values of these speeds depend on the type of stator winding, which can be *symmetrical* or *asymmetrical* (Fig. 8.4). These windings can be distinguished from each other by the Fourier spectrum of the MMF produced by each of them [91]. The symmetrical winding produces MMF harmonics which are odd multiples of the number of pole pairs p. With the asymmetrical winding the number of pole pairs cannot be distinguished.

8.4 Fault-tolerant PM brushless machines

First prototypes of fault-tolerant PM brushless machines of *modular construction* were designed and tested in Germany in the late 1980s [23, 24]. There are many potential faults that can occur in a motor and associated power electronics [93]; however, the most common faults are

- stator winding open-circuit
- stator winding short-circuit
- inverter switch open-circuit (similar to winding open-circuit)
- inverter switch short-circuit (similar to winding short-circuit)
- d.c. link capacitor failure

The most successful approach is a multiphase machine — solid state converter system in which each phase may be regarded as a single module. The machine must produce rated torque with any single phase fault. Thus, the machine must be overrated by a fault-tolerant rating factor k_{fault} that depends on the number of stator independent phase windings m_1, i.e.,

$$k_{fault} = \frac{m_1}{m_1 - 1} \qquad (8.8)$$

Eqn (8.8) shows that a three-phase machine ($m_1 = 3$) must be overrated by 50%, if the two remaining phases are to make up the short fall from one lost phase. The higher the number of stator phases, the lower the penalty for fault tolerance. Differences between standard PM brushless machine and fault-tolerant PM brushless machine are summarized in Table 8.1.

Table 8.1. Differences between standard and fault-tolerant PM brushless machine

Quantity	Standard Machine	Fault-Tolerant Machine
Number of phases	3	more than 3
Construction of stator	one cylindrical unit	modular construction
Construction of winding	distributed in slots	concentrated around teeth (one coil per slot)
Phase reactance per unit	less than 1	1
Mutual inductance	up to 50% of the phase self-inductance	less than 5% of the phase self-inductance
Short circuit current	higher than rated current	the same as rated current
Participation of harmonics in torque production	fundamental harmonic only	higher harmonics can be engaged
Interaction of phase currents	current in one phase affects currents in remaining phases	current in one phase has almost no influence on other phases
Slot-opening permeance harmonics in the air gap	low harmonics content (large number of slots)	high harmonic content (concentrated coils)
Losses in PMs	problematic only at high speeds	significant edddy current losses in PMs at low speeds

8.5 Surface PM versus salient-pole rotor

The PM rotor design has a fundamental influence on the *output power–to–volume* ratio.

Two motors of surface PM and salient-pole construction, as shown in Fig. 8.5, have been investigated [10]. As the rare earth PMs are rather expensive the power density must be maximized.

Tests made on two 50-kW, 200-V, 200-Hz, 6000-rpm motors designed according to Fig. 6.5 show the following [10]:

- the salient pole motor causes greater space harmonic content in the air gap than the surface PM rotor motor,
- the synchronous reactances of the surface PM rotor motor ($X_{sd} = X_{sq} = 0.56\ \Omega$) are smaller than those of the salient rotor motor ($X_{sd} = 1.05\ \Omega$ and $X_{sq} = 1.96\ \Omega$),
- the subtransient reactances in the d-axis are $X''_{sd} = 0.248\ \Omega$ for the surface PM rotor and $X''_{sd} = 0.497\ \Omega$ for the salient pole rotor motor, which results in different commutation angles (21.0^0 versus 29.8^0),
- the rated load angle of the surface PM rotor motor is smaller than that of the salient pole motor (14.4^0 versus 36.6^0),
- the relatively large load angle of the salient pole motor produces large torque oscillations of about 70% of the average torque as compared with only 35% for the surface PM rotor motor,

(a)

(b)

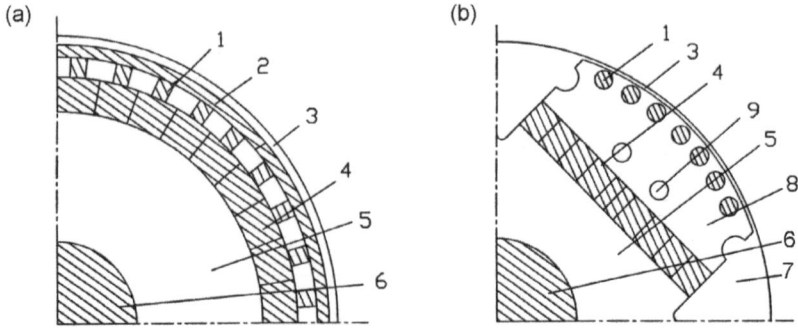

Fig. 8.5. Cross sections of large PM motors: (a) surface PM rotor, (b) salient-pole rotor: 1 — damper bar, 2 — bandage (retaining sleeve), 3 — air gap, 4 — PM, 5 — rotor hub, 6 — rotor shaft, 7 — gap between poles, 8 — pole shoe, 9 — axial bolt.

- the output power is 42.9 kW for the salient pole motor and 57.4 kW for the surface PM rotor motor,
- the surface PM rotor motor has better efficiency than the salient pole motor (95.3% versus 94.4%),
- the volume of PMs is proportional to the output power and is 445 cm^3 for the salient pole motor and 638 cm^3 for the surface PM rotor motor.

The stator dimensions and the apparent power have been kept the same for the two tested motors.

The air gap field of the salient pole motor is of rectangular rather than sinusoidal shape and produces additional higher harmonic core losses in the stator teeth [10]. The air gap field harmonics due to the stator slots and current carrying winding produce eddy current losses in the rotor pole faces and stator core inner surface. In the surface PM motor the field harmonics induce high-frequency eddy current losses in the damper. The damper is usually made from a copper cylinder and has axial slots to reduce the eddy current effect.

8.6 Electromagnetic effects

8.6.1 Armature reaction

The action of the armature currents in the phase windings causes a cross field in the air gap. This implies a distortion of the PM excitation field. The resulting flux induces a proportional EMF in the armature phase conductors. At some points in the air gap the difference between the d.c. link voltage and the induced EMF may decrease significantly, thus reducing the rate of increase of the armature current in the corresponding phase [199].

The armature reaction also shifts the magnetic neutral line of the resultant flux distribution by a distance dependent on the armature current. The displacement between the current and flux distribution contributes to the decrease in the electromagnetic torque [199]. Moreover, the distorted excitation field and flux in the d-axis produces noise, vibration and torque ripple.

The armature reaction together with the commutation effect can produce dips in the phase current waveforms, which in turn reduce the mean value of the armature current and electromagnetic torque.

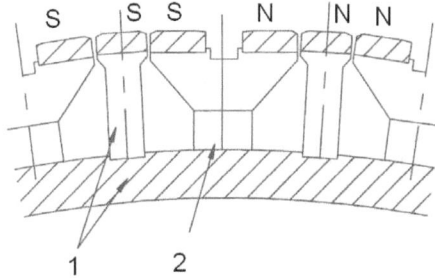

Fig. 8.6. Rotor of a large a.c. motor with surface PMs and nonferromagnetic parts for reducing the armature reaction according to Siemens: 1 — ferromagnetic core, 2 — nonferromagnetic core.

The influence of the armature reaction on the electromagnetic torque can be minimized by an increase in the air gap or using an anisotropic material with a large reluctance in the q-axis of the magnetic circuit. An interesting construction of the rotor magnetic circuit shown in Fig. 8.6 has been proposed by Siemens [23, 24]. The nonmagnetic parts suppress the cross-field of the armature currents and thus reduce the effect of armature reaction. Consequently, the inverter can better be utilized because the EMF waveform is less distorted.

8.6.2 Damper

As in synchronous machines with electromagnetic excitation, the damper reduces the flux pulsation, torque pulsation, core losses and noise. On the other hand, the damper increases the losses due to higher harmonic induced currents as it is designed in the form of a cage winding or high conductivity cylinder.

A damper can minimize dips in the current waveform due to armature reaction and commutation and increase the electromagnetic torque. The damper bars reduce the inductance of the armature coil with which it is aligned and reduces the effects of the firing of the next phase. Thus, the addition of the complete damper (cage with several bars per pole) can increase significantly the output power.

It is known from circuit theory that the self-inductance of a magnetically coupled circuit decreases when a short-circuited coil is magnetically coupled to it. The same happens to the mutual inductance between two circuits magnetically coupled. Obviously, if the coupling between the short-circuited coil and the magnetic circuit changes, the self and mutual inductances also change.

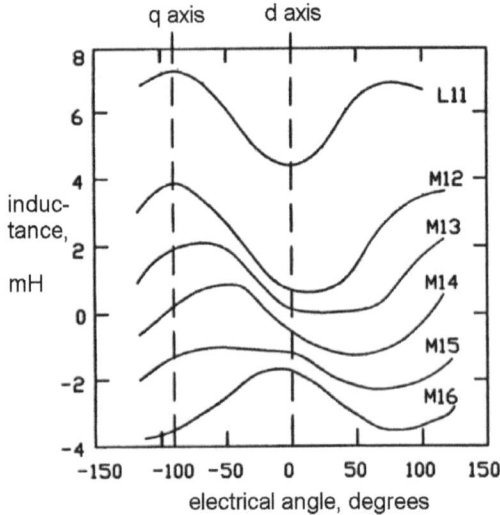

Fig. 8.7. Self and mutual inductances of a 6-phase, 75-kW, 60-Hz, 900-rpm, 8-pole PM motor as functions of the rotor position angle between the axis of each phase and the d-axis of the motor [129, 199].

Fig. 8.7 shows the test results on a 6-phase, 75-kW, 60-Hz, 900-rpm PM motor prototype [129, 199]. The angle between the d and q-axis is equal to 22.5^0 geometrical (8 poles), which corresponds to 90^0 electrical. The self inductance L_{11} and mutual inductances M_{12}, M_{13}, ..., M_{16} have been plotted against the angle between the axis of each phase and the d-axis of the motor.

The mutual inductance M_{14} between two phases shifted by 90^0 electrical with no damper would be nearly zero, since the magnetic circuit is almost isotropic. However, the damper introduces a magnetic coupling between the phases 1 and 4, and M_{14} varies with the angle. The voltage induced in the damper by one of the phases produces a magnetic flux that links this phase with the other phases so that the mutual inductance is greater than zero.

All the inductances plotted in Fig. 8.7 would also vary with the rotor angle if the magnetic circuit was made of anisotropic material and the damper was removed. In this case the self-inductance L_{11} would have its maximum value at the zero angle, i.e., when the center axis of the phase 1 and the d-axis coincide [199].

8.6.3 Winding losses in large motors

Suppose that the phase armature current has a square-wave shape with a flat-topped value $I_a^{(sq)}$. Such a rectangular function can be resolved into Fourier series:

- for a 120^0 square wave

$$i_a(t) = \frac{4}{\pi} I_a^{(sq)} (\cos \frac{\pi}{6} \sin \omega t + \frac{1}{3} \cos 3 \frac{\pi}{6} \sin 3\omega t$$

$$+ \frac{1}{5} \cos 5 \frac{\pi}{6} \sin 5\omega t + \ldots \frac{1}{n} \cos n \frac{\pi}{6} \sin n\omega t)$$

- for a 180^0 square wave

$$i_a(t) = \frac{4}{\pi} I_a^{(sq)} (\sin \omega t + \frac{1}{3} \sin 3\omega t + \frac{1}{5} \sin 5\omega t + \ldots \frac{1}{n} \sin n\omega t)$$

where $\omega = 2\pi f$ and n are the higher time harmonics. The armature winding losses can be represented as a sum of higher harmonic losses. For a three-phase winding the time harmonics $n = 5, 7, 11, 13, \ldots$ are only present. Thus

$$\Delta P_a = m_1 \frac{1}{2} R_{1dc} (I_{am1}^2 k_{1R1} + I_{am5}^2 k_{1R5} + \ldots I_{amn}^2 k_{1Rn})$$

where I_{man} are the amplitudes of higher time harmonic currents. For a 120^0 square wave

$$\Delta P_a = m_1 [I_a^{(sq)}]^2 R_{1dc} k_{1R1} \frac{8}{\pi^2} [\cos^2(\frac{\pi}{6}) + \frac{1}{5^2} \cos^2(5 \frac{\pi}{6}) \frac{k_{1R5}}{k_{1R1}}$$

$$+ \frac{1}{7^2} \cos^2(7 \frac{\pi}{6}) \frac{k_{1R7}}{k_{1R1}} + \ldots \frac{1}{n^2} \cos^2(n \frac{\pi}{6}) \frac{k_{1Rn}}{k_{1R1}}] \tag{8.9}$$

and for a 180^0 square wave

$$\Delta P_a = m_1 [I_a^{(sq)}]^2 R_{1dc} k_{1R1} \frac{8}{\pi^2} \left(1 + \frac{1}{5^2} \frac{k_{1R5}}{k_{1R1}} + \frac{1}{7^2} \frac{k_{1R7}}{k_{1R1}} + \ldots \frac{1}{n^2} \frac{k_{1Rn}}{k_{1R1}}\right) \tag{8.10}$$

The coefficients k_{1R1}, k_{1R5}, $k_{1R7}, \ldots k_{1Rn}$ are the coefficients of skin effect for time harmonics according to eqn (B.8), i.e.,

$$k_{1Rn} = \varphi_1(\xi_n) + \left(\frac{m_{sl}^2 - 1}{3} - \frac{m_{sl}^2}{16}\right) \Psi_1(\xi_n) \tag{8.11}$$

where ξ_n is according to eqn (B.7) for $f = nf$. For $\xi_n > 1.5$ the coefficients $\varphi_1(\xi_n) \approx \xi_n$ and $\Psi_1(\xi_n) \approx 2\xi_n$, thus

$$k_{1Rn} = \xi_n + 2\left(\frac{m_{sl}^2 - 1}{3} - \frac{m_{sl}^2}{16}\right)\xi_n = \left(\frac{1 + 2m_{sl}^2}{3} - \frac{m_{sl}^2}{8}\right)\xi_n \qquad (8.12)$$

and

$$\frac{k_{1Rn}}{k_{1R1}} \approx \sqrt{n} \qquad (8.13)$$

since $\xi_n/\xi_1 \approx \sqrt{n}$. The armature winding losses can be expressed with the aid of the coefficient K_{1R}, i.e.,

$$\Delta P_a = m_1 [I_a^{(sq)}]^2 R_1 K_{1R} \qquad (8.14)$$

where

- for a 120^0 square wave

$$K_{1R} = k_{1R1}\frac{8}{\pi^2}[\cos^2(\frac{\pi}{6}) + \cos^2(5\frac{\pi}{6})\frac{1}{5\sqrt{5}} + \cos^2(7\frac{\pi}{6})\frac{1}{7\sqrt{7}} + \ldots$$

$$+ \cos^2(n\frac{\pi}{6})\frac{1}{n\sqrt{n}}]$$

- for a 180^0 square wave

$$K_{1R} = k_{1R1}\frac{8}{\pi^2}\left(1 + \frac{1}{5\sqrt{5}} + \frac{1}{7\sqrt{7}} + \ldots + \frac{1}{n\sqrt{n}}\right)$$

Adding the higher harmonic terms in brackets up to $n = 1000$, the sum of the series is equal to 1.0 for a 120^0 square wave and 1.3973 for a 180^0 square wave. The coefficient of skin effect for a three-phase d.c. brushless motor with rectangular armature conductors is

- for a 120^0 square wave

$$K_{1R} \approx 0.811\left[\varphi_1(\xi_1) + \left(\frac{m_{sl}^2 - 1}{3} - \frac{m_{sl}^2}{16}\right)\psi_1(\xi_1)\right] \qquad (8.15)$$

- for a 180^0 square wave

$$K_{1R} \approx 1.133\left[\varphi_1(\xi_1) + \left(\frac{m_{sl}^2 - 1}{3} - \frac{m_{sl}^2}{16}\right)\psi_1(\xi_1)\right] \qquad (8.16)$$

The winding losses for a 120^0 square wave are only $0.811/1.133 = 0.716$ of those for a 180^0 square wave assuming the same frequency, flat-top current and winding construction.

8.6.4 Minimization of losses

The large volume of ferromagnetic cores, supporting steel elements as bolts, fasteners, clamps, bars, ledges, spiders, etc., high magnetic flux density in the air gap and intensity of leakage fields including higher harmonic fields contribute to large power losses in magnetic circuits and structural ferromagnetic components. The core losses are minimized by using very thin stator laminations (0.1 mm) of good quality electrotechnical steel. It is also recommended to use laminations for the rotor spider instead of solid steel. To avoid local saturation, careful attention must be given to uniform distribution of the magnetic flux.

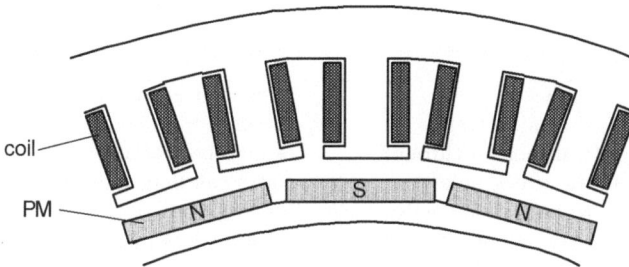

Fig. 8.8. Concentrated non-overlapping stator winding (with one slot coil pitch).

For multiphase modular motors the armature winding losses can be reduced by rearranging the stator winding connections. To avoid excessive higher harmonic losses, both MMF space distribution and armature current waveforms should be close to sinusoidal, or at least higher harmonics of low frequency should be reduced.

Multiphase modular motors can also be designed with stator coil span equal to one slot instead of one full pitch. Such a winding is called *winding with concentrated non-overlapping coils* (Fig. 8.8). The concentrated non-overlapping winding must meet condition (6.18). The end connections are very short, the winding losses are significantly reduced and the slot fill factor is high.

8.7 Cooling

The degree of utilization of active materials is directly proportional to the intensity of cooling. The efficiency of PM motors is high, for $2p > 4$ the stator yoke is thin and the rotor losses are small, so that PM brushless motors can achieve good performance with indirect conduction cooling, i.e., the stator winding heat is conductively transferred throughout the stator yoke to the external surface of frame. Direct liquid cooling of stator conductors allows for significant increase of the power density due to high heat transfer rates. On the other hand, it requires hollow stator conductors and hydraulic installation support.

The rotor gets hot due to heat rejected from the stator through the air gap and eddy current losses in PMs and ferromagnetic core. Although the rotor cooling systems are simpler, careful attention must be given to the temperature of PMs as their remanent flux density B_r and coercivity H_c decreases, diminishing the machine performance.

Table 8.2 summarizes some typical approaches to PM motors cooling [223].

Table 8.2. Large PM motor cooling options [223].

Approach	Description	Comments
Stator (armature)		
Indirect conduction	Armature heat is conductively transported through the stator yoke to to heat exchange surface; external surface can be cooled by surrounding air, water or oil	• coolant is completely isolated from conductor • high heat transfer rates can be achieved due to thin yoke of some PM motors • pod motors (ship propulsion) cooled by water
Direct conduction	Winding is flooded with dielectric fluid; heat conduction across winding insulation	• very high heat transfer rates are achieved • dielectric fluid must be used • slot fill factor compromised
Direct cooling	Winding is internally cooled; direct fluid contact	• highest heat transfer • dielectric fluid must be used • slot fill factor compromised • complex fluid manifolding
Rotor		
Passive air	Rotor cavity air circulation driven by rotor motion	Simplest, requires shaft rotation, good match for quadratic loads
Forced air	Rotor cavity air circulation driven by separate blower	Extra motor with associated manifolding requested
Open-loop spray	Liquid spray on rotor surface	Wet rotor, requires pump system
Closed loop internal	Liquid (oil) circulated internal to rotor	Highest heat transfer, requires rotating shaft seals

8.8 Construction of motors with cylindrical rotors

First prototypes of rare-earth PM motors rated at more than 1 MW for ship propulsion were built in the early eighties. This section contains a review of constructions and associated power electronics converters for large PM motors designed in Germany [23, 24, 132, 276]. Although some construction technologies and control techniques are now outdated, the experience and data base obtained have become a foundation for design of modern large PM brushless motors.

8.8.1 Motor with reduced armature reaction

One of the first large PM brushless motors rated at 1.1 MW, 230 rpm was built by Siemens in Nuremberg, Germany [23, 24]. The rotor construction with $2p = 32$ poles is shown in Fig. 8.6. There are nonferromagnetic poles with large reluctance in the q-axis to reduce the armature reaction. To minimize the cogging torque, the rotor was made with skewed poles. The stator (inner diameter 1.25 m, stack length 0.54 m) with a six-phase detachable winding consists of 8 replaceable modules (each module covering four pole pitches) to provide on-board repairing. A single module contains 2×6 coils located in 24 slots of a single-layer, full pitch winding. The one-module winding is shown in Fig. 8.9.

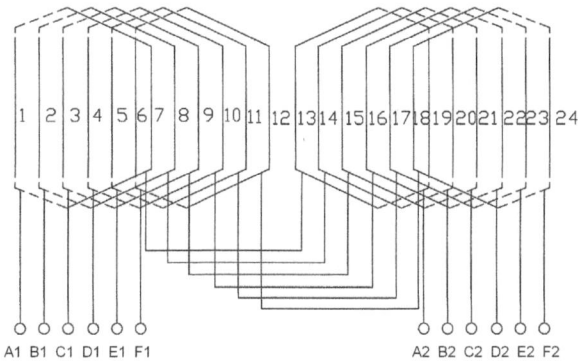

Fig. 8.9. Stator winding of one module of a 6-phase, 1.1-MW large PM motor manufactured by Siemens.

The six phase has been chosen since an even number of phases enables the division into two equal redundant systems (two line converters) and the duration of the trapezoidal current can be maximized to 150^0 or even 180^0 with respect to the achievable commutation time [23, 24]. Each of the phase

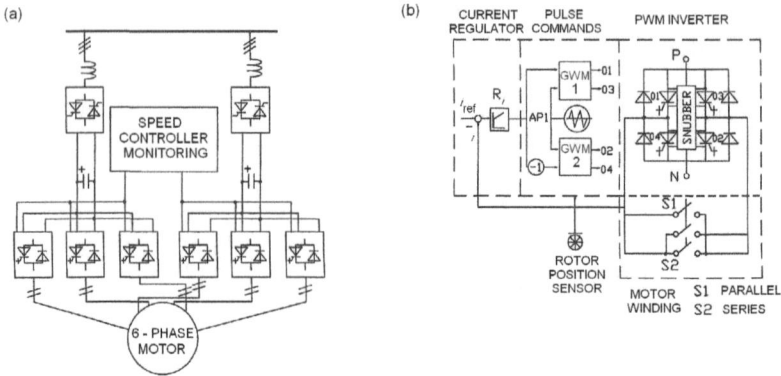

Fig. 8.10. Large variable-speed drive with six-phase PM motor manufactured by *Siemens*: (a) power circuit, (b) inverter module for phase control current.

windings consists of two halves of identical coils being connected in series (switch S2 closed in Fig. 8.10b) at low speeds, i.e., less than 55% of the rated speed and in parallel (switch S1 closed in Fig. 8.10b) at high speeds (reduction of synchronous reactance).

A separate PWM inverter is assigned per phase as shown in Fig. 8.10a. Three inverters in parallel are connected to a constant d.c. link. Two line converters are fed from 660-V three-phase on-board supply. The inverter module consists of a four-quadrant bridge which controls the phase current according to the set value which is dependent on the rotor angle. GTO thyristors and diodes with blocking capacity of 1.6 kV and a permissible turn-off current of 1.2 kA have been used. At 230 rpm and $f = 61.33$ Hz the commutation angle is rather large. Through the application of PWM, the armature current is kept quasi-constant within approximately 120^0 and coincides with the induced voltage e_f. The total current flow duration is about 160^0. A water cooling system with axial ducts is used for the stator.

The instantaneous electromagnetic power and developed torque per phase can be found as a product of EMFs and armature current according to eqns (8.6) and (8.7).

The large PM brushless motor shows about 40% less mass, frame length and volume, and 20 to 40% less power loss as compared with a d.c. commutator motor [23, 24].

8.8.2 Motors with modular stators

A large PM motor built by ABB in Mannheim, Germany, in co-operation with Magnet Motor GmbH at Starnberg [23, 24] is shown in Fig. 8.11. The stator consists of $Z = N_c = 88$ identical elements and $Z/2$ stator sections (two teeth or two elements screwed together). The number of stator slots is $s_1 = Z$. The

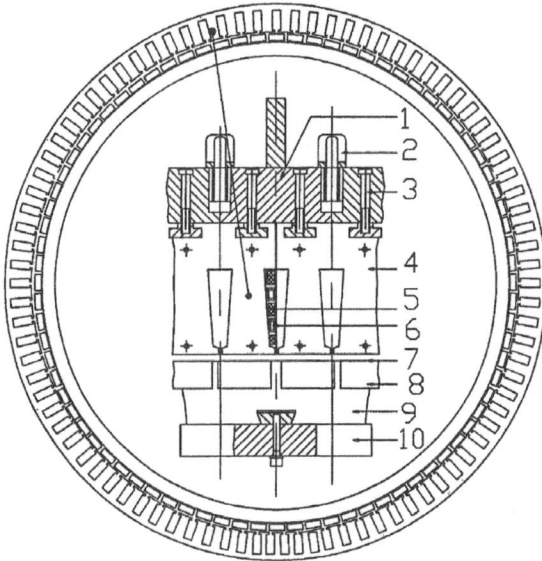

Fig. 8.11. Cross section of a large PM motor built by ABB and Magnet Motor GmbH: 1 — stator frame section, 2 — ledges clamping the stator sections, 3 — bolt for mounting the stator element, 4 — stator core, 5 — stator winding, 6 — cooling duct, 7 — fiberglass bandage, 8 — PMs, 9 — rotor yoke, 10 — rotor ring (spider).

number of stator concentrated coils $N_c = s_1 = Z$. The number of stator phases $m_1 = 11$; i.e., each phase section (pole pair) is formed by $p_a = Z/m_1 = 8$ elements per phase which have a spatial angular displacement of $360^0/8 = 45^0$. The coil span is $Z/(2m_1)$ and end connections are very short. According to eqn (6.18)

$$\frac{N_c}{GCD(N_c, 2p)} = \frac{88}{GCD(88, 80)} = 11 = m_1$$

where $k = 1$. The stator inner diameter is 1.3 m and the stack length is 0.5 m. To reduce the core loss, laminations thickness of 0.1 mm has been used. A direct water cooling system with hollow stator conductors has been designed. The rotor consists of $2p = 80$ rare-earth PM poles secured by a fiberglass bandage.

The principle of operation can be explained with the aid of Fig. 8.12 in which a five-phase, 12-pole, and 10-element motor is shown. According to eqn (6.18)

$$\frac{N_c}{GCD(N_c, 2p)} = \frac{10}{GCD(10, 12)} = 5 = m_1$$

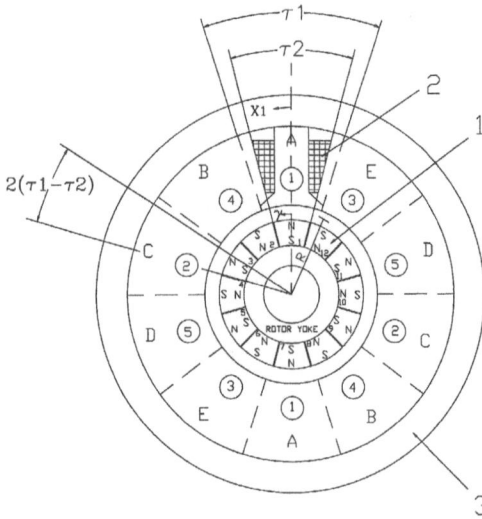

Fig. 8.12. Cross section of simple multiphase PM motor with $m_1 = 5$, $2p = 12$, and $Z = 10$: 1 — PM, 2 — stator element (tooth and winding), 3 — stator yoke.

where $k = 1$. For an odd number of phases the number of stator coils is $N_c = Z = p_a m_1$, where p_a is the number of stator (armature) pole pairs per phase. The stator coil pitch is $\tau_1 = 2\pi/s_1$ and the rotor pole pitch is $\tau_2 = 2\pi/(2p) = \pi/p$. After m_1 angular distances $(\tau_1 - \tau_2)$, where m_1 is the number of phases, the phase B is under the same conditions as phase A; i.e., the feasible number of phases can be determined on the basis of the following equality [24]:

$$m_1(\tau_1 - \tau_2) = \tau_2$$

Since $m_1 p(\tau_1 - \tau_2) = p\tau_2$, the same condition can be expressed in electrical degrees as $m_1 p(\tau_1 - \tau_2) = \pm\pi$. Thus

$$\frac{s_1}{2p} = \frac{m_1}{m_1 \pm 1} \tag{8.17}$$

The "+" sign is for $2p < Z$ and the "−" sign is for $2p > Z$. Indeed, for the motor with $s_1 = 88$, $2p = 80$, $m_1 = 11$

$$\frac{s_1}{2p} = \frac{88}{80} = 1.1 \qquad\qquad \frac{m_1}{m_1 - 1} = \frac{11}{11 - 1} = 1.1$$

and for the motor with $s_1 = 10$, $2p = 12$, $m_1 = 11$

$$\frac{s_1}{2p} = \frac{10}{12} = 0.833 \qquad\qquad \frac{m_1}{m_1 + 1} = \frac{5}{5 + 1} = 0.833$$

The harmonic torque equation can be derived on the basis of an analysis of space and time harmonics of the exciting magnetic flux density and stator line current density [23, 24].

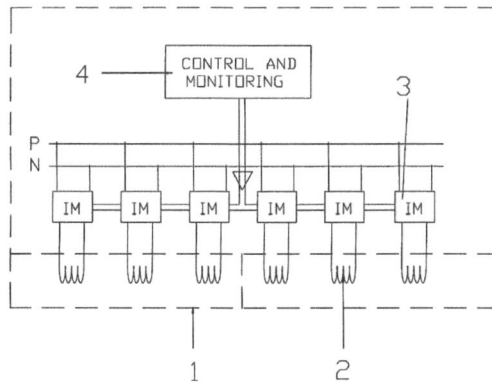

Fig. 8.13. Basic control unit: 1 — stator element, 2 — winding subelement, 3 — inverter module, 4 — microprocessor.

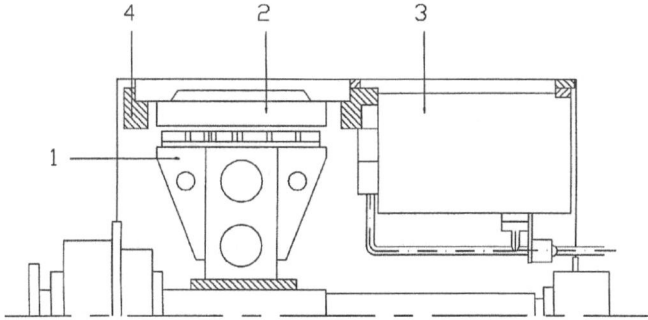

Fig. 8.14. Longitudinal section of a 1.5-MW PM motor for ship propulsion designed by *ABB*: 1 — rotor, 2 — armature, 3 — electronics, 4 — housing.

The large PM motor has a simple construction, compact stator winding with short overhangs, and reduced magnetic couplings between phases. The last enables each phase winding to be energized individually with no influence of switching of the adjacent phases.

Each stator element can be fed from its own inverter; therefore the inverters can be small. To obtain a better modular drive system, the winding of each stator element has been divided into three identical subelements as shown in Fig. 8.13. Each subelement is fed from a small 5-kVA PWM inverter module IM and, in total, $88 \times 3 = 264$ inverter modules are used. Using this advanced modular technology a highly reliable drive has been achieved. Owing to symmetry, the failure of one inverter module causes the switch-off of seven further modules. These modules are in a stand-by mode for substituting other failing inverter modules. In this way 97% of the rated output power can be obtained with the failure of up to eight inverter modules. The basic control unit (Fig. 8.13) is assigned to two adjacent stator elements. It contains six inverter modules and a command device with a microprocessor.

The motor efficiency at rated speed is as high as 96% and exceeds 94% in the speed range from rated speed to 20% of the rated speed. At low speed, the winding losses can be minimized by connecting the stator winding subelements in series [23, 24].

A similar 1.5-MW PM brushless motor (Fig. 8.14), the so-called "MEP motor" (multiple electronic PM motor) developed for submarine and surface ship propulsion has been described in [276]. The rotor consists of a steel spider construction carrying a laminated core with surface SmCo PMs mounted on it. Both rotor and stator stacks consist of 0.1 mm thick laminations. The stator is water cooled and the rotor is cooled by natural convection. All rotor parts including PMs are protected against corrosion by fiberglass bandages and epoxy paint. The motor insulating system is designed according to class F. The power electronic modules are mechanically integrated in the motor housing and connected to the corresponding stator units. Hall sensors are used to provide speed and position signals to the power electronics converters. To improve the motor self-starting capability, the numbers of stator and rotor poles are different. The speed is in the range of 0 to 180 rpm, the d.c. inverter input voltage is in the range of 285 to 650 V, the torque is 76 kNm, number of armature poles (teeth) $Z = 112$, number of rotor PM poles is $2p = 114$, number of power electronics modules is 28, diameter of machine is 2.25 m, length 2.3 m, air gap (mechanical clearance) 4 mm and mass 22,000 kg [276]. Very high efficiency can be obtained in the speed range from 10 to 100% of the rated speed.

8.8.3 Study of large PM motors with different rotor configurations

A 3.8-MW, 4-pole, SmCo PM, load commutated, inverter-fed synchronous motor has been investigated by AEG, Berlin, Germany [132].

The first rotor is designed with surface PMs according to Fig. 8.15a. It does not have a damper. A retaining sleeve (bandage and epoxy resin) is used to protect the PMs against centrifugal forces. The air gap is limited by mechanical constraints and rotor surface losses.

The calculated subtransient inductance is equal to that of the synchronous inductance. Both amount to 40%. This high value of subtransient inductance is due to the absence of a damper. A PM synchronous motor with such a

PROTECTIVE CYLINDER
(BANDAGE AND EPOXY RESIN)

(a)   (b)

PM   DAMPER

                       PM

g   2g

(c)

DAMPER BAR
POLE SHOE
PM

2g

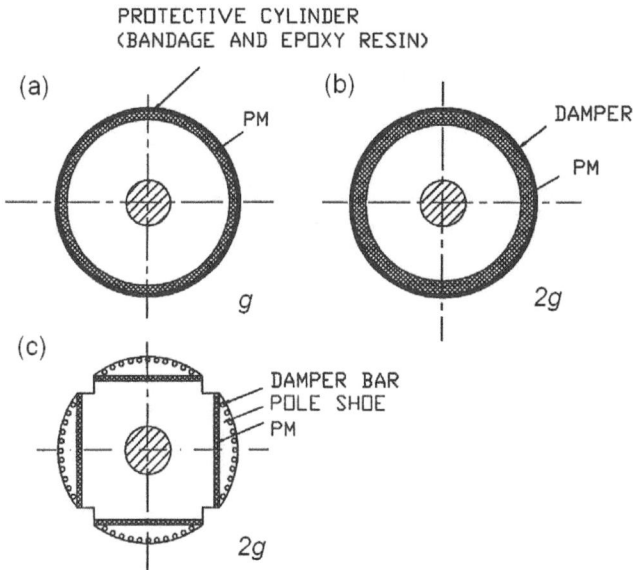

Fig. 8.15. Rotors of large synchronous motors studied by AEG: (a) with surface PMs without damper, (b) with surface magnets and damper cylinder, (c) with salient poles and cage winding.

rotor design and parameters is not suitable for operation as a load commutated PM synchronous motor (large subtransient inductance). A forced commutated converter (PWM inverter) is the most suitable power source in this case. An increase in the air gap of 1.5 times causes a decrease in both synchronous and subtransient inductances to 30%. On the other hand, this design requires 30% more SmCo material to keep the air gap flux density at the previous value.

To feed the PM synchronous motor from a load commutated CSI, the subtransient inductance must be reduced to about 10%. This can be achieved with the use of a damper. Fig. 8.15b shows the cross section of the designed PM rotor with an outer damper conductive cylinder. The air gap g is twice as much as that in Fig. 8.15a in order to minimize the losses in the damping cylinder. The required PM material amounts to twice that required in the first design without the damper. A calculated subtransient inductance equal to about 6% is very suitable for load commutated converters. To decrease the surface losses in the damping cylinder, grooves could be made on the conductive cylinder surface. This requires a thicker cylinder, which increases the PM material volume required to obtain the same air gap magnetic flux density.

A reduction in the amount of PM material with a low subtransient inductance is achieved using the configuration shown in Fig. 8.15c. Here, internal

Table 8.3. Comparison of large variable speed PM motor drives

Parameter	Load commutated thyristor CSI (Fig. 1.4b)	Forced commutated VSI with GTOs (Fig. 1.4c)	Load commutated thyristor VSI (similar to Fig. 1.4d)
Power factor $\cos\phi$	leading $\phi = 30^0$	$\cos\phi = 1$	leading $\phi = 5\ldots10^0$
Damper	with	without	without/with
Subtransient inductance	6%	40%	20 to 30%
Motor volume or mass for the same torque	100%	80%	85%
Rotor construction	relatively complicated	simple	simple
Mass of SmCo PMs	100%	50%	70%
Inverter	CSI simple thyristor bridge with d.c. link reactor	VSI GTO bridge with reverse current diodes and d.c. link capacitor	VSI thyristor bridge with antiparallel diodes and d.c. link capacitor
Converter volume	relatively small	relatively large	medium
Current harmonics	significant	small	medium
Motor harmonic losses	relatively high	low	medium
Torque ripple	significant	low	medium
Four-quadrant operation	possible with any additional elements	possible but regenerative capability necessary	possible but regenerative capability necessary
Starting	with pulsed link current	without any problem	with pulsed link current and low starting torque
Speed range	1:10	1:1000	1:5

PMs magnetized radially and pole shoes are used. The pole shoes are not only used to protect the PMs against demagnetization, but also to carry the bars of the damper windings. The volume of PM material is reduced to about 70% of that required for a damping cylinder, but it is still more than in the case of surface PMs without dampers. The subtransient inductance amounts to 6%.

The comparison of discussed above large variable speed PM motor drives is summarized in Table 8.3 [132].

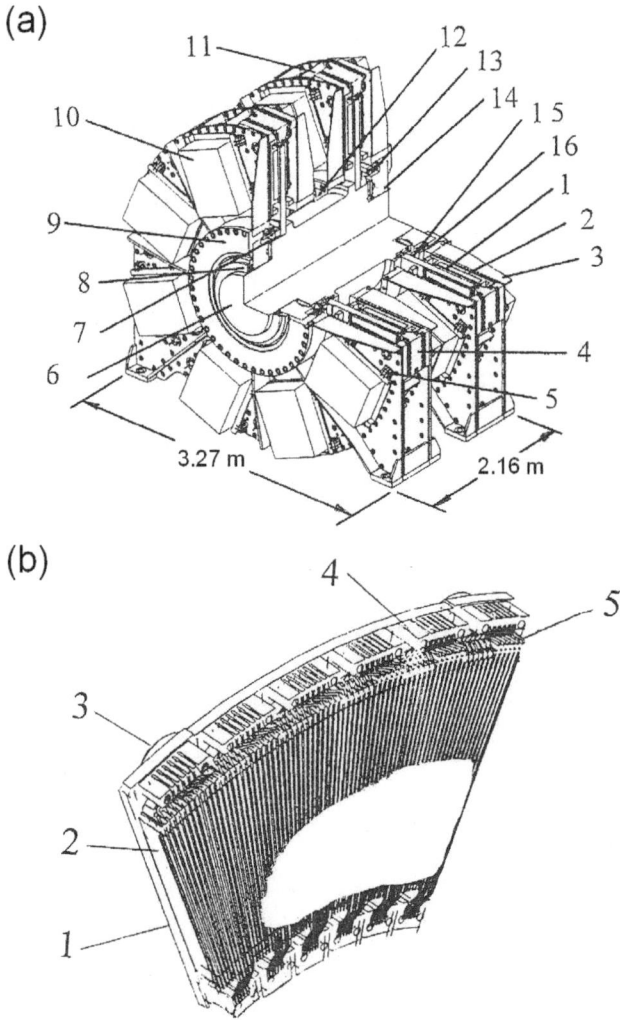

Fig. 8.16. Double disk PM brushless motor: (a) cutaway isometric view. 1 — PMs, 2 — stator assembly, 3 — housing, 4 — shock snubber, 5 — shock mount, 6 — rotor shaft, 7 — rotor disk clamp, 8 — shaft seal assembly, 9 — bearing retainer, 10 — stator segment, 11 — center frame housing, 12 — spacer housing, 13 — rotor disk, 14 — bearing assembly, 15 — rotor seal runner, 16 — rotor seal assembly; (b) stator segment. 1 — cold plate, 2 — slotted core, 3 — winding terminals, 4 — end connection cooling block, 5 — slot with conductors. Courtesy of Kaman Aerospace EDC, Hudson, MA, U.S.A.

Table 8.4. Design data of large axial flux PM brushless motors manufactured by Kaman Aerospace EDC, Hudson, MA, U.S.A.

Quantity	PA44-5W-002	PA44-5W-001	PA57-2W-001
Output power P_{out}, kW	336	445	746
Rated speed, rpm	2860	5200	3600
Maximum speed, rpm	3600	6000	4000
Efficiency at rated speed	0.95	0.96	0.96
Torque at rated speed, Nm	1120	822	1980
Stall torque, Nm	1627	1288	2712
Mass, kg	195	195	340
Power density, kW/kg	1.723	2.282	2.194
Torque density, Nm/kg	5.743	4.215	5.823
Diameter of frame, m	0.648	0.648	0.787
Length of frame, m	0.224	0.224	0.259
Application	Drilling industry, Traction		General purpose

8.9 Construction of motors with disk rotors

Stators of large axial flux PM brushless motors with disk type rotors usually have three basic parts [54]:

- aluminum cold plate
- bolted ferromagnetic core
- polyphase winding

The *cold plate* is a part of the frame and transfers heat from the stator to the heat exchange surface. The slots are machined into a laminated core wound in a continuous spiral in the circumferential direction. The copper winding, frequently a Litz wire, is placed in slots and then impregnated with a potting compound. The construction of a double disk motor developed by Kaman Aerospace EDC, Hudson, MA, U.S.A. is shown in Fig. 8.16 [54]. Specifications of large axial flux motors manufactured by Kaman are given in Table 8.4.

8.10 Transverse flux motors

8.10.1 Principle of operation

In a transverse flux motor (TFM) the electromagnetic force vector is perpendicular to the magnetic flux lines. In all standard or longitudinal flux motors the electromagnetic force vector is parallel to the magnetic flux lines. The TFM can be designed as a single-sided (Fig. 8.17a) or double-sided machine

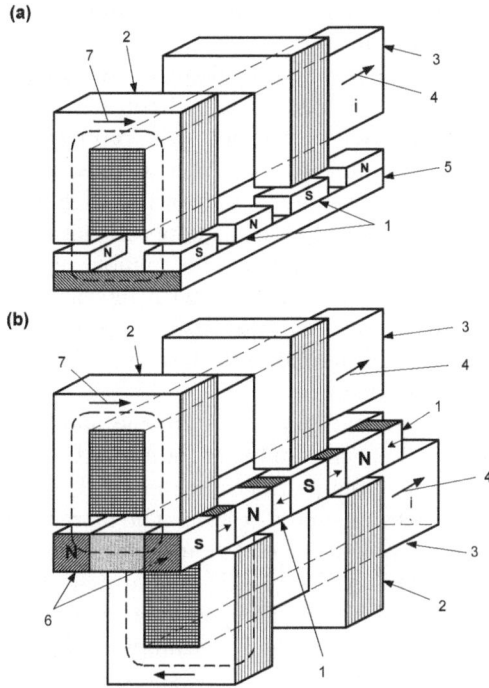

Fig. 8.17. PM transverse flux motor: (a) single sided, (b) double sided. 1 — PM, 2 — stator core, 3 — stator winding, 4 — stator current, 5 — rotor yoke, 6 — mild steel poles shoes, 7 — magnetic flux.

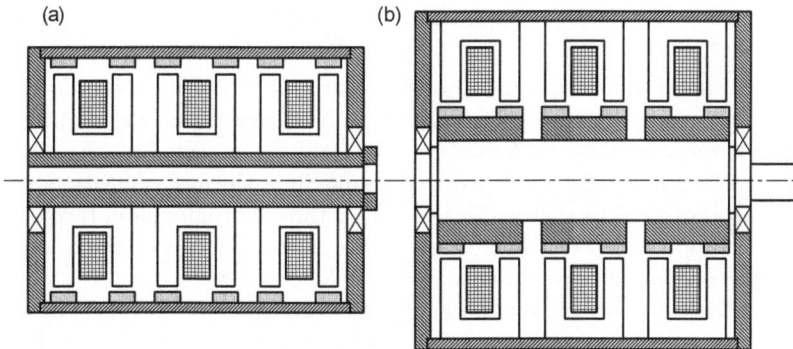

Fig. 8.18. Three-phase TFM consisting of three single phase units with: (a) internal stator, (b) external stator.

(Fig. 8.17b). Single-sided machines are easier to manufacture and have better prospects in practical applications.

The stator consists of a toroidal single-phase winding embraced by U-shaped cores. The magnetic flux in U-shaped cores is perpendicular to the stator conductors and direction of rotation. The rotor consists of surface or buried PMs and a laminated or solid core. A three-phase machine can be built of three of the same single phase units as shown in Fig. 8.18. The magnetic circuits of either stator or rotor of each single-phase unit should be shifted by $360^0/(pm_1)$ mechanical degrees where p is the number of the rotor pole pairs and m_1 is the number of phases. A TFM with internal stator (Fig. 8.18a) has a smaller external diameter. It is also easier to assemble the winding and internal stator cores. On the other hand, the heat transfer conditions are worse for internal than for external stator.

If the number of the rotor PM poles is $2p$, the number of the stator U-shaped cores is equal to p, i.e., the number of the stator U-shaped cores is equal to the number of the rotor pole pairs p. Each of the U-shaped cores creates one pole pair with two poles in axial direction. The more the poles, the better utilization and smoother operation of the machine. The power factor also increases with the number of poles. TFMs have usually from $2p = 24$ to 72 poles. The input frequency is higher than power frequency 50 or 60 Hz and the speed at an increased frequency is low. For example, a TFM with $2p = 36$ fed with 180 Hz input frequency operates at the speed $n_s = f/p = 180/18 = 10$ rev/s $= 600$ rpm.

Specifications of small two-phase and three-phase TFMs manufactured by Landert-Motoren AG, Bülach, Switzerland are shown in Table 8.5 [266].

The peak value of the line current density of a single phase is [146]

$$A_m = \frac{\sqrt{2}I_a N_1}{2\tau} = \frac{p\sqrt{2}I_a N_1}{\pi D_g} \tag{8.18}$$

where I_a is the stator (armature) *rms* current, N_1 is the number of turns per phase, τ is the stator pole pitch and D_g is the average air gap diameter. At constant *ampere turns-to-diameter* ratio the line current density can be increased by increasing the number of pole pairs. Since the force density (shear stress) is proportional to the product $A_m B_{mg}$, the electromagnetic torque of the TFM is proportional to the number of pole pairs. The higher the number of poles, the higher the torque density of a TFM.

Since at large number of poles and increased frequency the speed is low and the electromagnetic torque is high, TFMs are inherently well-suited propulsion machines to gearless electromechanical drives. Possible designs of magnetic circuits of single-sided TFMs are shown in Fig. 8.19. In both designs the air gap magnetic flux density is almost the same. However, in the TFM with magnetic shunts the rotor can be laminated radially [146].

Table 8.5. TFMs manufactured by Landert-Motoren AG, Bülach, Switzerland.

	SERVAX MDD1-91-2	SERVAX MDD1-91-3	SERVAX MDD1-133-2	SERVAX MDD1-133-3
Number of phases	2	3	2	3
Continuous torque (no active cooling)				
• at standstill, Nm	3.5	4.5	12	16
• at 300 rpm, Nm	2.5	3.3	8	10
• at 600 rpm, Nm	1.5	2	5	7
Efficiency				
• at 300 rpm	0.60	0.65	0.68	0.76
• at 600 rpm	0.65	0.68	0.70	0.80
EMF constant, V/rpm	0.07	0.07	0.16	0.15
Torque constant, Nm/A	1.8	2.7	2.8	4
Rotor	external			
Outer diameter, mm	91	91	133	133
Protection	IP54			
Class of insulation	F			
Cooling	IC410			

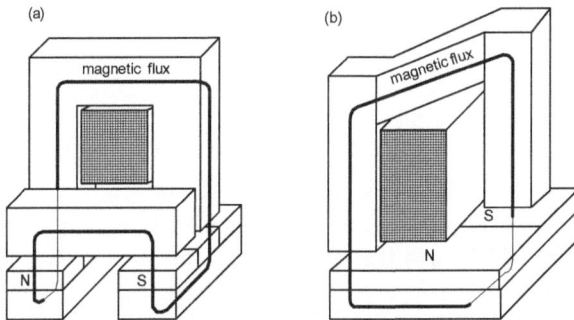

Fig. 8.19. Practical single-sided TFMs: (a) with magnetic shunts and surface PMs, (b) with twisted stator cores and surface PMs.

8.10.2 EMF and electromagnetic torque

According to eqn (5.6) the first harmonic of the magnetic flux per pole pair per phase excited by the PM rotor of a TFM is

$$\Phi_{f1} = \frac{2}{\pi}\tau l_p B_{mg1} \tag{8.19}$$

where $\tau = \pi D_g/(2p)$ is the pole pitch (in the direction of rotation), l_p is the axial length of the stator pole shoe (Fig. 8.20) and B_{mg1} is the first harmonic of the air gap peak magnetic flux density. With the rotor spinning at constant speed $n_s = f/p$, the fundamental harmonic of the magnetic flux is

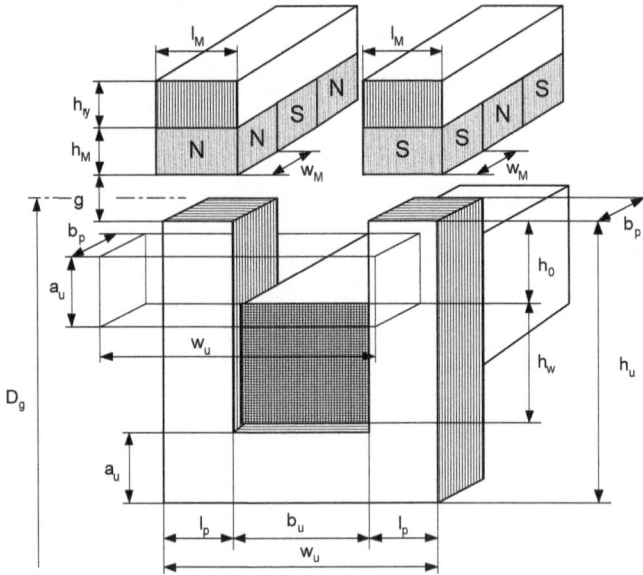

Fig. 8.20. Dimensions of U-shaped stator core and coil.

$$\phi_{f1} = \Phi_{f1}\sin(\omega t) = \frac{2}{\pi}\tau l_p B_{mg1}\sin(\omega t) = \frac{2}{\pi}\tau l_p k_f B_{mg}\sin(\omega t) \qquad (8.20)$$

where the form factor $k_f = B_{mg1}/B_{mg}$ of the excitation field is given by eqns (5.2) and (5.23) in which b_p is the width of the stator pole shoe (salient pole stator). The approximate air gap magnetic flux density B_{mg} can be found using eqn (2.14) both for the magnetic circuit shown in Fig. 8.19a and Fig. 8.19b. The instantaneous value of the sinusoidal EMF at no load induced in N_1 armature turns by the rotor excitation flux Φ_{f1} is

$$e_f = N_1 p\frac{d\Phi_{f1}}{dt} = \omega N_1 p\Phi_{f1}\cos(\omega t) = 2\pi f N_1 p\Phi_{f1}\cos(\omega t)$$

where p is the number of the stator pole pairs (U-shaped cores). The peak value of EMF is $2\pi f n_1 p\Phi_{f1}$. Thus, the *rms* value of EMF is

$$E_f = \frac{2\pi f N_1 p\Phi_{f1}}{\sqrt{2}} = \pi\sqrt{2}\,N_1 p^2\Phi_{f1}n_s \qquad (8.21)$$

or

$$E_f = 2\sqrt{2}f N_1 p\tau l_p k_f B_{mg} \qquad (8.22)$$

The electromagnetic power

$$P_{elm} = m_1 E_f I_a \cos \Psi = 2\sqrt{2}\ m_1 f N_1 p \tau l_p k_f B_{mg} I_a \cos \Psi \qquad (8.23)$$

where Ψ is the angle between the current I_a and EMF E_f. The electromagnetic torque developed by the TFM is

$$T_d = \frac{P_{elm}}{2\pi n_s} = \frac{m_1}{2\pi n_s} E_f I_a \cos \Psi = \frac{m_1}{\sqrt{2}} N_1 p^2 \Phi_{f1} I_a \cos \Psi \qquad (8.24)$$

As in the case of other motors, the EMF and electromagnetic torque can be brought to simpler forms

$$E_f = k_E n_s \qquad \text{and} \qquad T_d = k_T I_a \qquad (8.25)$$

Assuming $\Phi_{f1} = const$, the EMF constant and torque constant are, respectively,

$$k_E = \pi\sqrt{2}\ N_1 p^2 \Phi_{f1} \qquad (8.26)$$

$$k_T = \frac{m_1}{2\pi} k_E \cos \Psi = \frac{m_1}{\sqrt{2}} N_1 p^2 \Phi_{f1} \cos \Psi \qquad (8.27)$$

For $I_{ad} = 0$ the total current $I_a = I_{aq}$ is torque-producing and $\cos \Psi = 1$.

8.10.3 Armature winding resistance

The armature winding resistance can be calculated approximately as

$$R_1 \approx k_{1R} \pi [D_g \pm g \pm (h_w + h_o)] \frac{N_1}{a_w \sigma_1 s_a} \qquad (8.28)$$

where k_{1R} is the skin-effect coefficient for resistance (Appendix B), h_w is the coil height, $h_o = h_u - h_w - a_u$ is the top portion of the "slot" not filled with conductors, a_w is the number of parallel wires, σ_1 is the conductivity of the armature conductor at a given temperature and s_a is the cross section of the armature single conductor. The "+" sign is for the external stator and the "−" sign is for the internal stator.

8.10.4 Armature reaction and leakage reactance

The *mutual reactance* corresponding to the *armature reaction reactance* in a synchronous machine can analytically be calculated in an approximate way. One U-shaped core (pole pair) of the stator can be regarded as an a.c. electromagnet with N_1 turn coil which, when fed with the sinusoidal current I_a, produces peak MMF equal to $\sqrt{2}I_a N_1$. The equivalent d-axis field MMF per pole pair per phase which produces the same magnetic flux density as the armature reaction MMF is

$$\sqrt{2}I_{ad}N_1 = \frac{B_{ad}}{\mu_0}g' = \frac{B_{ad1}}{\mu_0 k_{fd}}g'$$

where g' is the equivalent air gap and $k_{fd} = B_{ad1}/B_{ad}$ is the d-axis form factor of the armature reaction according to eqn (5.24). Thus, the d-axis armature current as a function of B_{ad1} is

$$I_{ad} = \frac{B_{ad1}}{k_{fd}\mu_0} \frac{g'}{\sqrt{2}N_1} \qquad (8.29)$$

At constant magnetic permeability, the d-axis armature EMF

$$E_{ad} = 2\sqrt{2}fN_1p\tau l_p B_{ad1} \qquad (8.30)$$

is proportional to the armature current I_{ad}. Thus, the d-axis armature reaction reactance is

$$X_{ad} = \frac{E_{ad}}{I_{ad}} = 4\mu_0 fN_1^2 p\frac{\tau l_p}{g'}k_{fd} \qquad (8.31)$$

Similarly, the q-axis armature reactance

$$X_{aq} = \frac{E_{aq}}{I_{aq}} = 4\mu_0 fN_1^2 p\frac{\tau l_p}{g'}k_{fq} \qquad (8.32)$$

The d and q-axis form factors of the armature reaction can be found in a similar way as in Section 5.6. Most TFMs are designed with surface configuration of PMs and $k_{fd} = k_{fq} = 1$, i.e., $X_{ad} = X_{aq}$.

Neglecting the saturation of the magnetic circuit, the equivalent air gap is calculated as

- for the TFM with magnetic shunts (Fig. 8.19a)

$$g' = 4\left(g + \frac{h_M}{\mu_{rrec}}\right) \qquad (8.33)$$

- for the TFM with twisted U-shaped cores (Fig. 8.19b)

$$g' = 2\left(g + \frac{h_M}{\mu_{rrec}}\right) \qquad (8.34)$$

where g is the mechanical clearance in the d-axis, h_M is the radial height of the PM (one pole) and μ_{rrec} is the relative recoil magnetic permeability of the PM. To take into account the magnetic saturation, the equivalent air gap g' should be multiplied by the saturation factor k_{sat} in the d-axis and k_{satq} in the q-axis.

The armature reaction inductances

$$L_{ad} = \frac{X_{ad}}{2\pi f} = \frac{2}{\pi}\mu_0 N_1^2 p\frac{\tau l_p}{g'}k_{fd} \qquad (8.35)$$

$$L_{aq} = \frac{X_{aq}}{2\pi f} = \frac{2}{\pi}\mu_0 N_1^2 p \frac{\tau l_p}{g'} k_{fq} \tag{8.36}$$

The leakage inductance of the stator winding is approximately equal to the sum of the "slot" leakage inductance and pole-top leakage reactance. The approximate equation is

$$L_1 \approx \mu_0 \pi [D_g \pm g \pm (h_w + h_o)] N_1^2 (\lambda_{1s} + \lambda_{1p}) \tag{8.37}$$

where h_w is the height of coil, $h_o = h_u - h_w - a_u$ is the top portion of the "slot" not filled with conductors, the "+" sign is for the external stator and the "−" sign is for the internal stator. The coefficients of leakage permeances are

• coefficient of "slot" leakage permeance

$$\lambda_{1s} = \frac{h_w}{3b_u} + \frac{h_o}{b_u} \tag{8.38}$$

• coefficient of pole-top leakage permeance

$$\lambda_{1p} \approx \frac{5g/b_u}{5 + 4g/b_u} \tag{8.39}$$

where $b_u = w_u - 2l_p$ (Fig. 8.20).

The leakage inductance according to eqn (8.37) is much smaller than that obtained from measurements and the FEM. Good results are obtained if the eqn (8.37) is multiplied by 3 [147]. For most TFMs $L_1 > L_{ad}$ and $L_1 > L_{aq}$.

The leakage inductance can also be estimated as a sum of three inductances, i.e., due to lateral leakage flux, "slot" leakage flux and leakage flux about the portion of the coil not embraced by the ferromagnetic core [18].

The synchronous reactances in the d and q axis according to eqn (5.15) are the sums of the armature reaction reactances (8.31), (8.32) and leakage reactance $X_1 = 2\pi f L_1$.

8.10.5 Magnetic circuit

Kirchhoff's equations for the MVD per pole pair are

• for the TFM with magnetic shunts (Fig. 8.19a)

$$4\frac{B_r}{\mu_0 \mu_{rrec}} h_M = 4\frac{B_{mg}}{\mu_0 \mu_{rrec}} h_M + 4\frac{B_{mg}}{\mu_0} g + \sum_i H_{Fei} l_{Fei}$$

• for the TFM with twisted U-shaped cores (Fig. 8.19b)

$$2\frac{B_r}{\mu_0 \mu_{rrec}} h_M = 2\frac{B_{mg}}{\mu_0 \mu_{rrec}} h_M + 2\frac{B_{mg}}{\mu_0} g + \sum_i H_{Fei} l_{Fei}$$

where $H_c = B_r/(\mu_0\mu_{rrec})$ and $\sum_i H_{Fei}l_{Fei}$ is the magnetic voltage drop in ferromagnetic parts of the magnetic circuit (stator and rotor cores). The above equations can be expressed with the aid of the saturation factor k_{sat} of the magnetic circuit

• for the TFM with magnetic shunts (Fig. 8.19a)

$$4\frac{B_r}{\mu_0\mu_{rrec}}h_M = 4\frac{B_{mg}}{\mu_0}\left(\frac{h_M}{\mu_{rrec}} + gk_{sat}\right) \tag{8.40}$$

where

$$k_{sat} = 1 + \frac{\sum_i H_{Fei}l_{Fei}}{4B_{mg}g/\mu_0} \tag{8.41}$$

• for the TFM with twisted U-shaped cores (Fig. 8.19b)

$$2\frac{B_r}{\mu_0\mu_{rrec}}h_M = 2\frac{B_{mg}}{\mu_0}\left(\frac{h_M}{\mu_{rrec}} + gk_{sat}\right) \tag{8.42}$$

where

$$k_{sat} = 1 + \frac{\sum_i H_{Fei}l_{Fei}}{2B_{mg}g/\mu_0} \tag{8.43}$$

Both eqns (8.40) and (8.42) give almost the same value of the air gap magnetic flux density, i.e.,

$$B_{mg} = \frac{B_r}{1 + (\mu_{rrec}g/h_M)k_{sat}} \tag{8.44}$$

Please note that the saturation factors k_{sat} expressed by eqns (8.41) and (8.43) are different.

8.10.6 Advantages and disadvantages

The TFM has several advantages over a standard PM brushless motor, i.e.,

(a) at low rotor speed the frequency in the stator (armature) winding is high (large number of poles), i.e., a low speed machine behaves as a high speed machine, which is the cause of better utilization of active materials than in standard (longitudinal flux) PM brushless motors for the same cooling system, i.e., higher torque density or higher power density;
(b) less winding and ferromagnetic core materials for the same torque;
(c) simple stator winding consisting of a single ring-shaped coil (cost effective stator winding, no end connection);
(d) unity winding factor ($k_{w1} = 1$);
(e) the more the poles, the higher the torque density, higher power factor and less the torque ripple;

(f) a three-phase motor can be made of three (or multiples of three) identical single-phase units;

(g) a three-phase TFM can be fed from a standard three-phase inverter for PM brushless motors using a standard encoder;

(h) the machine can operate as a low speed generator with high frequency output current.

Although the stator winding is simple, the motor consists of a large number of poles ($2p \geq 24$). There is a double saliency (the stator and rotor) and each salient pole has a separate "transverse flux" magnetic circuit. Careful attention must be given to the following problems:

(a) to avoid a large number of components, it is necessary to use radial laminations (perpendicular to the magnetic flux paths in some portions of the magnetic circuit), sintered powders or hybrid magnetic circuits (laminations and sintered powders);

(b) the motor external diameter is smaller in the so-called "reversed design," i.e., with external PM rotor and internal stator;

(c) the TFM uses more PM material than an equivalent standard PM brushless motor;

(d) the power factor decreases as the load increases and special measures must be taken to improve the power factor;

(e) as each stator pole faces the rotor pole and the number of stator and rotor pole pairs is the same, special measures must be taken to minimize the cogging torque.

8.11 Applications

8.11.1 Ship propulsion

The *ship electric propulsion* scheme shown in Fig. 1.25 requires large electric motors. For example, a 90,000 gt *cruise ship* employs two 19.5 MW synchronous motors. Low speed PM brushless motors offer significant savings in mass (up to 50%) and efficiency (2 to 4% at full load and 15 to 30% at partial load) as compared to high speed synchronous motors with electromagnetic excitation and reduction gears. Fig. 8.21 shows the most powerful PM brushless motor in the world for advanced ship propulsion rated at 36.5 MW and 127 rpm.

An optimum undisturbed water inflow to the *propeller* and consequently reduced propeller pressure pulses (causing vibration and noise) and increased propulsion efficiency can be achieved with the aid of *pod propulsor* (Fig. 8.22). Reduction of vibration and noise considerably enhances passenger comfort. The propeller acts as a tractor unit located in front of the pod. The pod can be rotated through 360^0 to provide the required thrust in any direction. This

Fig. 8.21. PM brushless motor rated at 36.5 MW and 127 rpm for advanced ship propulsion. Photo courtesy of DRS Technologies, Parsippany, NJ, U.S.A.

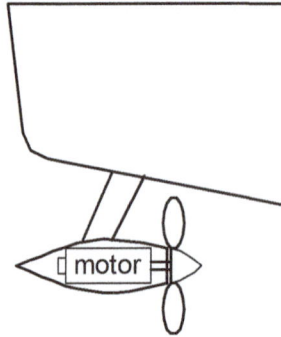

Fig. 8.22. Pod propulsor with electric motor.

eliminates the requirement for stern tunnel thrusters and ensures that ships can maneuver into ports without tug assistance.

Rolls-Royce, Derby, U.K. has been developing a 20 MW TFM for naval electric ship propulsion (Fig. 8.23) [220]. The TFM has been considered to be the optimum propulsion motor to meet the high efficiency and power density requirements. Each phase consists of two rotor rims with buried PMs and two armature coils (outer phase and inner phase) [220]. An 8-phase TFM emits a low noise and has significant reverse mode capability as all eight phases are

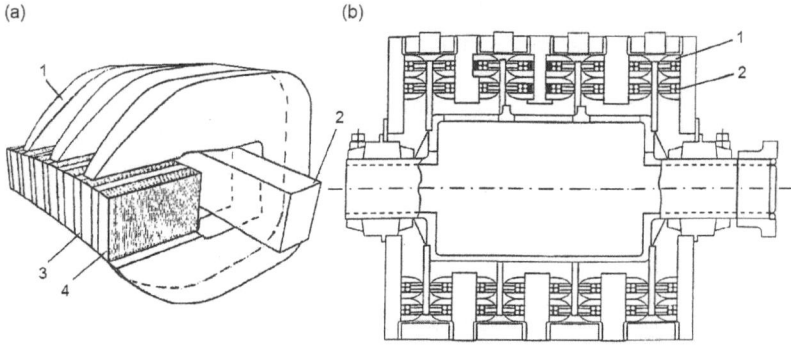

Fig. 8.23. 20-MW TFM for electric ship propulsion: (a) construction; (b) longitudinal section. 1 — U-shaped twisted core, 2 — stator winding, 3 — PM, 4 — mild steel pole. Courtesy of Rolls-Royce, Derby, U.K.

Table 8.6. Specification data of a 20 MW TFM prototype for naval ship propulsion. Courtesy of Rolls-Royce, Derby, U.K.

Rated power, MW	20	
Rated speed, rpm	180	
Number of poles	130	
Rated frequency, Hz	195	
Number of phases	8	
Supply voltage (d.c. link), V	5000	
Number of rotor disks	4	
Number of rotor rims per disk	4	
PM material	NdFeB	
Force density at rated load, kN/m^2	120	
Outer diameter, m	2.6	
Overall length, m	2.6	
Shaft diameter, m	0.5	
Estimated overall mass, t	39	
Power density at rated load, kW/kg	0.513	
	Outer phase	Inner phase
Mean rim diameter, m	2.1	1.64
Number of turns per coil	10	12
Number of coils per phase	2	2
Coil current (rms), A	750	500
Peak coil current, A	1000	670
Power per phase of solid state converter, KW	3200	1800
Converter type	PWM IGBT VSI (isolated phases)	

truly independent from one another. The stator components are supported by a water-cooled aluminum frame. The specifications of the TFM are shown in Table 8.6 [220].

To date, TFMs are predicted primarily for a conventional (inboard) fit for a surface ship or submarine [33]. The use of the TFM for podded electromechanical drives is highly attractive due to the low mass of the motor in comparison with other PM brushless motors.

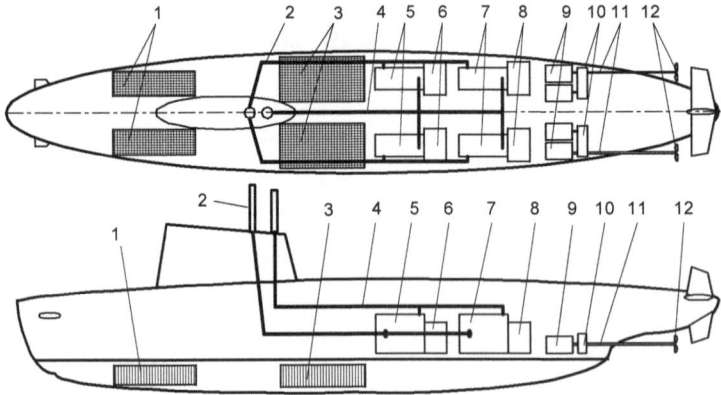

Fig. 8.24. Diesel-electric submarine: 1 — forward battery cells, 2 — snorkel (air intake), 3 — after battery cells, 4 — exhaust, 5 — diesel engines (compartment 1), 6 — generators (compartment 1), 7 — diesel engines (compartment 2), 8 — generators (compartment 2), 9 — electric propulsion motors, 10 — reduction gear, 11 — propulsion shafts, 12 — propeller.

8.11.2 Submarine propulsion

A typical "fleet" type *submarine* with four diesel engines and four electric motors is shown in Fig. 8.24. The diesel engines are coupled to electric generators which supply power to large electric propulsion motors which turn the propellers and charge batteries. Approximately, a 100 m long submarine uses four 1.2 MW diesel engines and 4 MW total power of electric propulsion motors. Diesel engines are divided between two compartments separated by a watertight bulkhead. If one room becomes flooded, the other two engines can still be operated.

When the submarine submerges below periscope depth, the diesel engines are shut down and the electric motors continue to turn the propellers being fed from batteries. The snorkel system permits a submarine to use its diesel engines for propulsion or charge batteries, while operating submerged at the periscope level. Both the air intake and exhaust are designed as masts which are raised to the position above the surface of the water.

Fig. 8.25. Large PM synchronous motor *Permasyn* for submarine propulsion. Courtesy of Siemens AG, Hamburg, Germany.

A large PM synchronous motor *Permasyn* with low noise level, high efficiency and small mass and volume specially designed for submarine propulsion at Siemens AG, Hamburg, Germany is shown in Fig. 8.25.

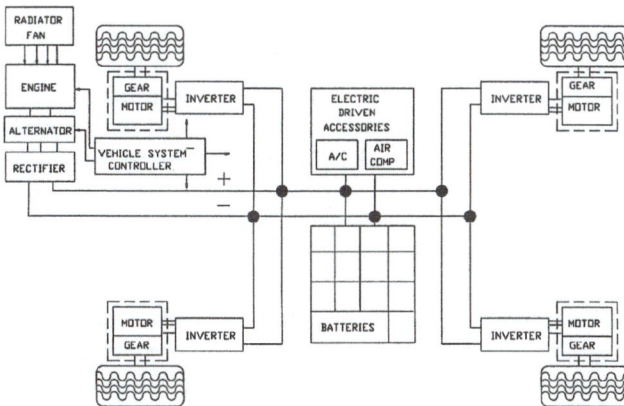

Fig. 8.26. A hybrid bus drive system with a.c. motors and reduction gears integrated into each of its four driven wheels.

8.11.3 Hybrid electric transit bus

Modern transit buses should, amongst other things, emit reduced pollution and be designed with low floors for easy access to people with physical problems. A *hybrid electric bus* with electric motors integrated in each of its four driven wheels can meet these requirements.

Table 8.7. Hybrid transit bus performance

Parameter	Conditions	Goal
Top speed	continuous operation, level road	93.6 km/h
Gradeability	16% grade	11.3 km/h
	2.5% grade	71.00 km/h
Startability	seated load mass	17% grade
	gross vehicle mass	13% grade
Maximum acceleration	0% grade	1.47 m/s^2
and deceleration	0% grade	1.27 m/s^2
(gross vehicle mass)		

The components included in the hybrid transit bus (Fig. 8.26) are brushless motors (induction, PM or switched reluctance) to supply or accept power from the wheels, power electronics converters, a battery for energy storage, and the auxiliary power unit consisting of a diesel engine, alternator, rectifier, and associated control [177]. The bus is a series hybrid with four independent brushless motors — one integrated into each driven wheel.

To minimize emissions and maximize fuel economy the electric drives are powered by a battery-assisted auxiliary power unit consisting of a down-sized combustion engine coupled to an alternator operating under closed-loop control. Each electromechanical drive consists of a 75-kW brushless oil-cooled motor integrated with a transmission into a compact wheel motor unit. Since the transmission is a simple single-speed device, and the integrated wheel motor units are located right at the drive wheels, torque is transmitted to the four rear drive wheels very efficiently and the need for an expensive and heavy multi-speed transmission and a bulky rear axle-differential assembly are eliminated. The output torque from each propulsion motor is controlled by a microprocessor-based electronics package containing both a d.c. — a.c. inverter and the actual motor-control circuitry. The auxiliary power unit supplies electric power to each a.c. integrated wheel motor in addition to charging the propulsion batteries and supplying power to operate electric accessories. In fact, the auxiliary power unit supplies average power to the drives, while the propulsion batteries fill in the demand peaks during vehicle acceleration and also receive power during regenerative braking. The auxiliary power unit is rated at 100 kW with controls to operate over a d.c. voltage range from 250 to 400 V. A battery with a capacity of 80 Ah and a total energy storage of about 25 kWh is required [177].

Table 8.8. Comparison of different 75 kW brushless motors for electrical vehicles. Courtesy of Voith Turbo GmbH & Co. KG, Heidenheim, Germany.

	IM	SRM	HSM	PMSM	PMMSM	TFM
Rotor	Internal Cu cage	Internal	External	Internal	Internal	External
Gear stages	1	2	1	1	1	1
Gear reduction ratio	6.22	12.44	6.22	6.22	6.22	6.22
Number of poles	2	6	20	24	40	44
Rated speed, rpm	940	1232	616	616	616	570
Rated frequency, Hz	49	82	103	123	205	209
Air gap, mm	1.00	1.00	2.00	2.00	3.00	1.2 to 2.0
Diameter, mm						
• inner	111	56	282	313	328	90
• gap	266	278	351	341	354	354
• outer	413	400	410	410	410	366
Stack length, mm	276	200	243	229	255	124
Stack + end connections, mm	397	350	285	265	295	212
Volume, 10^{-3} m^3	53.2	44.0	37.6	35.0	38.9	22.3
Mass of active parts, kg	272	147	106	79	71	73
PM mass, kg	—	—	2.7	4.7	7.0	11.5
Efficiency	0.900	0.930	0.932	0.941	0.949	0.976
Inverter power, kVA	396	984	254	361	385	455

The hybrid bus uses its electronic drive controls to operate the a.c. motors as generators during deceleration or while maintaining speed on a downhill grade to recover a portion of its kinetic energy. The recovered energy is routed to the propulsion battery, thus lowering fuel consumption and emissions.

The performance of a 12.2-m long hybrid electric transit bus with a seated load mass of 15,909 kg and gross vehicle mass of 20,045 kg are shown in Table 8.7 [177].

Table 8.8 shows a comparison of different 75-kW brushless electric motors for propulsion of hybrid electric buses [193].

The induction motor (IM) has a rotor cage winding. The switched reluctance motor (SRM) has 6 poles in the stator (2 poles per phase) and 4 poles in the rotor.

The hybrid synchronous motor (HSM) has both PM and electromagnetic excitation. The PMs are located on the rotor tooth faces and the d.c. excitation winding is located in rotor slots. The electromagnetic excitation is used to boost the starting and allows a real field weakening in the upper speed range.

The PM synchronous motor (PMSM) has been designed with surface configuration of rotor PMs.

The polyphase PM modular synchronous motor (PMMSM) has the stator coil span equal to only one tooth instead of full pole pitch; i.e., it has concentrated non-overlapping coils (Fig. 8.8). Thus, the end connections are extremely short and the winding power losses are substantially reduced. The simple coil shape is easy to manufacture using automated techniques and allows for a high slot fill factor.

The double-sided TFM has U-shaped stator cores made of powder materials, buried magnets and external rotor [193].

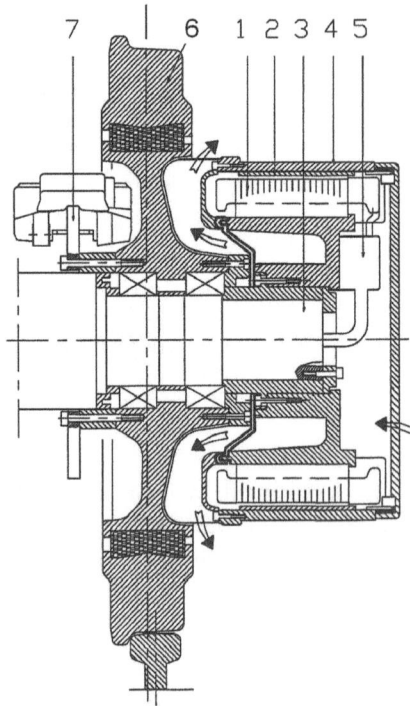

Fig. 8.27. Gearless motor wheel for a light rail vehicle with PM brushless motor: 1 — stator, 2 — external rotor with PMs, 3 — axle of the wheel, 4 — rotor enclosure, 5 — terminal board, 6 — rim of the wheel, 7 — brake.

8.11.4 Light rail system

The most important advantages of a direct electromechanical drive with gearless PM brushless motor over a geared motor are:

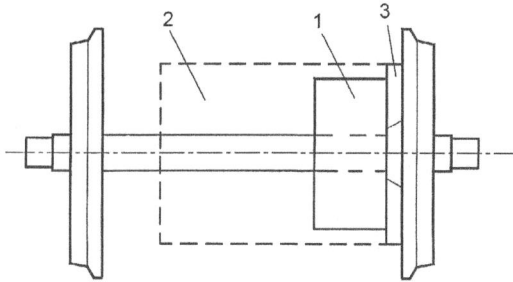

Fig. 8.28. Comparison of direct wheel electromechanical drives with brushless motors. 1 — PM TFM, 2 — induction motor, 3 — clutch.

- the gravity center of the bogie can be lowered;
- the wheel diameter can be reduced as motors are removed from the trolleys and gearboxes eliminated;
- it is easy to design a steerable bogie for negotiating sharp curves;
- electromechanical drives with gearless motors require limited maintenance (no oil);
- the noise is reduced.

Typical full load mass of a street car is 37 t and wheel diameter is 0.68 m. Four gearless motors with rated torque of 1150 Nm can replace the traditional propulsion system. The motorwheel is shown in Fig. 8.27 [39]. The motor is built in the wheel in order to minimize the overall dimensions and to improve the heat transfer. To achieve a speed of 80 km/h the rated power of the motor should be about 75 kW. The external rotor with PMs is integral with the wheel and the stator (armature) is integral with the spindle. Representatively, the armature diameter is 0.45 m, the external motor diameter is 0.5 m, the armature stack length is 0.155 m, the air gap magnetic flux density is 0.8 T and the armature line current density is 45 kA/m [39]. The motor is located externally while the disk brake is located internally.

The size of the direct wheel drive with PM TFM as compared with an equally rated induction motor is shown in Fig. 8.28 [305].

In the 1990s the Railway Technical Research Institute (RTRI) in Tokyo (Kokubunji), Japan, carried out research in traction PM brushless motors for *narrow gauge express train* to reach a maximum speed of 250 km/h. Japan has about 27,000 route km of narrow 1067 mm track and over 2000 route km of standard 1435 mm *Shinkansen* track. The target speed increase on narrow gauge tracks in Japan is over 160 km/h at the beginning of the 21st century (a speed record of 256 km/h on 1067 mm track was achieved in 1973 in the Republic of South Africa).

Several motor designs and drive configurations have been studied. The specifications of the RMT1A motor with surface PMs rated at 80 kW, 580 V,

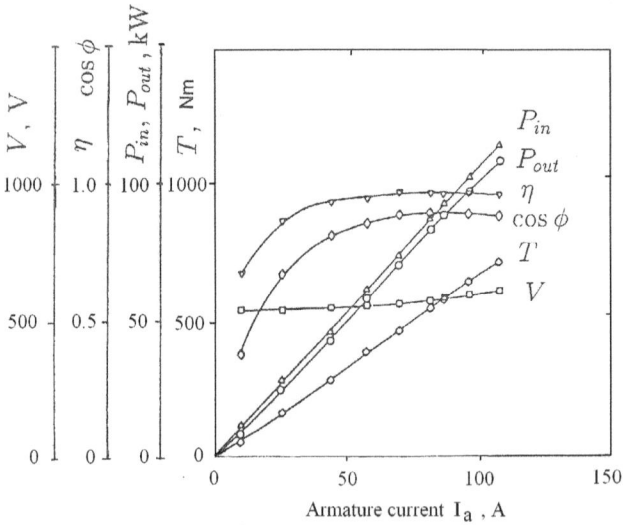

Fig. 8.29. Steady-state characteristics of RMT1A PM brushless motor at constant speed $n = 1480$ rpm [210].

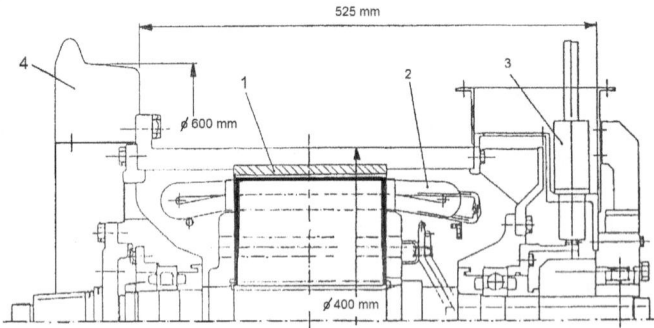

Fig. 8.30. RMT1A traction PM brushless motor: 1 — surface PMs, 2 — internal stator, 3 — position sensors, 4 — wheel. Courtesy of K. Matsuoka and K. Kondou [210], Railway Technical Research Institute, Kokubunji, Japan.

1480 rpm motor are given in Table 8.9 and steady state load characteristics are plotted in Fig. 8.29 [210]. The gearless wheel-mounted motor with internal stator (armature) is shown in Fig. 8.30 [210].

Table 8.9. Specification data of RMT1 PM brushless motor for narrow gauge express trains developed by K. Matsuoka and K. Kondou [210], Railway Technical Research Institute, Kokubunji, Japan.

Rated continuous power, kW	80
Rated continuous torque, Nm	516
Rated voltage, V	580
Rated current, A	86
Rated speed, rpm	1480
Rated frequency, Hz	74
Number of poles, $2p$	6
Torque constant, Nm/A	6
Cooling	Forced ventilation
Mass without axle, kg	280
Power density, kW/kg	0.286
Torque density, Nm/kg	1.843

Numerical examples

Numerical example 8.1

The speed of a large, three-phase, 24-pole, brushless motor is 2000 rpm. The full pitch armature winding has been designed from AWG copper conductors with square cross section. The height of a single conductor is $h_c = 3.264$ mm, the number of conductors per slot arranged above each other in two layers is $m_{sl} = 8$ and the conductivity of copper at 75^0C is $\sigma_1 = 47 \times 10^6$ S/m. The ratio of the *end connection–to–effective core length* is $l_{1e}/L_i = 1.75$. Find the coefficient of skin effect.

Solution

The frequency of current in the armature conductors is

$$f = \frac{2000}{60} \times 12 = 400 \text{ Hz}$$

The reduced height of the armature conductor calculated on the basis of eqn (B.7) is

$$\xi_1 \approx h_c\sqrt{\pi\mu_0\sigma_1}\sqrt{f}$$

$$= 0.003264\sqrt{\pi \times 0.4\pi \times 10^{-6} \times 47 \times 10^6}\sqrt{400} = 0.888$$

The coefficients $\varphi_1(\xi_1)$ and $\Psi_1(\xi_1)$ calculated on the basis of eqns (B.5) and (B.6) are

$$\varphi_1(\xi_1) = 1.05399 \qquad\qquad \Psi_1(\xi_1) = 0.20216$$

The coefficient of skin effect according to eqns (8.15) and (8.16) is

- for a 120^0 square wave

$$K_{1R} = 0.811 \left[1.05399 + \left(\frac{8^2 - 1}{3} - \frac{8^2}{16} \right) 0.20216 \right] = 3.642$$

- for a 180^0 square wave

$$K_{1R} = 1.133 \left[1.05399 + \left(\frac{8^2 - 1}{3} - \frac{8^2}{16} \right) 0.20216 \right] = 5.088$$

Since the skin effect is only in that part of a conductor that lies in the slot, the practical coefficient of skin effect will be smaller, i.e.,

- for a 120^0 square wave

$$K'_{1R} = \frac{K_{1R} + l_{1e}/L_i}{1 + l_{1e}/L_i} = \frac{3.642 + 1.75}{1 + 1.75} = 1.961$$

- for a 180^0 square wave

$$K'_{1R} = \frac{5.088 + 1.75}{1 + 1.75} = 2.486$$

The coefficient of skin effect is too high and either stranded round conductors or rectangular conductors with transpositions must be used. The winding has been incorrectly designed since at 400 Hz the winding losses exceed 1.961 (120^0 square wave) and 2.486 (180^0 square wave) times the losses calculated for d.c. current.

Numerical example 8.2

A three-phase, 50-kW, 6000-rpm, 200-Hz, 346-V (line-to-line), Y-connected PM synchronous motor has stator winding impedance $\mathbf{Z}_1 = R_1 + jX_1 = 0.027 + j0.026$ Ω, inner stator diameter $D_{1in} = 0.18$ m, effective length of stator core $L_i = 0.125$ m, number of stator turns per phase $N_1 = 32$, stator winding factor $k_{w1} = 0.96$ and air gap magnetic flux density $B_{mg} = 0.7$ T. The PMs are uniformly distributed on the rotor surface. The total air gap (PMs, damper, bandage, mechanical clearance) multiplied by the Carter's coefficient k_C and saturation factor k_{sat} is $g_t = gk_Ck_{sat} + h_M/\mu_{rrec} = 15$ mm. The stator core losses are equal to 2% of the output power, the additional losses are equal to 1% of the output power and the rotational losses are equal to 436 W. Find the output power, efficiency and power factor at load angle $\delta = 14.4^0$.

Solution

This motor is very similar to that described in [10]. The number of pole pairs is $p = f/n_s = (200/6000) \times 60 = 2$. The pole pitch

$$\tau = \frac{\pi D_{1in}}{2p} = \frac{\pi \times 0.18}{4} = 0.1414 \text{ m}$$

For the surface PM rotor, the form factors of the armature field $k_{fd} = k_{fq} = 1$. According to eqns (5.31), (5.32) and (5.33) the armature reaction reactances are

$$X_{ad} \approx X_{aq}$$

$$= 4 \times 3 \times 0.4\pi \times 10^{-6} \times 200 \frac{(32 \times 0.96)^2}{2\pi} \frac{0.1414 \times 0.125}{0.015} = 0.534 \ \Omega$$

According to eqns (5.15) the synchronous reactances are

$$X_{sd} \approx X_{sq} = 0.026 + 0.534 = 0.56 \ \Omega$$

The magnetic flux excited by the rotor is calculated on the basis of eqn (5.6)

$$\Phi_f = \frac{2}{\pi} 0.1414 \times 0.125 \times 0.7 = 0.007876 \text{ Wb}$$

The form factor of the excitation field has been assumed $k_f \approx 1$.
    Through the use of eqn (5.5) the EMF induced by the rotor excitation flux is

$$E_f = \pi\sqrt{2} \times 200 \times 32 \times 0.96 \times 0.007876 = 215 \text{ V}$$

The phase voltage is $V_1 = 346/\sqrt{3} \approx 200$ V.
    The armature current in d-axis — eqn (5.40)

$$I_{ad} = \frac{200(0.56 \times 0.9686 - 0.027 \times 0.2487) - 215 \times 0.56}{0.56 \times 0.56 + 0.027^2} = -42.18 \text{ A}$$

where $\cos\delta = \cos 14.4^0 = 0.9686$ and $\sin\delta = \sin 14.4^0 = 0.2487$.
    The armature current in q-axis — eqn (5.41)

$$I_{aq} = \frac{200(0.027 \times 0.9686 + 0.56 \times 0.2487) - 215 \times 0.027}{0.56 \times 0.56 + 0.027^2} = 86.79 \text{ A}$$

The current in armature winding according to eqn (5.43) is

$$I_a = \sqrt{(-42.18)^2 + 86.79^2} = 96.5 \text{ A}$$

The angle Ψ between E_f and I_a (Fig. 5.4)

$$\Psi = \arctan \frac{I_{ad}}{I_{aq}} = \arctan \frac{(-42.18)}{86.79} = \arctan(-0.486) = -25.9^0$$

Since $E_f > V_1$, the d-axis armature current I_{ad} is negative and the angle Ψ is negative. This means that the motor is overexcited (Fig. 5.4b). The input power on the basis of eqn (5.44)

$$P_{in} = 3V_1(I_{aq} \cos \delta - I_{ad} \sin \delta)$$

$$= 3 \times 200[86.79 \times 0.9686 - (-42.18) \times 0.2487] = 56,733 \text{ W} \approx 56.7 \text{ kW}$$

The input apparent power

$$S_{in} = m_1 V_1 I_a = 3 \times 200 \times 96.5 = 57,900 \text{ VA} = 57.9 \text{ kVA}$$

Power factor $\cos \phi$

$$\cos \phi = \frac{P_{in}}{S_{in}} = \frac{56,733}{57,900} = 0.9798 \qquad \text{and} \qquad \phi = 11.52^0$$

The stator (armature) winding losses according to eqn (B.12)

$$\Delta P_a = m_1 I_a^2 R_1 = 3 \times 96.5^2 \times 0.027 = 754.3 \text{ W}$$

The input power is the sum of the output power and losses, i.e.,

$$P_{in} = P_{out} + \Delta P_a + 0.02 P_{out} + 0.01 P_{out} + \Delta P_{rot}$$

where $\Delta P_{Fe} = 0.02 P_{out}$ and $\Delta P_{str} = 0.01 P_{out}$. Thus, the output power is

$$P_{out} = \frac{1}{1.03}(P_{in} - \Delta P_a - \Delta P_{rot})$$

$$= \frac{1}{1.03}(56,733 - 754.3 - 436) = 53,924.9 \text{ W} \approx 54 \text{ kW}$$

The stator core losses

$$\Delta P_{Fe} = 0.02 \times 53,924.9 = 1078.5 \text{ W}$$

The additional losses

$$\Delta P_{str} = 0.01 \times 53,924.9 = 539.2 \text{ W}$$

Total losses at $\delta = 14.4°$

$$\sum \Delta P = \Delta P_a + \Delta P_{Fe} + \Delta P_{str} + \Delta P_{rot}$$

$$= 754.3 + 1078.5 + 539.2 + 436 = 2808 \text{ W}$$

The efficiency

$$\eta = \frac{53,924.9}{56,733} = 0.95 \qquad \text{or} \qquad 95\%$$

Methods of calculating the power losses are given in Appendix B and in many publications, e.g., [10, 185].

Numerical example 8.3

A three-phase, 160-kW, 180-Hz, 346-V (line-to-line), Y-connected, 3600-rpm, 300-A, surface-type PM synchronous motor has the following armature winding parameters: number of slots $s_1 = 72$, coil pitch $w_c/\tau = 10$, a single-layer armature winding and number of turns per phase $N_1 = 24$. The armature resistance is $R_1 = 0.011$ Ω, the d-axis synchronous inductance is $L_{sd} = 296 \times 10^{-6}$ H and the q-axis synchronous inductance is $L_{sq} = 336 \times 10^{-6}$ H. The stator inner diameter is $D_{1in} = 0.33$ m and the effective length of the stator core is $L_i = 0.19$ m. The air gap magnetic flux density is $B_{mg} = 0.65$ T, the core and additional losses are $\Delta P_{Fe} + \Delta P_{str} = 0.02 P_{out}$, and the rotational losses are $\Delta P_{rot} = 1200$ W.

(a) Find the electromagnetic torque, power factor, efficiency and shaft torque at the load angle $\delta = 26°$;
(b) Redesign this motor into a six-phase motor keeping similar ratings.

<u>Solution</u>

This three-phase motor is very similar to that described in the paper [9].

(a) Three-phase motor

For $f = 180$ Hz and $n_s = 3600$ rpm $= 60$ rev/s, the number of pole pairs is $p = 180/60 = 3$ and the number of poles $2p = 6$. The pole pitch $\tau = \pi \times 0.33/6 = 0.1728$ m. The pole pitch expressed in number of slots is $72/6 = 12$. The phase voltage is $V_1 = 346/\sqrt{3} = 200$ V.

Synchronous reactances

$$X_{sd} = 2\pi f L_{sd} = 2\pi 180 \times 296 \times 10^{-6} = 0.3348 \ \Omega$$

$$X_{sq} = 2\pi f L_{sq} = 2\pi 180 \times 333 \times 10^{-6} = 0.3766 \ \Omega$$

The rotor magnetic flux according to eqn (5.6)

$$\Phi_f = \frac{2}{\pi} \times 0.1728 \times \times 0.19 \times 0.65 = 0.013586 \text{ Wb}$$

The form factor of the excitation field has been assumed $k_f \approx 1$.

For the number of slots per pole per phase $q_1 = s_1/(2pm_1) = 72/(6 \times 3) = 4$ the winding factor according to eqns (A.1) to (A.3) is

$$k_{w1} = k_{d1} k_{p1} = 0.9578 \times 0.966 = 0.925$$

where the distribution factor

$$k_{d1} = \frac{\sin[\pi/(2m_1)]}{q_1 \sin[\pi/(2m_1 q_1)]} = \frac{\sin[\pi/(2 \times 3)]}{4 \sin[\pi/(2 \times 3 \times 4)]} = 0.9578$$

and the pitch factor

$$k_{p1} = \sin\left(\frac{w_c}{\tau} \frac{\pi}{2}\right) = \sin\left(\frac{10}{12} \frac{\pi}{2}\right) = 0.966$$

The EMF excited by the rotor flux calculated on the basis of eqns (5.5)

$$E_f = \pi\sqrt{2} \times 180 \times 24 \times 0.925 \times 0.013586 = 241.2 \text{ V}$$

The armature currents for $\delta = 26^0$ ($\sin \delta = 0.4384$, $\cos \delta = 0.8988$) according to eqns (5.40) to (5.43)

$$I_{ad} = \frac{200(0.3766 \times 0.8988 - 0.011 \times 0.4384) - 241.2 \times 0.3766}{0.3348 \times 0.3766 + 0.011^2} = -191.6 \text{ A}$$

$$I_{aq} = \frac{200(0.011 \times 0.8988 + 0.3348 \times 0.4384) - 241.2 \times 0.011}{0.3348 \times 0.3766 + 0.011^2} = 226.9 \text{ A}$$

$$\Psi = \arctan \frac{I_{ad}}{I_{aq}} = \arctan \frac{-191.6}{226.9} = \arctan(-0.8444) = -40.18^0$$

$$I_a = \sqrt{(-191.6)^2 + 226.9^2} = 296.98 \text{ A}$$

The input active power according to eqn (5.44)

$$P_{in} = 3 \times 200[226.9 \times 0.8988 - (-191.6)0.4384] = 172,560.8 \text{ W} \approx 172.6 \text{ kW}$$

The input apparent power

$$S_{in} = 3 \times 200 \times 296.98 = 178,188 \text{ VA} \approx 178.2 \text{ kVA}$$

The power factor

$$\cos\phi = \frac{P_{in}}{S_{in}} = \frac{172,560.8}{178,188.0} = 0.97 \qquad \text{and} \qquad \phi = 14.1^0$$

The electromagnetic power according to eqn (5.45)

$$P_{elm} = 3[226.9 \times 241.2 + (-191.6)226.9(0.3348 - 0.3766)]$$

$$= 169,657.8 \text{ W} \approx 169.7 \text{ kW}$$

The developed torque according to eqns (5.19) and (5.46)

$$T_d = T_{syn} + T_{drel} = \frac{P_{elm}}{2\pi n_s} = m_1\frac{I_{aq}E_f}{2\pi n_s} + m_1\frac{I_{ad}I_{aq}}{2\pi n_s}(X_{sd} - X_{sq})$$

$$= 3\frac{226.9 \times 241.2}{2\pi \times 60} + 3\frac{(-191.6) \times 226.9}{2\pi \times 60}(0.3348 - 0.3766)$$

$$= 435.5 + 14.5 = 450 \text{ Nm}$$

The reluctance torque $T_{drel} = 14.5$ Nm is very small compared with the synchronous torque $T_{ds} = 435.5$ Nm.

The armature winding losses according to eqn (B.12)

$$\Delta P_a = 3 \times 296.98^2 \times 0.0011 = 2910.4 \text{ W}$$

The rotational loss is $\Delta P_{rot} = 1200$ W. The core and additional losses are equal to $0.02P_{out}$. Thus, the following equation can be written

$$P_{in} = P_{out} + \Delta P_{rot} + \Delta P_{str} + \Delta P_{Fe} + \Delta P_a$$

or

$$P_{in} = P_{out} + 0.02P_{out} + \Delta P_{rot} + \Delta P_a$$

The output power

$$P_{out} = \frac{1}{1.02}(P_{in} - \Delta P_{rot} - \Delta P_a)$$

$$= \frac{1}{1.02}(172,560.8 - 1200 - 2910.4) = 165,147.4 \text{ W} \approx 165.1 \text{ kW}$$

The efficiency

$$\eta = \frac{165,147.4}{172,560.8} = 0.957 \qquad \text{or} \qquad 95.7\%$$

The shaft torque

$$T_{sh} = \frac{P_{out}}{2\pi n_s} = \frac{165,147.4}{2\pi 60} = 438.1 \text{ Nm}$$

(a) Six-phase motor

The number of slots per pole per phase will be twice as small, i.e.,

$$q_1 = \frac{72}{6 \times 6} = 2$$

For the same coil pitch ($w_c/\tau = 10/12$ or 10 slots) the pitch factor remains the same. i.e., $k_{p1} = 0.966$ while the distribution factor on the basis of eqn (A.6) will increase, i.e.,

$$k_{d1} = \frac{\sin[\pi/(2 \times 6)]}{2\sin[\pi/(2 \times 6 \times 2)]} = 0.9916$$

The winding factor according to eqn (A.1)

$$k_{w1} = 0.9916 \times 0.966 = 0.958$$

The number of turns per phase should be the same, i.e., $N_1 = 24$. This is because the air gap magnetic flux density must remain the same, i.e., $B_{mg} = 0.65$ T to obtain the same torque as that developed by the three-phase motor. The magnetic flux excited by the rotor is the same, i.e., $\Phi_f = 0.013586$ Wb.

The number of coils for a single-layer winding is equal to half of the number of slots, i.e., $c_1 = s_1/2 = 36$. The number of coils per phase will be 6. If there are two parallel conductors $a_w = 2$, the number of conductors per coil is

$$N_{1c} = \frac{a_w N_1}{(c_1/m_1)} = \frac{2 \times 24}{36/6} = 8$$

In the previous three-phase motor with 12 coils per phase the number of parallel conductors should have been $a_w = 4$ so that $N_{1c} = 8$, since the armature current was approximately twice as high. In the six-phase motor the cross section of armature conductor is twice as small because there must be space for twice as many conductors in each coil to accommodate the coils in the same slots.

The armature resistance will remain approximately the same as the number of coils in which the same span per phase has been reduced twice, i.e., the length of conductor has halved and the cross section of the conductor has also halved. This means that $R_1 \approx 0.011 \ \Omega$.

The synchronous reactances of the six-phase motor can be estimated on the assumption that the armature reaction reactances are predominant. Analyzing eqns (5.31) and (5.33), only the number of phases and winding factor will change. The saturation factor of the armature circuit is approximately the same because the magnetic flux density has not been changed. The synchronous reactance $X_s^{(6)}$ of the six phase motor will increase to

$$X_s^{(6)} \approx \frac{6}{3}\left(\frac{0.958}{0.925}\right)^2 X_s^{(3)} = 2.145 X_s^{(3)}$$

Thus

$$X_{sd} = 2.145 \times 0.3348 = 0.7182 \ \Omega \quad \text{and} \quad X_{sq} = 2.145 \times 0.3766 = 0.8078 \ \Omega$$

The phase voltage should be kept the same, i.e., $V_1 = 200$ V. For the star connection, the line voltage according to eqn (8.3) is

$$V_{1L} = 2V_1 \sin\left(\frac{360^0}{2m_1}\right) = 2 \times 200 \times \sin\left(\frac{360^0}{2 \times 6}\right) = 200 \text{ V}$$

For a six-phase system and star connection, the line voltage is equal to the phase voltage. Similarly, for a six phase system and a polygon connection, the line current is equal to the phase current.

The EMF induced by the rotor excitation flux, given by eqn (5.5), will increase due to an increase in the winding factor, i.e.,

$$E_f = \pi\sqrt{2}180 \times 24 \times 0.958 \times 0.013586 = 249.8 \text{ V}$$

For the same load angle $\delta = 26^0$, the new armature currents are $I_{ad} = -99.15$ A, $I_{aq} = 107.18$ A, and $I_a = 141.6$. The armature current has decreased by more than half. The angle between the q-axis and the armature current I_a is $\Psi = -42.8^0$.

The input active power absorbed by the six-phase motor is $P_{in} = 167,761.4$ W, the input apparent power is $S_{in} = 175,212$ VA and the power factor is $\cos\phi = 0.9575$ ($\phi = 16.7^0$). There is a slight decrease in the power factor as compared with the three-phase motor.

The electromagnetic power is $P_{elm} = 166,354.4$ W, the developed torque is $T_d = 441.3$ Nm ($T_d = 450$ Nm for the three-phase motor). The synchronous torque is $T_{ds} = 426.1$ Nm and the reluctance torque is $T_{drel} = 15.15$ Nm. Although the number of phases has doubled, the armature winding losses are halved since the armature current has been reduced more than twice, i.e.,

$$\Delta P_a = 6 \times 146.01^2 \times 0.011 = 1407.1 \text{ W}$$

The output power

$$P_{out} = \frac{1}{1.02}(167,761.4 - 1200 - 1407.1) = 161,916.1 \text{ W}$$

The efficiency

$$\eta = \frac{161,916.1}{167,761.4} = 0.9652 \qquad \text{or} \qquad 96.52\%$$

The increase in the efficiency is only 0.86% as compared with the three-phase motor. The shaft torque

$$T_{sh} = \frac{161,916.1}{2\pi \times 60} = 429.5 \text{ Nm}$$

Table 8.10. Design data of a single-sided three-phase TFM with external rotor

Number of stator phases	$m_1 = 3$
Rated speed, rpm	480
Rated armature current, A	$I_a = 22.0$
Number of pole pairs (U-shaped or I-shaped cores per phase)	$p = 18$
Number of rotor poles	$2p = 36$
Average air gap diameter, m	$D_g = 0.207$
Air gap in the d-axis, mm	$g = 0.8$
Stator pole shoe thickness (circumferential), m	$b_p = 0.012$
Axial width of U-shaped core, mm	$w_u = 52$
Axial width of I-shaped core, mm	$w_u = 52$
Height of U-shaped core, mm	$h_u = 52$
Axial width of U-shaped core pole, mm	$l_p = 15$
Width of leg and height of yoke of U-shaped core, mm	$a_u = 15$
Height of I-shaped core, mm	15
I-shaped core thickness (circumferential), mm	12
Number of turns per phase	$N_1 = 70$
Number of parallel conductors	$a_w = 1$
Conductivity of stator wire at 20^0C, S/m	$\sigma_1 = 57 \times 10^6$
Height of the stator wire, mm	$h_c = 2.304$
Width of the stator wire, mm	$w_c = 2.304$
Height of PM, mm	$h_M = 4.0$
Width of PM (circumferential direction), mm	$w_M = 17.0$
Length of PM (axial direction), m	$l_M = 17.0$
Height of the rotor laminated yoke, mm	$h_{ry} = 8.0$
Remanent magnetic flux density, T	$B_r = 1.25$
Coercive force, A/m	$H_c = 923 \times 10^3$
Rotational loses at 480 rpm, W	$\Delta P_{rot} = 200$
Stacking coefficient	$k_i = 0.95$
Temperature of stator windings, ^{0}C	75
Temperature of rotor, ^{0}C	40

Numerical example 8.4

The design data of a single-sided three phase TFM with external rotor and surface PMs are given in Table 8.10. The stator windings are Y-connected and the stator magnetic circuit is composed of U-shaped cores with I-shaped magnetic flux shunts between them (Fig. 8.19a). Both U-shaped and I-shaped cores are made of cold-rolled electrotechnical steel. The rotor PMs are glued to laminated rings (two rings per phase) of the same axial width as PMs. Find the steady state performance of the TFM for $I_{ad} = 0$. The saturation of the magnetic circuit and influence of the armature reaction flux on the rotor excitation flux can be neglected.

Solution

The average pole pitch

$$\tau = \frac{\pi D_g}{2p} = \frac{\pi \times 207}{36} = 18.06 \text{ mm}$$

The motor external diameter without the rotor external cylinder (supporting structure)

$$D_{out} = D_g + g + 2(h_M + h_{ry}) = 207 + 0.8 + 2(4 + 8) = 231.8 \text{ mm}$$

The length of the motor without the rotor bearing disks

$$L = m_1 w_u + (0.6 \dots 1.0 l_p)(m_1 - 1) = 3 \times 52 + 0.8 \times 15 \times 2 = 180 \text{ mm}$$

The air gap magnetic flux density at no load for unsaturated magnetic circuit ($k_{sat} = 1$) can be found on the basis of eqn (8.44), i.e.,

$$B_{mg} = \frac{1.25}{1 + 1.078 \times 0.8/4.0} = 1.0283 \text{ T}$$

where

$$\mu_{rrec} = \frac{1}{\mu_0} \frac{B_r}{H_c} = \frac{1}{0.4\pi \times 10^{-6}} 1.25923 \times 10^3 = 1.078$$

The magnetic flux densities in the U-shaped and I-shaped cores are approximately the same. These values secure low core losses at increased frequency. The magnetic flux density in the rotor laminated ring

$$B_{ry} = \frac{1}{2} B_{mg} \frac{w_M}{h_{ry}} = \frac{1}{2} 1.0283 \frac{17}{8} = 1.0925 \text{ T}$$

Armature winding current density

$$J_a = \frac{I_a}{a_w h_c w_c} = \frac{22}{1 \times 2.304 \times 2.304} = 4.14 \text{ A/mm}^2$$

Winding packing factor (fill factor)

$$\frac{N_1 a_w h_c w_c}{h_w b_u} = \frac{70 \times 1 \times 2.304 \times 2.304}{22 \times 22} = 0.768$$

where

$$h_w = h_u - 2a_u = 52 - 2 \times 15 = 22 \text{ mm}$$

$$b_u = w_u - 2l_p = 522 \times 15 = 22 \text{ mm}$$

The winding packing (space) factor, depending on the insulation thickness, for low voltage TFMs and square wires is approximately 0.75 to 0.85.

Reluctance of the air gap

$$R_g = \frac{g}{\mu_0 b_p l_p} = \frac{0.0008}{0.4\pi \times 10^{-6} \times 0.015 \times 0.012} = 3.5368 \times 10^6 \; 1/\mathrm{H}$$

Reluctance of the PM

$$R_{PM} \approx \frac{h_M}{\mu_0 \mu_{rrec} b_p l_p} = \frac{0.004}{0.4\pi \times 10^{-6} \times 1.078 \times 0.015 \times 0.012} = 16.4 \times 10^6 \; 1/\mathrm{H}$$

Reluctance per pole pair with ferromagnetic laminations being neglected

$$R_{pole} = 4(R_g + R_{PM}) = 4(3.5368 + 16.4)10^{-6} = 79.745 \times 10^6 \; 1/\mathrm{H}$$

Armature reaction (mutual) inductance in the d-axis calculated on the basis of reluctance

$$L_{ad} = pN_1^2 \frac{1}{R_{pole}} = 18 \times 70^2 \frac{1}{79.745 \times 10^6} = 0.001106 \; \mathrm{H}$$

Line current density

$$A = \frac{I_a N_1}{2\tau} = \frac{22 \times 70}{2 \times 0.01806} = 42,635.6 \; \mathrm{A/m}$$

Peak value of line current density

$$A_m = \sqrt{2}A = \sqrt{2}42,635.6 = 60,295.9 \; \mathrm{A/m}$$

Tangential force density (shear stress)

$$f_{sh} = \alpha_{PM} B_{mg} A = 0.623 \times 1.0283 \times 42,635.6 = 27,313.7 \; \mathrm{N/m^2}$$

where the PM coverage coefficient

$$\alpha_{PM} = \frac{b_p l_p}{w_M l_M} = \frac{12 \times 15}{17 \times 17} = 0.623$$

Pole shoe-to-pole pitch ratio

$$\alpha_i = \frac{b_p}{\tau} = \frac{12.0}{18.06} = 0.664$$

Form factor of the excitation field according to eqn (5.23)

$$k_f = \frac{4}{\pi} \sin \frac{0.664 \times \pi}{2} \approx 1.1$$

Peak value of the fundamental harmonic of the air gap magnetic flux density

$$B_{mg1} = k_f B_{mg} = 1.1 \times 1.0283 = 1.131 \text{ T}$$

Fundamental harmonic of the magnetic flux according to eqn (8.19)

$$\Phi_{f1} = \frac{2}{\pi} 0.01806 \times 0.015 \times 1.131 = 1.952 \times 10^{-4} \text{ Wb}$$

No-load EMF at rated speed $n_s = 480/60 = 8$ rev/s according to eqn (8.21)

$$E_f = \pi\sqrt{2} \times 70 \times 18^2 \times 1.952 \times 10^{-4} \times 8 = 157.3 \text{ V}$$

Coefficient of the slot leakage permeance according to eqn (8.38)

$$\lambda_{1s} = \frac{22}{3 \times 22} + \frac{15}{22} = 1.015$$

where

$$h_o = 52 - 22 - 15 = 15 \text{ mm}$$

Coefficient of pole-top leakage permeance according to eqn (8.39)

$$\lambda_{1p} = \frac{5 \times (0.8/22)}{5 + 4 \times (0.8/22)} = 0.0353$$

Leakage inductance according to eqn (8.37) multiplied by 3 (correction factor)

$$L_1 \approx 3 \times 0.4\pi \times 10^{-6} \times \pi [0.207 - 0.0008 - (0.022 + 0.015)] \times 70^2 \times (1.015 + 0.0353)$$

$$= 0.010313 \text{ H}$$

Equivalent air gap according to eqn (8.33)

$$g' = 4 \left(0.0008 + \frac{0.004}{1.078} \right) = 0.01804 \text{ mm}$$

Armature reaction (mutual) inductance in the d-axis according to eqn (8.35) for $k_{fd} = k_{fq} = 1$

$$L_{ad} = \frac{2}{\pi} 0.4\pi \times 10^{-6} \times 70^2 \times 18 \frac{0.01806 \times 0.015}{0.01804} 1.0 = 0.0010596 \text{ H}$$

is almost the same as that calculated on the basis of reluctances. For surface PMs, the armature reaction inductance in the q-axis $L_{aq} = L_{ad} = 0.0010596$ H.

Synchronous inductances

$$L_{sd} = L_1 + L_{ad} = 0.010313 + 0.0010596 = 0.011372 \text{ H}$$

$$L_{sq} = L_1 + L_{aq} = 0.010313 + 0.0010596 = 0.011372 \text{ H}$$

Synchronous reactances

$$X_{sd} = 2\pi f L_{sd} = 2\pi \times 144 \times 0.011372 = 10.29 \ \Omega \quad \text{and} \quad X_{sq} = X_{sd} = 10.29 \ \Omega$$

where the input frequency

$$f = n_s p = \frac{480}{60} 18 = 144 \text{ Hz}$$

Electromagnetic power at $I_{ad} = 0$

$$P_{elm} = m_1 E_f I_a = 3 \times 157.3 \times 22 = 10,381.8 \text{ W}$$

Electromagnetic torque at $I_{ad} = 0$

$$T_d = \frac{10381.8}{2\pi \times 8} = 206.54 \text{ Nm}$$

where $n_s = 480$ rpm $= 8$ rev/s. EMF constant

$$k_E = \frac{E_f}{n_s} = \frac{157.3}{480} = 0.3277 \text{ V/rpm}$$

Torque constant at $I_{ad} = 0$ ($\cos \Psi = 1$) according to eqn (8.27)

$$k_T = \frac{3}{2\pi} \times (0.3277 \times 60) \times 1 = 9.39 \text{ Nm/A}$$

Average length of turn for armature system with I-shaped shunts

$$l_{1av} \approx \pi(D_g - h_u) = \pi(0.207 - 0.052) = 0.487 \text{ m}$$

Conductivity of copper wire at 75°C

$$\sigma_1 = \frac{57 \times 10^6}{1 + 0.00393(75 - 20)} = 46.87 \times 10^6 \text{ S/m}$$

Reduced thickness of conductor and Field's functions for resistance according to eqns (B.5), (B.6) and (B.7)

$$\xi_1 \approx h_c \sqrt{\pi \times f \times \mu_0 \times \sigma_1} = 0.002304 \sqrt{\pi \times 144 \times 0.4 \times \pi \times 10^{-6} \times 46.87 \times 10^6}$$

$$= 0.002304 \times 163.23 = 0.376$$

$$\varphi_1(\xi_1) = 0.376 \frac{\sinh(2 \times 0.376) + \sin(2 \times 0.376)}{\cosh(2 \times 0.376) - \cos(2 \times 0.376)} = 1.00178$$

$$\Psi_1(\xi_1) = 2 \times 0.376 \frac{\sinh(0.376) - \sin(0.376)}{\cosh(0.376) + \cos(0.376)} = 0.00666$$

Number of conductor layers

$$m_{sl} \approx \frac{h_u - 2a_u - 0.003}{a_w h_c} = \frac{0.052 - 2 \times 0.015 - 0.003}{1 \times 0.002304} = 8$$

Skin effect coefficient for the armature resistance according to eqn (B.10)

$$k_{1R} \approx 1.00178 + \frac{8^2 - 1}{3} 0.00666 = 1.1416$$

Armature winding resistance

$$R_1 = k_{1R} \frac{N_1 l_{1av}}{a_w \sigma_1 h_c w_c} = 1.1416 \frac{70 \times 0.487}{1 \times 46.87 \times 10^6 \times 0.2304 \times 0.2304} = 0.1564 \; \Omega$$

The increase of the resistance due to the skin effect is approximately 4%. Since $\xi_1 < 1$, the skin effect has practically no influence on the slot leakage inductance.

Stator winding losses

$$\Delta P_a = 3 \times 0.1564 \times 22^2 = 227.1 \; \text{W}$$

Mass of the stator U-shape cores per phase

$$m_U = 7700 p b_p k_i [w_u h_u - (h_u - a_u) b_u]$$

$$= 7700 \times 18 \times 0.012 \times 0.95 [0.052 \times 0.052 - (0.052 - 0.015) \times 0.022] = 2.986 \; \text{kg}$$

Mass of the stator I-shaped cores

$$m_I = 7700 p b_p k_i w_u a_u = 7700 \times 18 \times 0.012 \times 0.95 \times 0.052 \times 0.015 = 1.232 \; \text{kg}$$

The magnetic flux density in the U-shaped core is $B_{1U} \approx B_{mg} = 1.0283$ T (uniform cross section). The corresponding specific core losses at 1.0283 T and 50 Hz are $\Delta p_{FeU} = 2.55$ W/kg. The magnetic flux density in the I-shaped core has been assumed $B_{FeI} \approx 1.1 B_{mg} = 1.131$ T (10% increase due to grooves for bandages). The corresponding specific core losses at 1.131 T and 50 Hz are $\Delta p_{FeI} = 3.25$ W/kg. The stator (armature) core losses (Appendix B, eqn (B.19))

$$\Delta P_{Fe} = m_1 (\Delta p_{FeU} m_U + \Delta p_{FeI} m_I) \left(\frac{f}{50} \right)^{1.333}$$

$$= 3(2.55 \times 2.986 + 3.25 \times 1.232) \left(\frac{144}{50}\right)^{1.333} = 206.5 \text{ W}$$

Additional losses

$$\Delta P_{str} \approx 0.0125 P_{elm} = 0.0125 \times 10,381.8 = 129.8 \text{ W}$$

Output power

$$P_{out} = P_{elm} - \Delta P_{rot} - \Delta P_{str} = 10,381.8 - 200 - 129.8 = 10,052.0 \text{ W}$$

Shaft torque

$$T_{sh} = \frac{P_{out}}{2\pi n_s} = \frac{10,052.0}{2\pi \times 8} = 200 \text{ Nm}$$

Input power

$$P_{in} = P_{elm} + \Delta P_a + \Delta P_{Fe} = 10,381.8 + 227.1 + 206.5 = 10,815.4 \text{ W}$$

Efficiency

$$\eta = \frac{10.052.0}{10.815.4} = 0.929$$

Phase input voltage ($I_a = I_{aq}$, $\psi = 0^0$) according to eqn (5.48)

$$V_1 = \sqrt{(E_f + I_a R_1)^2 + (I_a X_{sq})^2}$$

$$= \sqrt{(157.3 + 22.0 \times 0.1564)^1 + (22.0 \times 10.29)^2} = 277.8 \text{ V}$$

The line-to-line voltage is $V_{1L-L} = \sqrt{3} \times 277.8 = 481$ V.
Power factor according to eqn (5.47)

$$\cos\phi \approx \frac{E_f + I_a R_1}{V_1} = \frac{157.3 + 22.0 \times 0.1564}{277.8} = 0.579$$

Mass of rotor yokes per phase

$$m_y = 2 \times 7700\pi (D_g + g + 2h_M + h_{ry}) h_{ry} l_M$$

$$= 2 \times 7700\pi (0.207 + 0.0008 + 2 \times 0.004 + 0.008) \times 0.008 \times 0.017 = 1.472 \text{ kg}$$

Mass of PMs

$$m_{PM} = 7500 N_{PM} h_M l_M w_M = 7500 \times 72 \times 0.004 \times 0.017 \times 0.017 = 0.624 \text{ kg}$$

where the number of PMs per phase $N_{PM} = 2 \times (2p) = 2 \times 36 = 72$. Mass of armature winding per phase

$$m_a = 8200 N_1 l_{1av} a_w h_c w_c$$

$$= 8200 \times 70 \times 0.487 \times 1 \times 0.002304 \times 0.002304 = 1.484 \text{ kg}$$

Mass of active materials per machine

$$m = m_1(m_U + m_I + m_a + m_{PM} + m_y)$$

$$= 3(2.986 + 1.232 + 1.484 + 0.624 + 1.472) = 23.39 \text{ kg}$$

Power and torque density

$$\frac{P_{out}}{m} = \frac{10,052}{23.39} = 429.7 \text{ W/kg} \qquad \frac{T_{sh}}{m} = \frac{200}{23.39} = 8.55 \text{ Nm/kg}$$

The steady state performance characteristics at $n_s = 480$ rpm are shown in Fig. 8.31. All values have been computed in the same way as those for the rated current.

If the negative component of d-axis current is injected, the input voltage decreases and the power factor increases. For example, at $\psi = -10^0$

$$I_{ad} = I_a \sin \psi = 22.0 \sin(-10^0) = -3.82 \text{ A}$$

$$I_{aq} = I_a \cos \psi = 22.0 \cos(-10^0) = 21.67 \text{ A}$$

Input voltage per phase obtained on the basis of phasor diagram

$$V_1 \approx \sqrt{(E_f + I_{ad}X_{sd})^2 + (I_{aq}X_{sq} - I_{ad}R_1)^2}$$

$$= \sqrt{(157.3 - 3.82 \times 10.29)^2 + (21.67 \times 10.29 + 3.82 \times 0.1564)^2} = 252.8 \text{ V}$$

The line-to-line voltage at $\psi = -10^0$ is $V_{1L-L} = \sqrt{3} \times 252.8 = 437.9$ V. The load angle

$$\delta = \sin^{-1} \frac{I_{aq}X_{sq} - I_{ad}R_1}{V_1} = \sin^{-1} \frac{21.67 \times 10.29 + 3.82 \times 0.1564}{252.8} = 62.18^0$$

Power factor

$$\cos \phi = \cos(\delta + \psi) = \cos(62.18^0 - 10^0) = 0.613$$

(a)

(b)

(c)

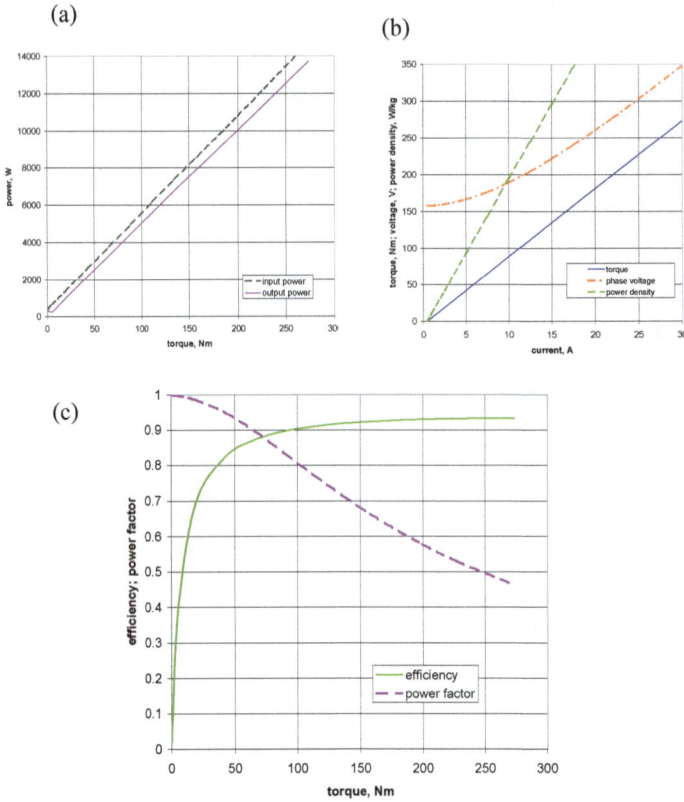

Fig. 8.31. Steady state characteristics at $n_s = 480$ rpm: (a) input and output power versus torque, (b) torque and phase voltage versus current, (c) efficiency and power factor versus torque. *Numerical example 8.4.*

Electromagnetic power for $X_{sd} = X_{sq}$ according to eqn (5.45)

$$P_{elm} = m_1 I_{aq} E_f = 3 \times 21.67 \times 157.3 = 10,226 \text{ W}$$

Input power

$$P_{in} = P_{elm} + \Delta P_{Fe} + \Delta P_a = 10,226.0 + 206.5 + 227.1 = 10,659.7 \text{ W}$$

Output power

$$P_{out} = P_{elm} - \Delta P_{rot} - \Delta P_{str} = 10,226.0 - 200.0 - 129.8 = 9896.2 \text{ W}$$

Shaft torque

$$T_{sh} = \frac{9896.2}{2\pi \times 8} = 196.9 \text{ Nm}$$

Efficiency

$$\eta = \frac{9896.2}{10,659.7} = 0.928$$

Although the power factor increases, both the shaft torque and efficiency slightly decrease.

9

High Speed Motors

High speed motors that develop *rotational speeds in excess of* 5000 *rpm* are necessary for centrifugal and screw compressors, grinding machines, mixers, pumps, machine tools, textile machines, drills, handpieces, aerospace, flywheel energy storages, etc. The actual trend in high speed electromechanical drives technology is to use PM brushless motors, solid rotor induction motors or switched reluctance motors. The highest efficiency and highest power density is achieved with PM brushless motors.

9.1 Why high speed motors?

The output equation (5.67) for a PM synchronous machine (Chapter 5) indicates that the developed power is basically proportional to the speed and the volume of the rotor, i.e., the output coefficient (5.70) itself is dependent upon both the rotor geometry and speed. Thus, eqn (5.70) can be rewritten to obtain the output power per rotor volume of a synchronous machine proportional to the synchronous speed n_s, peak value of the air gap magnetic flux density B_{mg} and peak value of the line current density A_m of the armature winding, i.e.,

$$\frac{P_{out}}{\pi D_{1in}^2 L_i} = \frac{0.5}{\epsilon} k_{w1} n_s B_{mg} A_m \eta \cos\phi \propto n_s B_{mg} A_m \qquad (9.1)$$

The higher the speed n_s the higher the power density of the machine. Increase in speed n_s (frequency) at the same output power P_{out} results in smaller volume and mass of the machine. Eqn (9.1) also says that the power density can be increased by applying high magnetic flux density B_{mg} in the air gap and high line current density A_m of the armature winding. High magnetic flux density B_{mg} can be achieved by using magnetic materials with high saturation magnetic flux density, e.g., cobalt alloys. High line current density A_m can be achieved by using intensive cooling systems, e.g., liquid cooling system.

Iron-cobalt (Fe-Co-V) alloys with Co contents ranging from 15 to 50% have the highest known saturation magnetic flux density, up to 2.4 T at room temperature. Fe-Co-V alloys are the natural choice for applications such as aerospace, where mass and space saving are of prime importance. Additionally, the Fe-Co-V alloys have the highest Curie temperature of any ferromagnetic alloy family and have found use in elevated temperature applications. The nominal composition, e.g., for Hiperco 50 from Carpenter, PA, U.S.A. is 49% Fe, 48.75% Co, 1.9% V, 0.05% Mn, 0.05% Nb and 0.05% Si. Similar to Hyperco 50 is Vacoflux 50 (50% Co) and Vacodur 50 cobalt-iron alloy from Vacuumschmelze, Hanau, Germany [295].

9.2 Mechanical requirements

The rotor diameter is limited by the *bursting stress* at the design speed. The rotor axial length is limited by its *stiffness* and the *first critical (whirling) speed*.

Since the centrifugal force acting on a rotating mass is proportional to the linear velocity squared and inversely proportional to its radius of rotation, the rotor must be designed with a small diameter and must have a very high mechanical integrity. The *surface linear speed* (tip speed) of the rotor

$$v = \pi(D_{1in} - 2g)n_s \qquad (9.2)$$

is an engineering measure of mechanical stresses on the rotor under action of the centrifugal forces. The maximum permissible surface linear speed depends on the rotor construction and materials.

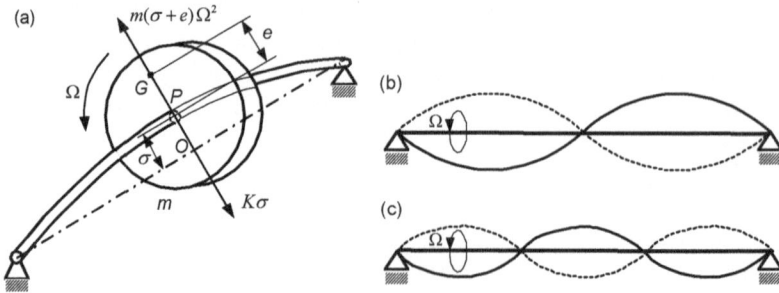

Fig. 9.1. Single mass flexible rotor with residual unbalance and possible modes of oscillations: (a) 1st mode; (b) 2nd mode; (c) 3rd mode. O – center of rotation, G – center of gravity, P – geometric center.

When the shaft rotates, centrifugal force will cause it to bend out. For a single rotating mass m (Fig. 9.1a) the first critical (whirling) rotational speed and first critical angular speed according to eqn (1.10), in which $i = 1$, are respectively

$$n_{cr} = \frac{1}{2\pi}\sqrt{\frac{K}{m}} \qquad\qquad \Omega_{cr} = \sqrt{\frac{K}{m}} \qquad (9.3)$$

The static deflection for $a_i = 0.5L$ according to eqn (1.11) takes the form

$$\sigma = \frac{mgL^3}{48EI} = \frac{mg}{K} \qquad (9.4)$$

where the stiffness

$$K = 48\frac{EI}{L^3} \qquad (9.5)$$

area moment of inertia I is given by eqn (1.12) for $i = 1$, EI is the bending stiffness and L is the bearing span (Fig. 9.1a). Distinguishing between the rotor stack (E, I, L_i) and shaft (E_{sh}, I_{sh}, L), the stiffness of the shaft with rotor stack is

$$K = 48\frac{EI}{L_i^3} + 48\frac{E_{sh}I_{sh}}{L^3} \qquad (9.6)$$

The modulus of elasticity of the laminated stack E is from 1 to 20% of the modulus of elasticity E_{sh} of the steel shaft. The stronger the clamping of laminations, the higher the modulus E.

Neglecting damping, the centrifugal force is $m\Omega^2(\sigma + e)$ and the restoring force (deflection force) is $K\sigma$, in which σ is the shaft deflection, e is imbalance distance (eccentricity) and $\sigma + e$ is the distance from the center of rotation to the center of gravity. From the force balance equation

$$K\sigma = m\Omega^2(\sigma + e) \qquad (9.7)$$

the deflection of shaft can be found as

$$\sigma = \frac{m\Omega^2 e}{K(1 - m\Omega^2/K)^2} = \frac{e}{(\Omega_{cr}/\Omega)^2 - 1} \qquad (9.8)$$

The shaft deflection $\sigma \to \infty$ if $\Omega = \Omega_{cr}$. No matter how small the imbalance distance e is, the shaft will whirl at the natural frequency. The mass rotates about the center of rotation O if $\Omega < \Omega_{cr}$. Point O and G are opposite each other. The mass rotates about the center of gravity G if $\Omega > \Omega_{cr}$. Point O approaches point G.

It is recommended that the synchronous (rated) speed of the motor should meet the following conditions [77]

- if $n_s < n_{cr}$ then

$$n_s > 0.75\frac{n_{cr}}{2p} \qquad \text{or} \qquad n_s < 1.33\frac{n_{cr}}{2p} \qquad (9.9)$$

- if $n > n_{cr}$ then

$$n_s > 1.33 n_{cr} \tag{9.10}$$

With the radial magnetic pull being included, the first critical speed is [110]

$$n_{cr} = \frac{1}{2\pi} \sqrt{\frac{K - K_e}{m}} \tag{9.11}$$

where K_e is the negative spring coefficient (stiffness) induced by the electromagnetic field (magnetic pull). This coefficient is given e.g., in [110].

9.3 Construction of high speed PM brushless motors

Design guidelines for high speed PM brushless motors include:

- compact design and high power density;
- minimum number of components;
- high efficiency over the whole range of variable speed;
- power factor close to unity over the whole range of speed and load;
- ability of the PM rotor to withstand high temperature;
- optimal *cost–to–efficiency* ratio to minimize the system *cost–output power* ratio;
- high reliability (failure rate less than 5% within 80,000 h);
- low total harmonics distortion (THD).

Given below are the fundamental issues, which are essential in electromagnetic, mechanical and thermal design of high speed PM brushless motors:

- Volume and mass: the higher the speed, the higher the power density;
- Power losses and efficiency: special attention must be given to windage and core losses;
- Laminations: cobalt alloy, non-oriented silicon steel or amorphous alloy laminations;
- Stator conductors: small diameter stranded conductors or Litz wires;
- Higher harmonics generated by the solid state converter: parasitical effect as losses, vibration and noise depend on the harmonic content;
- Cooling system: intensive air or oil cooling system;
- PM excitation system: if the temperature of the hot end of the shaft exceeds 150°C, SmCo PMs must be used;
- Rotor tensile hoop stresses: properly selected rotor diameter, *rotor diameter–to–length* ratio and rotor retaining sleeve (material and thickness);
- Thermal compatibility of rotor materials: thermal expansions of the rotor retaining sleeve and rotor core produce compressing stresses on PMs, fluctuating with the temperature;

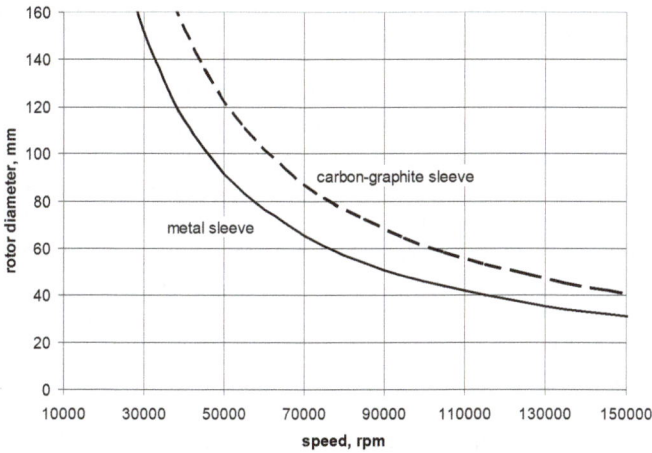

Fig. 9.2. Maximum rotor outer diameters for metal and carbon-graphite sleeves as functions of design speeds.

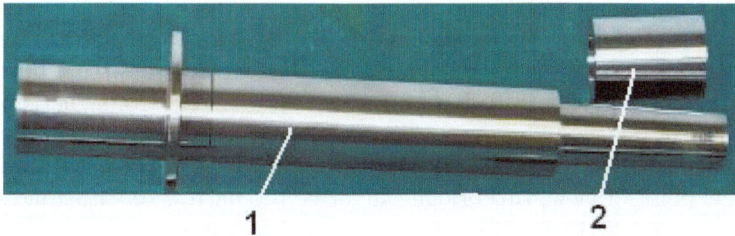

Fig. 9.3. Rotor of a 100 kW, 70,000 rpm PM brushless motor for an oil free compressor. 1 — PM rotor with a retaining sleeve, 2 — foil bearing journal sleeve. Photo courtesy of Mohawk Innovative Technology, Albany, NY, U.S.A.

Table 9.1. Characteristics of retaining sleeves for high speed PM brushless machines

Quantity	non-magnetic metal sleeve	non-metalic wound sleeve
Material	titanium alloys stainless steels Inconel 718 (NiCoCr based alloy)	carbon-graphite carbon fiber glass fiber
Maximum temperature, °C	290	180
Maximum surface linear speed, m/s	240	320

- Rotor dynamics: the first critical speed of the rotor should be much higher or much lower than the rated speed.

The stator core is stacked of slotted or slotless laminations. For input frequencies 400 Hz and lower, 0.2-mm thick laminations are used. For higher frequencies, 0.1-mm laminations are necessary. Vacuum impregnated coils made of stranded conductors are inserted into slots. To minimize the space harmonics, the stator winding is made as a double layer winding with shorted coils. For very high speeds and low voltages, when the EMF induced in single turn stator coils is too high, small number of coils, single layer winding or parallel paths must be used. Hollow conductors and direct water cooling are too expensive for machines rated below 200 kW. The stator volume is affected by winding losses and heat dissipation.

PM rotor designs include bread loaf (Fig. 5.1g), surface-type (Fig. 5.1c, 5.1h), inset-type (Fig. 5.1d) or interior-type rotors (Fig. 5.1b, 5.1i, 5.1j). All surface-type PM rotors are characterized by minimal leakage flux. Bread loaf surface-type PM rotors provide, in addition, the highest magnetic flux density in the air gap (large volume of PM material). All surface-type, including bread loaf and inset-type PM rotors, can be used only with an external rotor *retaining sleeve* (can). In the case of an interior-type PM rotors the retaining sleeve is not necessary, but the ferromagnetic bridge in the rotor core between neighboring PMs must be very carefully sized. From an electromagnetic point of view, this bridge should be very narrow to obtain full saturation, preventing the circulation of leakage flux between neighboring rotor poles. From a mechanical point of view, this bridge cannot be too narrow to withstand high mechanical stresses. In practice, interior-type PM rotors without retaining sleeves can be used at speeds not exceeding 6000 rpm.

Good materials for retaining sleeves are nonferromagnetic and have high permissible stresses, low specific density and good thermal conductivity. If the magnetic saturation effect is used effectively, a thin steel sleeve in low power machines can sometimes be better than a sleeve made of nonferromagnetic material. Typical materials, maximum operating temperatures and maximum surface linear speeds are given in Table 9.1. Maximum rotor outer diameters for metal and carbon-graphite sleeves as functions of design speeds are plotted in Fig. 9.2. A PM rotor with a metal retaining sleeve for a 110 kW, 70,000 rpm brushless motor is shown in Fig. 9.3. Reinforced plastics and brass can also be used for retaining sleeves.

To increase the electromagnetic coupling between the magnets and the stator, the air gap should be made as small as mechanically possible. However, the use of a small air gap increases the tooth ripple losses in the retaining sleeve, if the sleeve is made of current-conducting material.

To minimize the losses in the retaining sleeve and PMs, torque ripple and acoustic noise, the stator slots should have very narrow slot openings or be closed (Fig. 6.16c). In the case of closed stator slots, the slot closing bridge should be highly saturated under normal operating conditions.

Active radial and axial magnetic bearings (Fig. 1.14) or air bearings are frequently used. High speed PM brushless motors integrated with magnetic bearings and solid state devices are used in gas compressors providing a true oil free system, reduced maintenance and high efficiency.

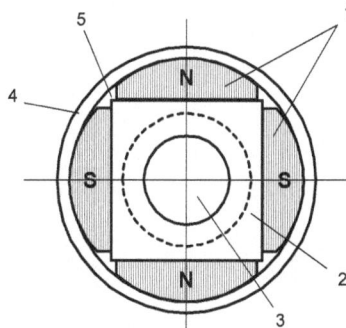

Fig. 9.4. Typical rotor of a four-pole high speed PM brushless motor with bread-loaf PMs and retaining sleeve: 1 — PM, 2 — rotor ferromagnetic core, 3 — shaft, 4 — nonmagnetic retaining sleeve, 5 — nonmagnetic material.

Typical rotor of a high speed PM brushless motor has bread loaf PMs secured by a metal or carbon-graphite retaining sleeve. A four-pole rotor with bread loaf PM is shown in Fig. 9.4. The 2D magnetic field distribution in the cross section of a four-pole PM brushless motor is plotted in Fig. 9.5. A six-pole rotor is shown in Fig. 9.6.

A PM rotor with retaining sleeve proposed by SatCon Technology Corp., Cambridge, MA, U.S.A., is shown in Fig. 9.7. The rotor is divided into segments. Within each segment a retaining sleeve holds the plastic bonded NdFeB PM magnetized radially. Although bonded NdFeB exhibits only half the remanent magnetic flux density of the sintered NdFeB, the lower electric conductivity limits the rotor eddy current losses at high speeds. A high speed motor developed by SatCon for centrifugal compressors rated at 21 kW, 47,000 rpm, 1567 Hz has the rotor diameter 46 mm, efficiency 93 to 95% and power factor $\cos \phi \approx 0.91$ [176]. The predicted reliability is 30,000 h lifetime and cost 13.4 $/kW (10 $/hp) for rotor and stator set (assembly separate) [176].

Specifications of small two-pole high speed motors with slotless windings manufactured by Koford Engineering, Winchester, OH, U.S.A. are given in Table 9.2. Slotless design reduces tooth ripple losses in the metal retaining sleeve and PMs and allows for reducing the stator outer diameter. These motors are designed either with Hall sensors or for sensorless controllers. High speed slotless motors listed in Table 9.2 can be used in laboratory pumps, aerospace applications, unmanned aircrafts, military robots, handpieces and medical instruments.

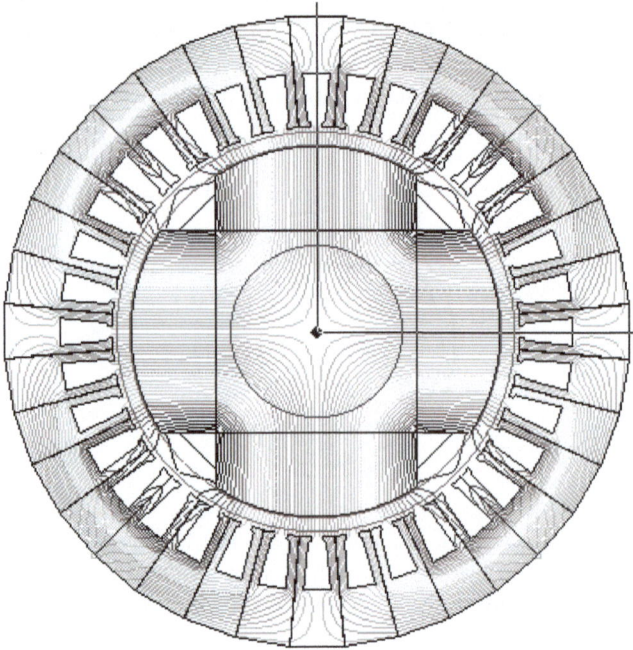

Fig. 9.5. Magnetic flux distribution in a cross section of a four-pole high speed PM brushless motor with ferromagnetic shaft.

Fig. 9.6. Rotor of a six-pole high speed PM brushless motor with metal retaining sleeve. Photo courtesy of Electron Energy Corporation, Landisville, PA, U.S.A.

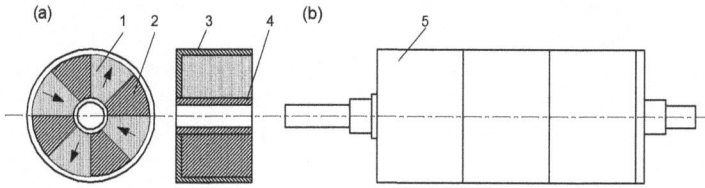

Fig. 9.7. PM rotor of segmented construction: (a) single segment, (b) three-segment rotor. 1 — PM, 2 — nonferromagnetic material, 3 — retaining sleeve, 4 — steel sleeve, 5 — complete rotor consisting of three segments. Courtesy of SatCon, Cambridge, MA, U.S.A.

Table 9.2. Specifications of small high speed slotless PM brushless motors manufactured by Koford Engineering, Winchester, OH, U.S.A.

Type of winding	101	102	103
Peak output power, W	109	181	22
Rated supply voltage, V	12	12	24
No load speed, rpm	155,000	200,000	48,900
Stall torque, $\times 10^{-3}$ Nm	28.8	36.8	17.6
Stall current, A	40	66	8
Continuous torque, $\times 10^{-3}$ Nm	4.45	2.54	5.72
No load current, A $\pm 50\%$	0.492	0.770	0.115
EMF constant, $\times 10^{-5}$ V $\pm 12\%$	7.74	6.00	49.10
Torque constant, $\times 10^{-3}$ Nm/A	0.72	0.558	4.38
Maximum efficiency	0.79	0.80	0.67
Winding resistance, $\Omega \pm 15\%$	0.3	0.18	3.0
Armature inductance, mH	0.13	0.10	2.4
Mechanical time constant, ms	17	17	9
Rotor moment of inertia, $\times 10^{-7}$ kgm^2	0.3		
Static friction, $\times 10^{-3}$ Nm	0.092		
Thermal resistance winding-to-frame, ^{0}C/W	7.0	7.0	8.0
Thermal resistance frame-to-ambient, ^{0}C/W	24.0		
Maximum winding temperature, ^{0}C	125		
External dimensions, mm	diameter 16, length 26		

9.4 Design of high speed PM brushless motors

The objective function is generally the maximum output power available from a particular high speed motor at given speed. The power is limited by the thermal and mechanical constraints.

In the design of high speed PM brushless motors the following aspects should be considered:

(a) *Mechanical design constraints* are important due to the high cyclic stress placed on the rotor components. Materials with high fatigue life are favored. Materials with low melting points, such as aluminum, should be avoided or restricted.

(b) *Capital and operational costs* are generally directly linked. The use of magnetic bearings over traditional rolling element bearings or oil lubricated bearings is a very important consideration. The capital cost of magnetic bearings is high, but the operational costs are less since the rotational loss and power consumption are reduced and there is no maintenance.

(c) *Dynamic analysis* of the rotor assembly, including shaft, core stack and bearing sleeves should be carried out with great detail using the 3D FEM simulation.

(d) *Static and dynamic unbalance.* Even a very small unbalance can produce high vibration. For example, a static unbalance of 0.05 N at a speed of 100,000 rpm produces an additional centrifugal force of more than 600 N.

Unbalance occurs when the center of gravity of a rotating object is not aligned with its center of rotation. *Static unbalance* is where the rotor mass center (principal inertia axis) is displaced parallel to the rotor geometric spin axis. *Dynamic unbalance* is where the rotor mass center is not coincidental with the rotational axis.

It is generally not difficult to design a high speed PM brushless motor rated at a few kWs and speed 7000 to 20,000 rpm with efficiency about 93 to 95%. The efficiency of high speed PM brushless motors rated above 80 kW and 70,000 to 90,000 rpm is over 96%. Core losses, windage losses and metal sleeve losses are high. Slotless stator, amorphous cores and foil bearings can increase the efficiency up to 98%.

Fig. 9.8a shows a 5-kW, 150,000 rpm motor with surface PMs and nonferromagnetic stainless steel retaining sleeve [283]. The input voltage is $V_1 = 200$ V, input frequency is $f = 2500$ Hz, number of poles $2p = 2$, the effective air gap is 6 mm, outer stator diameter is 90 mm, thickness of stator laminations is 0.1 mm, stator winding d.c. resistance per phase is $R_1 = 0.093$ Ω, and stator winding leakage inductance is $L_1 = 0.09$ mH. The slot ripple losses can be substantially reduced by expanding the air gap. High energy NdFeB magnets are therefore required.

At the rated speed of 150,000 rpm, the rotor surface speed will reach nearly 200 m/s and the resultant stress is calculated as high as 200 N/mm^2. Because this far exceeds the allowable stress of the magnets (80 N/mm^2), to prevent the magnets from exfoliating, initially, a nonferromagnetic stainless steel sleeve was shrunk on the PMs to retain them [283]. Although the stainless steel has low electric conductivity, the losses occurred in a relatively thick can were still quite large at the speeds over 100,000 rpm. Nonconductive fiber-reinforced plastic was then used [283].

To provide a high frequency ($f = 2500$ Hz) and realize a compact power circuit, a quasi-current source inverter (CSI) has been employed. This inverter consists of a diode rectifier, a current-controlling d.c. chopper and a voltage type inverter (Fig. 9.8b) [283]. To improve the input power factor, the large

(a)

(b)

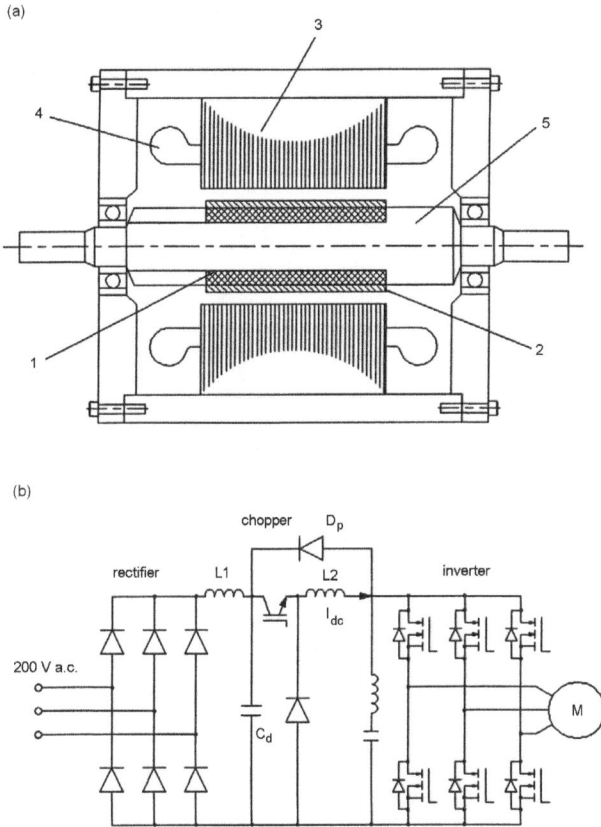

Fig. 9.8. High speed, 5-kW, 150,000 rpm PM brushless motor: (a) longitudinal section, (b) circuit configuration of the quasi-current source inverter. 1 — PMs, 2 — retaining sleeve, 3 — stator core, 4 — stator winding, 5 — shaft; D_p on = inverter operates as a VSI, D_p off = inverter operates as a CSI [283].

electrolytic capacitor C_d of the filter has been replaced with a substantially reduced film capacitor, which is enabled by the appropriate current control of the chopper.

9.5 Ultra high speed motors

A concept of an *ultra high speed PM synchronous motor* rotating at 500,000 is shown in Fig. 9.9 [74]. The rotor consists of a cylinder of rare-earth PM material magnetized radially and reinforced by a nonferromagnetic high tensile material, such as stainless steel, forming an external can (retaining sleeve). The stator has three teeth (salient poles), three slots, and three or six coils.

In assessing the power limit at a particular speed it is important to establish the optimum relationship between the rotor diameter and the outer stator diameter, and the complex coupling of the mechanical bursting and whirling constraints with the winding loss and power development. Optimum geometry prefers a small rotor diameter in relation to the stator outer diameter [74].

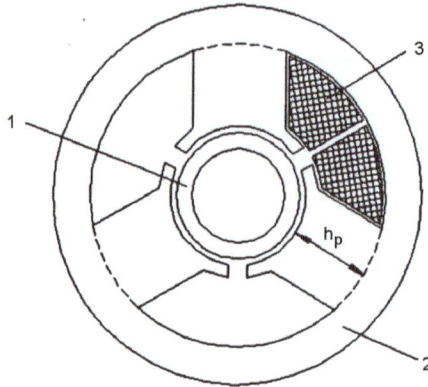

Fig. 9.9. Cross section of high speed PM synchronous motor with cylindrical PM and three stator slots: 1 — PM, 2 — stator core, 3 — stator winding [74].

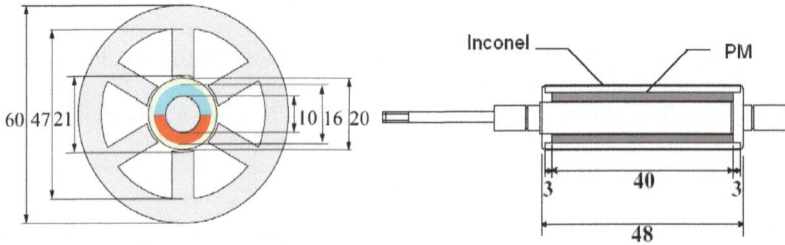

Fig. 9.10. Ultra high speed 5 kW, 240,000 rpm PM brushless motor. All dimensions are in millimeters. Courtesy of University of Nagasaki, Japan [234].

Small size ultra high speed two-pole PM brushless motor rated at 5 kW, 240,000 rpm and 4 kHz is shown in Fig. 9.10. The directly coupled motor is used to start up a gas turbine of a co-generation system. After the gas turbine is ignited and becomes the source of mechanical power, the motor operates as a generator. The two-pole rotor with surface PMs is equipped with a 2-mm thick metal retaining sleeve. The stator outer diameter is 60 mm, stator stack length is 40 mm and the rated voltage is 200 V. An oil circulation system is used for the shaft bearing and the cooling of the motor. This system consists of a filter, reserve tank and trochoid pump. The motor is circulated by the oil and the oil is cooled by the circulated water [234]. The control electronics consists of a control unit with DSP TMS320C32. The power supply unit consists of a PWM controlled voltage source inverter (VSI).

9.6 Applications

9.6.1 High speed aerospace drives

High speed PM brushless motors are used in the following aerospace electromechanical drives:

- electric fuel pumps;
- electric actuation systems for flight controls;
- electric cabin air compressors;
- nitrogen generation systems;
- compartment refrigeration units;
- supplemental cooling units.

The concept of more electric aircraft (MEA), which removes hydraulic, pneumatic and gearbox driven subsystems in favor of electrical driven subsystems, has necessitated the development of high performance compact lightweight motor drives and starter/generator systems [306]. Table 9.3 identifies the numbers and power of motor drive systems necessary to support a generic MEA fighter [306].

Aerospace electromechanical drive systems require high power density (small size motors), high reliability, low EMI and RFI interference level, high efficiency, precise speed control, high starting torque, fast acceleration and linear torque–speed characteristics. Constructional design features of aircraft electromechanical drives demand special optimized design and packaging, coolant passing through the motor and compatibility of materials with the coolant.

An exemplary magnetic circuit of a d.c. PM brushless motor used in aircraft technology is shown in Fig. 9.11 [216]. To keep high magnetic flux density in the teeth, cobalt alloy laminations with saturation magnetic flux density close to 2.4 T are frequently used. There are symmetrical axial channels in the rotor stack. The number of channels is equal to the number of rotor poles. The laminated rotor with axial channels has several advantages, such as improved power–mass ratio, improved heat transfer and cooling system, lower moment of interia (higher mechanical time constant) and lower losses in the rotor core.

In high speed motors the core losses form a large portion of the total losses. These losses can be minimized amongst others by designing laminated cores with uniform distribution of magnetic flux. Fig. 9.11 shows the magnetic flux distribution in the cross section area of the magnetic circuit [216]. The

Table 9.3. Electrical power requirements for MEA fighter aircraft [306]

System description	Continuous maximum total power kW	Number of motor drives	Largest motor drive kW
Flight controls	80	28	50
Environmental control system	40	10	10
Fuel management system	35	10	9
Pneumatic system	30	2	15
Landing system	30	20	5
Miscellaneous other	20	10	1
Engine starter/generator system	125 per channel	6	125
Total number of power conditioning units		86	

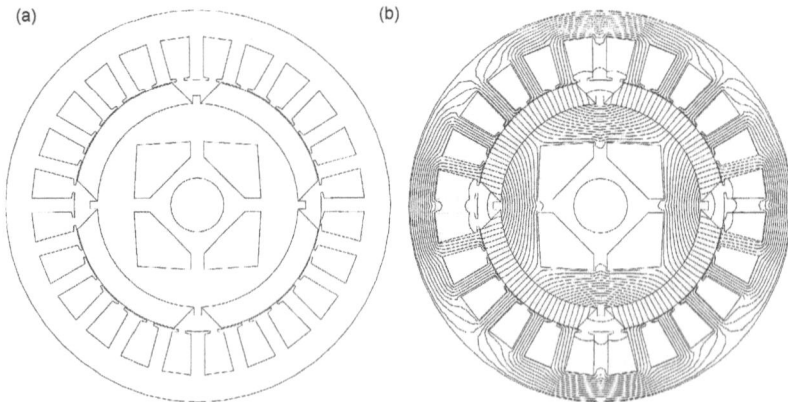

Fig. 9.11. Cross section of the magnetic circuit of a PM d.c. brushless motor with laminated rotor and axial channels for aerospace: (a) outline, (b) magnetic flux distribution [216].

magnetic flux in the rotor is uniformly distributed due to the axial channels. To avoid saturation of stator teeth, wider teeth are required [216]. For a PM brushless motor rated at 200 W efficiency of 90% is achievable, i.e., 9% higher than that of a motor with solid rotor core [216].

The PMs in the form of rectangular blocks or segments are secured on the laminated hub and embraced using a nonferromagnetic retaining sleeve (metal or carbon-graphite). The thickness of the sleeve is designed to be sufficient to protect PMs against centrifugal forces and provide the required air gap (mechanical clearance).

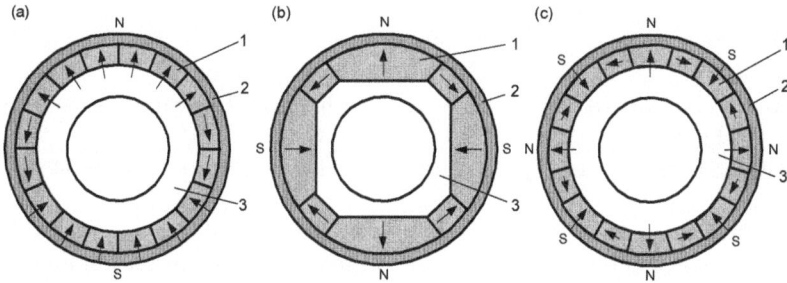

Fig. 9.12. Rotors of aerospace PM brushless motors with magnets arranged in Halbach cylinder: (a) two-pole rotor; (b) four-pole rotor; (c) eight-pole rotor.

Other rotor constructions for aerospace PM brushless motors are shown in Fig. 9.12. Halbach cylinder adds from 3 to 5% to the air gap magnetic flux density. Rotors shown in Fig. 9.12 are sometimes equipped with an inner heat exchanger for intensive air cooling.

Most brushless d.c. motors in aircraft electromechanical drive systems use encoders or Hall sensors, but significant developments are coming up in favor of search coil techniques [216]. Sensorless control methods increase electronics complexity, but improve reliability and allow the motor to operate at elevated temperature.

Aerospace brushless d.c. motors are easily integrated with their drive electronics in compact packages (integrated electromechanical drives). This reduces the number of sub-assemblies, connecting wires, production cost, space and mass. The motor is fully controllable and protected against faults. Life is limited by bearings only.

One of the main problems in designing electric motors for aerospace and defense is that the space envelope is often small and the motor must be very small. For example, an aerospace PM brushless motor rated at 90 kW, 32,000 rpm, 95% efficiency, measures 125 mm in its outer stack diameter and 100 mm stack length (without winding overhangs).

9.6.2 High speed spindle drives

Owing to the improvements in power electronics and control techniques the *electrical spindle drives* have shown a significant evolution. At present time very high speed spindle drives acquire the speed range between 10,000 and 100,000 rpm [32]. Moreover, there is demand to increase the speed limit up to 300,000 rpm for special applications. This evolution has been a consequence of the high speed metal cutting, milling and grinding machine tools used in manufacturing processes, e.g., light alloys for aerospace applications. Spindle drives are used by commercial aircraft manufacturers in both drilling rivet holes and milling stringers, spars, and precision components. The main demands for high speed electrical spindles are [32]:

- higher "power speed" product compared with the values for standard applications, particularly for milling and grinding machines;
- increased bearing robustness as a result of high mechanical stresses during the machining processes;
- position control at zero speed in order to allow for automatic changing of tools;
- high efficiency cooling system to reach the highest output power–to–volume ratio compatible with the actual magnetic and insulation material technology;
- suitable lubrication system in order to get high quality behavior and minor friction problems;
- capability to work in different positions;
- duty cycle requirements in proportion to the large size of the pieces to work.

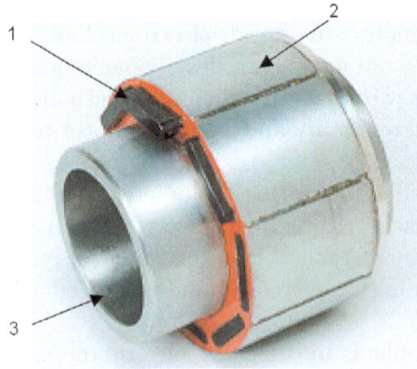

Fig. 9.13. Rotor with embedded PMs for a brushless spindle motor. 1 — PM, 2 — laminated core, 3 — inner sleeve. Photo courtesy of Mitsubishi, Nagoya, Japan [171].

PM brushless spindle motors for machine tools in comparison with their induction counterparts display much higher efficiency in the low-speed, high-torque range. This makes it possible to reduce the size of the spindle unit and to simplify the cooling structures, taking advantage of the lower heat generation [171]. Fig. 9.13 shows a prototype of an embedded PM rotor [171]. The brushless motor with embedded PM rotor can produce both synchronous and reluctance torque.

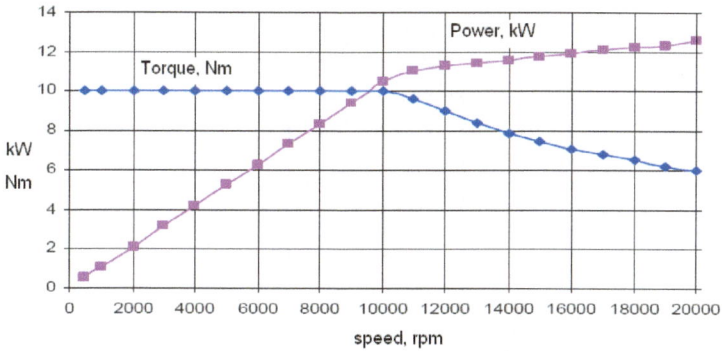

Fig. 9.14. Steady-state characteristics of a 12 kW, 500 to 20,000 rpm PM brushless spindle motor (SD60124 type) manufactured by Fischer Precise, Racine, WI, U.S.A.

Fig. 9.14 shows output power–speed and torque–speed characteristics of a 12 kW, four-pole, 500 to 20,000 rpm, 65 A PM brushless spindle motor with liquid water cooling system. The length of the motor housing is 516.5 mm and outer diameter of housing is 119 mm. The length of the stator stack with winding overhangs is about one quarter of the housing length.

9.6.3 Flywheel energy storage

A *flywheel energy storage* (FES) system draws electrical energy from a primary source, such as the utility grid, and stores it in a high-density rotating flywheel. The flywheel system is actually a kinetic, or mechanical battery, spinning at very high speeds ($> 20,000$ rpm) to store energy that is instantly available when needed. Upon power loss, the motor driving the flywheel acts as a generator. As the flywheel continues to rotate, this generator supplies power to the customer load.

Advanced FES systems have rotors made of high strength *carbon-composite filaments* that spin at speeds from 20,000 to over 50,000 rpm in a vacuum enclosure and use magnetic bearings. Composites are desirable materials for flywheels due to their light weight and high strength. Lightness in high speed rotors is good from two standpoints: the ultra-low friction bearing assemblies are less costly and the inertial loading which causes stress in the material at high rotational speeds is minimized. High strength is needed to achieve maximum rotational speed.

Flywheel circumferential speed is higher than that given in Table 9.1. For example, for diameter $D = 0.5$ m and 50,000 rpm the linear surface speed is 1308 m/s.

The kinetic energy stored in a rotating homogeneous disk with its outer diameter D, thickness t and density ρ is

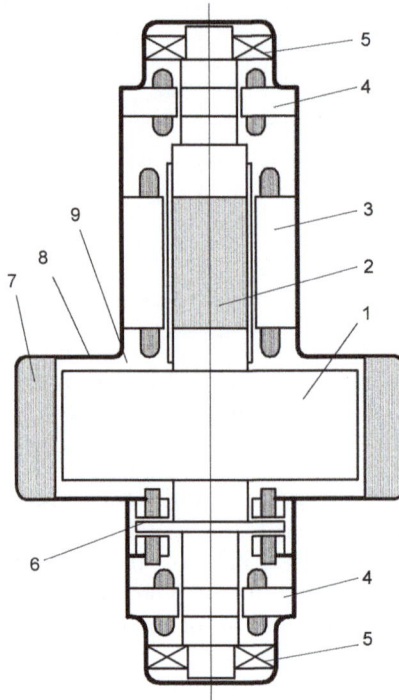

Fig. 9.15. Construction of a typical FES with PM brushless machine. 1 — flywheel, 2 — rotor of PM brushless motor/generator, 3 — stator of PM brushless motor/generator, 4 — radial magnetic bearing, 5 — backup bearing, 6 — thrust magnetic bearing, 7 — burst shield, 8 — vacuum containment, 9 — vacuum.

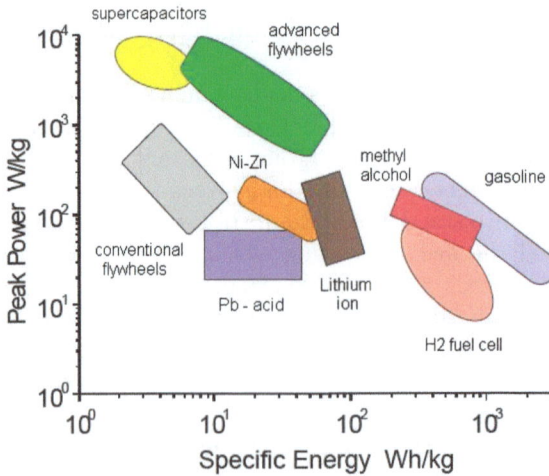

Fig. 9.16. Comparison of energy storage systems.

Table 9.4. Comparison of mechanical properties of metals and composite materials

Material	Specific mass density ρ kg/m^3	Young modulus GPa	Poisson ratio	Failure stress σ_f MPa	Material index σ_f/ρ MNm/kg
Steel AISI 4340	7800	190 to 210	0.27 to 0.30	1800	0.22
Alloy AlMnMg	2700	70 to 79	0.33	600	0.22
Titanium TiAl6Zr5	4500	105 to 120	0.34	1200	0.27
Glass fiber reinforced polymer (GFRP) 60% volume of electrical grade glass (E-Glass)	2550	70 to 80	0.22	1600	0.60
Carbon fiber reinforced polymer (CFRP) 60% volume of high tenacity (HT) carbon	1500	225 to 240	0.28 to 0.36	2400	1.60

$$E_k = \frac{1}{2}J\Omega^2 = \frac{\pi}{64}\rho t D^4 \Omega^2 \tag{9.12}$$

where the moment of inertia and mass of disk are, respectively

$$J = \frac{\pi}{32}\rho t D^4 \qquad\qquad m = \frac{\pi}{4}\rho t D^2 \tag{9.13}$$

The energy density of a flywheel is the first criterion for the selection of a material, i.e.,

$$\frac{E_{kin}}{m} = \frac{1}{16}D^2\Omega^2 \tag{9.14}$$

As the angular speed Ω increases, the energy stored in the flywheel and radial tensile stress due to centrifugal forces increase too. The maximum tensile stress on a spinning disk is

$$\sigma_{max} = \frac{3+\nu}{8}\rho\frac{D^2}{4}\Omega^2 \approx \frac{1}{8}\rho D^2 \Omega^2 \tag{9.15}$$

where ν is Poisson's ratio. In eqn (9.15) $\nu \approx 1/3$. The maximum stress $\sigma_{max} < \sigma_f$ where σ_f is the failure stress. An appropriate safety factor should also be included. Putting $D^2\Omega^2 = 8\sigma_f/\rho$ from eqn (9.15) into eqn (9.14)

$$\frac{E_{kin}}{m} = \frac{1}{2}\frac{\sigma_f}{\rho} \tag{9.16}$$

The maximum energy density of a rotating mass m is dependent on only the failure stress σ_f of the material and its specific mass density ρ. The best materials for high performance flywheels are those with high value of the *failure stress–to–mass density* ratio (material index) [19]. Composite materials such as, for example, glass fiber reinforced polymer (GFRP) or carbon fiber reinforced polymer (CFRP) are much better than metals or metal alloys (Table 9.4).

Fig. 9.15 shows a typical FES with PM brushless machine and magnetic bearings. To minimize tooth ripple losses in retaining sleeve and PMs, a slotless stator is sometimes used. When the mechanical energy is stored in the flywheel, the PM brushless machine operates as a motor. When the energy stored in the flywheel is utilized as electrical energy, the PM brushless machine operates in generating mode.

Flywheels have a much higher power density than batteries (Fig. 9.16), typically by a factor of 5 to 10. While batteries can supply backup power for a significantly longer period than a flywheel and consume less standby power, most other characteristics favor a flywheel, i.e.,

- the design life for a flywheel is typically about 20 years, while most batteries in uninterruptible power supply (UPS) applications will last only 3 to 5 years;
- batteries must be kept within a narrow operating temperature range about room temperature, while flywheels tolerate normal outdoor ambient temperature conditions;
- frequent cycling has little impact on flywheel life, while frequent cycling significantly reduces battery life;
- flywheel reliability is 5 to 10 times greater than a single battery string or about equal to two battery strings operating in parallel;
- flywheels are more compact, using only about 10 to 20% of the space required to provide the same power output from batteries;
- flywheel maintenance is generally less frequent and less complicated than for batteries;
- flywheels avoid battery safety issues associated with chemical release.

9.6.4 Dental handpieces

Electric motor driven *dental handpieces* are rapidly replacing traditional air turbine driven handpieces. Air turbine driven high-speed handpieces typically develop speeds between 250,000 and 420,000 rpm and relatively low torque. Electric motor driven handpieces have speeds up to 200,000 rpm and relatively high torque. This implies that air turbine driven handpieces are faster than electric handpieces. However, when a bur in an air turbine driven handpiece contacts material to be cut, the speed will drop by as much as 40% or more (depending on the hardness of the material) because the air pressure is insufficient to maintain the speed of the turbine under high load [191]. An electric motor driven handpiece offers smooth, constant torque that does not vary as the bur meets resistance. Because of the absence of air, electric handpieces are quieter and the chance of air embolism in a surgical site is eliminated [191]. Electric motor driven handpieces offer cutting power from 33 to 45 W (greater than air turbine driven handpieces).

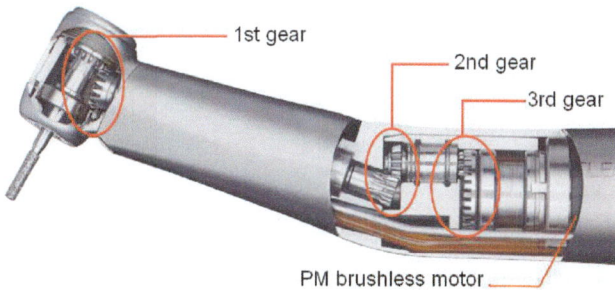

Fig. 9.17. Dental handpiece with PM brushless motor. Courtesy of Kavo Dental, Lake Zurich, IL, U.S.A.

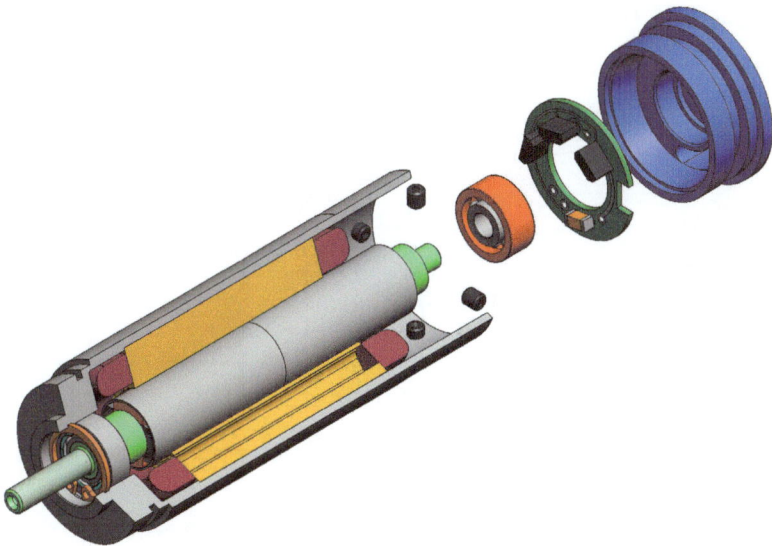

Fig. 9.18. Small 13-mm diameter, 73.6-mNm, 50-V d.c. BO512-050 PM brushless motor (Table 9.5). Photo courtesy of Portescap^{TM}, A Danaher Motion Company, West Chester, PA, U.S.A.

Table 9.5. Small-diameter brushless slotted BO512-050 PM brushless motors for dental and power surgical instruments manufactured by Portescap[TM], A Danaher Motion Company, West Chester, PA, U.S.A. [219].

	Motor type	
Specifications	Design A	Design B
Diameter of housing, mm	12.7	
Length of housing, mm	47.0	
Rated d.c. voltage, V	50	50
Peak torque (torque constant × peak current), mNm	72.3	35.7
Continuous stall torque, mNm	8.98	8.98
Maximum continuous current, A	1.33	0.67
Peak current, A	10.7	2.73
No–load speed at 24 V, rpm	70,600	36,500
Resistance line-to-line, Ω	4.64	18.3
Inductance line-to-line, mH	0.22	0.89
EMF constant, V/krpm	0.708	1.37
Torque constant mNm/A	6.76	13.1
Rotor moment of inertia, 10^{-8}kgm^2	4.94	4.94
Mechanical time constant, ms	5.02	5.02
Electrical time constant, ms	0.05	0.05
Thermal resistance, ^{0}C/W	15.9	15.9
Maximum continuous power dissipation, W	8.18	8.18
Mass, g	44	44

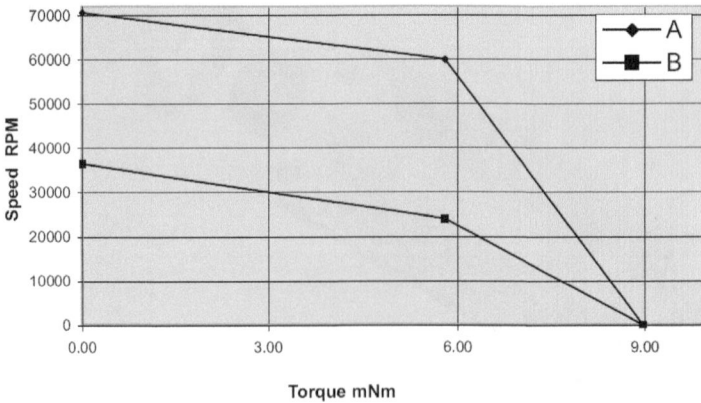

Fig. 9.19. Speed-torque characteristics of BO512-050 PM brushless motor (Table 9.5). Photo courtesy of Portescap[TM], A Danaher Motion Company, West Chester, PA, U.S.A.

In electric handpieces, the bur is connected through gears in the head of the handpiece to a central drive shaft that is physically turned by the motor (Fig. 9.17). Electric handpieces have typically a step up 5:1 gear ratio. The *Electro torque* handpiece (Fig. 9.17) from Kavo Dental, Lake Zurich, IL U.S.A. with PM brushless motor offers a speed range of 2,000 to 40,000 rpm at the motor shaft. When used in combination with their 25LPA handpiece (5:1 gear ratio), speeds can be increased to 200,000 rpm.

Small 13-mm diameter, four-pole, 73.6-mNm, 50-V d.c. slotted PM brushless motor for dental instruments and power surgical devices is shown in Fig. 9.18. It can withstand in excess of 1000 autoclave[1] cycles. The speed-torque characteristic is shown in Fig. 9.19.

To reduce core losses and temperature of the stator, low loss ferrmagnetic alloys are used for the magnetic circuit, e.g., Megaperm© 40 L with specific losses 0.2 W/kg at 1 T and 50 Hz [221] or amorphous alloys.

Electric dental handpieces show the following advantages [66]:

- high torque with very little stalling;
- quiet and smooth operation with a reduced potential for hearing damage and a less irritating sound for patients;
- low levels of vibration;
- precision cutting with high concentricity;
- one electric motor runs several handpiece attachments (high speed and low speed);
- low-speed attachments for the electric motor allow for easy cutting of dentures, temporary resin restorations, orthodontic appliances, occlusal splints, plaster or stone.

The disadvantages of high-speed electric dental handpieces include [66]:

- higher price and weight compared with high-speed air-rotor handpieces;
- the heads on the contrangles are comparatively large;
- because of high torque, the dentist inadvertently may place excessive load on a tooth during cutting;
- infection control measures must be observed carefully to avoid damaging the handpiece with repeated sterilizations.

9.6.5 Sheep shearing handpieces

The force required to move a handpiece through wool can be reduced by limiting the number of teeth on the cutter and increasing the stroke of the cutter [256]. The overall dimensions and temperature rise require a built-in drive and motor efficiency of at least 96% [256]. A feasibility study has indicated that a 2-pole PM brushless d.c. motor can meet these requirements

---

[1] Autoclave is an apparatus (as for sterilizing) using superheated high-pressure steam cycles.

[256]. The most promising appeared to be a slotless geared drive with an amorphous stator core and unskewed winding (Fig. 9.20). To minimize the motor losses the controller supplies sinusoidal voltages to match the back EMF waveforms and a position sensor has a phase-locked loop to allow the phase of the supply to be electronically adjusted.

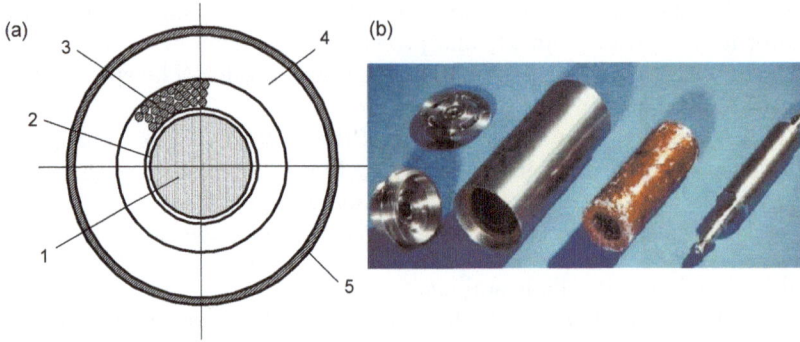

Fig. 9.20. PM brushless motor for sheep shearing handpiece: (a) cross section; (b) disassembled motor. 1 — two-pole NdFeB rotor, 2 — air gap (mechanical clearance), 3 — slotless armature winding, 4 — slotless amorphous laminations, 5 — housing. Courtesy of CSIRO, Newcastle, New South Wales, Australia.

The stator winding losses can be reduced by applying a high magnetic flux density in the air gap (NdFeB PMs) and axial straight conductors rather than skewed conductors. The core losses can be minimized by reducing the flux change frequency leading to the choice of a two-pole rotor. Further reduction can be achieved by using an amorphous magnetic alloy instead of silicon steel or by eliminating the stator teeth and using a slotless core. The stray losses in the armature winding can be reduced by stranding the conductors. The eddy current losses in the frame can be reduced by ensuring the armature stack is radially not too thin, so that the external magnetic field is low.

The measured data of an experimental three-phase, two-pole, $13,300$ rpm, 150-W brushless d.c. motor with NdFeB PMs and slotless amorphous stator core (Metglas 2605-S2) for sheep shearing are: full load phase current $I_a = 0.347$ A, total losses 6.25 W, phase winding resistance $R_1 = 7$ Ω at 21^0C, mean magnetic flux density in the stator core 1.57 T and full load efficiency $\eta = 0.96$ [256].

Numerical examples

Numerical example 9.1

Find the outer diameter and stack length of a 100 kW, 60,000 rpm PM brushless motor. The efficiency should be $\eta = 96\%$, power factor $\cos\phi = 0.95$, stator winding factor $k_{w1} = 0.96$, EMF–to-voltage ratio $\epsilon = 0.85$, peak value of the air gap magnetic flux density $B_{mg} = 0.72$ T, peak value of the stator line current density $A_m = 120,000$ A/m, number of poles $2p = 4$, air gap (mechanical clearance) $g = 1.5$ mm, rotor length equal to stator stack length $L_i = 1.5D$, distance between bearings $L \approx 1.8L_i$ and shaft diameter $d_{sh} = 0.35D$. Assume that the shaft specific mass density is $\rho_{sh} = 7700$ kg/m³, rotor mean mass specific density $\rho = 7200$ kg/m³, shaft modulus of elasticity $E_{sh} = 200$ GPa and rotor modulus of elasticity $E = 8$ GPa.

Solution

For $L_i \approx 1.5D_{1in}$, the stator inner diameter as determined by eqn (9.1) is

$$D_{1in} = \sqrt[3]{\frac{\epsilon P_{out}}{0.75\pi k_{w1}n_s A_m \eta \cosh\phi}}$$

$$= \sqrt[3]{\frac{0.85 \times 100\ 000}{0.75\pi 0.96 \times (60\ 000/60)0.72 \times 120\ 000 \times 0.96 \times 0.95}} = 0.078\text{m} = 78\text{ mm}$$

The rotor outer diameter

$$D_{2out} = D_{1in} - 2g = 78 - 2 \times 1.5 = 75 \text{ mm}$$

The rotor surface linear speed

$$v = \pi \times 0.075\frac{60\ 000}{60} = 236 \text{ m/s}$$

The linear surface speed is acceptable both for metal and carbon-graphite sleeves. The length of the rotor stack equal to the length of stator stack $L_i = 1.5D_{1in} = 1.5 \times 78 = 117$ mm, the distance between bearings is $L \approx 1.8L_i = 1.8 \times 116 = 211$ mm and the diameter of shaft $d_{sh} = 0.35D = 0.35 \times 75 = 26$ mm. The mass of the rotor and mass of the rotor shaft are, respectively,

$$m = \rho\frac{\pi}{4}(D_{2out}^2 - d_{sh}^2)L_i = 7200\frac{\pi}{4}(0.075^2 - 0.026^2)0.117 = 3.28 \text{ kg}$$

$$m_{sh} = \rho_{sh}\frac{\pi}{4}d_{sh}^2 L_i = 7700\frac{\pi}{4}0.026^2 \times 0.211 = 0.88 \text{ kg}$$

The second area moment of inertia of the rotor stack and shaft are, respectively,

$$I = \pi \frac{D_{2out}^4 - d_{sh}^2}{64} = \pi \frac{0.075^4 - 0.026^4}{64} = 1.54 \times 10^{-6} \text{ m}^4$$

$$I_{sh} = \pi \frac{d_{sh}^2}{64} = \pi \frac{0.026^4}{64} = 0.023 \times 10^{-6} \text{ m}^4$$

The stiffness of the rotor according to eqn (9.6)

$$K = 48 \frac{8 \times 10^9 \times 1.54 \times 10^{-6}}{0.117^3} + 48 \frac{200 \times 10^9 \times 0.023 \times 10^{-6}}{0.211^3} = 391.5 \times 10^6 \text{ N/m}$$

The first critical speed

$$n_{cr} = \frac{30}{\pi} \sqrt{\frac{K}{m + m_{sh}}} = \frac{30}{\pi} \sqrt{\frac{391.6 \times 10^6}{3.28 + 0.88}} = 92,590 \text{ rpm}$$

The rotor speed is less than the critical speed, i.e., $n_s < n_{cr}$. In this case, inequality (9.9) must apply, i.e., $0.75 n_{cr}/(2p) = 0.75 \times 92\,590/4 = 17\,360 < n_s$ rpm. Thus, $60\,000 > 0.75 n_{cr}/(2p)$ and from a mechanical point of view the rotor has been sized correctly. In industrial design of electric motors, the first critical speed should be verified with the aid of structural FEM computations.

Fig. 9.21. Cross section and magnetic flux distribution of a 100-kW, four-pole, 40,000-rpm high speed PM brushless motor.

Numerical example 9.2

For the 10-kW, four-pole, 40,000 rpm high speed PM brushless motor shown in Fig. 9.21 find the losses in retaining sleeve, windage losses and bearing losses at the temperature $\vartheta = 100^0$ C. The dimensions of the machine are as follows: rotor outer diameter $D_{2out} = 50$ mm, shaft diameter $d_{sh} = 12$ mm, stack length $L_i = 50$ mm, non-magnetic air gap (including retaining sleeve) $g = 3$ mm, stator slot opening $b_{14} = 3$ mm, thickness of retaining sleeve $d_{sl} = 1.8$ mm. The electric conductivity of retaining sleeve at 20^0C is $\sigma_{sl} = 0.826 \times 10^6$ S/m, the mean specific mass density of the rotor is 7200 kg/m^3 and the specific mass density of the rotor shaft is 7800 kg/m^3. The peak value of the air gap magnetic flux density is $B_{mg} = 0.71$ T. The number of stator slots is $s_1 = 6$. The axial velocity of cooling air is $v_{ax} = 10$ m/s.

<u>Solution</u>

<u>Losses in retaining sleeve</u>

The stator inner diameter

$$D_{1in} = D_{2out} + 2g = 50 + 2 \times 3 = 56 \text{ mm}$$

The stator slot pitch

$$t_1 = \frac{\pi 56}{6} = 29 \text{ mm}$$

Coefficient taking into account eddy currents in the sleeve in tangential direction according to eqn (B.21)

$$k_r = 1 + \frac{1}{\pi}\frac{29}{50} = 1.187$$

The auxiliary function according to eqn (B.24)

$$u = \frac{3}{2 \times 3} + \sqrt{1 + \left(\frac{3}{2 \times 3}\right)^2} = 1.618$$

The auxiliary function according to eqn (A.28) for calculation of Carter's coefficient

$$\gamma_1 = \frac{4}{\pi}\left\{\left(\frac{3}{2 \times 3}\right)\arctan\left(\frac{3}{2 \times 3}\right) - \ln\left[\sqrt{1 + \left(\frac{3}{2 \times 3}\right)^2}\right]\right\} = 0.153$$

Carter's coefficient according to eqn (A.27)

$$k_C = \frac{29}{1 - 0.153 \times 3} = 1.016$$

The magnitude of magetic flux density due to slot openings according to eqns (B.22) and (B.23)

$$B_{msl} = \frac{1 + 1.618^2 - 2 \times 1.618}{1 + 1.618^2} 1.016 \frac{2}{\pi} 0.71 = 0.048 \text{ T}$$

Electric conductivity of sleeve at $\vartheta = 100^0 C$

$$\sigma_{sl} = \frac{0.826 \times 10^6}{1 + 0.007(100 - 20)} = 0.5295 \times 10^6 \text{ S/m}$$

Losses in retaining sleeve due to slot ripple as given by eqn (B.20)

$$\Delta P_{sl} = \frac{\pi^3}{2} B_{msl}^2 \left(\frac{40\ 000}{60}\right)^2 0.5295 \times 10^6 \times 0.003 \times 0.05 \times 1.187 = 114.5 \text{ W}$$

<u>Windage losses</u>

The air density at $\vartheta = 100^0 C$ according to eqn (B.37)

$$\rho = -10^{-8}100^3 + 10^{-5}100^2 - 0.0045 \times 100 + 1.2777 = 0.918 \text{ kg/m}^3$$

The dynamic viscosity of air at $\vartheta = 100^0 C$ according to eqn (B.39)

$$\rho = -2.1664 \times 10^{-11}100^2 + 4.7336 \times 10^{-8}100 + 2 \times 10^- - 5 = 0.2452 \times 10^{-6} \text{ Pa s}$$

The angular speed of the rotor

$$\Omega = 2\pi \frac{40\ 000}{60} = 4188.8 \text{ rad/s}$$

Reynolds number according to eqn (B.43)

$$Re = \frac{1}{2} \frac{0.918 \times 4188.8 \times 0.05 \times (0.003 - 0.0018)}{0.2452 \times 10^{-6}} = 47 \times 10^4$$

Since $Re = 47 \times 10^4 > 10^4$, the friction coefficient according to the second equation (B.38) is

$$c_f = 0.0325 \frac{[2(0.003 - 0.0018)/0.05)]^{0.3}}{(47 \times 10^4)^{0.2}} = 0.00096$$

The friction losses in the air gap due to the resisting drag torque according to eqn (B.36)

$$\Delta P_a = \frac{1}{16} 0.00096\pi 0.918 \times 4188.8^3 \times 0.05^4 \times 0.05 = 4.0 \text{ W}$$

Reynolds number for rotating disc according to eqn (B.46)

$$R_{ed} = \frac{1}{4}\frac{0.918 \times 4188.8 \times 0.05^2}{0.2452 \times 10^{-6}} = 0.98 \times 10^5$$

Since $Re = 0.98 \times 10^5 < 3 \times 10^5$, the friction coefficient for rotating disks according to the first equation (B.45) is

$$c_{fd} = \frac{3.87}{(0.98 \times 10^5)^{0.5}} = 0.012$$

The friction losses for the flat circular surface of the rotor (two sides) due to the resisting drag torque according to eqn (B.44)

$$\Delta P_{ad} = \frac{1}{64}0.012 \times 0.918 \times 4188.8^3(0.05^5 - 0.012^5) = 4.1 \text{ W}$$

The mean tangential velocity of the cooling air in the air gap is $v_t \approx 0.5v = 0.5 \times 104.72 = 52.36$ m/s. The axial velocity of the air flow through the air gap is $v_{ax} = 10$ m/s. The losses due to axial cooling medium flow according to eqn (B.47) are

$$\Delta P_c = \frac{1}{12}\pi 0.918 \times 52.36 \times 10 \times 4188.8(0.056^3 - 0.05^2) = 26.7 \text{ W}$$

Thus, the total windage losses given by eqn (B.35) are

$$\Delta P_{wind} = 4.0 + 4.1 + 26.7 \approx 34.8 \text{ W}$$

Bearing losses

The mass of the rotor assuming that the length of the shaft is $2L_i$

$$m_{rot} = 7200\frac{\pi D_{2out}^2}{4}L_i + 7800\frac{\pi d_{sh}^2}{4}(2L_i)$$

$$= 7200\frac{\pi 0.05^2}{4}0.05 + 7800\frac{\pi 0.012^2}{4}(2 \times 0.05) = 0.795 \text{ kg}$$

Assuming $k_{fb} \approx 2$, the bearing friction losses calculated with the aid of eqn (B.31) are

$$P_{fr} = 2 \times 0.795 \times 40\,000 \times 10^{-3} = 63.6 \text{ W}$$

Numerical example 9.3

Calculate the performance of the 10 kW, 40,000 rpm high speed PM brushless motor shown in Fig 9.21. The motor is fed from a PWM voltage source inverter. Dimensions of the motor, number of poles and number of stator slots are given in Numerical Example 9.2. The number of turns per coil is $N_c = 15$, coil pitch expressed in the number of slots is $w_{sl} = 1$, magnetic flux density in the air gap $B_{mg} = 0.71$ T, PM pole shoe arc–to–pole pitch $\alpha_i = 0.833$, d.c. link voltage $V_{dc} = 376$ V, amplitude modulation index $m_a = 1.0$, stator winding resistance per phase of hot machine $R_1 = 0.0154$ Ω, d-axis synchronous reactance $X_{sd} = 1.526$ Ω, q-axis synchronous reactance $X_{sq} = 1.517$ Ω, load angle $\delta = 15.7^0$ and core losses $\Delta P_{Fe} = 197$ W.

Solution

The number of slots per pole $Q_1 = 6/4 = 1.5$, number of slots per pole per phase $q_1 = 6/(4 \times 3) = 0.5$, number of turns per phase $N_1 = (s_1/m_1)N_c = (6/3)15 = 30$, frequency $f = pn_s = 2(40\ 000/60) = 1333.3$ Hz, stator inner diameter $D_{1in} = D_{2out} + 2g = 50 + 2 \times 3 = 56$ mm, pole pitch $\tau = \pi D_{1in}/(2p) = \pi 56/4 = 44$ mm and the rotor surface linear speed $v = \pi D_{2out} n_s = \pi 0.05 \times 40\ 000/60 = 104.7$ m/s. The line–to–line a.c. voltage according to eqn (6.9) is

$$V_{1L} \approx 0.612 \times 1.0 \times 376 \approx 230 \text{ V}$$

Thus, the phase voltage $V_1 = 230/\sqrt{3} \approx 133$ V. The stator winding distribution factor for such a motor $k_{d1} = 1$ and coil span measured in the number of slots $w_{sl} = 1$. The stator winding pitch factor according to eqn (A.3) is $k_{p1} = \sin[\pi w_{sl}/(2Q_1)] = \sin[\pi \times 1/(2 \times 1.5)] = 0.866$ and the resultant winding factor according to eqn (A.1) is $k_{w1} = 1 \times 0.866 = 0.866$. The magnetic flux and phase EMF according to eqns (5.6) and (5.5) are, respectively,

$$\Phi_f = \frac{2}{\pi}0.71 \times 0.044 \times 0.05 = 0.994 \times 10^{-5} \text{ Wb}$$

$$E_f = \pi\sqrt{2}1333.3 \times 30 \times 0.866 \times 10.944 \times 10^{-5} = 153 \text{ V}$$

Because $E_f > V_1 = 133$, the machine operates as an overexcited motor. The stator winding current is calculated on the basis of eqns (5.40), (5.41) and (5.43), i.e.,

$$I_{ad} = \frac{133(1.517 \cos 15.7^0 - 0.0154 \sin 15.7^0) - 153 \times 1.517}{1.526 \times 1.517 + 0.0154} = -16.9 \text{ A}$$

$$I_{aq} = \frac{133(0.0154 \cos 15.7^0 + 1.526 \sin 15.7^0) - 153 \times 0.0154}{1.526 \times 1.517 + 0.0154} = 23.5 \text{ A}$$

$$I_a = \sqrt{(-16.9)^2 + 23.5^2} = 28.9 \text{ A}$$

Now the EMF can be verified with the aid of eqn (5.39), i.e.,

$$E_f = V_1 \cos\delta + I_{ad}X_{sd} - I_{aq}R_1 = 133\cos 15.7^0 + |-16.9| \times 1.526 - 23.5 \times 0.0154 = 153 \text{ V}$$

The input power can be calculated with the aid of eqn (5.44) or eqn (5.3), i.e.,

$$P_{in} = 3 \times 133(23.5\cos 15.7^0 - (-16.9)\sin 15.7^0) = 10\ 826 \text{ W}$$

or

$$P_{in} = 3[23.5 \times 153 + 0.0154 \times 28.9^2 + (-16.9) \times 23.5(1.526 - 1.517)]$$

$$= 10\ 826 \text{ W}$$

The input apparent power

$$S_{in} = 3 \times 133 \times 28.9 = 11\ 494 \text{ VA}$$

The power factor

$$\cos\phi = \frac{10\ 826}{11\ 531} = 0.94$$

The angle between the current I_a and q-axis for overexcited motor

$$\Psi = \arccos\phi + \delta = \arccos 0.94 + 15.7 = 35.3^0$$

The stator winding losses according to eqn (B.12)

$$\Delta P_a = 3 \times 28.9^2 \times 0.0154 = 38.4 \text{ W}$$

The output (shaft) power

$$P_{out} = P_{in} - \Delta P_a - \Delta P_{Fe} - \Delta P_{sl} - \Delta P_{wind} - \Delta P_{fr}$$

$$= 10\ 826 - 197 - 114.5 - 34.8 - 63.6 = 10\ 378 \text{ W}$$

The efficiency

$$\eta = \frac{10\ 378}{10\ 826} = 0.959$$

The losses in retaining sleeve $\Delta P_{sl} = 114.5$ W, windage losses $\Delta P_{wind} = 34.8$ and bearing friction losses $\Delta P_{fr} = 63.6$ W have been calculated in Numerical example 9.2. The shaft torque

$$T_{sh} = \frac{10\ 826}{2\pi 40\ 000/60} = 2.5 \text{ Nm}$$

Numerical example 9.4

Compare the performance of two disk-type flywheels made of steel and GFRP (E-glass) and driven by PM brushless motors. Both disks have thickness $t = 45$ mm. Other specifications are as follows:

(a) Steel flywheel: outer diameter $D = 0.32$ m, 40,000 rpm, specific mass density $\rho = 7800$ kg/m^3, Poisson's ratio $\nu = 0.30$;
(b) Steel flywheel: outer diameter $D = 0.46$ m, 50,000 rpm, specific mass density $\rho = 2550$ kg/m^3, Poisson's ratio $\nu = 0.22$;

Solution

Steel flywheel

The mass of the disk according to eqn (9.13)

$$m = \frac{\pi}{4} 7800 \times 0.045 \times 0.32^2 = 28.32 \text{ kg}$$

The moment of inertia of the steel disk according to eqn (9.13)

$$J = \frac{\pi}{32} 7800 \times 0.045 \times 0.32^4 = 0.369 \text{ kg}^2$$

The angular speed of the disk

$$\Omega = 2\pi \frac{40\ 000}{60} = 4188.8 \text{ rad/s}$$

The kinetic energy of the steel flywheel according to eqn (9.12)

$$E_k = \frac{1}{2} 0.369 \times 4188.8^2 = 3\ 169\ 963 \text{ J} = 3.17 \text{ MJ}$$

The energy density

$$e_k = \frac{3\ 169\ 963}{28.23} = 112\ 294 \text{ J/kg} \approx 112 \text{ kJ/kg}$$

The radial tensile stress according to eqn (9.15)

$$\sigma = \frac{3 + 0.3}{8} 7800 \frac{0.32^2}{4} 4188.8^2 = 1\ 445\ 225\ 912 \text{ Pa} = 1.445 \text{ GPa}$$

The tensile stress–to–density ratio

$$\frac{\sigma}{\rho} = \frac{1\ 445\ 225\ 912}{7800} = 185\ 185.4 \text{ Nm/kg} \approx 0.185 \text{ MNm/kg}$$

The so called *shape factor* of the flywheel

$$k_{sh} = e_k \frac{\rho}{\sigma} = 112\ 294 \frac{7800}{1\ 445\ 225\ 912} = 0.606$$

GFRP flywheel

For the GFRP disk, the parameters calculated identically as for steel disk are: $m = 19.1$ kg, $J = 0.504$ kgm^2, $\Omega = 5236$ rad/s. $E_k = 6.9$ MJ, $e_k = 362.6$ kJ/kg, $\sigma = 1.488$ GPa, $\sigma/\rho = 0.584$ MNm/kg and $k_{sh} = 0.606$.

In spite of larger diameter, the GFRP flywheel is lighter, can spin at higher speed, stores more energy (3.17 MJ for steel flywheel and 6.9 MJ for GFRP flywheel) and has higher material index σ/ρ (0.185 MNm for steel flywheel and 0.584 MNm/kg for GFRP flywheel). The values of material index do not exceed the maximum values given in Table 9.4.

Brushless Motors of Special Construction

10.1 Single-phase motors

Many industrial and domestic applications, computers, office equipment and instruments require small single-phase auxiliary electric motors rated up to 200 W fed from 50 or 60 Hz single-phase mains. Applications, such as fans, sound equipment and small water pumps were a decade ago the exclusive domain of the very low-efficiency shaded pole induction motor. Owing to cost-effective magnets and power electronics, PM brushless motors can successfully replace cage induction motors. Higher efficiency and smaller sizes of PM brushless motors reduce the energy consumption, volume and mass of apparatus.

10.1.1 Single-phase two-pole motors with nonuniform air gap

A salient pole PM synchronous motor can be designed with nonuniform air gap as a self-starting motor. With regard to the stator magnetic circuit, two types of construction can be identified: U-shaped, two-pole asymmetrical stator magnetic circuit (Fig. 10.1) and two-pole symmetrical stator magnetic circuit (Fig. 10.2). In both asymmetrical and symmetrical motors, the nonuniform air gap can be smooth (Figs 10.1a and 10.2a) or stepped (Figs 10.1b and 10.2b). The leakage flux of a U-shaped stator is higher than that of a symmetrical stator. Nonuniform air gap, i.e., wider at one edge of the pole shoe than at the opposite edge, provides the starting torque as a result of misalignments of the stator and rotor field axes at zero current state.

The *rest angle* θ_0 of the rotor is the angle between the center axis of the stator poles and the axis of the PM rotor flux. These motors are self-starting only when, with the armature current $I_a = 0$, the angle $\theta_0 > 0$. The largest starting torque is achieved when the rest angle $\theta_0 = 90^0$. The motor constructions shown in Fig. 10.1 limit the rest angle to $\theta_0 \leq 5 \ldots 12^0$, which consequently results in a small starting torque. The motors shown in Fig. 10.2 can theoretically achieve θ_0 close to 90^0 [8].

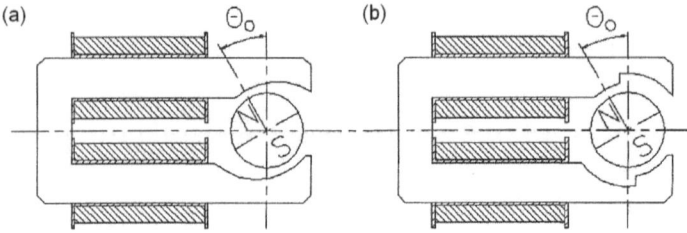

Fig. 10.1. Single-phase PM synchronous motor with asymmetrical stator magnetic circuit: (a) smooth nonuniform air gap, (b) stepped nonuniform air gap.

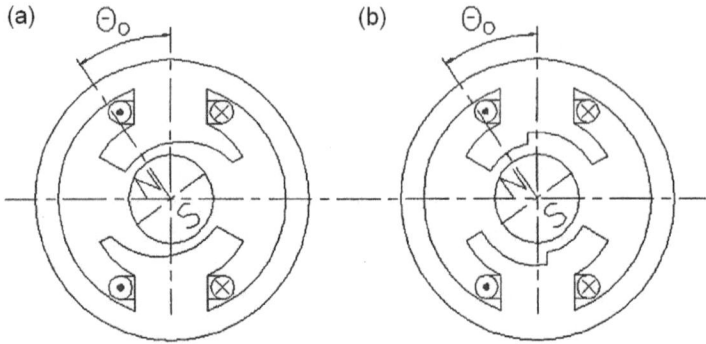

Fig. 10.2. Single-phase PM synchronous motor with symmetrical stator magnetic circuit: (a) smooth nonuniform air gap, (b) stepped nonuniform air gap.

With zero current in the stator winding, the rest angle $\theta_0 > 0$ as the attractive forces between PM poles and the stator stack align the rotor center axis with the minimum air gap (minimum reluctance). After switching on the stator voltage the stator magnetic flux will push the PM rotor towards the center axis of the stator poles. The rotor oscillates with its *eigenfrequency*. If this eigenfrequency is close enough to the stator winding supply frequency, the amplitude of mechanical oscillations will increase and the motor will begin to rotate continuously. The eigenfrequency depends on the moment of inertia of the rotor and mechanical parameters. Larger motors with lower eigenfrequency thus require lower supply frequencies. It is important to know the dynamic behavior of these motors at the design stage to ensure the desired speed characteristics.

The advantages of these motors are their simple mechanical construction and relatively high efficiency at small dimensions and rated power. Owing to a simple manufacturing process, it is easier to fabricate the asymmetrical stator than the symmetrical stator. The symmetrical design can achieve almost the maximum possible starting torque (θ_0 close to 90^0), while the asymmetrical design has a relatively small starting torque. It should be noted that the direction of rotation in both designs cannot be predetermined.

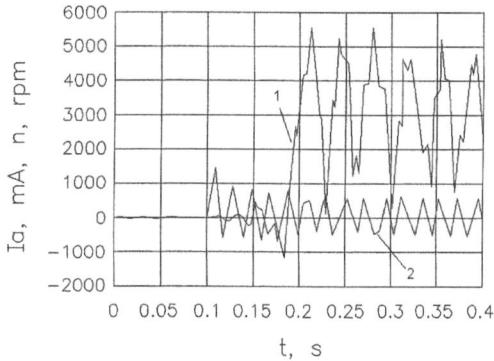

Fig. 10.3. Speed and current versus time in self-starting mode for a two-pole synchronous motor with oscillatory starting: 1 — speed, 2 — current.

The disadvantage is the limited size of the motor if it is to be utilized in its self-starting mode. Fig. 10.3 shows the oscillations of speed and current during start-up of a lightly loaded small motor [285]. If the load increases, the oscillations of speed decrease more quickly than those in Fig. 10.3.

The electromagnetic torque developed by the motor can be found by taking the derivative of the magnetic coenergy with respect of the rotor angular position θ — eqn (3.75). If the magnetic saturation is neglected, the torque developed by the motor with $p = 1$ is [8]

$$T_d = i_a \Psi_f \sin \theta - T_{drelm} \sin[2(\theta - \theta_0)] \tag{10.1}$$

where the electrodynamic torque is given by the first term, the reluctance torque is given by the second term, $\Psi_f = N_1 \Phi_f$ is the peak linkage flux, θ is the angle of rotation related to the d-axis and T_{drelm} is the peak value of the reluctance torque. When considering characteristics of the motor coupled to the machine to be driven, the following differential equations for the electrical and mechanical balance can be written [7]:

$$V_{1m} \sin(\omega t + \varphi_v) = i_a R_1 + L_1 \frac{di_a}{dt} + \frac{d\theta}{dt} \Psi_f \sin \theta \tag{10.2}$$

$$J_s \frac{d^2\theta}{dt^2} = i_a \Psi_f \sin \theta - T_{drelm} \sin[2(\theta - \theta_0)] - T_{sh} \tag{10.3}$$

where V_{1m} is the peak terminal voltage, φ_v is the voltage phase angle, i_a is the instantaneous stator (armature) current, R_1 is the stator winding resistance, L_1 is the stator winding inductance, J_s is the inertia of the drive system together with the load and T_{sh} is the external load (shaft) torque.

To simplify the analysis, it is advisable to calculate the characteristics for an unloaded motor, i.e., $T_{sh} = 0$, with the reluctance torque neglected $T_{drelm} = 0$. The equation of mechanical balance (1.16) at no load takes the following simple form [7]:

$$J_s \frac{d^2\theta}{dt^2} = i_a \Psi_f \sin\theta \qquad (10.4)$$

Solving eqn (10.4) for $i_a(t)$ and inserting into eqn (10.2), and then differentiating for θ gives [7]:

$$\frac{V_{1m}}{2}\sin(\omega t + \varphi_v) = \frac{V_{1m}}{4}[\sin(\omega t + \varphi_v + 2\theta) + \sin(\omega t + \varphi_v - 2\theta)]$$

$$+ \frac{J_s R_1}{\Psi_f}\frac{d^2\theta}{dt^2}\sin\theta + \frac{J_s L_1}{\Psi_f}\left(\frac{d^3\theta}{dt^3}\sin\theta - \frac{d^2\theta}{dt^2}\frac{d\theta}{dt}\cos\theta\right)$$

$$+ \frac{d\theta}{dt}\frac{\Psi_f}{4}[3\sin\theta - \sin(3\theta)] \qquad (10.5)$$

Eqn (10.5) is used to determine whether or not the motor can run-up and pull into synchronism, and how silently it can operate [7].

10.1.2 Single-phase multi-pole motors with oscillatory starting

Single-phase PM motors rated up to a few watts can also have a cylindrical magnetic circuit, multi-pole stator and rotor and the self-starting ability as a result of mechanical oscillations. The magnetic circuit of such a motor is schematically sketched in Fig. 10.4a. The PM rotor is shaped as a *six-point star*. Each pole is divided into two parts. Distances between two parts of poles are not equal: the distance between every second pole is equal to 2τ; for the remaining poles the distance is 1.5τ. The stator toroidal winding is fed from a single-phase source and creates 36 poles. The number of poles is equal to the number of teeth. Under the action of a.c. current the stator poles change their polarity periodically and attract the points of rotor star alternately in one or the other direction. The rotor starts to oscillate. The amplitude of mechanical oscillations increases until the rotor is pulled into synchronism.

The direction of rotation is not determined. To obtain the requested direction, the motor must be equipped with a *mechanical blocking system* to oppose the undesirable direction of rotation, e.g., spiral spring, locking pawl, etc.

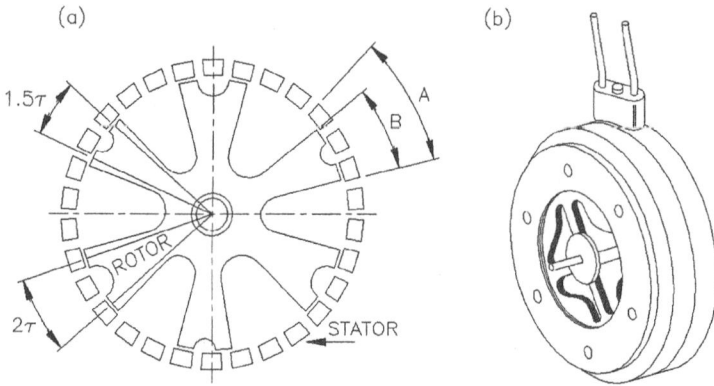

Fig. 10.4. Single-phase multi-pole PM synchronous motor with oscillatory starting: (a) magnetic circuit, (b) general view.

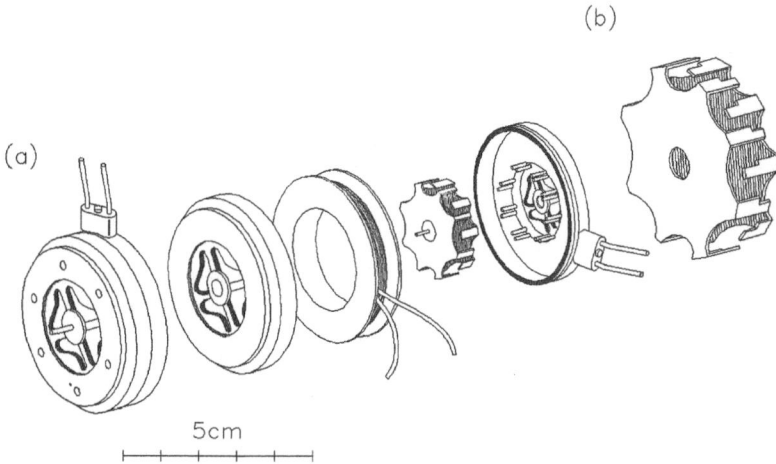

Fig. 10.5. Single phase multi-pole PM synchronous motor with oscillatory starting manufactured by AEG: (a) dismantled motor, (b) rotor construction.

Owing to constant speed, these motors have been used in automatic control systems, electric clocks, sound and video equipment, movie projectors, impulse counters, etc.

The PM single-phase motor with oscillatory starting for impulse counters manufactured by AEG has a disk rotor magnetized axially as shown in Fig. 10.5. Mild steel *claw poles* are placed at both sides of the disk-shaped PM (Fig. 10.6). To obtain the requested direction of rotation each pole is asymmetrical.

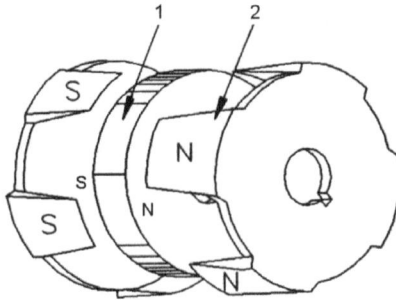

Fig. 10.6. PM claw-type rotor. 1 — disk-shaped PM magnetized axially, 2 — mild steel pole.

The stator consists of a toroidal single-phase winding with inner salient poles, the number of which is equal to the number of rotor poles. The stator poles are distributed asymmetrically pair by pair. One pole of each pair is shorted by a ring. Polarity of each pole pair is the same and changes according to the stator a.c. current. The oscillatory starting is similar to that of the motor shown in Fig. 10.4. Specification data of SSLK–375 AEG motor is: $P_{out} \approx 0.15$ W, $P_{in} \approx 2$ W, $V_1 = 110/220$ V, $2p = 16$, $f = 50$ Hz, $n_s = 375$ rpm, $\eta = 7.5\%$, starting torque $T_{st} = 0.785$ Nm with regard to 1 rpm, shaft torque at synchronous speed $T_{sh} = 1.47$ Nm. This motor can be used for measurement of d.c. pulses. The stator winding fed with a d.c. pulse can produce a magnetic field in the air gap which turns the rotor by one pole pitch, i.e. each pulse turns the rotor by 22.5^0 since the motor has 16 poles and $360^0/16 = 22.5^0$. Thus, one full revolution corresponds to 16 pulses. Such motors are sometimes equipped with a mechanical reducer. In this case the angle of rotation is determined by the gear ratio.

PM multi-pole synchronous motors with oscillatory starting are characterized by high reliability, small dimensions, relatively good power factor and poor efficiency as compared with their low rated power and price.

10.1.3 Single-phase cost-effective PM brushless motors

Single-phase and two-phase PM brushless motors are used in many applications including computer fans, bar code scanners (Table 10.1), bathroom equipment, vision and sound equipment, radio-controlled toys, mobile phones (vibration motors), automobiles, etc.

A single-phase PM brushless motor behaves as a self-starting motor if the input voltgae is controlled with regard to the rotor position. The motor is self-starting independently of the North or South rotor pole facing Hall sensor which is placed in the q-axis of the stator. The desired direction of rotation can be achieved with the aid of the electronic circuitry which produces the stator field with a proper phase shift related to the rotor position [151].

Table 10.1. Specifications of bar code scanner PM brushless motors manufactured by Sunonwealth Electric Machine Industry, Kaohsiung, Taiwan.

Model	BC105007	M11CBC03	M11BBC01	M12DBC02
Diameter, mm	19	19	17.8	32
Length, mm	11	11	8.8	19.3
Voltage, V	4.8	5	3.5	12
No-load speed, rpm	7600	8200	9000	9900
No-load current, mA	87	109	47	200
Starting torque, 10^{-3} Nm	0.54	0.59	0.31	8.33
Type of bearings	sleeve	ball	sleeve	ball
Acoustic noise, dBA	45 (5 cm)	45 (5 cm)	50 (5 cm)	50 (10 cm)
Operating temperature, ^{0}C	5 to 70	5 to 70	−10 to 60	−10 to 60

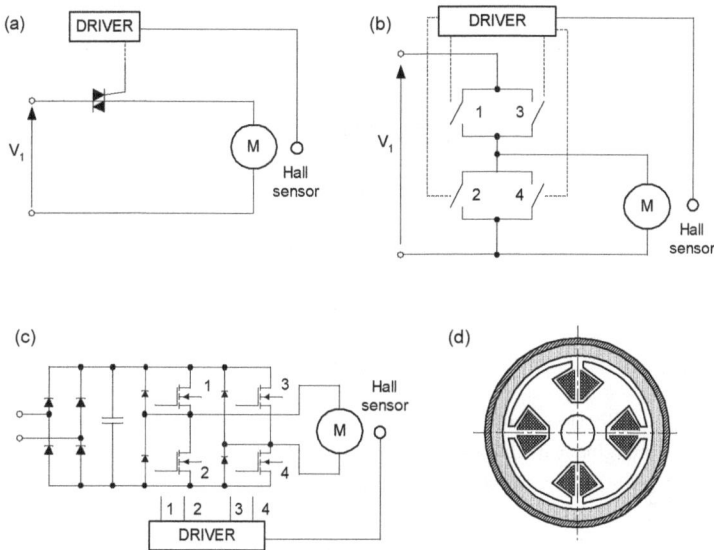

Fig. 10.7. Single-phase PM brushless motor drives: (a) triac converter; (b) four-switch converter; (c) full-bridge converter; (d) cost effective single phase PM brushless motor.

The simplest electronics circuitry with a triac is shown in Fig. 10.7a. The triac is switched on only when the supply voltage and the motor EMF are both positive or negative [151]. The task of the electronics circuitry is to feed the motor at starting and at transient operation, e.g., sudden overload. The motor is also equipped with a synchronization circuit which bypasses the power electronics converter and connects the motor directly to the mains at rated speed [151].

In the line-frequency variable-voltage converter shown in Fig. 10.7b the switches 1 and 3 are in on-state when the supply voltage and the motor EMF are both positive or negative [151]. The switches 2 and 4 conduct the current when the supply voltage and motor EMF have opposite signs and the voltage across the motor terminals is zero. Converters according to Figs 10.7a and 10.7b produce pulse voltage waveforms, low frequency harmonics (torque ripple) and transient state at starting is relatively long.

The d.c. link power electronics converter according to Fig. 10.7c provides the starting transient independent of the supply frequency and reduces the harmonic distortion of the inverter input current. The motor is supplied with a square voltage waveform at variable frequency.

The cost effective single-phase brushless motor has a salient pole stator and ring-shaped PM. Fig. 10.7d shows a four pole motor with external rotor designed for cooling fans of computers and instruments (see also Fig. 6.41).

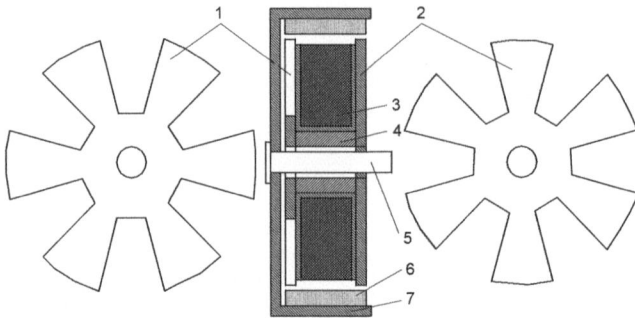

Fig. 10.8. Cost-effective single-phase PM TFM. 1 — stator left star-shaped steel disk, 2 — stator right star-shaped steel disk, 3 — stator winding (single coil), 4 — steel bush, 5 — shaft, 6 — ring PM magnetized radially, 7 — steel hub.

The cost-effective single phase PM brushless motor can also be designed as a TFM with star-shaped stator core embracing the single-coil winding [156]. In the single phase motor shown in Fig. 10.8 the stator has six teeth, i.e., half of the rotor PM poles. The stator left and right star-shaped steel disks are shifted by 30 mechanical degrees, i.e., one pole pitch to face rotor poles of opposite polarity. In this way closed transverse magnetic flux paths are created. The cogging torque can be reduced by proper shaping of the stator salient poles.

10.2 Actuators for automotive applications

PM brushless actuators can provide high torque density and conversion of electric energy into mechanical energy at high efficiency. Most rotary electromechanical actuators for motor vehicles must meet the following requirements [240, 241, 242]:

- working stroke less than one full revolution (less than 360^0)
- high torque
- symmetrical performance both in left and right directions of rotation
- the same stable equilibrium position for the PM cogging torque and electromagnetic torque.

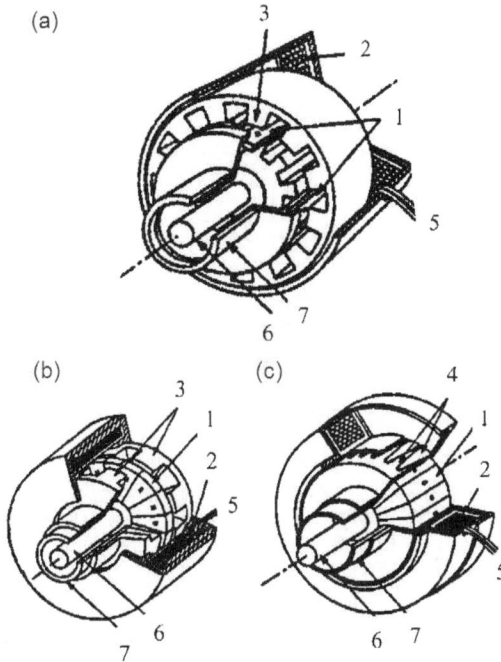

Fig. 10.9. Rotary PM actuators for automotive applications: (a) cylindrical, (b) disk type, (c) claw pole. 1 — PM, 2 — excitation coil, 3 — toothed magnetic circuit, 4 — claw poles, 5 — terminal leads, 6 — torsion bar, 7 — pinion. Courtesy of Delphi Technology, Shelby, MI, U.S.A.

Basical topologies of PM brushless actuators are shown in Fig. 10.9 [240, 241, 242]. The stator consists of an external magnetic circuit and excitation coil. The rotor is comprised of two mechanically coupled toothed ferromagnetic structures and a multipole PM ring or disk inserted between them (cylindrical and disk actuators) or below them (claw pole actuator). The number of salient teeth of the magnetic parts is equal to half of the PM poles and depends on the required angle of limited angular motion. The limited angular motion of the PM with respect of the toothed ferromagnetic parts is performed in addition to the full rotary motion of the rotor [241]. This feature requires an additional air gap between the stator and rotor which deteriorates the performance. Since the torque is proportional to the number of independent magnetic circuit sections equal to the number of PM pole pairs, multipole rotary actuators provide the so-called "gearing effect".

With the increase of the PM energy and applied magnetic field strength the magnetic flux density distribution in the air gap changes from sinusoidal to trapezoidal when the saturation level is approached. The actuators shown in Fig. 10.9 have been successfully implemented in General Motor's *Magnasteer* power steering assist system [240, 241, 242]. Both sintered and die quench (Table 2.4) NdFeB magnets have been used.

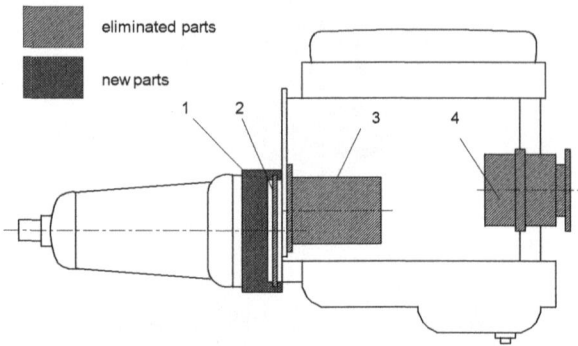

Fig. 10.10. Direct drive ISG replaces the classical generator (alternator), starter, flywheel, pulley and belt. 1 — ISG, 2 — flywheel, 3 — classical starter, 4 — classical generator (alternator).

10.3 Integrated starter-generator

The *integrated starter-generator* (ISG) replaces the conventional starter, generator and flywheel of the engine, integrates starting and generating in a single electromechanical device and provides the following auxiliary functions:

- automatic vehicle start-stop system which switches off the combustion engine at zero load (at traffic lights) and automatically restarts engine in less than 0.3 s when the gas pedal is pressed;
- pulse-start acceleration of the combustion engine to the required cranking idle speed and only then the combustion process is initiated;
- boost mode operation, i.e., the ISG operates as electric auxiliary motor to shortly drive or accelerate the vehicle at low speeds;

- regenerative mode operation, i.e., when the vehicle brakes, the ISG operates as an electric generator, converts mechanical energy into electrical energy and helps to recharge the battery;
- active damping of torsional vibration which improves driveability.

Automatic vehicle start-stop system and pulse-start acceleration improves the fuel economy up to 20% and reduces emissions up to 15%. The rotor of the ISG, like a flywheel, is axially fastened on the crankshaft between the combustion engine and clutch (transmission) as shown in Fig. 10.10. Because the application of ISG eliminates the traditional generator (alternator), starter, flywheel, pulley and generator belt, the number of components of the vehicle propulsion system is reduced.

The ISG rated between 8 and 20 kW and operates on a 42 V electrical system. Both induction and PM brushless machines can be used. To reduce the cost of ISGs, ferrite PMs are economically justified. The ISG is a flat machine with the outer stator diameter approximately 0.3 m and number of poles $2p \geq 10$. Buried magnet rotor topology according to Fig. 5.1j is recommended. For example, for a three-phase 12 pole machine the number of the stator slots is 72. The ISG is interfaced with the vechicle electrical system with the aid of 42 V d.c. six switch inverter.

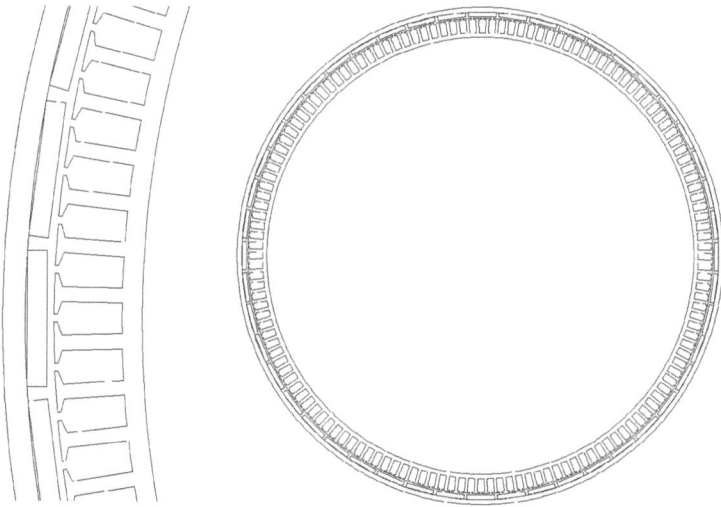

Fig. 10.11. Outline of large diameter PM brushless motor with $2p = 48$ and $s_1 = 144$.

Table 10.2. Comparison of large-diameter 20 Nm, 800 rpm PM brushless, induction and switched reluctance motors.

Specifications	PM brushless motor	Cage induction motor	Switched reluctance motor
Number of poles	48	48	48 (stator)
Number of stator slots	144	144	48
Number of rotor slots	–	131	36
Outer diameter, mm	468	540	540
Inner diameter, mm	411	470	470
Axial width of stack, mm	6	6	6
Peak torque, Nm	36.7	38.0	37.0
Current at peak torque, A	11.0	28.6	12.0
Current density at peak torque, A/mm^2	10.6	11.0	11.6
Continuous torque, Nm	20.0	22.5	19.0
Current density a continuous torque, A/mm^2	6.0	6.5	19.0
Efficiency at peak torque, %	85.5	70.9	77.2
Efficiency at continuous torquem %	85.0	65.7	78.3
Motor mass, kg	2.41	7.15	4.15
Peak torque density, Nm/kg	15.2	5.3	8.9

10.4 Large diameter motors

A thin annular ring with a large diameter–to–length ratio, allows the motor to be wrapped around a driven shaft. The direct connection of the motor to the load effectively eliminates problems of torsional resonances due to couplings. The absence of gears removes errors caused by friction and backlash, creating a high performance positioning system. Direct integration with the shaft also results in minimal volume of the motor. A large-diameter PM brushless motor with inner stator is shown in Fig. 10.11.

The outer diameters are up to 850 mm, the outer diameter–to–axial stack width ratio is 20 to 80 and number of poles $2p$ is from 16 to 64. Comparison of large diameter 20 Nm, 800 rpm PM brushless, induction and switched reluctance motors is given in Table 10.2. The PM brushless motor has the highest efficiency, lowest stator winding current density, lowest mass and highest torque density.

The direct-drive large-diameter PM brushless motor is ideally suited to high acceleration applications requiring improved response for rapid start–stop action. Applications include semiconductor manufacturing, laser scanning and printing, machine tool axis drives, robot bases and joints, coordinate measuring systems, stabilized gun platforms and other defense force equipment.

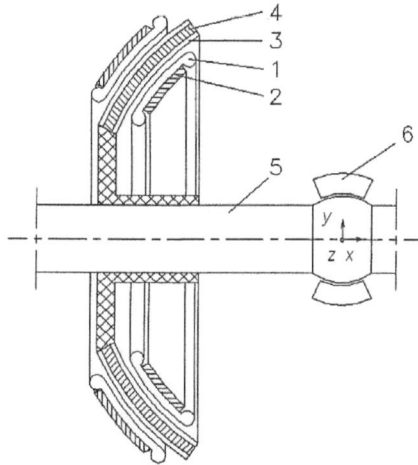

Fig. 10.12. Spherical three-axis motor with a PM rotor: 1 — armature winding, 2 — armature core, 3 — PMs, 4 — rotor core, 5 — shaft, 6 — spherical bearing. Courtesy of TH Darmstadt, Germany [11, 12].

10.5 Three-axis torque motor

The *three-axis torque motor* can be designed as a PM or reluctance *spherical motor*. It can be used, e.g., in airborne telescopes. This motor has double-sided stator coils, similar to disk motors. There are usually slotless stator coils to reduce torque pulsations. The rotor can rotate 360^0 around the x-axis, and by only a few degrees in the y- and z-axis.

Fig. 10.12 shows an example of a three-axis torque motor, designed and tested at the Technical University of Darmstadt [11, 12]. The rotor segments cover a larger angle than the stator segments. The overhang ensures a constant torque production for the range of y- and z-axis movement — 10^0 for this motor (Fig. 10.12).

In designs with a complete spherical rotor three stator windings shaped as spherical sections have to envelop the rotor. The three windings correspond to the three spatial axes and are thus aligned perpendicularly to each other. These three stator windings can be replaced with just one winding when the spherical rotor and stator have a diameter smaller than the diameter of the sphere, as shown in Fig. 10.12. The stator winding can then be divided into four sections with each section being controlled separately. Depending on the

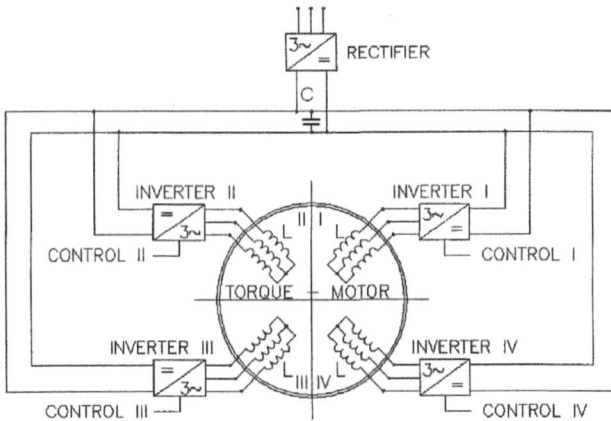

Fig. 10.13. Four inverters suppling the winding sections of a spherical motor. Courtesy of TH Darmstadt, Germany [11, 12].

number of the winding sections activated and on the direction of the MMF wave, each of the three torques may be produced. The four inverters supplying the winding sections are shown in Fig. 10.13.

Owing to the 3D spherical structure, the performance calculation of this type of motor is usually done using the FEM.

10.6 Slotless motors

Cogging effect can be eliminated if PM brushless motors are designed without stator slots, i.e., the winding is fixed to the inner surface of the laminated stator yoke. Sometimes, a toroidal winding, e.g., coils wound around the cylindrical stator core, are more convenient for small motors. In addition to zero cogging torque, *slotless PM brushless motors* have the following advantages over slotted motors:

- higher efficiency in the higher speed range which makes them excellent small power high speed motors (Table 9.2);
- lower winding cost in small sizes;
- higher winding–to–frame thermal conductivity;
- smaller eddy current losses in rotor retaining ring and/or PMs due to larger air gap and less higher harmonic contents in the air gap magnetic flux density distribution;
- lower acoustic noise.

The drawbacks include lower torque density, more PM material, lower efficiency in the lower speed range and higher armature current. With the increase in the total air gap (mechanical clearance plus winding radial thickness) the air gap magnetic flux density according to eqn (2.14) decreases and consequently so does the electromagnetic torque. The volume (radial height) of PMs must significantly increase to keep the torque close to that of an equivalent slotted motor [149].

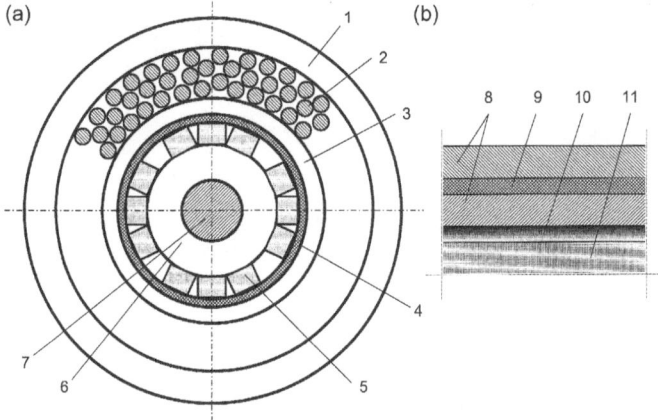

Fig. 10.14. Construction of a high speed slotless PM brushless motor: (a) cross section; (b) rotor composite retaining sleeve (bandage). 1 — stator yoke (back iron), 2 — stator conductors, 3 — air gap, 4 — retaining sleeve, 5 — PM, 6 — rotor core, 7 — shaft, 8 — carbon fiber, 9 — carbon woven, 10 — glass fiber [97].

A high speed slotless PM brushless motor is shown in Fig. 10.14 [97]. PMs are protected against centrifugal forces with the aid of a composite retaining sleeve. Two rings made of GFRP can be inserted into the stator core, one ring at each end of the stator. The outer diameter of the GFRP ring is equal to the stator core inner diameter and the inner diameter of the GFRP ring is determined by the mechanical clearance between the stator and rotor. Then, holes are drilled in GFRP rings at the conductor positions and conductors are threaded through these holes [97].

The armature reaction reactances X_{ad} and X_{aq} of slotless motors with surface configuration of PMs can be calculated with the aid of eqns (5.31) and (5.33), in which $g' = g + h_M/\mu_{rrec}$ and $q'_q = g_q$, provided that both g and g_q comprise the radial thickness of the armature winding. Owing to large nonferromagnetic gap, the magnetic saturation both in the d and q axis can be neglected.

Applications of slotless motors include medical equipment (handpieces, drills and saws), robotic systems, test and measurement equipment, pumps, scanners, data storage, semiconductor handling. Specifications of small two-pole high speed motors with slotless armature winding are given in Table 9.2.

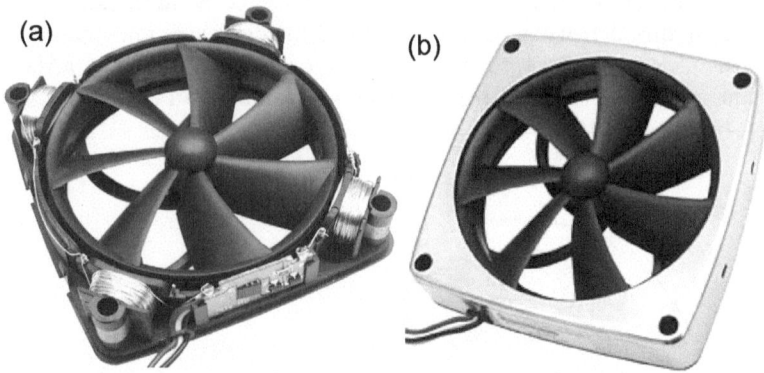

Fig. 10.15. Construction of a TDF: (a) PM brushless motor, (b) TDF for CPU coolers. Courtesy of C.J. Chen, CEO, Yen Sun Technologies, Kaohsiung, Taiwan.

10.7 Tip driven fan motors

In the *tip driven fan* (TDF) the outer stator of PM brushless motor consists of four portions, one at each corner (Fig. 10.15a). A cylindrical PM embraces the tips of fan blades. In comparison with traditional fans, the motor hub area is reduced by 75%, the air flow is increased by 30% and the efficiency of heat dissipation is increased by 15%. Typical specifications of a TDF for CPU coolers (Fig. 10.15b) are: dimensions 75 mm × 75 mm × 75 mm, number of poles $2p = 12$, power consumption 2 W, rated current 0.17 A, rated speed 4500 rpm, rated voltage 12 C d.c. maximum air flow 0.014 m^3/s, noise level 34 dB(A). Corner-located PM brushless motors rated at 120 to 600 W and speed 2700 rpm are used in cooling TDFs with volume flow 0.335 to 0.8 m^3/s (ABB, Flakt Oy, Sweden).

Numerical examples

Numerical example 10.1

Find the performance characteristics of the single-phase two-pole PM synchronous motor with oscillatory starting shown in Fig. 10.16a. The input voltage is $V_1 = 220$ V, input frequency is $f = 50$ Hz, conductor diameter is $d_a = 0.28$ mm, number of turns per coil is $N_1 = 3000$, number of coils is $N_c = 2$ and NdFeB PM diameter is $d_M = 23$ mm.

Solution

For the class F enamel insulation of the 0.28-mm conductor, its diameter with insulation is 0.315 mm. If the conductors are distributed in 20 layers ($20 \times 0.315 = 6.3$ mm), 150 turns in each layer ($150 \times 0.315 = 47.25$ mm), they require a space of $6.3 \times 47.25 = 297.7$ mm^2. Assuming a spool thickness of 1 mm, 9 layers of insulating paper of thickness 0.1 mm (every second layer) and an external protective insulation layer of 0.6 mm, the dimensions of the coil are: length $47.25 + 2 \times 1.0 \approx 50$ mm and thickness $6.3 + 9 \times 0.1 + 0.6 = 7.8$ mm plus the necessary spacing, in total, the radial thickness will be about 8.6 mm.

The average length of the stator turn on the basis of Fig. 10.16a is $l_{1av} = 2(9 + 8.6 + 20 + 8.6) = 92.4$ mm.

The resistance of the stator winding is calculated at a temperature of 75^{0}C. The conductivity of copper at 75^0 C is $\sigma_1 = 47 \times 10^6$ S/m. The cross section area of a conductor $s_a = \pi d_a^2/4 = \pi \times (0.28 \times 10^{-3})^2)/4 = 0.0616 \times 10^{-6}$ m^2.

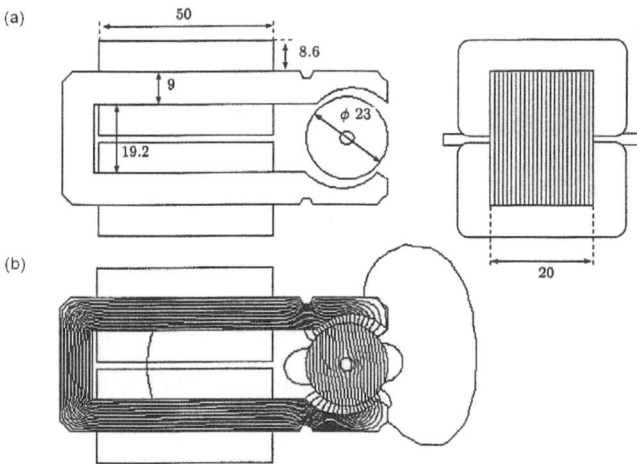

Fig. 10.16. Single-phase PM synchronous motor with asymmetrical stator magnetic circuit and smooth nonuniform air gap: (a) dimensions, (b) magnetic flux plot in rotor's rest position of $\theta_0 = 5^0$ (the rest angle is shown in Fig. 10.1).

The total winding resistance of the two coils is

$$R_1 = \frac{2l_{1av}N_1}{\sigma_1 s_a} = \frac{2 \times 0.0924 \times 3\,000}{47 \times 10^6 \times 0.0616 \times 10^{-6}} = 191.5\Omega$$

The calculation of the winding inductance and reluctance torque is done using a two-dimensional finite element model and the energy/current perturbation method (Chapter 3, Section 3.12.2). Fig. 10.16b shows a flux plot of the motor in its rest position at an angle of 5^0. The reluctance torque and total torque as functions of rotor angle are shown in Fig. 10.17a. The stator current at $V_1 = 220$ V is about 0.26 A which corresponds to the current density $J_a =$

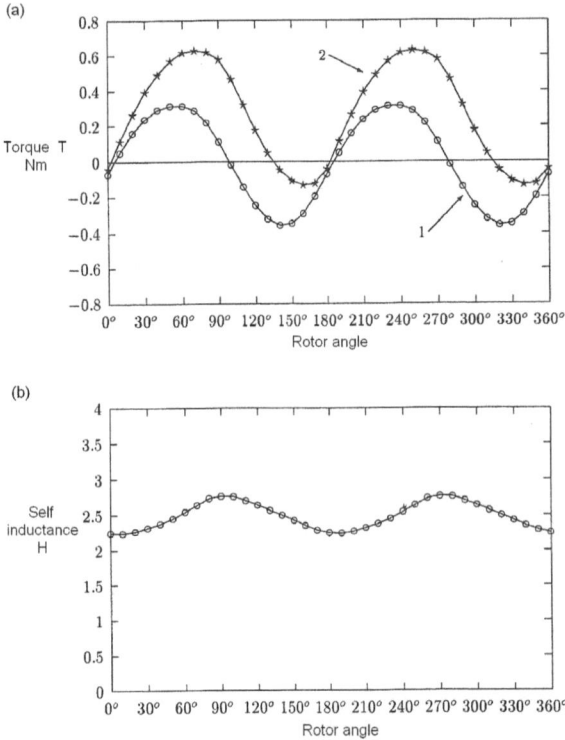

Fig. 10.17. Characteristics of a single-phase two-pole PM synchronous motor: (a) torques versus rotor angle; (b) self-inductance versus rotor angle at $I_a = 0.275$ A. 1 — reluctance torque at $I_a = 0$; 2 — total torque at $I_a = 0.26$ A.

$0.26/0.0616 = 4.22$ A/mm^2. The self-inductance as a function of the rotor angle is plotted in Fig. 10.17b.

Numerical example 10.2

A 1.5 kW, 1500 rpm, 50 Hz PM brushless motor has the stator inner diameter $D_{1in} = 82.5$ mm and air gap (mechanical clearance) in the d-axis $g = 0.5$ mm. The height of surface NdFeB PMs is $h_M = 5$ mm, remanence $B_r = 1.25$ T and coercivity $H_c = 965$ kA/m. The standard slotted stator has been replaced by a slotless stator with inner diameter $D'_{in} = 94$ mm (inner diameter of winding) and winding thickness $t_w = 6.5$ mm. The rotor yoke diameter has proportionally been increased to keep the same mechanical clearance $g = 0.5$ mm between magnets and stator winding.

Neglecting the saturation of magnetic circuit and armature reaction find the torque of the slotless motor. How should the rotor be redesigned to obtain 80% of the torque of the standard slotted motor?

Solution

The shaft torque of the slotted motor

$$T_{sh} = \frac{1500}{2\pi(1500/60)} = 9.55 \text{ Nm}$$

The magnetic flux density in the air gap of the slotted motor according to eqn (2.14)

$$B_g \approx \frac{1.25}{1 + 1.03 \times 0.5/5} = 1.133 \text{ T}$$

where the relative recoil magnetic permeability $\mu_{rrec} = 1.25/(0.4\pi \times 10^{-6} \times 965,000) = 1.03$. The magnetic flux density of a slotless motor

$$B'_g \approx \frac{1.25}{1 + 1.03 \times 7/5} = 0.512 \text{ T}$$

where the total nonferromagnetic air gap $t_w + g = 6.5 + 0.5 = 7$ mm.

The shaft torque of the slotless motor using PMs with the same height

$$T'_{sh} \approx \frac{D'_{in}}{D_{in}} \frac{B'_{mg}}{B_{mg}} T_{sh} = \frac{94}{82.5} \frac{0.512}{1.133} 9.55 = 4.917 \text{ Nm}$$

The slotless motor can develop 80% torque of the slotted motor if the magnetic flux density will increase to

$$B''_{mg} = \frac{T''_{sh}}{T_{sh}} \frac{D_{1in}}{D'_{1in}} B_{mg} = 0.8 \frac{82.5}{94} 1.133 = 0.7955 \text{ T}$$

To obtain this value of the magnetic flux density the height of the PM according to eqn (2.14) must increase to

$$h'_M = \frac{B''_{mg} g' \mu_{rrec}}{B_r - B_{mg}"} = \frac{0.7955 \times 7 \times 1.03}{1.25 - 0.7955} \approx 12.6 \text{ mm}$$

The volume and cost of PMs will approximately increase $h'_M/h_M = 12.6/5.0 = 2.52$ times.

Stepping Motors

11.1 Features of stepping motors

A rotary *stepping motor* is defined as a singly-excited motor converting electric pulses into discrete angular displacements. It has salient poles both on the stator and rotor but only one polyphase winding. Normally, the stator carries the winding which is sometimes called a *control winding*. The input signal (pulse) is converted directly into a requested shaft position, without any rotor position sensors or feedback. Stepping motors are highly reliable and low-cost electrical machines compatible with modern digital equipment.

Stepping motors find their applications in speed- and position-control systems without expensive feedback loops (open-loop control). Examples are computer peripherals (printers, scanners, plotters), camera, telescope and satellite dish positioning systems, robotic arms, numerically controlled (NC) machine tools, etc. A typical control circuit of a stepping motor (Fig. 1.7) consists of an *input controller*, *logic sequencer* and *power driver*. Output signals (rectangular pulses) of a logic sequencer are transmitted to the input terminals of a power driver which distributes them to each of the phase windings (commutation). Stepping motors can be classified in three ways:

- with active rotor (PM rotor)
- with reactive rotor (reluctance type)
- hybrid motors.

A stepping motor should meet the following requirements: very small step, bidirectional operation, noncumulative positioning error (less than ±5% of step angle), operation without missing steps, small electrical and mechanical time constants, can be stalled without motor damage. Stepping motors provide very high torque at low speeds, up to 5 times the continuous torque of a d.c. commutator motor of the same size or double the torque of the equivalent brushless motor. This often eliminates the need for a gearbox. The most important advantages of stepping motors are

- rotational speed proportional to the frequency of input pulses
- digital control of speed and position
- open-loop control (except for special applications)
- excellent response to step commands, acceleration and deceleration
- excellent low speed–high torque characteristics
- very small steps without mechanical gears (it is possible to obtain one step per 24 h)
- simple synchronization of a group of motors
- long trouble-free life.

On the other hand, stepping motors can show the tendency of losing synchronism, resonance with multiples of the input frequency, and oscillations at the end of each step. The efficiency and speed of stepper motors are lower than those of brushless motors.

Typical applications of stepping motors are printers, plotters, X–Y tables, facsimile machines, barcode scanners, image scanners, copiers, medical apparatus, and others.

11.2 Fundamental equations

11.2.1 Step

The *step* of a rotary stepping motor is the angular displacement of the rotor due to a single input pulse, i.e.,

- for a PM stepping motor

$$\theta_s = \frac{360^0}{2pm_1} \qquad \text{or} \qquad \theta_s = \frac{\pi}{pm_1} \qquad (11.1)$$

- for a reluctance stepping motor

$$\theta_s = \frac{360^0}{s_2m_1n} \qquad \text{or} \qquad \theta_s = \frac{2\pi}{s_2m_1n} \qquad (11.2)$$

where p is the number of rotor pole pairs, m_1 is the number of stator phases, s_2 is the number of rotor teeth, $n = 1$ for a symmetrical commutation, and $n = 2$ for an asymmetrical commutation.

11.2.2 Steady-state torque

Under constant current excitation ($f = 0$) the steady-state synchronizing torque T_{dsyn} developed by the motor varies as a function of mismatch angle $p\theta$ (electrical degrees) between the direct axis of the rotor and the axis of the stator MMF, i.e.,

$$T_{dsyn} = T_{dsynm} \sin(p\theta) \tag{11.3}$$

where T_{dsynm} is the maximum synchronizing torque for $p\theta = \pm 90^0$.

11.2.3 Maximum synchronizing torque

The maximum synchronizing torque is proportional to the stator MMF iN and the rotor magnetic flux Φ_f, i.e.,

$$T_{dsynm} = piN\Phi_f = pi\Psi_f \tag{11.4}$$

where $\Psi_f = N\Phi_f$ is the peak linkage flux. The synchronizing torque increases with the number of the rotor pole pairs p.

11.2.4 Frequency of the rotor oscillations

The frequency of the rotor oscillations can be found through the solution of a differential equation of the rotor motion (Section 11.9). The approximate analytical solution gives the following result:

$$f_0 = \frac{1}{2\pi} \sqrt{\frac{pT_{dsynm}}{J}} \tag{11.5}$$

where J is the moment of inertia of the rotor. In practical stepping motors where the torque-frequency curves are nonlinear (Section 11.10.3), the main resonance frequency is slightly lower than f_0.

11.3 PM stepping motors

In a *stepping motor with an active rotor* the rotor PMs produce an excitation flux Φ_f. The two-phase control winding located on the stator salient poles receives input rectangular pulses, as in Fig. 11.1. The *synchronizing torque* is produced similarly to that in a synchronous motor. The commutation algorithm is $(+A) \rightarrow (+B) \rightarrow (-A) \rightarrow (-B) \rightarrow (+A)$.... Since the stator control winding consists of two phases (two poles per phase), the value of step for this motor is $\theta_s = 360^0/(pm_1) = 360^0/(1 \times 2) = 90^0$. It means that the rotor turns by 90^0 after each input pulse as a result of the synchronizing torque. This motor is characterized by a four-stroke commutation (four strokes per each full revolution). The value of step can also be written as $\theta_s = 360^0/(2pm_1) = 360^0/(kp) = 360^0/(4 \times 1) = 90^0$, where $k = 2m_1$ is the number of strokes per revolution.

The claw-pole *canstack motor* shown in Fig. 11.2 is a type of PM stepping motor [70]. It is essentially a low-cost, low-torque, low-speed machine ideally suited to applications in fields such as computer peripherals, office automation,

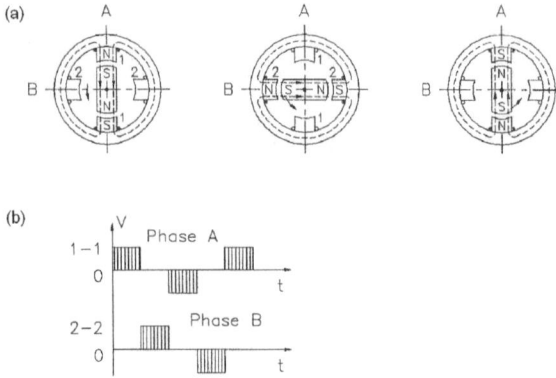

Fig. 11.1. Principle of operation of a stepping motor with active rotor: (a) rotor positions under action of input pulses, (b) phase voltage waveforms.

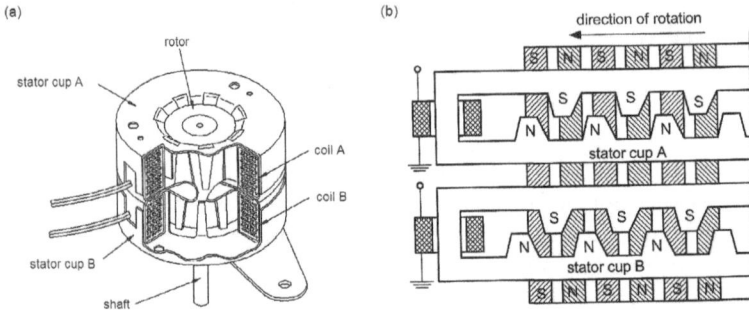

Fig. 11.2. Canstack PM stepping motor: (a) general view; (b) principle of operation.

valves, fluid metering and instrumentation. The stator looks similar to a metal can with punched teeth drawn inside to form claw-poles. In a two-phase motor, two stator phases (windings A and B and two claw-pole systems) produce heteropolar magnetic fluxes. The rotor consists of two multipole ring PMs mounted axially on the same shaft. The rotor pole pitches of two ring magnets are aligned with each other while the stator pole pitches are shifted by a half of the rotor pole pitch. The number of poles is such that the canstack motors have steps angles in the range of 7.5 to 20^0. In two-phase configuration the control current waveforms are as in Fig. 11.1b.

Fig. 11.2b shows how the canstack motor operates. With phase A being excited with positive current the rotor moves left to reach an equilibrium position. Then, the current in phase A is turned off and phase B is energized with positive current. The rotor turns further in the same direction since the stator claw-poles of phase B are shifted by the half of the rotor pole pitch. In

Table 11.1. Specifications of canstack stepping motors manufactured by Thomson Airpax Mechatronics LLC, Cheshire, CT, U.S.A.

Specifications	Unipolar 42M048C 1U	2U	Bipolar 42M048C 1B	2B
Step angle, degrees	7.5			
Step angle accuracy, %	±5			
d.c. operating voltage, V	5	12	5	12
Number of leads	6	6	4	4
Winding resistance, Ω	9.1	52.4	9.1	52.4
Winding inductance, mH	7.5	46.8	14.3	77.9
Holding torque (two phases on), 10^{-2} Nm	7.34		8.75	
Detent torque, 10^{-2} Nm	0.92			
Rotor moment of inertia, 10^{-7} kgm^2	12.5			
Rotation	bidirectional			
Ambient operating temperature, $^{\circ}$C	−20 to +70			
Mass, kg	0.144			
Dimensions, mm	42 diameter, 22 length			

the next step the phase B is turned off and phase A excited with a negative current which further advances the rotor movement to the left until it reaches the next equilibrium position. The commutation algorithm according to Fig. 11.1b provides continuous four stroke rotation.

The canstack construction results in relatively large step angles, but its overall simplicity ensures economic high-volume production at very low cost. Drawbacks include resonance effects and relatively long setting times, rough performance at low speed unless a microstepping drive is used and liability to undetected position loss as a result of operating in open loop. The current of the motor shown in Fig. 11.2 is practically independent of the load conditions. Winding losses at full speed are relatively high and can cause excessive heating. At high speeds the motor can be noisy.

Specifications of selected canstack stepping motors manufactured by Thomson Airpax Mechatronics LLC, Cheshire, CT, U.S.A. are given in Table 11.1.

11.4 Reluctance stepping motors

A *stepping motor with a reactive rotor* (reluctance stepping motor) produces a *reluctance torque*. The magnetic flux paths tend to penetrate through the magnetic circuit with minimal reluctance. The reluctance torque tends to align the rotor with the symmetry axis of the stator salient poles.

A reluctance stepping motor has a rotor made of mild steel and a salient pole stator with a multiphase control winding. The motor shown in Fig. 11.3 has a three-phase stator winding. For certain time intervals two neighboring

(a)

(b)

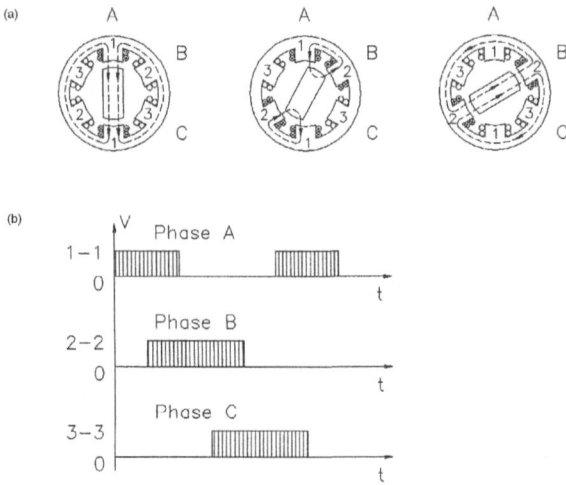

Fig. 11.3. Principle of operation of a reluctance stepping motor (with reactive rotor): (a) rotor positions under action of input pulses, (b) phase voltage waveforms.

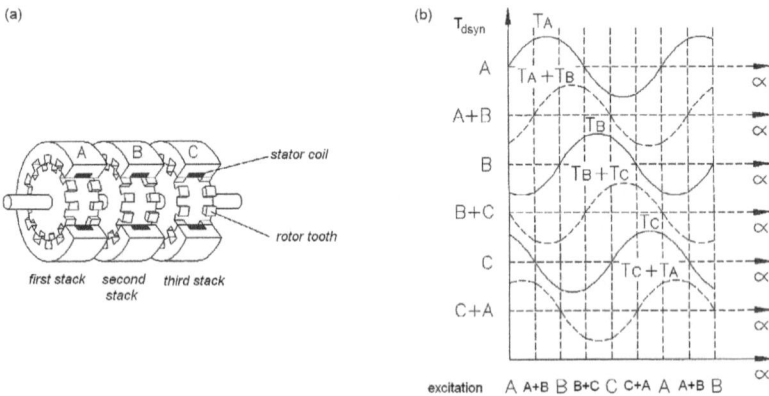

(a)

(b)

Fig. 11.4. Three-stack stepping motor: (a) construction, (b) torque as a function of the angular displacement.

phases are fed simultaneously. The commutation algorithm is: $A \rightarrow A + B \rightarrow B \rightarrow B + C \rightarrow C \rightarrow C + A \rightarrow A \ldots$. This is an asymmetrical, six-stroke commutation with a step value of $\theta_s = 360^0/(s_2 m_1 n) = 360^0/(2 \times 3 \times 2) = 30^0$.

Reluctance stepping motors are not sensitive to current polarity (no PMs) and so require a different driving arrangement from the other types.

In order to increase the output torque without degrading acceleration performance, parallel stators and rotors are added. A *three-stack stepping motor* has three stators mutually shifted from each other, and three rotors (Fig.

11.4a). This makes it possible to obtain a large number of steps with small angular displacement when three (or more) stators are subsequently fed with input pulses. The distribution of the static torque for subsequent stators fed with pulses according to the sequence $A, A + B, B, B + C, C, C + A, \ldots$ is shown in Fig. 11.4b. The aggregates $T_A + T_B$, $T_B + T_C$, and $T_C + T_A$ are the resultant torques produced by two stators. The *torque–to–inertia* ratio remains the same.

11.5 Hybrid stepping motors

A *hybrid stepping motor* is a modern stepping motor which is becoming, nowadays, more and more popular in industrial applications. The name is derived from the fact that it combines the operating principles of the PM and reluctance stepping motors. Most hybrid motors are two-phase, although five-phase versions are available. A recent development is the "enhanced hybrid" stepping motor which uses flux focusing magnets to give a significant improvement in performance, albeit at extra cost.

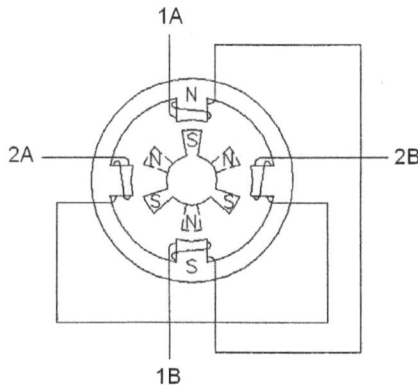

Fig. 11.5. Simple 12 step/rev hybrid stepping motor.

The operation of the hybrid stepping motor is easily understood when looking at a very simple model (Fig. 11.5) which produces 12 steps per revolution. The rotor of this machine consists of two star-shaped mild steel pieces with three teeth on each. A cylindrical PM magnetized axially is placed between the mild steel pieces making one rotor end a North pole and the other a South pole. The teeth are offset at the North and South ends as shown in Fig. 11.5. The stator magnetic circuits consists of a cylindrical yoke having four poles which run the full length of the rotor. Coils are wound on the stator poles and connected together in pairs.

11.5.1 Full stepping

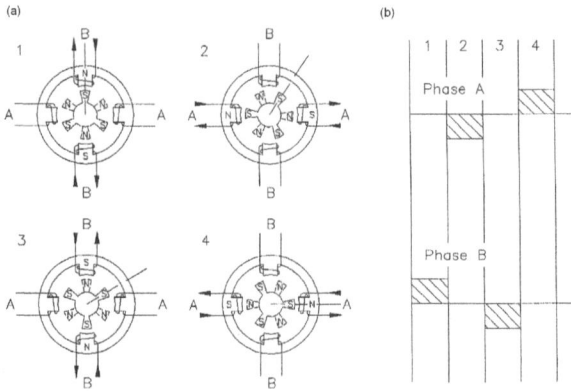

Fig. 11.6. Full stepping, one phase on: (a) rotor positions; (b) phase voltage waveforms.

With no current flowing in any of the stator windings, the rotor will tend to take up one of the rest equilibriun positions shown in Fig. 11.5. This is because the rotor PM is trying to minimize the reluctance of the magnetic flux path from one end to the other. This will occur when a pair of North and South pole rotor teeth are aligned with two of the stator poles. The torque tending to hold the rotor in one of these positions is usually small and is called the *detent torque*. The motor will have twelve possible detent positions.

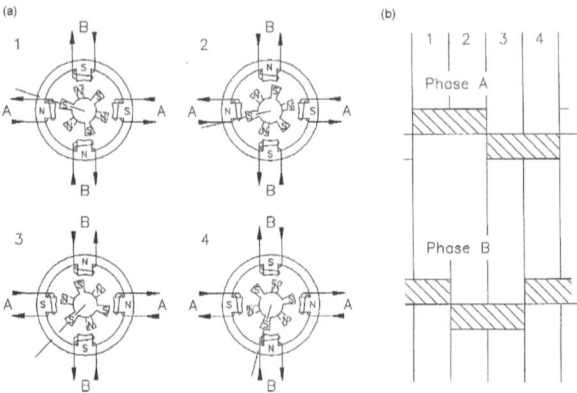

Fig. 11.7. Full stepping, two phases on: (a) rotor positions; (b) phase voltage waveforms.

If current is now passed through one pair of stator windings (Fig. 11.6a) the resulting North and South stator poles will attract teeth of the opposite polarity on each end of the rotor. There are now only three stable positions for the rotor, the same as the number of the rotor tooth pairs. The torque required to deflect the rotor from its stable position is now much greater, and is referred to as the *holding torque*. By switching the current from the first to the second set of stator windings, the stator field rotates through 90^0 and attracts a new pair of rotor poles. This results in the rotor turning through 30^0, corresponding to one full step. Returning to the first set of stator windings but energizing them in the opposite direction, the stator field is moved through another 90^0 and the rotor takes another 30^0 step. Finally the second set of windings are energized in the opposite direction to give a third step position. Now, the rotor and the stator field go back to the first condition, and after these four steps the rotor will have moved through two tooth pitches or 120^0. The motor performs 12 steps/rev. Reversing the sequence of current pulses the rotor will be moving in the opposite direction.

If two coils are energized simultaneously (Fig. 11.7), the rotor takes up an intermediate position since it is equally attracted to two stator poles. Under these conditions a greater torque is produced because all the stator poles are influencing the rotor. The motor can be made to take a full step simply by reversing the current in one set of windings; this causes a 90^0 shifting of the stator field as before. In fact, this would be the normal way of driving the motor in the full-step mode, always keeping two windings energized and reversing the current in each winding alternately. When the motor is driven in its full-step mode, energizing two windings or "phases" at a time (Fig. 11.7b), the torque available on each step will be the same (subject to very small variations in the motor and drive characteristics).

11.5.2 Half stepping

By alternately energizing one winding and then two windings (Fig. 11.8), the rotor moves through only 15^0 at each stage and the number of steps per revolution is doubled. This is called *half stepping*, and most industrial applications make use of this stepping mode. Although there is sometimes a slight loss of torque, this mode results in much better smoothness at low speeds and less overshoot at the end of each step.

In the half-step mode two phases are energized alternately and then only one phase (Fig. 11.8b). Assuming the drive delivers the same winding current in each case, this will cause greater torque to be produced when there are two windings energized. In other words alternate steps will be strong and weak. This is not as much of a problem as it may appear to be. The available torque is obviously limited by the weaker step, but there will be a significant improvement in low speed smoothness over the full-step mode.

Approximately equal torque on every step can be produced by using a higher current level when there is only one winding energized. The motor

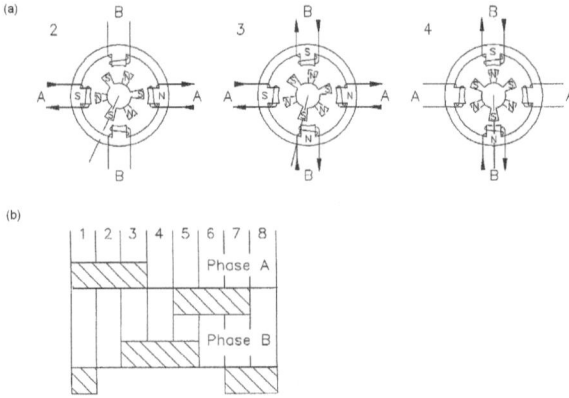

Fig. 11.8. Half stepping: (a) rotor positions; (b) phase current waveforms.

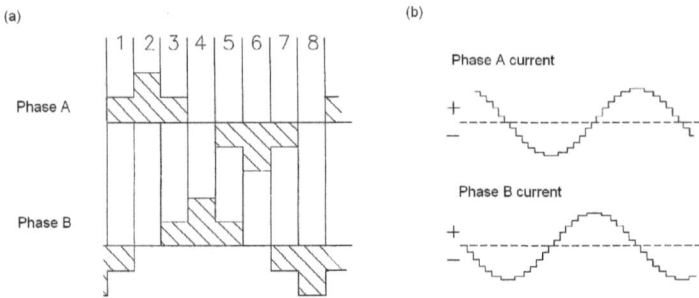

Fig. 11.9. Current waveforms to produce approximately equal torque: (a) half step current, profiled, (b) waveforms of a microstep motor.

must be designed to withstand thermally the rated current in two phases. With only one phase energized, the same total power will be dissipated if the current is increased by 40% [70]. Using this higher current in the one-phase-on state produces approximately equal torque on alternate steps (Fig. 11.9a).

11.5.3 Microstepping

Energizing both phases with equal currents produces an intermediate step position halfway between the one-phase-on positions. If the two-phase currents are unequal, the rotor position will be shifted towards the stronger pole. This effect is utilized in the *microstepping drive,* which subdivides the basic motor step by proportioning the current in the two windings. In this way the step size is reduced and the low-speed smoothness is dramatically improved. High-resolution microstep drives divide the full motor step into as many as 500 microsteps, giving 100,000 steps per revolution. In this situation the current pattern in the windings closely resembles two sinewaves with a 90^0 phase shift between them (Fig. 11.9b). The motor is now being driven very much as though it is a conventional a.c. two-phase synchronous motor. In fact, the stepping motor can be driven in this way from a 50- or 60-Hz sinewave source by including a capacitor in series with one phase.

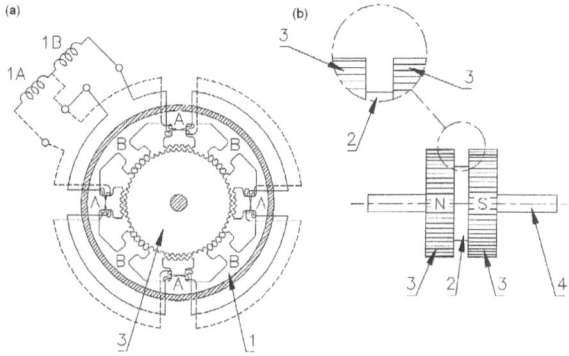

Fig. 11.10. Hybrid stepping motor performing 200 steps per revolution: (a) cross section (only the winding of phase A is shown), (b) rotor. 1 — stator core, 2 — PM, 3 — ferromagnetic disk with teeth, 4 — shaft.

11.5.4 Practical hybrid motor

The practical hybrid stepping motor operates in the same way as the simple model shown in Fig. 11.5. A larger number of teeth on the stator and rotor gives a smaller basic step size. The stator has a two-phase winding. Each phase winding consists of two sections. The stator shown in Fig. 11.10 has 8 poles each with 5 teeth, making a total of 40 teeth [70]. If a tooth is placed in each of the zones between the stator poles, there would be a total of $s_1 = 48$ teeth. The rotor consists of an axially magnetized PM located between two ferromagnetic disks (Fig. 11.10) with $s_2 = 50$ teeth per disk, two more than the number of uniformly distributed stator teeth [70]. There is a half-tooth displacement between the two sections of the rotor.

If rotor and stator teeth are aligned at 12 o'clock, they will also be aligned at 6 o'clock. At 3 and 9 o'clock the teeth will be misaligned. However, due to the displacement between the sets of rotor teeth, alignment will occur at 3 and 9 o'clock at the other end of the rotor.

The windings are arranged in sets of four, and wound such that diametrically opposite poles are the same. So referring to Fig. 11.5, the North poles at 12 and 6 o'clock attract the South-pole teeth at the front of the rotor; the South poles at 3 and 9 o'clock attract the North-pole teeth at the back. By switching current to the second set of coils, the stator field pattern rotates through 45^0 but to align with this new field the rotor only has to turn through $\theta_s = 360^0/(100 \times 2) = 1.8^0$ where $2p = 100$ (two disks with 50 teeth each) and $m_1 = 2$. This is equivalent to one quarter of a tooth pitch or 7.2^0 on the rotor, giving 200 full steps per revolution.

There are as many detent positions as there are full steps per revolution, normally 200. The detent positions correspond to rotor teeth being fully aligned with stator teeth. When the power is applied, there is usually current in both phases (zero phase state). The resulting rotor position does not correspond with a natural detent position, so an unloaded motor will always move by at least one half a step at power-on. Of course, if the system is turned off other than at the zero phase state, or if the motor is moved in the meantime, a greater movement may be seen at power-up.

For a given current pattern in the windings, there are as many stable positions as there are rotor teeth (50 for a 200-step motor). If a motor is desynchronized, the resulting positional error will always be a whole number of rotor teeth or a multiple of 7.2^0. A motor cannot "miss" individual steps — position errors of one or two steps must be due to noise, spurious step pulses or a controller fault.

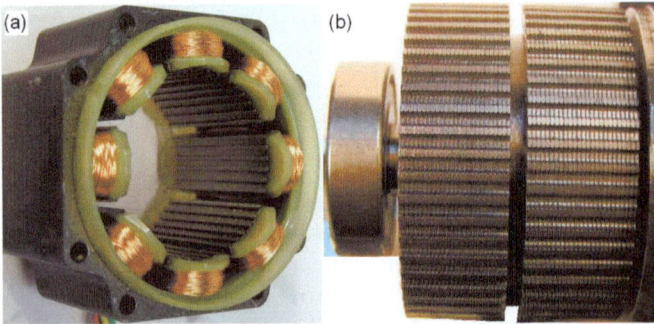

Fig. 11.11. Hybrid stepping motor for inkjet printer: (a) stator; (b) rotor.

The stator and rotor tooth pitches should meet the condition [126]:

$$t_1 = \frac{t_2}{180^0 - t_2} 180^0 \tag{11.6}$$

where $t_1 = 360^0/s_1$ and $t_2 = 360^0/s_2$ are the stator and rotor tooth pitches in degrees, respectively.

Table 11.2 specifies design data of selected two-phase hybrid stepping motors manufactured by MAE, Offanengo, Italy. Fig. 11.11 shows a two-phase hybrid stepping motor used for moving the print head assembly of an inkjet printer.

Table 11.2. Specifications of two phase hybrid stepping motors manufactured by MAE, Offanengo, Italy.

Specifications	0150AX08 0150BX08	0100AX08 0100BX08	0033AX04 0033BX04	0220AX04 0220BX04
Step angle, degrees	1.8			
Step angle accuracy, %	5			
Rated phase current, A	1.5	1.0	0.33	2.2
Maximum applicable voltage, V	75			
Number of leads	8	8	4	4
Phase resistance, Ω	1.5	3.4	33.8	0.7
Phase inductance, mH	1.5	3.8	54.6	1.2
Holding torque (unipolar, two phases on), 10^{-2} Nm	25	27	–	–
Holding torque (bipolar, two phases on), 10^{-2} Nm	33	34	32	31
Detent torque, 10^{-2} Nm	3.4			
Rotor moment of inertia, 10^{-7} kgm^2	56			
Insulation class	B			
Rotation	bidirectional			
Ambient operating temperature, ^{0}C	-20 to $+40$			
Mass, kg	0.34			
Dimensions, mm	57.2 diameter, 40 length			

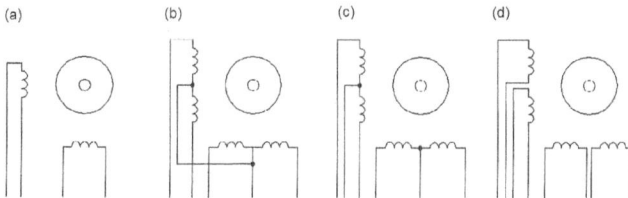

Fig. 11.12. Two-phase winding lead configuration: (a) 4-lead, (b) 5-lead, (c) 6-lead, (d) 8-lead.

11.5.5 Bipolar and unipolar motors

In *bipolar* PM and hybrid stepping motors the two phase windings are made with no center taps with only 4 leads (Fig. 11.12a).

To reverse the current without increasing the number of semiconductor switches, the so called *unipolar* or *bifilar* windings with center tap on each of the two windings are used in PM and hybrid stepping motors (Fig. 11.12b to d). The center taps are typically wired to the positive supply, and the two ends of each winding are alternately grounded to reverse the direction of the magnetic field provided by that winding. A bifilar winding consists of two parallel windings wound in opposite directions. If all the windings are brought out separately, there will be a total of 8 leads (Fig. 11.12d). Although this configuration gives the greatest flexibility, there is a lot of motors produced with only 6 leads, one lead serving as a common connection to each winding. This arrangement limits the range of applications of the motor since the windings cannot be connected in parallel.

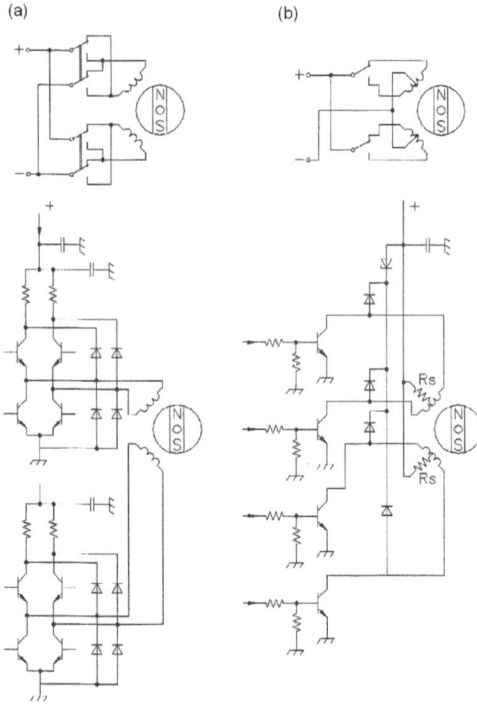

Fig. 11.13. Control methods of two-phase stepping motors: (a) bipolar control, (b) unipolar control.

11.6 Motion control of stepping motors

The control circuit turns the current in each winding on and off and controls the direction of rotation. The basic control methods of stepping motors are [309]:

(a) bipolar control when the whole phase winding is switched on (Fig. 11.13a);
(b) unipolar control when only half of the phase winding is switched on at the same time (Fig. 11.13b).

Control methods affect the torque–step frequency characteristics. In a *bipolar control* the torque is high because the whole winding carries the current. The disadvantage is that to reverse the current, eight semiconductor switches are necessary. A *unipolar control* method is simple and fewer semiconductor switches are needed. On the other hand, the full torque is never produced since only half of the winding carries the currrent. The diodes in Fig. 11.13 are used to protect the switches against reverse voltage transients.

For unipolar control where each winding draws under 0.5 A the following IC Darlington arrays can drive multiple motor windings directly from logic inputs: ULN200X family from Allegro Microsystems, DS200X from National Semiconductor or MC1413 from Motorola. Examples of bipolar ICs (H-bridge drivers) are: IR210X from International Rectifier, L298 from SGS-Thompson (up to 2 A) and LMD18200 from National Semiconductor (up to 3A).

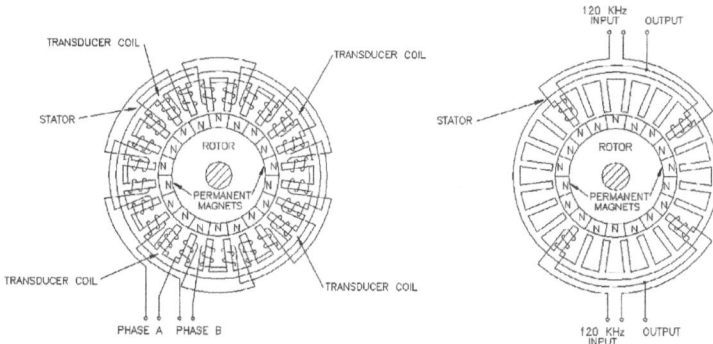

Fig. 11.14. Two-phase PM stepping motor with shaft position transducers: (a) power windings, (b) transducer arrangement.

11.7 PM stepping motors with rotor position transducers

Stepping motors with PM surface rotors usually develop higher torque than hybrid stepping motors. On the other hand, position resolution of stepping motors with surface configuration of PMs is less than that for hybrids. Rotor position transducers can improve the resolution.

Fig. 11.14 shows a two-phase PM stepping motor called *sensorimotor* with 24 poles uniformly spaced around the stator and 18 surface PMs on the rotor [154]. The stator two-phase winding uses only 20 poles whilst the other four are used in a shaft position transducer which measures the inductance. Under the motor operation the inductance of each coil varies as a sine wave.

One cycle of inductance values corresponds to a 20^0 shaft rotation and one shaft rotation produces 18 sensing cycles. The transducer coils are positioned on the stator in such a way as to obtain maximum inductance for one coil of a pair and minimum for the other coil (six pole-spacing between coils or 90^0). Each pair of coils is connected in series across a 120-kHz square-wave source. Voltage signals at the center points between each coil pair are utilized for sensing the shaft position. Pairs of coils produce a.c. square wave signals shifted by 90^0 from each other. A position detector converts the two signals into corresponding d.c. signals by a sampling technique. The resulting values feed an analog–to–digital converter that produces a hexadecimal signal representing the shaft position. Position resolution up to 1:9000 is realistic. The operation of the sensorimotor is similar to that of a d.c. brushless motor.

11.8 Single-phase stepping motors

Single-phase stepping motors are widely used in watches, clocks, timers, and counters. The stator made of a U-shaped laminated core with two poles and a yoke is common. The winding with concentrated parameters is located on the yoke. The air gap is nonuniform as shown in Fig. 11.15. The rotor is a two-pole PM cylinder. For an unexcited stator winding the stable rest position of the magnetic poles of the rotor is in the narrowest parts of the air gap as in Fig. 11.15a or b. If the winding is excited to produce flux in the direction shown in Fig. 11.15a, the rotor will rotate clockwise through 180^0 from position (a) to (b). The magnetic polarities of the stator poles (narrowest part of the air gap) and PM repel each other. To turn the rotor from position (b) to (a), the stator winding must be excited as indicated in Fig. 11.15b.

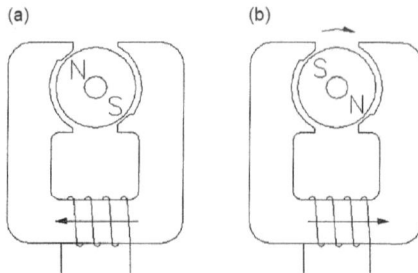

Fig. 11.15. Detent positions (a) and (b) of a single phase stepping motor.

A *wrist-watch stepping motor* operates on the same principle (Fig. 11.16). The rare-earth PM rotor with its diameter of about 1.5 mm is inside the stator mild steel core. Two external bridges are highly saturated when the winding is excited so that the flux passes through the rotor. The uniform air

(a)

coil

stator

(b)

(c)

pulse width

1 s

rotor

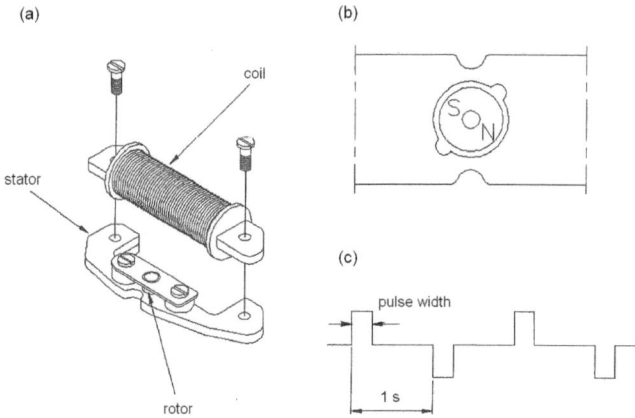

Fig. 11.16. A single-phase wrist-watch stepping motor: (a) general view, (b) air gap and stator poles, (c) input voltage waveform.

gap is created by two internal slots. The stator voltage waveform consists of short rectangular pulses (a few ms) of positive and negative polarities (Fig. 11.16c). A circuit for memorizing the rotor position and excitation in the correct polarity is used in order not to miss the first step when one starts a watch after synchronizing it with a standard time.

11.9 Voltage equations and electromagnetic torque

The EMFs induced in the phase windings of a two-phase stepping motor are

- in phase A

$$e_A = N\frac{d\phi}{dt} = N\frac{d\phi}{d\theta}\frac{d\theta}{dt} = -pN\Phi_f \sin(p\theta)\frac{d\theta}{dt} \tag{11.7}$$

- in phase B

$$e_B = -pN\Phi_f \sin[p(\theta - \gamma)]\frac{d(\theta - \gamma)}{dt} \tag{11.8}$$

The instanteneous value of the linkage flux

$$N\phi = \Psi_f \cos(p\theta) = N\Phi_f \cos(p\theta) \tag{11.9}$$

The peak magnetic linkage flux $\Psi_f = N\Phi_f$ is for $\theta = 0$ (Fig. 11.17). Thus, the voltage balance equations for the stator phase windings are

$$v = Ri_A + L\frac{di_A}{dt} + M\frac{di_B}{dt} - \frac{d}{dt}[N\Phi_f \cos(p\theta)] \tag{11.10}$$

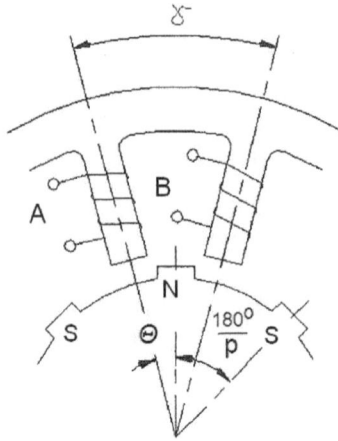

Fig. 11.17. Model of a PM stepping motor for the electromagnetic analysis.

$$v = Ri_B + L\frac{di_B}{dt} + M\frac{di_A}{dt} - \frac{d}{dt}N\Phi_f \cos[p(\theta - \gamma)] \qquad (11.11)$$

where v is the d.c. input voltage, L is the self-inductance of each phase, M is the mutual inductance between phases and R is the stator-circuit resistance per phase. It has been assumed that L and M are independent of α.

The electromagnetic torque developed by the current i_A in phase winding A (Fig. 11.17) is [157]

$$T_{dsyn}^{(A)} = -pi_A\Psi_f \sin(p\theta) = -pi_A N\Phi_f \sin(p\theta) \qquad (11.12)$$

where p is the number of rotor pole pairs, N is the number of stator turns per phase, Φ_f is the peak flux per pole pair produced by PMs, and θ is the rotational angle. The magnetic flux Φ_f can be found on the basis of the demagnetization curve and permeances (Chapter 2) or by using the FEM (Chapter 3). The peak flux linkage is $\Psi_f = N\Phi_f$. In a similar way, the electromagnetic torque produced by the current i_B can be found as [157]

$$T_{dsyn}^{(B)} = -pi_B N\Phi_f \sin[p(\theta - \gamma)] \qquad (11.13)$$

where γ is the space angle between center axes of the neighboring stator phase windings (Fig. 11.17). For a single-phase excitation $\gamma = 0$ and $T_{dsyn}^{(B)} = T_{dsyn}^{(A)}$ (both poles are coincident). Eqns (11.12) and (11.13) are also valid for a hybrid motor, but p must be replaced by the number of rotor teeth s_2 [174].

The torque produced by a two-phase stepping motor is the superposition of torques produced by each phase alone, i.e.,

$$T_{dsyn} = T_{dsyn}^{(A)} + T_{dsyn}^{(B)} = -pNi_A\Phi_f \sin(p\theta) - pNi_B\Phi_f \sin[p(\theta - \gamma)] \qquad (11.14)$$

Neglecting the spring constant, the torque balance eqn (1.16) has the following form

$$J\frac{d^2\theta}{dt^2} + D\frac{d\theta}{dt} \pm T_{sh} = T_{dsyn} \qquad (11.15)$$

where J is the moment of inertia of all rotating masses including the rotor, D is the damping coefficient taking into account the a.c. component of mutual attraction between the rotor flux and the stator field, air friction, eddy-current, and hysteresis effects.

Eqns (11.7), (11.10) and (11.11) are nonlinear equations and can be solved only numerically. It is also possible to solve them analytically by making simplifications [174].

11.10 Characteristics

11.10.1 Torque–angle characteristics

The steady-state *torque–angle characteristic* is the relation between the external torque of the excited motor and the angular displacement of the rotor (Fig. 11.18a). The maximum steady-state torque is termed the *holding torque* which corresponds to the angle θ_m. At displacements larger than θ_m, the steady-state torque does not act in a direction towards the original equilibrium position, but in the opposing direction towards the next equilibrium position. The holding torque is the maximum torque that can be applied to the shaft of an excited motor without causing continuous motion.

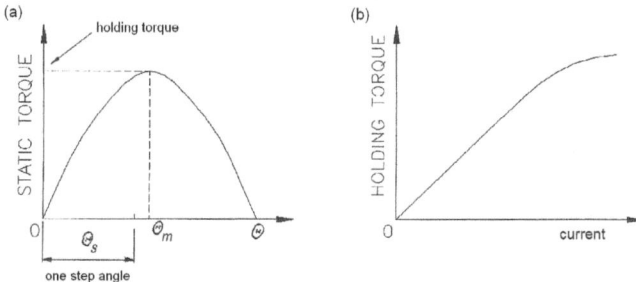

Fig. 11.18. Steady-state characteristics: (a) torque-angle, (b) torque-current.

11.10.2 Torque–current characteristics

The electromagnetic torque developed by a stepping motor is proportional to the stator input current (Fig. 11.18b). The graph of the holding torque plotted against the stator current per phase is called the *torque–current characteristic*.

11.10.3 Torque–frequency characteristics

The performance of the stepping motor is best described by the *torque–frequency characteristic* (Fig. 11.19). The frequency is equal to the number of steps per second. There are two operating ranges: the *start-stop (or pull in) range* and the *slew (or pull out) range*.

The torque of the stepping motor decreases with an increase in frequency, which can be attributed to the following factors: (a) at constant power the torque is inversely proportional to the frequency, (b) the action of the damping torque set up due to the rotating EMF, (c) the stator winding EMF becomes close to the supply voltage, and current through the stator windings does not have enough time to reach the steady-state value within a step period, which reduces the resultant stator flux.

The *maximum starting frequency* is defined as the maximum control frequency at which the unloaded motor can start and stop without losing steps.

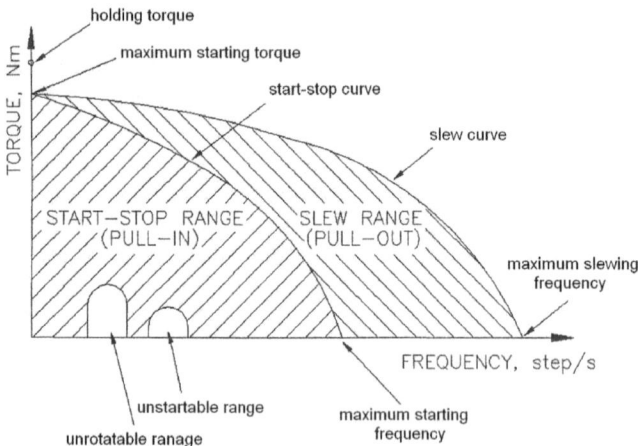

Fig. 11.19. Torque–frequency characteristics of a stepping motor.

The *maximum slewing frequency* is defined as the maximum frequency (stepping rate) at which the unloaded motor can run without losing steps.

The *maximum starting torque* or *maximum pull-in torque* is defined as the maximum load torque with which the energized motor can start and synchronize with the pulse train of a very low frequency (a few Hz), without losing steps, at constant speed.

The *start-stop range* is the range where the stepping motor can start, stop and reverse the direction by applying pulses at constant frequency without losing step. If an inertial load is added, this speed range is reduced. So the start-stop speed range depends on the load inertia. The upper limit to the start-stop range is typically between 200 and 500 full step/s (1.0 to 2.5 rev/s) [70].

To operate the motor at faster speeds it is necessary to start at a speed within the start-stop range and then accelerate the motor into the slew region. Similarly, when stopping the motor it must be decelerated back into the start-stop range before the clock pulses are terminated. Using acceleration and deceleration "ramping" allows much higher speeds to be achieved, and in industrial applications the useful speed range extends to about 10,000 full step/s (3000 rpm) [70]. Continuous operation at high speeds is not normally possible with a stepping motor due to rotor heating, but high speeds can be used successfully in positioning applications.

The torque available in the slew range does not depend on load inertia. The torque-frequency curve is normally measured by accelerating the motor up to speed and then increasing the load until the motor stalls. With a higher load inertia, a lower acceleration rate must be used but the available torque at the final speed is unaffected.

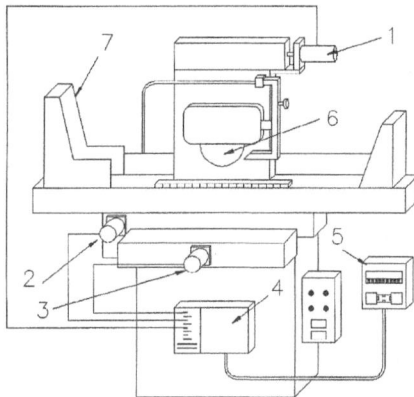

Fig. 11.20. A modern surface-grinding machine: 1,2,3 — stepping motors, 4 — motion controller, 5 — control panel, 6 — grinding wheel, 7 — safety guard. Courtesy of Parker Hannifin Corporation, Rohnert Park, CA, U.S.A.

11.11 Applications

Stepping motors are used in numerically controlled (NC) machine tools, robots, manipulators, computer printers, electronic typewriters, X–Y plotters, telefaxes, scanners, clocks, cash registers, measuring instruments, metering pumps, remote control systems, and in many other machines.

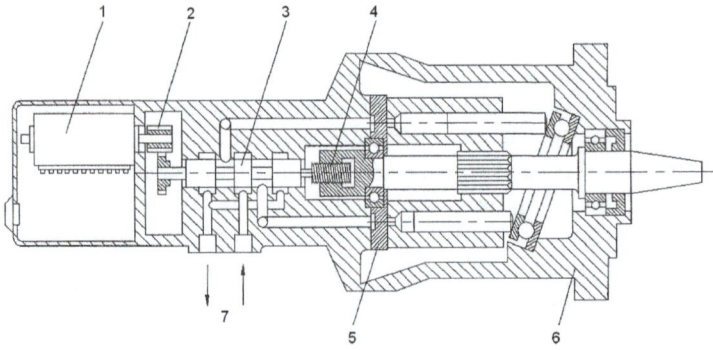

Fig. 11.21. Electrohydraulic stepping motor: 1 — electric stepping motor, 2 — tooth gear, 3 — four-edge control slider, 4 — screw, 5 — oil distribution panel, 6 — hydraulic motor with axial piston system, 7 — inlet and outlet of oil.

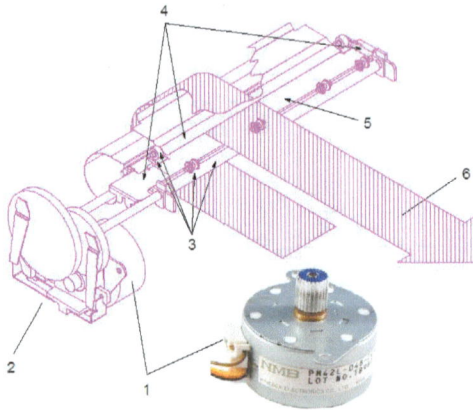

Fig. 11.22. Paper feeder of an inkjet printer. 1 — stepping motor, 2 — main drive system, 3 — pinch/drive rollers, 4 — paper control shims, 5 — print zone, 6 — media direction [44].

Fig. 11.20 shows an automatic *surface grinding machine* with three stepping motors. In digitally-controlled industrial drives, where large torques are required, the so-called *electrohydraulic stepping motors* (Fig. 11.21) have been used since 1960 [309].

Inkjet printers have made rapid technological advances in recent years. An inkjet printer is any printer that places extremely small droplets of ink onto paper to create an image. The three-color printer has been around for several years now and has succeeded in making color inkjet printing an affordable option. The *print head* is the main part of the printer. It contains a series of nozzles that spray the ink onto the paper.

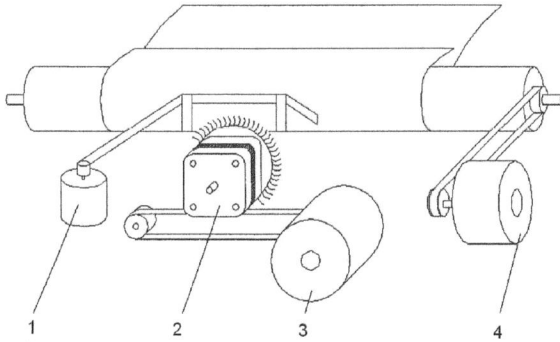

Fig. 11.23. Electronic typewriter with four stepping motors for: 1 — ribbon transportation, 2 — daisy wheel, 3 — carriage, 4 — paper feed.

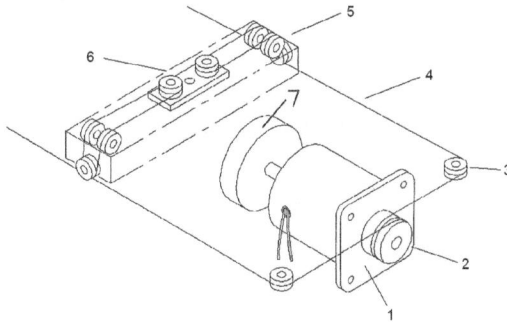

Fig. 11.24. Driving mechanism of X–Y plotter using stepping motors: 1 — stepping motor, 2 — sheave pulley, 3 — plastic pulleys on ball bearings, 4 — nylon-coated stainless twisted fiber cable, 5 — Y arm envelope, 6 — pulling mounting block on pen carriage, 7 — damper.

The print head stepping motor moves the print head assembly (print head and ink cartridges) back and forth across the paper while printing. A belt attaches the print head to the stepper motor and a stabilizer bar is used to stabilize the print head. Another stepping motor drives the *paper feeder* and the paper rollers (Fig. 11.22). This stepping motor synchronizes the movement of the paper with that of the print head so that the images that are printing appear on the correct place of the page.

The *electronic typewriter* with daisy wheel shown in Fig. 11.23 uses four stepping motors. By the end of the 1980s, word processor applications on personal computers had largely replaced the tasks previously accomplished with typewriters. Electronic typewriters can still serve many useful functions in todays offices, e.g., can improve efficiency and enhance precision and control

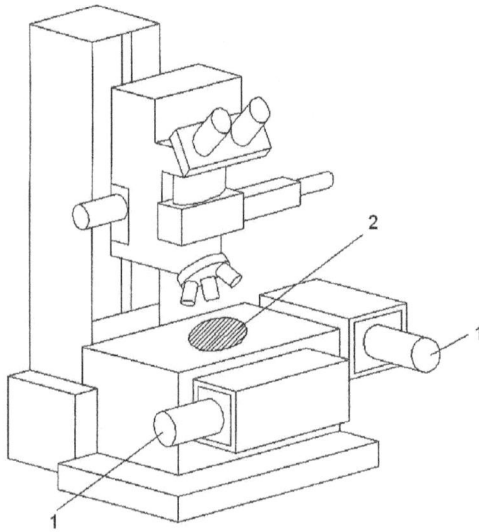

Fig. 11.25. Stepping motors for specimen positioning in an optical microscope: 1 — microstepping motor, 2 — specimen. Courtesy of Parker Hannifin Corporation, Rohnert Park, CA, U.S.A.

when it comes to labels, envelopes, printed forms and other items that a computer struggles to print on.

A *plotter* is a vector graphics printing device. Early plotters worked by placing the paper over a roller which moved the paper back and forth for X motion, while the pen moved back and forth on a single arm for Y motion. In the 1980s, the small and lightweight Hewlett Packard 7470 plotter used an innovative "grit wheel" mechanism which moved only the paper. The Hewlett Packard X–Y plotter driving mechanism is shown in Fig. 11.24 [238]. Pen plotters have essentially become obsolete and have been replaced by inkjet printers.

An *optical microscope* (Fig. 11.25) used in medical research laboratories requires a sub-micron positioning in order to automate the visual inspection process [70]. This can be done with the aid of high-resolution microstepping motors (Fig. 11.9).

Numerical examples

Numerical example 11.1

A two-phase, two-pole PM stepping motor (Fig. 11.1a) has $N = 1200$ turns per phase and is loaded with an external torque $T_{sh} = 0.1$ Nm. The motor is fed with rectangular waveforms (Fig. 11.1b) the magnitude of which is $i_A = i_B = 1.5$ A. Only one phase, A or B, is fed at a time. Find the peak magnetic flux Φ_f required to accelerate an inertial load $J = 1.5 \times 10^{-4}$ kgm^2 from $\Omega_1 = 50$ to $\Omega_2 = 250$ rad/s during $\Delta t = 0.08$ s.

Assumption: The stator field reaction on the rotor magnetic field, leakage flux, and damping coefficient are neglected.

Solution

1. The angular acceleration

$$\frac{d\Omega}{dt} = \frac{\Omega_2 - \Omega_1}{\Delta t} = \frac{250 - 50}{0.08} = 2500 \text{ rad/s}^2$$

2. The electromagnetic torque required — eqn (11.15)

$$T_{dsyn} = J\frac{d\Omega}{dt} + T_{sh} = 1.5 \times 10^{-4} \times 2500 + 0.1 = 0.375 + 0.1 = 0.475 \text{ Nm}$$

The first term is predominant and the torque developed by a stepping motor is mainly dependent on the dynamic torque (torque required to accelerate the inertia of the rotor and other rotating masses).

3. The magnetic flux. The maximum flux linkage is at $p\theta = 0$. The step angle for $2p = 2$ and $m_1 = 2$ according to eqn (11.1) is $\theta_s = 90^0$. It means that for a four-stroke commutation the necessary torque developed by the motor should be calculated for $\theta = \theta_s = 90^0$. Thus, the magnetic flux required is

$$\Phi_f = \frac{T_{dsyn}}{pNi_A \sin(p\theta)} = \frac{0.475}{1 \times 1200 \times 1.5 \sin(1 \times 90^0)} = 2.64 \times 10^{-4} \text{ Wb}$$

If, for example, the dimensions of PM are $w_M = 25$ mm, $l_M = 35$ mm, and the leakage flux is neglected, the air gap magnetic flux density should be

$$B_g \approx \frac{\Phi_f}{w_M l_M} = \frac{2.64 \times 10^{-4}}{25 \times 10^{-3} \times 35 \times 10^{-3}} \approx 0.3 \text{ T}$$

Anisotropic barium ferrite can be used for the rotor construction. To design the magnetic circuit follow all steps of Example 2.2 or 3.2.

Numerical example 11.2

The hybrid stepping motor shown in Fig. 11.10 has the rotor outer diameter $D_{2out} = 51.9$ mm, stator outer diameter $D_{1out} = 92$ mm and air gap $g = 0.25$ mm. The stator has 8 poles each with 5 teeth and the rotor has $s_2 = 50$ teeth on each ferromagnetic disk.

Calculate the ideal stator tooth pitch. Find the magnetic flux distribution, using the FEM, for one rotor disk with the stator and rotor teeth aligned and with the teeth totally misaligned. Assume that there is no current in the stator winding.

Solution

The rotor tooth pitch, with 50 teeth, is $t_2 = 360^0/50 = 7.2^0$. The most suitable stator tooth pitch is defined by eqn (11.6), i.e.,

$$t_1 = \frac{7.2^0}{180^0 - 7.2^0} 180^0 = 7.5^0$$

(a)   (b)

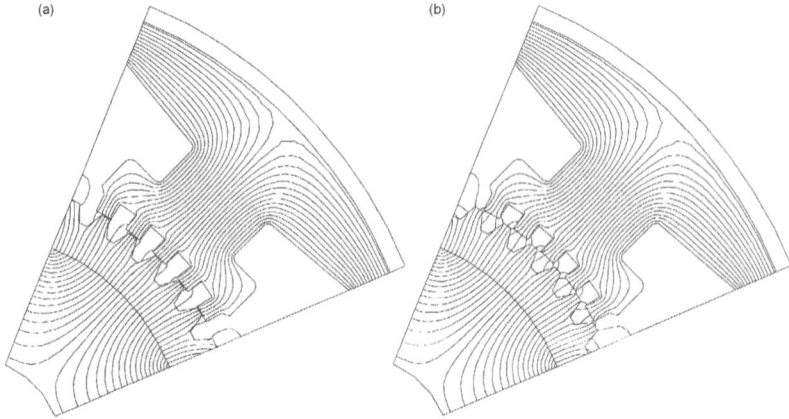

Fig. 11.26. Magnetic flux distribution of hybrid stepping motor with: (a) stator and rotor teeth aligned and (b) stator and rotor teeth misaligned. *Numerical example 11.2.*

A 2D cross-sectional FEM analysis of one pole of the motor is sufficient to show the flux distribution. The stator teeth are rectangular in shape while the rotor teeth are trapezoidal. Fig. 11.26a shows the magnetic flux distribution with the stator and rotor teeth aligned and Fig. 11.26b shows the magnetic flux distribution with the teeth misaligned.

Numerical example 11.3

A PM stepping micromotor according to [286] is shown in Fig. 11.27a. The four-pole cylindrical rotor made of isotropic barium ferrite has the outer diameter of 1.0 mm, inner diameter of 0.25 mm and length of 0.5 mm. The stator winding consists of a single coil with $N = 1000$ turns made of copper conductor with its diameter $d = 25 \times 10^{-6}$ m = 25 μm. The average length of the stator turn is 3.4 mm. The motor is fed with 3-V rectangular pulses. The frequency of the input pulses is variable; however, 100 Hz has been assumed as the rated frequency.

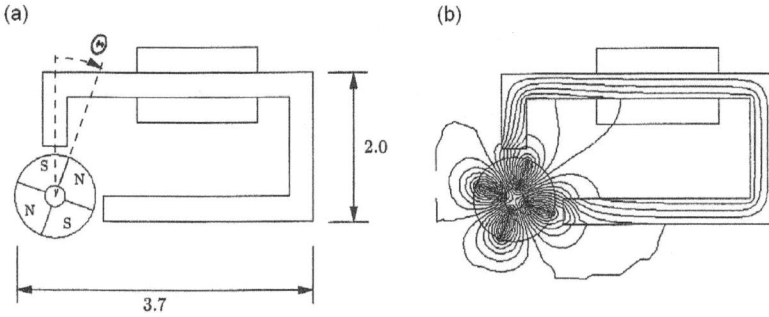

Fig. 11.27. PM stepping micromotor: (a) outline, (b) magnetic flux distribution in the longitudinal section at $\theta = 20°$ and $i_a = 0.02$ A. *Numerical example 11.3.*

Find the magnetic flux distribution in the longitudinal section of the motor and the steady-state synchronizing torque as a function of the rotor position.

Solution

The cross section area of the stator conductor is $s = \pi d^2/4 = \pi(25 \times 10^{-6})^2/4 = 490.87 \times 10^{-12}$ m^2.

The stator winding resistance at 75°C according to eqn (B.1)

$$R = \frac{1000 \times 0.0034}{47 \times 10^6 \times 490.87 \times 10^{-12}} = 147.4 \ \Omega$$

The magnitude of the current pulse

$$i = I_m = \frac{V}{R} = \frac{3}{147.4} = 0.02 \text{ A}$$

The current density

$$J = \frac{0.02}{490.87 \times 10^{-12}} = 40.74 \times 10^6 \text{ A/m}^2$$

as in other types of electric micromotors is very high.

A 2D FEM model has been built. The magnetic flux distribution at $\theta = 20°$ and $i = 0.02$ A is plotted in Fig. 11.27b.

The synchronizing torque is calculated using the Maxwell stress tensor line integral method. The PM rotor is shifted in increments of 5° until a full electrical cycle is completed. Fig. 11.28 shows the reluctance torque of the unexcited motor and the total synchronizing torque of the excited motor which is the sum of the reluctance and electromagnetic synchronous torque. The zero rotor angle $\theta = 0$ corresponds to the q-axis of the rotor aligned with the center axis of the vertical stator pole.

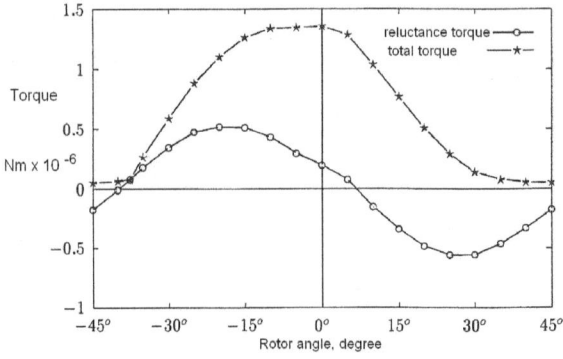

Fig. 11.28. Synchronizing torque and reluctance torque as functions of the rotor position. *Numerical example* 11.3.

Micromotors

12.1 What is a micromotor?

A *micromachine* is defined as a very small electromechanical apparatus (1µm to 1 cm) for conversion of electrical energy into mechanical energy or/and vice versa. Most micromotors, microgenerators or microactuators have external dimensions in the millimeter or submillimeter range; however, the developed torque or thrust should be high enough to overcome the losses.

Practical rotary micromotors, microgenerators and microactuators operate as *electromagnetic* or *electrostatic* devices, which use forces arising from the field energy changes.

Electrostatic micromachines are micromachined out of silicon. Silicon is unquestionably the most suitable material. It has a modulus of elasticity of 110.3 GPa, about 52 to 60% of that of carbon steel, higher yield-strength than stainless steel, low specific mass density of 2330 kg/m^3, higher strength–to–mass ratio than aluminum, high thermal conductivity of 148 W/(m K) and a low thermal expansion coefficient about 2.8×10^{-6} 1/K. Although silicon is difficult to machine using normal cutting tools, it can be chemically etched into various shapes.

The manufacturing and development infrastructures for silicon are already well established. The raw materials are well known. Their availability and processing expertise have the experience of more than 40 years of development in the IC industry. All this infrastructure and know-how have been implemented to advanced *micro electromechanical systems* (MEMS), while enjoying the advantages of batch processing, i.e., reduced cost and increased throughput.

Most of the applications have focused on *electrostatic microdrives* having typical rotor diameters of 100 µm. Electrostatic micromotors have been fabricated entirely by planar IC processes within the confines of a *silicon wafer*. This is done by selectively removing wafer material and has been used for many years for most silicon pressure sensors. Over the last two decades, surface micromachining, silicon fusion bonding, and a process called LIGA-based on a combination of deep-etch X-ray lithography, electroforming and molding processes[1] – have also evolved into major micromachining techniques. These methods can be complemented by standard IC processing techniques such as ion implantation, photolithography, diffusion, epitaxy, and thin-film deposition [46].

12.2 Permanent magnet brushless micromotors

The *magnetic micromotor* is an attractive option in applications with dimensions above 1 mm and where high voltages, needed in electrostatic micromotors, are unacceptable or unattainable. PM brushless micromotors dominate for rotor dimensions above 1 mm as micromachines of both cylindrical or disk construction. High energy rare earth PMs are used for rotors. The magnets move synchronously with the rotating magnetic fields produced by very small copper conductor coils or current gold paths on silicon substrates [300].

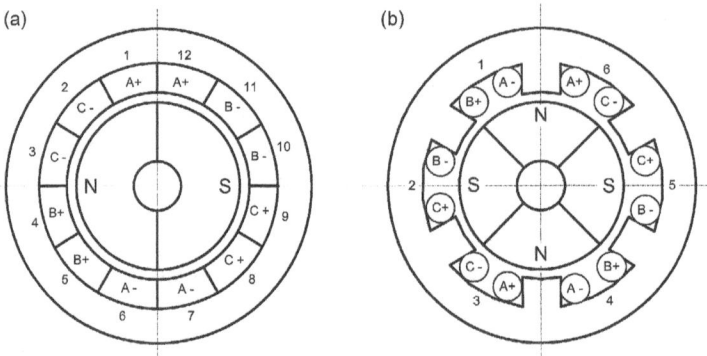

Fig. 12.1. Cross sections of cylindrical brushless PM micromotors with stator outer diameter greater than 2 mm: (a) with slotless stator; (b) with slotted stator.

12.2.1 Cylindrical micromotors

Cylindrical micromotors are usually designed as three-phase or two-phase PM brushless machines with a small number of rotor poles. The stator can be slotless (Fig. 12.1a) or slotted (Fig. 12.1b). Typically, the resistance of the stator winding is much higher than the synchronous reactances, i.e., $R_1 >> X_{sd}$ and $R_1 >> X_{sq}$, especially in slotless machines, in which the winding inductance is very small. This is why in overexcited micromotors (negative d-axis current) the phase voltage $V_1 > E_f$ (Fig. 12.2).

---

[1] The term LIGA is an acronym for the German terms meaning litography (*Litographie*), electroforming (*Galvanoformung*) and molding (*Abformung*).

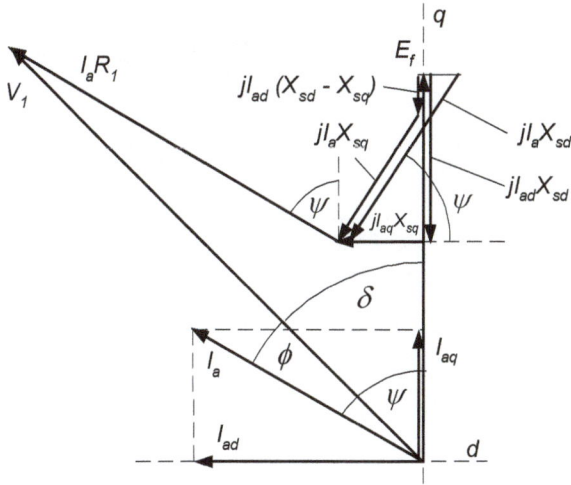

Fig. 12.2. Phasor diagram of an overexcited micromotor in which $R_1 \gg X_{sd}$, $R_1 \gg X_{sq}$, $V_1 > E_f$ and $\psi = \phi + \delta$.

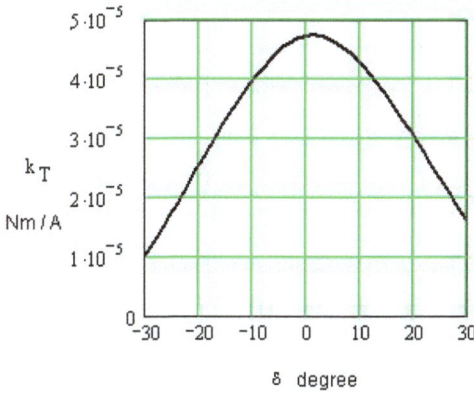

Fig. 12.3. Torque constant k_T as a function of the load angle δ for a 0.1 to 0.2 W, 80,000 rpm, 1 V PM brushless micromotor.

According to Fig. 12.2 eqns (5.58) and (5.59) take the following form:

$$V_1 \cos \delta = E_f - I_{ad} X_{sd} + I_a R_1 \cos(\phi + \delta) \tag{12.1}$$

$$= E_f - I_a X_{sd} \sin(\phi + \delta) + I_a R_1 \cos(\phi + \delta)$$

$$V_1 \sin \delta = I_a X_{sq} \cos(\phi + \delta) + I_a R_1 \sin(\phi + \delta) \tag{12.2}$$

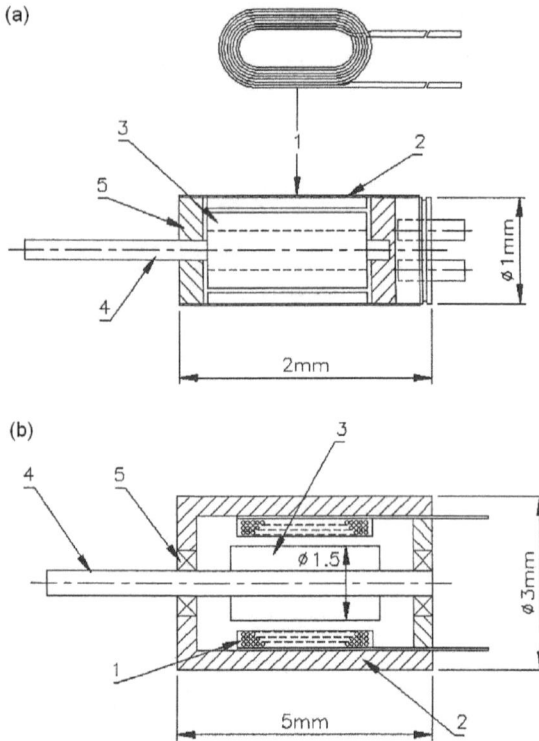

Fig. 12.4. Construction of cylindrical brushless PM micromotors: (a) synchronous micromotor designed at Eindhoven University of Technology, Netherlands, (b) 4-mW, 2-V micromotor designed by Toshiba. 1 — stator coil, 2 — stator yoke, 3 — rotor PM, 4 — stainless steel shaft, 5 — bearing.

$$X_{sd} = \frac{E_f - V_1 \cos\delta + I_a R_1 \cos(\phi + \delta)}{I_a \sin(\phi + \delta)} \tag{12.3}$$

$$X_{sq} = \frac{V_1 \sin\delta - I_a R_1 \sin(\phi + \delta)}{I_a \cos(\phi + \delta)} \tag{12.4}$$

The angle between the armature current I_a and q-axis is $\Psi = \phi + \delta$. The torque constant (6.21) for a given type of PM brushless micromotor is very sensistive to the load angle δ (Fig. 12.3).

Cross sections of magnetic circuits of cylindrical micromotors with stator outer diameter in the range of a few millimeters are shown in Fig. 12.1. The micromotor shown in Fig. 12.1a has a slotless stator winding with distributed parameters. Such a winding in the shape of a thin-wall cylinder is fabricated using round copper magnet wires encapsulated in a resin. Then, two halves

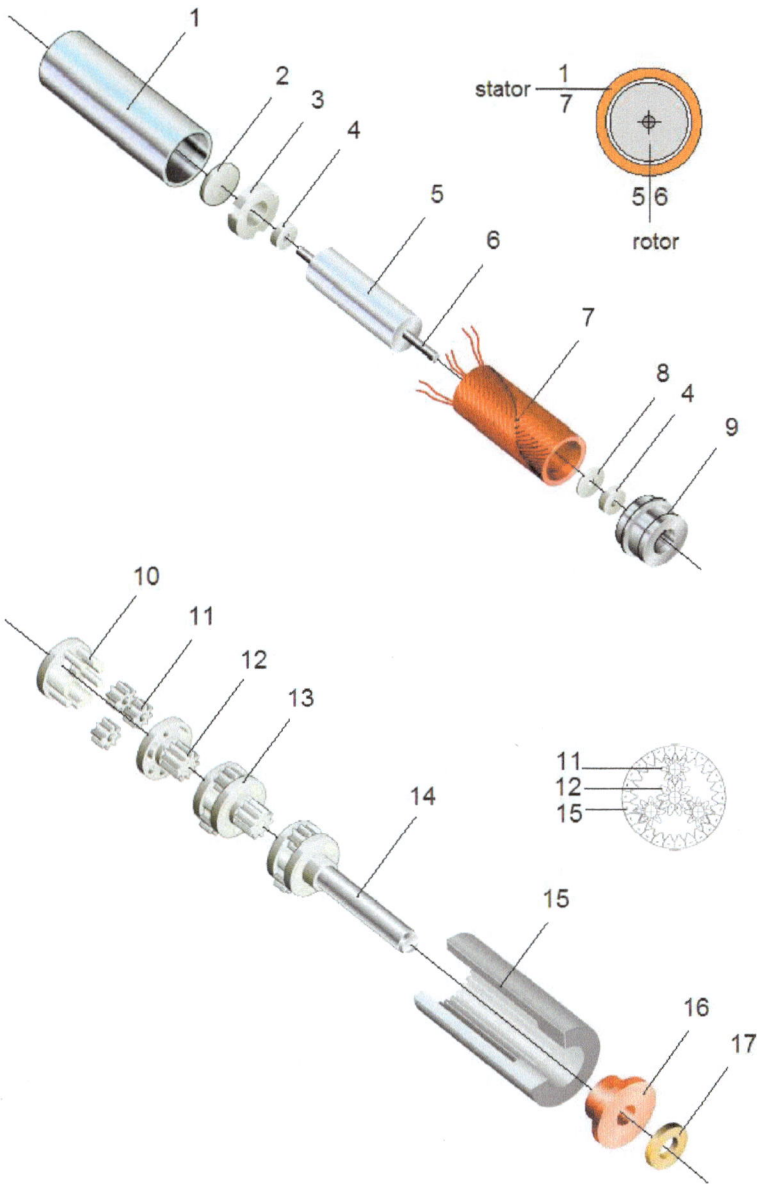

Fig. 12.5. Expanded view of the smallest in the world electromechanical drive system with PM brushless micromotor and microplanetary gearhead. 1 — housing (enclosure) of micromotor, 2 — end cap, 3 — bearing support, 4 — bearing of micromotor, 5 — PM, 6 — shaft, 7 — armature winding, 8 — washer, 9 — end cover, 10 — satellite carrier, 11 — satellite gear, 12 — sun gear, 13 — planetary stage, 14 — output shaft, 15 — housing of microplanetary gearhead, 16 — bearing cover, 17 — retaining ring. The outer diameter of housing is 1.9 mm. Source: Faulhaber Micro Drive Systems and Technologies - Technical Library, Croglio, Switzerland.

Table 12.1. PM brushless micromotors manufactured by Minimotor SA, Faulhaber Group, Croglio, Switzerland

Specifications	Motor type		
	001B	006B	012B
Diameter of housing, mm	1.9	6.0	6.0
Length of housing, mm	5.5	20.0	20.0
Rated voltage, V	1.0	6.0	12.00
Rated torque, mNm	0.012	0.37	0.37
Stall torque, mNm	0.0095	0.73	0.58
Maximum output power, W	0.13	1.56	1.58
Maximum efficiency, %	26.7	57.0	55.0
No–load speed, rpm	100 000	47 000	36 400
No–load current, A	0.032	0.047	0.016
Resistance line-to-line, Ω	7.2	9.1	59.0
Inductance line-to-line, μH	3.9	26.0	187.0
EMF constant, mV/rpm	0.00792	0.119	0.305
Torque constant mNm/A	0.0756	1.13	2.91
Rotor moment of inertia, gcm^2	0.00007	0.0095	0.0095
Angular acceleration, rad/s^2	1350×10^3	772×10^3	607×10^3
Mechanical time constant, ms	9.0	6.0	6.0
Temperature range, $^{\circ}$C	-30 to $+125$	-20 to $+100$	-20 to $+100$
Mass, g	0.09	2.5	2.5

of the stator ferromagnetic ring are placed around the stator winding. The stator core can be made of very thin laminations (permalloy, amorphous alloy or cobalt alloy). The rotor PM is of cylindrical shape with round hole for the shaft. To obtain adequate rotor stiffness, the shaft behind the rotor magnet is of the same diameter as the magnet.

In the micromotor shown in Fig. 11.1b the stator coils are fabricated together with the stator ferromagnetic poles and then placed inside the stator ring-shaped core (yoke). The outer diameter of the stator is bigger than the diameter of the slotless micromotor shown in 12.1a.

Examples of prototypes of low speed cylindrical PM brushless mircomotors are shown in Fig. 12.4 [139]. The stator (armature) coils consist of a few turns of flat copper wire (typical thickness 35 μm).

Specifications of very small PM brushless micromotors for surgical devices, motorized catheters and other clinical engineering devices are listed in Table 12.1. The smallest in the world electromechanical drive system with PM brushless micromotor and microplanetary gearhead is shown in Fig. 12.5. The rotor has a 2-pole NdFeB PMs on a continuous spindle. The maximum output power is 0.13 W, no-load speed 100 000 rpm, maximum current 0.2 A (thermal limit) and maximum torque 0.012 mNm (Table 12.1) [289].

12.2.2 Fabrication of magnetic micromotors with planar coils

Planar coils can be fabricated, e.g., by local electroplating[2] of gold. A large cross section of the gold lines is necessary to keep the power consumption and thermal load at acceptable levels. For NdFeB magnets with a typical dimension of 1 mm, forces of 150 μN, torques of 100 nNm and maximum speeds of 2000 rpm can be achieved [300]. The PMs are guided in channels or openings in the silicon itself or in additional glass layers.

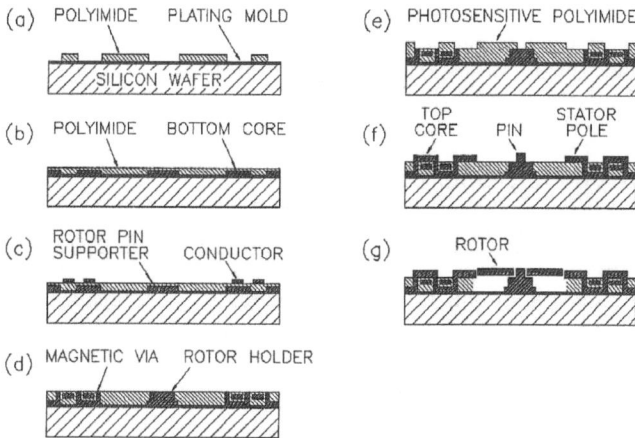

Fig. 12.6. Fabrication sequence for a magnetic micromotor: (a)polyimide deposition, dry etching; (b) bottom core electroplating; (c) patterning of conductor; (d) magnetic via hole and rotor pin electroplating; (e) photosensitive polyimide deposition and developing; (f) top core and stator pole electroplating; (g) rotor and stator microassembly [65].

The fabrication process of a *magnetic micromotor* (Fig. 12.6) would start with a silicon wafer as substrate, onto which silicon nitride is deposited. Cr–Cu–Cr layers are deposited onto this substrate using electron-beam evaporation, to form an electroplating seed layer. Polyimide is then spun on the wafer to build electroplating molds for the bottom magnetic core. A 40 μm thick polyimide layer is built-up. After curing, the holes which contain bottom magnetic cores are etched until the copper seed layer is exposed. The electroplating forms are then filled with NiFe permalloy [65].

The stator winding can also be made of interleaved, electroplated copper coils that are dielectrically isolated from a 1-mm thick NiFeMo substrate by a 5 μm polyimide (high temperature engineering polymer) layer (Fig. 12.7).

---

[2] Electroplating is the process of using electrical current to reduce metal cations in a solution and coat a conductive object with a thin layer of metal.

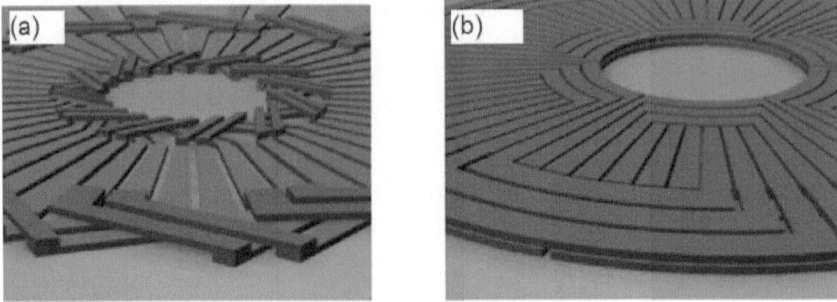

Fig. 12.7. Stator winding patterns for 8-pole micromachine: (a) 2 turns per pole; (b) 3-turns pole [16]. Photo courtesy of Georgia Institute of Technology, Atlanta, GA, U.S.A.

12.2.3 Disk-type micromotors

A disk-shaped rare earth PM, magnetized radially, which rotates on the silicon chip surface driven by four surrounding planar coils [300] is shown in Fig. 12.8. The planar coils on silicon substrate generate a rotating field which is predominantly horizontal, while the magnet is held in position by the guide hole in the glass sheet.

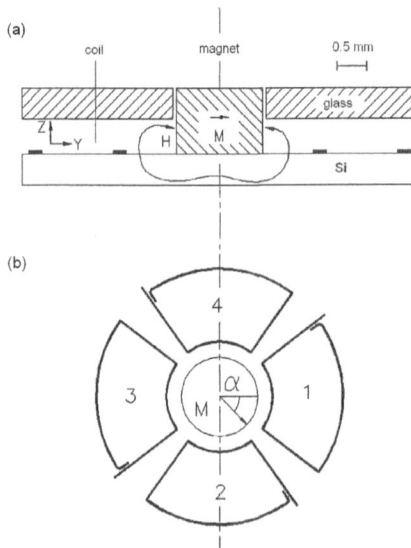

Fig. 12.8. Micromotor with rotating PM [300].

According to [300], a very small PM with its height of 1.0 mm and diameter of 1.4 mm can be used in a planar coil micromotor. The diameter of the

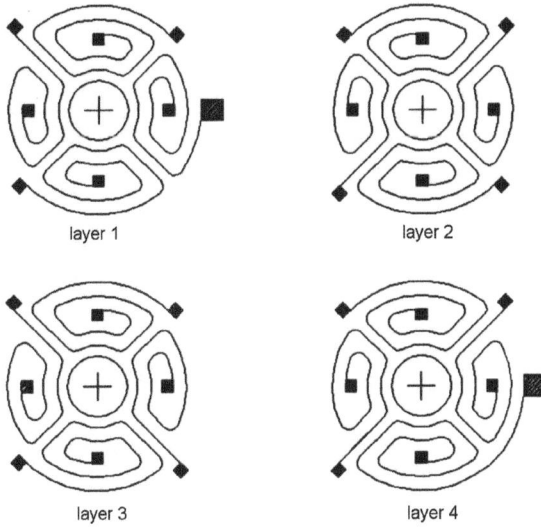

Fig. 12.9. An example of four layer winding for a PM micromotor. Courtesy of TU Berlin, Germany, [140].

guide hole is 1.4 mm. Each planar coil extends over an angle of 80^0 and has a resistance of 1.4 Ω. The synchronous speed of 2000 rpm was obtained at the current of 0.5 A. This current in two opposite coils generates a mean lateral magnetic field of 90 A/m. For a perpendicular orientation of field and magnetization, a maximum torque of 116 nNm is produced by the motor shown in Fig. 12.8.

Owing to their simple construction and advances in IC manufacture, it is possible that the size of these PM micromotors can be reduced further. PMs with a diameter of 0.3 mm have already been manufactured [300].

Etched windings are also used in some slotless axial flux micromachines [138, 139, 140]. The advantage of the slotless winding design is the elimination of the cogging torque, the tooth saturation and tooth losses. The disadvantage is that the coils are stressed by the electromagnetic forces and by the mechanical vibration. Thus, these micromotors are not robust enough for all applications.

A four-layer etched winding is shown in Fig. 12.9 [140]. The conducting material for the prototypes is gold with some addition of palladium. Extremely high accuracy is necessary in the *etching process of multilayer windings* using the thick film technology. For substrate, different ceramic materials and glass have been used. The cross section of the conductors is from 3750 to 7500 μm^2 and the distance between conducting paths varies from 150 μm to 200 μm. The current density in a multilayer etched winding is very high: from 1000

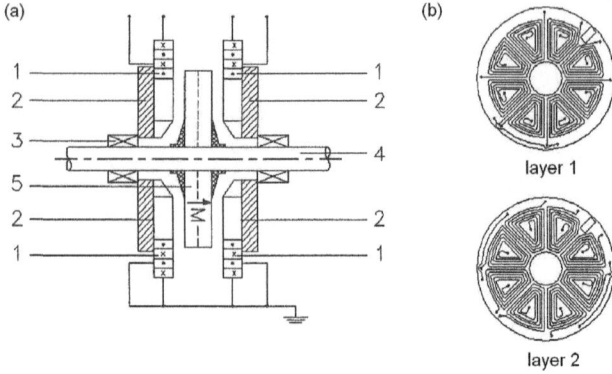

Fig. 12.10. Etched winding disk type PM motor: (a) schematic, (b) layout of etched windings. 1 — windings (2+2 phases), 2 — back iron, 3 — bearing, 4 — shaft, 5 — PM. Courtesy of TU Berlin, Germany [138].

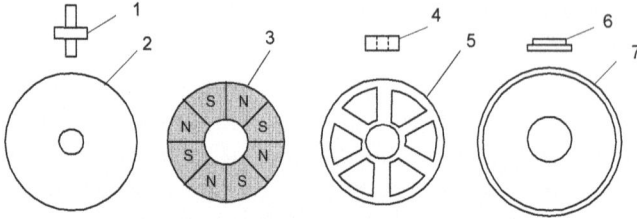

Fig. 12.11. Construction of penny-motor: 1 — shaft, 2 — soft steel yoke cover, 3 — PM ring, 4 — ball bearing, 5 — stator winding, 6 — flange, 7 — bottom steel yoke.

to 10,000 A/mm^2 [138, 140]. To reduce the nonferromagnetic air gap, a thin ferromagnetic liquid layer with $\mu_r \approx 10$ can be added [138].

Fig. 12.10 shows a disk type 8-pole prototype micromotor with etched winding [138]. The dimensions of the NdFeB PM are: thickness 3 mm, outer diameter 32 mm and inner diameter 9 mm. Although the stator laminated core reduces losses, it is difficult to manufacture. In a four-layer etched stator winding as in Fig. 12.10b the current conducting path has a width of 0.4 mm and a thickness of 0.1 mm. The electric time constant of the stator winding per phase is $L_1/R_1 = 0.7$ μs. The four-phase stator winding is fed by four transistors. Magnetoresistive position sensors have been used. The sensor PM has the dimensions $3 \times 3 \times 1$ mm. The torque of 0.32 mNm at a speed of 1000 rpm and input voltage 3.4 V has been developed.

Ultra-flat PM micromotor, the so-called *penny-motor*, is shown in Fig. 12.11 [178]. The thickness is 1.4 to 3.0 mm, outer diameter about 12 mm, torque constant up to 0.4 μNm/mA and speed up to 60,000 rpm. A 400 μm eight-pole PM and three-strand 110 μm disk shaped litographically produced stator winding have been used [178]. Plastic bound NdFeB magnets are a cost-effective solution. However, the maximum torque is achieved with sintered NdFeB magnets. A miniature ball bearing has a diameter of 3 mm. Penny-motors find applications in miniaturized hard disk drives, cellular phones as vibration motors, mobile scanners and consumer electronics.

12.3 Applications

Micromotors and microactuators are used in high precision manufacturing, glass-fiber and laser mirror adjustments, military and aerospace industry, medical engineering, bioengineering, and microsurgery. By inserting a micro-motor intravascularly, surgery can be done without large openings of vessels.

12.3.1 Motorized catheters

A brushless motor with planetary gearhead and outer diameter below 2 mm has many potential applications such as *motorized catheters*[3], minimally inva-sive surgical devices, implantable drug-delivery systems and artificial organs [289]. An ultrasound catheter shown in Fig. 12.12a consists of a catheter head with an ultrasound transducer on the motor/gearhead unit and a catheter tube for the power supply and data wires. The site to be examined can be reached via cavities like arteries or the urethra[4]. The supply of power and data to and from the transmit/receive head is provided via slip rings.

(a)   (b)

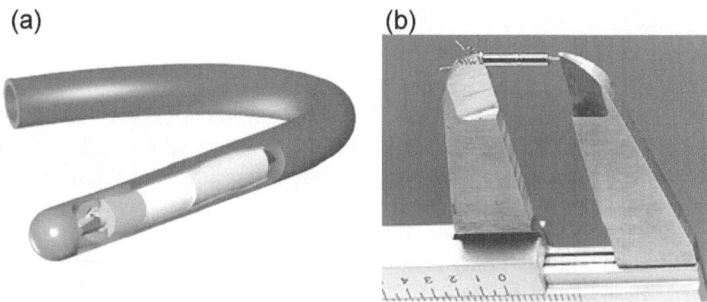

Fig. 12.12. Ultrasound motorized catheter: (a) general view; (b) 1.9-mm diame-ter PM brushless motor (Table 12.1). Source: Faulhaber Micro Drive Systems and Technologies - Technical Library, Croglio, Switzerland [289].

---

[3] Catheter is a tube that can be inserted into a body cavity, duct or vessel.

[4] Urethra is a tube which connects the urinary bladder to the outside of the body.

The stator of the brushless motor is a coreless type with skewed winding. The outer diameter of the motor is 1.9 mm, length of motor alone is 5.5 mm and together with gearhead is 9.6 mm (Figs 12.5 and 12.12b, Table 12.1). The high-precision rotary speed setting allows analysis of the received ultrasound echoes to create a complex ultrasound image.

12.3.2 Capsule endoscopy

Capsule endoscopy helps doctors to evaluate the condition of the small intestine. This part of the bowel cannot be reached by traditional upper *endoscopy*[5] or by *colonoscopy*. [6] The most common reason for doing capsule endoscopy is to search for a cause of bleeding from the small intestine. It may also be useful for detecting polyps, inflammatory bowel disease (Crohn' s disease), ulcers, and tumors of the small intestine.

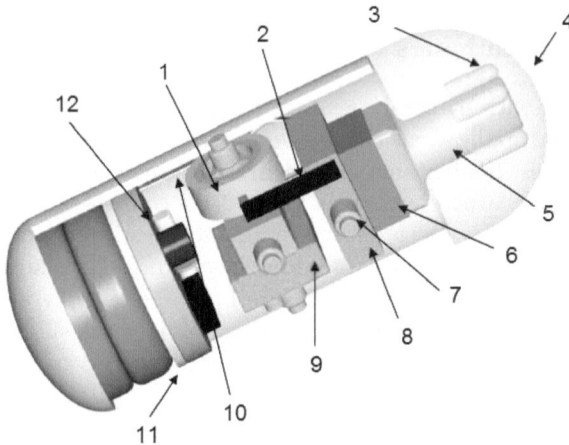

Fig. 12.13. Intracorporeal video probe IVP2. 1 — Q-PEM motor, 2 — transmission shaft, 3 — LEDs, 4 — transparent cover, 5 — optics, 6 — CMOS sensor, 7 — fixation point, 8 — camera chip, 9 — localization chip, 10 — electrical wires, 11 — batteries, 12 — data transmission chip. Courtesy of Scuola Superiore Sant'Anna, Pisa, Italy.

Approximately the size of a large vitamin, the *capsule endoscope* includes a miniature color video camera, a light, a battery and transmitter. Images

---

[5] Endoscopy is the examination and inspection of the interior of body organs, joints or cavities through an endoscope. An endoscope is a device that uses fiber optics and powerful lens systems to provide lighting and visualization of the interior of a joint.

[6] Colonoscopy is a procedure that enables a gastroenterologist to evaluate the appearance of the inside of the colon (large bowel) by inserting a flexible tube with a camera into the rectum and through the colon.

Fig. 12.14. Camera tilting mechanism with wobble motor [14].

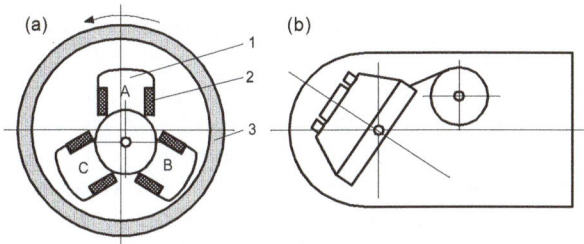

Fig. 12.15. Wobble motor: (a) cross section; (b) cam mechanism. 1 — stator core, 2 — stator coil, 3 — PM rotor wheel.

captured by the video camera are transmitted to a number of sensors attached to the patient's torso and recorded digitally on a recording device similar to a walkman or beeper that is worn around the patient's waist.

In next generation capsule endoscopy, e.g., *Sayaka capsule endoscope* [260], a *tiny stepper motor* rotates the camera as the capsule passes through the digestive tract, allowing it to capture images from every angle. Sayaka capsule is characterized by a double structure made up of an outer and an inner capsule. Whereas the outer capsule traverses through the gastrointestinal tract, the inner capsule alone spins. The spinning is produced by a small PM stepping motor with a stepping angle of 7.5^0. This stepping rotation is necessary to prevent fluctuation or blurring in the images. An 8-hour, 8-m passage from entrance to exit will yield 870,000 photos, which are then combined by software to produce a high-resolution image.

The autonomous *intracorporeal video probe* (IVP) system contains a CMOS image sensor with camera, optics and illumination, transceiver, system control with image data compression unit and a power supply [14]. The optical part is located on a tiltable plate, which is driven by a *wobble motor* (Fig. 12.13). The basic concept is to use a frontal view system with a vision angle up to 120^0 and a tilting mechanism able to steer the vision system (optics, illumination and image sensor) between about $\pm30^0$ in one plane (Fig. 12.14). By exploiting this technique, the device will perform an optimal view between

± 90 degrees in the xy plane. The tilting mechanism consists of a wobble motor (Fig. 12.15a) and simple mechanical parts, such as one cam and one shaft fixed to the vision system (Int. Patent Publication WO2006/105932). The cam system transforms the rotational action of the motor in a linear action to the shaft (Fig. 12.15b). The so–called Q-PEM stepping motor can be controlled with a precision of 340 steps per revolution. The motor with outer diameter of 4 mm and thickness of 3 mm draws about 100 mW of power.

Numerical examples

Numerical example 12.1

A three-phase, 0.15 W, 80,000 rpm, two-pole slotless PM brushless micromotor has a single layer stator winding consisting of two full pitch stator coils per phase. There are $N_1 = 8$ turns per phase wound with $d_a = 0.1024$ mm diameter wire. The number of parallel wires is $a_w = 2$, the d.c. bus voltage is $V_{dc} = 1.0$ V, the amplitude modulation index $m_a = 0.96$, the armature current $I_a = 0.448$ A and input power $P_{in} = 0.455$ W at the load angle $\delta = 3^0$. The peak value of the magnetic flux density in the air gap at no load is $B_{mg} = 0.65$ T, stator core losses $\Delta P_{Fe} = 0.1$ W and the stator winding phase resistance $R_1 = 0.28$ Ω at 75^0 C. The dimensions of the motor (Fig. 12.1a) are as follows: rotor outer diameter $D_{2out} = 1$ mm, shaft diameter inside the PM $d_{sh} = 0.4$ mm, air gap (mechanical clearance) $g = 0.2$ mm, radial thickness of the winding $h_w = 0.2$ mm, stator outer diameter $D_{1out} = 2.4$ mm and the length of the stator stack equal to the axial length of PM is $L_i = 4$ mm. Find the synchronous reactances in the d and q-axis, output power, efficiency, power factor, shaft torque, EMF constant and torque constant. The armature reaction, windage losses and losses in PMs (slotless stator) can be neglected.

<u>Solution</u>

The total number of coils is $N_c = 2 \times 3 = 6$. For a single layer winding the number of "slots" is $2N_c = 2 \times 6 = 12$. The number of "slots" per pole per phase is $q_1 = 2N_c/(2pm_1) = 12/(2 \times 3) = 2$. The winding factor according to eqns (A.1), (A.6) and (A.3) is $k_{w1} = 0.966 \times 1 = 0.966$. The stator core inner diameter $D_{1in} = D_{2out} + 2(g + h_w) = 1.0 + 2(0.2 + 0.2) = 1.8$ mm and the pole pitch $\tau = \pi \times 1.8/2 = 2.83$ mm. The input frequency is $f = 1 \times 80\ 000/60 = 1333.3$ Hz. Thus, the fundamental harmonic of the magnetic flux without armature reaction according to eqn (5.6) is

$$\Phi_{f1} = \frac{2}{\pi} 0.65 \times 0.00283 \times 0.004 = 4.68^{-6}\ \text{Wb}$$

and the stator EMF per phase excited by the rotor magnetic flux according to eqn (5.5) is

$$E_f = \pi\sqrt{2} \times 1333.3 \times 8 \times 0.966 \times 4.68^{-6} = 0.214 \text{ V}$$

The line-to-line and phase a.c. voltages according to eqn (6.9) are, respectively,

$$V_{1L} \approx 0.612 \times 0.96 \times 1.0 = 0.588 \text{ V}; \qquad V_1 = \frac{V_{1L}}{\sqrt{3}} = 0.339 \text{ V}$$

The EMF $E_f < V_1$ and the voltage drop across the armature resistance $I_a R_1 = 0.448 \times 0.28 = 0.126$ V is large compared with E_f, V_1 and inductive voltage drops, which is illustrated by the phasor diagram shown in Fig. 12.2. The current density in the armature winding

$$j_a = \frac{I_a}{0.25\pi d_a^2 a_w} = \frac{0.448}{0.25\pi 0.1024^2 \times 2} = 27.64 \times 10^6 \text{ A/m}^2$$

is much higher than that in normal-sized air cooled PM brushless motors. The power factor and angle between current and voltage are, respectively,

$$\cos\phi = \frac{0.455}{3 \times 0.339 \times 0.448} = 0.999; \qquad \phi = \arccos\phi = 2.72^0$$

The input apparent power is

$$S_{in} = \frac{0.455}{0.999} = 0.456 \text{ VA}$$

The angle between the armature current and q-axis according to phase diagram shown in Fig. 12.2 is

$$\Psi = \phi + \delta = 2.72 + 3.0 = 5.72^0$$

The d and q axis synchronous reactances are calculated on the basis of eqns (12.3) and (12.4), i.e.,

$$X_{sd} = \frac{0.214 - 0.339\cos 3.0 + 0.448 \times 0.28\cos(2.72 + 3.0)}{0.448\sin(2.72 + 3.0)} = 0.012 \text{ }\Omega$$

$$X_{sq} = \frac{0.339\sin 3.0 - 0.448 \times 0.28\sin(2.72 + 3.0)}{0.448\cos(2.73 + 3.0)} = 0.012 \text{ }\Omega$$

The d and q-axis armature currents (overexcited machine) are, respectively,

$$I_{ad} = -I_a\sin\Psi = -0.448\sin 5.72 = -0.045 \text{ A};$$

$$I_{aq} = I_a\cos\Psi = -0.448\cos 5.72 = 0.446 \text{ A}$$

Now, the input power can be verified with the aid of eqn (5.3), i.e.,

$$P_{in} = 3[0.446 \times 0.214 + 0.28 \times 0.448^2 + (-0.045)0.446(0.012 - 0.012)] = 0.455 \text{ W}$$

The armature winding losses are

$$\Delta P_a = 3 \times 0.448^2 \times 0.28 = 0.169 \text{ W}$$

The mass of the rotor with shaft is

$$m_r \approx 7700 \frac{\pi D_{2out}}{4}(2L_i) = 7700 \frac{\pi 0.001^2}{4} 2 \times 0.004 = 0.00005 \text{ kg} = 50 \text{ mg}$$

The bearing friction losses according to eqn (B.31) in which $k_{fb} = 2.5$ is

$$\Delta P_{fr} = 2.5 \times 0.00005 \times 80\,000 \times 10^{-3} = 0.0097 \text{ W}$$

The output power is

$$P_{out} = P_{in} - \Delta P_a - \Delta P_{Fe} - \Delta P_{fr} = 0.455 - 0.169 - 0.1 - 0.0097 \approx 0.177 \text{ W}$$

The efficiency is

$$\eta = \frac{0.177}{0.455} \approx 0.388$$

The shaft torque is

$$T_{sh} = \frac{0.177}{2\pi 80\,000/60} = 21.1 \times 10^{-6} \text{ Nm} = 0.0211 \text{ mNm}$$

The EMF constant and torque constant are, respectively,

$$k_E = \frac{0.214}{80\,000} = 2.678 \times 10^{-6} \text{ V/rpm};$$

$$k_T = \frac{21.1 \times 10^{-6}}{0.448} = 47.12 \times 10^{-6} \text{ Nm/A} = 0.04712 \text{ mNm/A}$$

Numerical example 12.2

A three-phase, 0.05 W, 10,000 rpm, four-pole PM brushless micromotor has a six-coil stator winding as shown in Fig. 12.1b. Concentrated-parameter coils are fabricated together with ferromagnetic cores and then inserted in the stator cylindrical yoke. There are 12 turns per coil. The cross section of armature conductor is 0.008107 mm^2. The peak value of the magnetic flux density in the air gap at no load is $B_{mg} = 0.725$ T, stator core losses $\Delta P_{Fe} = 0.02$ W, stator winding resistance at 75^{0}C is $R_1 = 1.276$ Ω, d-axis synchronous inductance $L_{sd} = 0.0058$ mH and q-axis synchronous inductance $L_{sq} = 0.0058$ mH. The dimensions of the motor (Fig. 12.1b) are as follows: rotor outer diameter $D_{2out} = 1$ mm, shaft diameter inside the PM $d_{sh} = 0.4$ mm, air gap (mechanical clearance) $g = 0.2$ mm, stator outer diameter $D_{1out} = 3.6$ mm and the length of the stator stack equal to the axial length of PM is $L_i = 7$ mm. For the d.c. bus voltage is $V_{dc} = 1.1$ V, amplitude modulation index $m_a = 1.0$ and load angle $\delta = 6^0$ find the armature current, input power, output power, power factor, efficiency, EMF constant and torque constant. The armature reaction, windage losses and losses in PMs (slotless stator) can be neglected.

Solution

The number of turns per phase is $N_1 = 2 \times 12 = 24$. The number of slots is the same as the number of coils, i.e., $s_1 = 6$. The number of slots per pole per phase is $q_1 = 6/(4 \times 3) = 0.5$. The winding factor according to eqns (A.1), (A.6) and (A.3) is $k_{w1} = 1 \times 1 = 1$. The stator core inner diameter $D_{1in} = D_{2out} + 2g = 1.0 + 2 \times 0.2 = 1.4$ mm and the pole pitch $\tau = \pi \times 1.4/4 = 1.13$ mm. The input frequency is $f = 1 \times 10\ 000/60 = 333.3$ Hz. The fundamental harmonic of the magnetic flux without armature reaction according to eqn (5.6) is $\Phi_{f1} = (2/\pi)0.725 \times 0.00113 \times 0.007 = 3.553^{-6}$ Wb. The stator EMF per phase excited by the rotor magnetic flux according to eqn (5.5) is $E_f = \pi\sqrt{2} \times 333.3 \times 24 \times 1.0 \times 3.553^{-6} = 0.1263$ V. The line-to-line a.c. voltage according to eqn (6.9) is $V_{1L} \approx 0.612 \times 1.0 \times 1.1 = 0.673$ V. The phase voltage is $V_1 = 0.673/\sqrt{3} = 0.389$ V. The synchronous reactances are $X_{sd} = 2\pi \times 333.3 \times 0.0058 \times 10^{-3} = 0.012$ Ω and $X_{sq} = 2\pi \times 333.3 \times 0.0058 \times 10^{-3} = 0.012$ Ω.

The armature currents I_{ad}, I_{aq} and I_a are calculated according to eqns (5.40), (5.41) and (5.43), respectively,

$$I_{ad} = \frac{0.389(0.012\cos 6.0 - 1.276\sin 6.0) - 0.1263 \times 0.012}{0.012 \times 0.012 + 1.276^2} = -0.03 \text{ A}$$

$$I_{aq} = \frac{0.389(1.276\cos 6.0 - 0.012\sin 6.0) - 0.1263 \times 1.276}{0.012 \times 0.012 + 1.276^2} = 0.204 \text{ A}$$

$$I_a = \sqrt{(-0.03)^2 + 0.204^2} = 0.206 \text{ A}$$

The EMF $E_f < V_1$ and the voltage drop across the armature resistance $I_a R_1 = 0.206 \times 1.276 = 0.263$ V is large compared with E_f, V_1 and inductive voltage drops $I_{ad}X_{sd} = 0.03 \times 0.012 = 0.00036$ V and $I_{aq}X_{sq} = 0.204 \times 0.012 = 0.00248$ V. The current density in the armature winding

$$j_a = \frac{0.206}{0.008107} = 25.47 \text{ A/mm}^2$$

is much higher than that in normal-sized air cooled PM brushless motors. The EMF can now be verified using the phasor diagram (Fig. 12.2), i.e.,

$$E_f = V_1 \cos \delta + |I_{ad}| X_{sd} - I_{aq} R_1$$

$$= 0.389 \cos 6.0 + |-0.03| \times 0.012 - 0.204 \times 1.276 = 0.126 \text{ V}$$

The input power according to eqn (5.44) is

$$P_{in} = 3 \times 0.389[0.204 \cos 6.0 - (-0.03) \sin 6.0] = 0.2405 \text{ W}$$

The input apparent power is

$$S_{in} = 3 \times 0.389 \times 0.206 = 0.2407 \text{ VA}$$

The power factor and angle between current and voltage are, respectively,

$$\cos \phi = \frac{0.2405}{0.2407} = 0.999$$

$$\phi = \arccos \phi = 2.327°$$

The angle between the armature current and q-axis according to phase diagram shown in Fig. 12.2 is

$$\Psi = \phi + \delta = 2.327 + 6.0 = 8.327°$$

The armature winding losses are

$$\Delta P_a = 3 \times 0.206^2 \times 1.276 = 0.163 \text{ W}$$

The mass of the rotor with shaft is

$$m_r \approx 7700 \frac{\pi D_{2out}}{4}(2L_i) = 7700 \frac{\pi 0.001^2}{4} 2 \times 0.007 = 0.00008 \text{ kg} = 80 \text{ mg}$$

The bearing friction losses according to eqn (B.31) in which $k_{fb} = 2.5$ is

$$\Delta P_{fr} = 2.5 \times 0.00008 \times 10\ 000 \times 10^{-3} = 0.0021 \text{ W}$$

The output power is

$$P_{out} = P_{in} - \Delta P_a - \Delta P_{Fe} - \Delta P_{fr} = 0.2405 - 0.163 - 0.02 - 0.0021 \approx 0.0553 \text{ W}$$

The efficiency is

$$\eta = \frac{0.0553}{0.2405} \approx 0.23$$

The shaft torque is

$$T_{sh} = \frac{0.0553}{2\pi 10\ 000/60} = 52.8 \times 10^{-6}\ \text{Nm} = 0.0528\ \text{mNm}$$

The EMF constant and torque constant are, respectively,

$$k_E = \frac{0.1263}{10\ 000} = 12.63 \times 10^{-6}\ \text{V/rpm};$$

$$k_T = \frac{52.8 \times 10^{-6}}{0.206} = 255.7 \times 10^{-6}\ \text{Nm/A} = 0.256\ \text{mNm/A}$$

13
Optimization

Optimization methods try to find the *maximum* or *minimum* of a function, where there may exist restrictions or *constraints* on the independent variables. Finding the maximum or minimum of a function which is also the global maximum or minimum of the function has considerable difficulty and is the complex part of any optimization method. In engineering, it is generally considered practical to search only for local solutions.

In the optimization of electrical machines the *objective function* and constraints can be computed using either the classical (circuital) approach or a numerical field computation approach, such as the FEM. The FEM is more accurate than the classical approach but requires substantially more sophisticated software and computational time. In numerical field computation problems the standard prerequisites of local optimization (convexity, differentiability, accuracy of the objective function) are usually not guaranteed. Deterministic optimization tools for solving local optimization, such as steepest-descent, conjugate gradient and quasi-Newton methods, are not ideally suited to numeric electromagnetic problems [130]. This is due to the difficulties with numerical calculation of derivatives because of discretization errors and numerical inaccuracy. Recently, noniterative optimization schemes have been proposed using artificial neural networks. Stochastic methods of optimization have, however, become more popular over the last few years for optimization of electrical machines due to their high probability of finding global minima [40] and their simplicity. Stochastic methods such as simulated annealing [259], genetic algorithm (GA) [40] and evolution strategies [170] have been successfully used for different aspects of electrical machine design.

The *population-based incremental learning* (PBIL) method [21] is a stochastic nonlinear programming method which shows many advantages over the present stochastic methods. Stochastic optimization has the disadvantage that it is not very efficient. This problem is compounded by the large computation time of the numeric field computations. A solution to this problem is the use of *response surface methodology* [37], which has been successfully used for optimization of e.g., PM d.c. commutator motors [40].

13.1 Mathematical formulation of optimization problem

The optimization of an electrical machine can be formulated as a general constrained optimization problem with more than one objective, i.e., minimization of costs, minimization of the amount of PM material, maximization of the efficiency and output power, etc. Finding the extremum (**Extr**) of vector-optimization problems is defined as

$$\text{Extr } F(\mathbf{x}) = \text{Extr } [f_1(\mathbf{x}), f_2(\mathbf{x}), \dots, f_k(\mathbf{x})] \tag{13.1}$$

where

$$F : \Re^n \to \Re^k \qquad g_i, h_j : \Re^n \to \Re \qquad \mathbf{x} \in \Re^n \qquad \mathbf{x} = (x_1, x_2, \dots, x_n)$$

subject to specified equality and inequality constraints:

$$g_i(\mathbf{x}) \le 0, \qquad i = 1, 2, \dots, m \tag{13.2}$$

$$h_j(\mathbf{x}) = 0, \qquad j = 1, 2, \dots, p \tag{13.3}$$

and specified limits for the independent variables:

$$\mathbf{x}_{min} \le \mathbf{x} \le \mathbf{x}_{max} \tag{13.4}$$

In eqns (13.1) to (13.3) $F(\mathbf{x})$ is the vector objective function with the objectives $f_i(\mathbf{x})$ to be minimized, $\mathbf{x}$ is the vector of design variables used in the optimization, g_i are the nonlinear inequality constraints, h_j are the equality constraints, and $\mathbf{x}_{min}$ and $\mathbf{x}_{max}$ are vectors of lower and upper bounds for the design variables.

In vector optimization problems there is a conflict between the individual objective functions $f_i(\mathbf{x})$ since there exists no solution vector $\bar{\mathbf{x}}$ for which all objectives gain their individual minimum. Vector optimization problems can be transformed from multi-objective optimization into single objective optimization using the method of objective weighting. Although objective weighting always leads to a noninferiority (Pareto-optimal) feasible solution [68], the estimation of the weighting factors and the optimization starting point are subjective choices and their influence can rarely be estimated in advance.

A more practical method of optimization is to minimize only one objective function while restricting the others with appropriate constraints. Most constraints will be upper or lower bound inequality constraints, which means constrained optimization procedures are required. The optimization is thus done for a feasible region, in which all the constraints are satisfied for the design variables.

13.2 Nonlinear programming methods

In *nonlinear programming* both the objective and constraint functions may be nonlinear. There is no general agreement to the best optimization method or approach [106]. The extremely vast subject of nonlinear programming has been divided into direct search methods, stochastic methods and gradient methods [38]. A short review of these three classes is described below.

Most numerical field problems have constraints of one form or another. A summary of the main constrained optimization techniques is also given below.

13.2.1 Direct search methods

Direct search methods are minimization techniques that do not require the explicit evaluation of any partial derivatives of the function, but instead rely solely on values of the objective function, plus information gained from earlier iterations. Direct search methods can loosely be divided into three classes: tabulation, sequential and linear methods [38].

Tabulation methods assume that the minimum lies within a known region. The methods of locating the optimum are: (i) evaluation of the function at *grid* points covering the region given by the inequalities, (ii) *random searching* assuming that the minimum would be found within a sufficiently large number of evaluations, or (iii) a generalized *Fibonacci search* finding the solution of the multivariate minimization problem by using a sequence of nested univariate searches [106].

Sequential methods investigate the objective function by evaluating the function at the vertices of some geometric configuration in the space of the independent variables. This method originated from the evolutionary operation (EVOP). EVOP is based on *factorial designs*. The objective function is evaluated at the vertices of a hypercube in the space of the independent variables. The vertex with the minimum function value becomes the center point of the next iteration and a new design is constructed about this point. This is a mutation type search mechanism to direct the search towards the optimal. *Fractional factorial experimentation* assumes systematic and symmetric vertices to reduce the number of objective function evaluations.

The simplex method evaluates the objective function of n independent variables at $n + 1$ mutually equidistant points, forming the vertices of a regular simplex. The vertex with the highest value is reflected in the centroid of the remaining n vertices, forming a new simplex. The size of the simplex is reduced if one vertex remains unchanged for more than M consecutive iterations, thus narrowing the search to the minimum.

Linear methods use a set of direction vectors to direct the exploration [107]. There is a large number of linear methods available such as:

(i) *The alternating variable search method* which considers each independent variable in turn and alters it until a minimum of the function is located, while the remaining $(n-1)$ variables remain fixed.

(ii) *The method of Hooke and Jeeves* which uses exploratory moves and pattern moves to direct the search towards the minimum by attempting to align the search direction with the principal axis of the objective function.

(iii) *Rosenbrock's method* which uses n mutually orthonormal direction vectors. Perturbations along each search direction are done in turn and if the result is no greater than the current best value, this trial point replaces the current point. This is repeated until the minimum is obtained.

(iv) *Davies, Swann and Campey method* uses n mutually orthonormal direction vectors and a linear univariate search algorithm on each direction in turn. After each stage is complete, the direction vectors are redefined.

(v) *Quadratic convergent methods* minimize functions which are quadratic in the independent variables making use of conjugate directions.

(vi) *Powell's method* is based on mutually conjugate directions and ensures that a direction is replaced only if by doing so a new set of direction vectors, at least as efficient as the current set, is likely to be obtained.

13.2.2 Stochastic methods

Simulated annealing (SA). This method generates a sequence of states based on an analogy from thermodynamics where a system is slowly cooled in order to achieve its lowest energy state. This is done using nature's own minimization algorithm based on the Boltzmann probability distribution. Thus, the design configuration is changed from one to two with objective functions f_1 and f_2 with a probability of $P = \exp[-(f_2 - f_1)/kT]$. If $f_2 > f_1$, the state with probability P is accepted. If $f_2 < f_1$, the probability is greater than one and the new state is accepted. At a given temperature, the configurations are arbitrarily changed using a random number generator and the designs are also changed with a dictated probability greater than one. The temperature is thus lowered for the next round of searches. This makes uphill excursions less likely and limits the search space. SA ultimately converges to the global optimum.

Multiple-restart stochastic hillclimbing (MRSH). This method initially generates a random list of solution vectors of the independent variables, using binary vectors. The solution vector corresponding to the minimum result of the objective function is used in an iterative loop. A bit of the solution vector is toggled and evaluated. The minimum for a sufficiently large number of iterations is assumed to be the minimum of the objective function.

Genetic algorithm (GA). This search method is rooted in the mechanisms of evolution and natural genetics. A GA combines the principles of survival of the fittest with a randomized information exchange. GAs generate a sequence of populations by using a selection mechanism, and use crossover as the search mechanism to guide the search towards the optimal solution.

Population-based incremental learning (PBIL). This is a combination of evolutionary optimization and hillclimbing [21]. PBIL is an abstraction of the GA that maintains the statistics of the GA, but abstracts away the crossover operation and redefines the role of the population.

13.2.3 Gradient methods

Gradient methods select the direction $\mathbf{s}_i$, of the n dimensional direction vector, using values of the partial derivatives of the objective function F with respect to the independent variables, as well as values of F itself, together with information gained from earlier iterations. The solution is thus improved, that is

$$F(\mathbf{x}_{i+1}) \leq F(\mathbf{x}_i) \qquad \mathbf{x}_{i+1} = \mathbf{x}_i + h_i \mathbf{s}_i \qquad (13.5)$$

where h_i is the step increment and $\mathbf{s}_i$ is the search direction. Types of gradient optimization methods are:

Methods of steepest descent use the normalized gradient vector at the current point to obtain a new point using a specified step length.

Newton's methods use a second order truncated Taylor series expansion of the objective function $F(\mathbf{x})$. The method requires zero, first and second derivatives of the function at any point.

Quasi-Newton methods use an approximation of the second derivative of the function which is updated after each iteration.

13.2.4 Constrained optimization techniques

Constrained optimization problems are generally transformed to unconstrained ones and are then optimized using one of the nonlinear programming methods described above. Some of the techniques used on constrained problems are:

Feasible direction method attempts to maintain feasiblilty by searching from one feasible point to another along feasible arcs. This method assumes a feasible point can be found when the procedure starts.

Penalty function transforms the optimization problem to include the constraints which enable F to be maintained while controlling constraint violations by penalizing them. Exact penalty function is similar to the classical penalty function except that the absolute value of the constraints are used [106].

Sequential unconstrained minimization technique is also similar to the classical penalty function except the penalty coefficient is increased after each step of the algorithm.

Augmented Lagrangian function or multiplier penalty function uses a Lagrangian function to which the penalty term is added. The problem is transformed to an augmented Lagrangian function which is minimized.

13.3 Population-based incremental learning

In terms of Darwinian models of natural selection and evolution, *life is a struggle in which only the fittest survive to reproduce.* GAs based on natural selection and genetic recombination were first proposed by Holland [150]. GAs generate a sequence of populations by using a *selection mechanism, crossover* and *mutation* as search mechanisms. In nature, competition for resources such as food means that the fittest individuals of a species dominate over the weaker ones. This natural phenomenon is called "the survival of the fittest." The fittest individuals thus get a chance to reproduce ensuring, implicitly, the survival of the fittest genes. The reproduction process combines the genetic material (chromosome) from the parents into a new gene. This exchange of part of the genetic material among chromosomes is called *crossover.*

GAs encode the solution as a population of binary strings. Each solution is associated with a fitness value determined from the objective function. The main operation in GA is crossover, although *mutation* plays the role of regenerating lost genetic material by causing sporadic and random alteration of the bits of the binary strings. GAs are generally characterized by their population size, crossover type, crossover rate and elitist selection. These control parameters affect how well the algorithm performs. The optimum set of parameters is dependent on the application being optimized.

PBIL is an abstraction of the GA that explicitly maintains the statistics contained in a GA's population, but abstracts away the crossover operation [21]. PBIL is in fact a combination of evolutionary optimization and hillclimbing. The PBIL algorithm uses a real valued probability vector which, when sampled, reveals a high evaluation solution vector with high probability.

The PBIL algorithm creates a probability vector from which samples are drawn to produce the next generation's population. As in the GA, the solution is encoded into a binary vector of fixed length. Initially the values of the probability vector are set to 0.5. A number of solution vectors, analogous to the population in GAs, are generated based upon the probabilities of the probability vector. The probability vector is pushed towards the generated solution vector with the highest evaluation (fitness value). This probability vector can thus be considered a prototype for high evaluation vectors for the function space being explored. Each bit of the probability vector is updated using

$$P_i = [P_i \times (1.0 - \delta l)] + (\delta l \times \sigma_i) \qquad (13.6)$$

where P_i is the probability of generating a one in the bit position i, σ_i is the ith position in the solution vector for which the probability vector is being changed and δl is the learning rate. The learning rate is the amount the probability vector is changed after each cycle. A new set of solution vectors is produced after each update of the probability vector. The entries in the probability vector start to drift towards either 0.0 or 1.0 as the search progresses to represent a high evaluation solution vector.

Mutation is used in PBIL for the same reasons as in the GA, to inhibit premature convergence. Mutations perturb the probability vector with a small probability in a random direction. PBILs are generally characterized by their number of samples, learning rate, number of vectors to update from and mutation rate. Fig. 13.1 shows a flow chart representation of PBIL.

PBIL has been shown to work as well, or better, than the GA. The main advantage of the PBIL over the GA is that since PBIL is characterized by fewer parameters and their values are less problem-related, as little problem-specific knowledge as possible is needed.

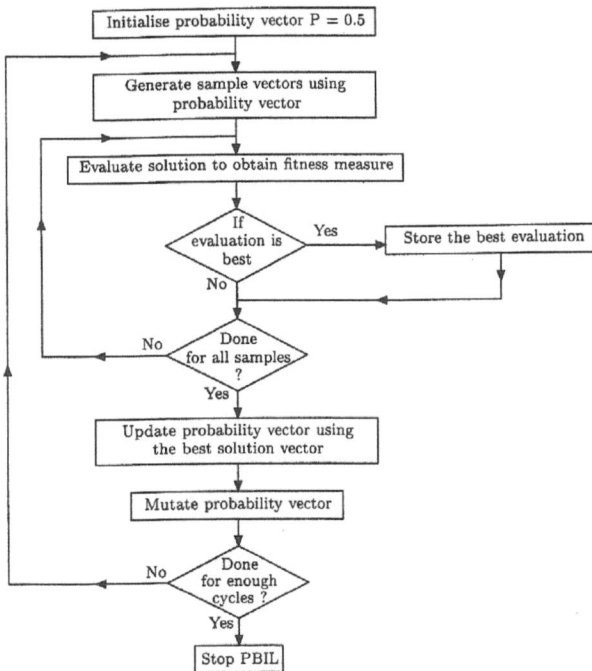

Fig. 13.1. Flow chart of PBIL algorithm.

13.4 Response surface methodology

When using a numeric field computation based on a FEM analysis, the objective function being optimized will require a lengthy computation time. Owing to this large computation time a faster approach than PBIL is required. *Response surface methodology* (RSM) is a collection of mathematical and statistical techniques useful for analyzing problems of several independent variables and model exploitation.

By careful design of the FEM experiments, RSM seeks to relate the output variable to the input variables that affect it. The computer experimental result y as a function of the input independent variables $x_1, x_2, \ldots, x_n$ is

$$y = f(x_1, x_2, \ldots, x_n) + \delta(x_1, x_2, \ldots, x_n) + \delta\epsilon \qquad (13.7)$$

where $\delta(\mathbf{x})$ is the bias error and $\delta\epsilon$ is a random error component [37]. If the expected result is denoted by $E(y) = S$, then the surface represented by $S = f(x_1, x_2, \ldots, x_n)$ is called the *response surface*.

The form of the relationship between the result and the independent variables is unknown. A polynomial expression is used as a suitable approximation for the true functional relationship between y and the independent variables. A polynomial expression of degree d can be thought of as a Taylor's series expansion of the true underlying theoretical function $f(\mathbf{x})$ truncated after terms of dth order. The assumption that low order polynomial models can be used to approximate the relationship between the result and the independent variables within a restricted region of the operability space is essential to RSM. Second order polynomial models are used to model the response surface.

$$f(x) = b_o + \sum_{i=1}^{n} b_i x_i + \sum_{i=1}^{n} b_{ii} x_i^2 + \sum_{i=1}^{n} \sum_{j=1}^{n} b_{ij} x_i x_j \qquad (13.8)$$

where $i < j$ and the coefficients b_o, b_i, b_{ii} and b_{ij} are found using the method of least squares. These unknown coefficients can be estimated most effectively if proper computer experimental designs are used to collect the data. Designs for fitting response surfaces are called *response surface designs*.

13.4.1 Response surface designs

An experimental design for fitting a second order model must have at least three levels of each factor, so that the model parameters can be estimated. Rotatable designs are the preferred class of second order response surface designs [37]. The variance of the predicted response at a point x, in a rotatable design, is a function only of the distance of the point from the design center and not a function of direction.

The collecting of the sample results is essential since a sufficiently accurate approximation has to be found with a minimum of experiments. If not all the factorial combinations are employed, the design is called an incomplete factorial design. The *Box-Behnken three level design* has been chosen for investigating the response surface. This is an incomplete factorial design which is a reasonable compromise between accuracy of the function and the required number of computations. The design generates second order rotatable designs or near-rotatable designs, which also possess a high degree of orthogonality.

13.4.2 Estimation of errors in response surface fitting

The errors in design of experimental methods are generally divided into two types [37]: (i) systematic or bias errors $\delta(\mathbf{x})$ which are the difference between the expected value of the response $E(y) = S$ and the approximate objective function $f(\mathbf{x})$, and (ii) the random or experimental error ϵ in sampling.

In numeric computer experiments, replicated experiments result in the same result, so random errors cannot be defined. Only the bias error from systematic departure of the fixed polynomial from the real response, due to an insufficient polynomial order, can be calculated. An estimate of the variance of the bias error is [217]

$$s_e^2 = \frac{1}{m-p} \sum_{i=1}^{m} (y_i - \hat{y}_i)^2 \qquad (13.9)$$

where s_e is the standard error of estimate, m is the number of observations, p is the number of coefficients in the polynomial, y_i is the observed response and $\hat{y}_i$ is the predicted response. The normalized error is [40]

$$\bar{\delta} = \frac{s_e}{y_o} \qquad (13.10)$$

where $y_o = (y_1 + y_2 + \ldots + y_n)/m$. The accuracy of the response surface $E(y)$ is varied by changing the size of the investigated region, since only second order polynomials are used.

13.5 Modern approach to optimization of PM motors

The *aim of the optimization procedure* is to minimize the cost of the active material of the motor, while ensuring a rated power and high efficiency. It is proposed to optimize a PM motor using the PBIL with RSM. The PBIL does not optimize the performance characteristics directly, but uses polynomial fits of the performance characteristics of the motor, created using RSM.

The motor characteristics are calculated using FEM (Chapter 3) and classical machine theory for a number of combinations of input parameters. Fig. 13.2 shows the logic sequence followed by the control program. The output characteristics and input parameters are used to fit polynomial equations of

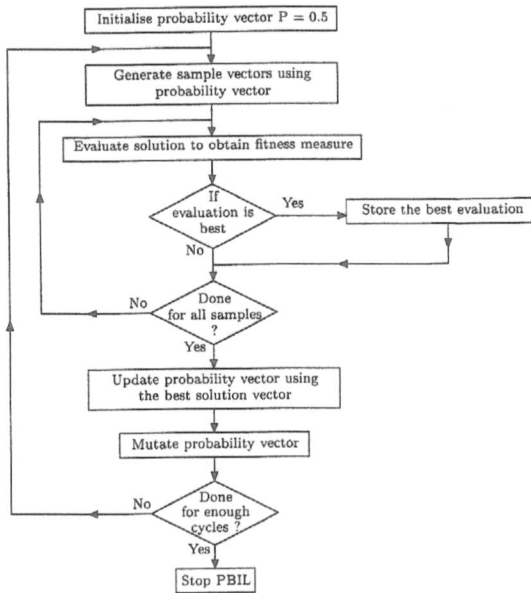

Fig. 13.2. Flow chart of the control program which produces the motor output characteristics.

second order to every output characteristic. These polynomials are used as objective functions and constraints in the PBIL optimization.

The following simplifications are made for the FEM: (a) the end leakage reactance is calculated using classical theory (Appendix A, eqn (A.10)) since it would be very difficult to calculate using a two-dimensional FEM; (b) the induced EMF and inductive reactance are assumed constant throughout the load range, and equal to the value obtained at rated current; (c) the model is independent of rotor position.

13.5.1 PM d.c. commutator motors

Consider a PM d.c. commutator motor with segmental magnets, as in Fig. 4.1 (Chapter 4). The optimization problem can be stated as minimize the cost

$$C(\mathbf{x}) = V_M c_{PM} + V_w c_w + V_c c_c \tag{13.11}$$

where V_M, V_w, V_c are the volume of the PM, copper conductors and steel core, respectively, and c_{PM}, c_w, c_c are the price per m^3 for the PM, copper conductors and electrotechnical steel, respectively. The optimization is subject to the constraints:

$$T_d(\mathbf{x}) \geq a \qquad \text{Nm}$$

$$P_{out}(\mathbf{x}) \geq b \qquad \text{W}$$

$$\eta(\mathbf{x}) \geq c$$

$$H_{max}(\mathbf{x}) \leq d \qquad \text{kA/m}$$

$$D_{out}(\mathbf{x}) \leq e \qquad \text{mm}$$

where T_d is the electromagnetic torque, P_{out} is the output power, η is the efficiency, H_{max} is the maximum magnetic field intensity of the PM and D_{out} is the outer stator diameter, a is the value of electromagnetic torque, b is the value of output power, c is the value of efficiency, d is the value of magnetic field intensity and e is the value of the stator outer diameter. These output characteristics are computed via postprocessing of the numeric field solution.

The electromagnetic torque is computed by using the Maxwell stress tensor according to eqn (3.71) multiplied by $D_{2out}/2$ (Chapter 3, Section 3.11), i.e.,

$$T_d = \frac{D_{2out} L_i}{2\mu_o} \oint_l B_n B_t \, dl \qquad (13.12)$$

where D_{2out} is the outer diameter of the rotor, L_i is the effective length of the rotor stack, B_n and B_t are the normal and tangential components of the flux density over a contour placed in the middle of the air gap.

The cost of active material, i.e., PMs, copper wire and electrotechnical steel is calculated according to eqn (13.11).

The maximum field strength H_{max} is computed from the FEM results with the armature current three times the rated current.

The power loss in the rotor steel and in the stator yoke are computed using the specific loss curves given by the manufacturers. An additional loss factor is multiplied to basic core loss results to account for higher harmonics in the magnetic flux.

The power balance is computed in order to evaluate the output power P_{out}, the speed n and the efficiency η. The electromagnetic power is

$$P_{elm} = V I_a - \sum R_a I_a^2 - \triangle P_{Fe} - I_a \triangle V_{br} \qquad (13.13)$$

where I_a is the armature current, $\sum R_a$ is the armature circuit resistance, $\triangle P_{Fe}$ are the armature core losses and $I_a \triangle V_{br}$ is the brush drop loss. The efficiency is then

$$\eta = \frac{P_{out}}{V I_a} \approx \frac{P_{elm}}{V I_a}$$

and speed $n = P_{elm}/(2\pi T_d)$.

13.5.2 PM synchronous motors

A surface PM rotor and a buried PM rotor synchronous motor have been considered. Only the rotor is being optimized since the stator of an existing induction motor has been used. The objective function attempts to minimize the volume of PM material used. The optimization problem can be expressed as: minimize $V_M(\mathbf{x})$ subject to the constraints

$$P_{elm}(\mathbf{x}) \geq P_{elm(d)} \qquad\qquad J_a(\mathbf{x}) \leq J_{ath}$$

$$\eta(\mathbf{x}) \geq \eta_d \qquad\qquad g \geq g_{min} \qquad (13.14)$$

$$h_M,\ w_M \geq h_{Mmin} \qquad\qquad D_{max} < D_{1in}$$

where V_M is the magnet volume, $P_{elm(d)}$ is the desired electromagnetic power, J_{ath} is the current density thermal limit at $P_{elm(d)}$, η_d is the desired electrical efficiency at $P_{elm(d)}$, g and h_M are the mechanical minimum sizes for the air gap and PM respectively, and D_{max} is the maximum diameter of the outer edge of the PMs.

The independent variables used in the surface PM rotor design are the air gap g, PM thickness h_M and the overlap angle β, and in the case of the buried PM rotor design the air gap g, PM thickness h_M and PM width w_M (Fig 13.3). The motor performance obtained from the FEM, using flux linkage and magnetizing reactances, is shown in Chapter 5, Examples 5.1 to 5.3.

The objective function is not easily expressed in terms of the independent variables. RSM is thus not used for modelling the objective function directly, but rather for modelling the performance characteristics of the PM synchronous motor, used in the constraints, in terms of the independent variables.

The characteristics for a number of combinations of the independent variables (factors) are calculated using the FEM. The factors range over the whole problem space and their combinations are determined by the Box-Behnken three level design method, with 3 factors and 3 levels the design needs 15 runs. Response surfaces are created for the output characteristics E_f, X_{sd}, X_{sq}, X_{ad} and X_{aq}. A second order polynomial is fitted to each of these five characteristics using the least squares method. These five polynomial equations are then used to calculate the electromagnetic power, efficiency and stator current in the optimization procedure. The PBIL optimization is used for the minimization of the PM volume with the constraints stated in eqn (13.14).

The volume of a surface magnet per pole is $V_M/(2p) = (0.25\beta\pi/360^0)$ $\times(D_{2out}^2 - D_{Min}^2)l_M$, where l_M is the magnet axial length, D_{2out} and D_{Min} are magnet diameters according to Fig. 13.3a. The volume of a buried magnet per pole is $V_M/(2p) = 2h_M w_M l_M$ (Fig. 13.3b).

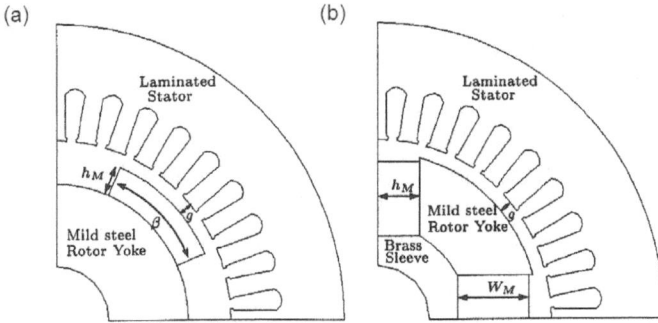

Fig. 13.3. Geometric layout of PM rotors showing design variables for: (a) surface PM rotor, (b) buried PM rotor.

Table 13.1. Optimization of surface PM rotor using RSM

P_{elm}	η	minimum $\cos\phi$	g	β	h_M	$\cos\phi$	PM volume per pole
kW	%		mm	deg	mm		mm^3
1.5	90	–	0.3	76.85°	0.77	0.754	4191
2.0	90	–	0.3	72.00°	1.15	0.871	5857
2.5	90	–	0.3	73.80°	1.75	0.974	9056
1.5	90	0.90	0.3	74.78°	1.46	0.900	7669
2.0	90	0.90	0.3	70.23°	1.38	0.900	6793
2.5	90	0.90	0.3	73.80°	1.75	0.974	9056
Final result							
2.2	90	0.90	0.3	71.40°	1.36	0.913	6838

Numerical examples

Numerical example 13.1

Optimization of a surface PM rotor. The stator of a commercial 380-V, 50-Hz, four-pole 1.5-kW induction motor has been used to design a synchronous motor with surface type PM rotor. Table 13.1 shows the results for different constraints in electromagnetic output power, efficiency and power factor. The results show that at low power ratings (1.5 kW) the design needs a substantial increase in PM material if the power factor is constrained to a minimum of 0.9 at rated power. At the high power rating of 2.5 kW the power factor well exceeds the design minimum.

The most feasible power rating for this motor should ensure that the stator winding current density remains well below the specification limits, but also maximizes the power output. A power rating of 2.2 kW is thus considered appropriate due to the excellent power factor at rated power and the maximum use of the stator windings. The stator used in this motor is from a 1.5 kW induction motor. The increase in power rating to 2.2 kW, when operating as a synchronous motor, is due to the higher stator winding current density possible in PM synchronous machines, and the improved efficiency and power factor. The stator winding current density has increased from 9.93 A/mm^2, for the induction motor, to 10.3 A/mm^2 for the PM synchronous motor (forced ventilation).

Higher time and space harmonics have been neglected and the effects of *cogging torque* have also been neglected throughout this optimization procedure. The cogging torque can significantly be reduced using an appropriate PM overlap angle [197]. Skewing of the stator teeth is not possible as an existing stator is being used. Using a PM $\beta = (k + 0.14)t_1$ where k is an integer and $t_1 = 360^0/s_1$ is the stator slot pitch (angle), the minimum cogging torque can be achieved [197]. The overlap angle is thus decreased from $\beta = 73.8^0$ to 71.4^0. The air gap and magnet thickness are again optimized for this fixed overlap angle. Table 13.1 shows the final optimized rotor details.

The performance of this optimized motor is compared with an initial surface PM design. Fig. 13.4 compares the electromagnetic power and Fig. 13.5 compares the efficiency for the two motors. The optimized surface PM synchronous motor has superior efficiency at the desired rated power with a reduction in the PM volume from 15,000 mm^3 per pole for the initial design (Example 5.1) to 6838 mm^3 per pole for the optimized motor.

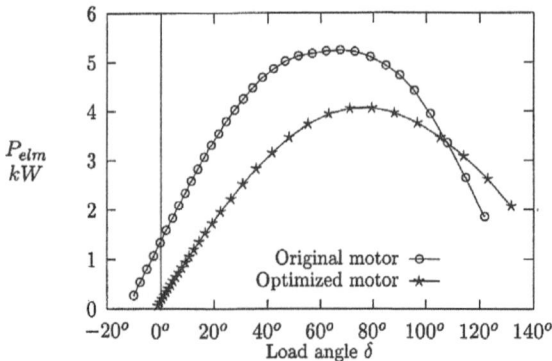

Fig. 13.4. Electromagnetic power versus load angle for surface PM motors. *Numerical example 13.1.*

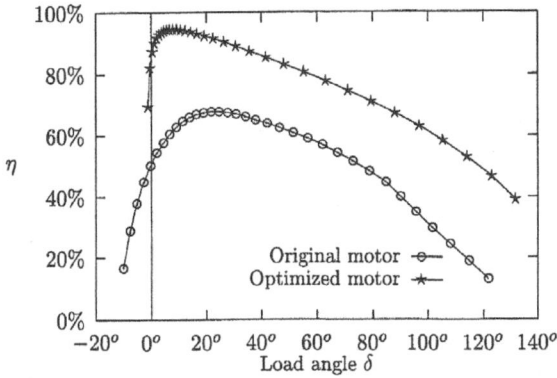

Fig. 13.5. Efficiency versus load angle for surface PM motors. *Numerical example 13.1.*

Numerical example 13.2

Optimization of a buried PM rotor. The stator of a commercial 380-V, 50-Hz, four-pole 1.5-kW induction motor has been used to design a synchronous motor with buried-type PM rotor. Table 13.2 shows the results for different constraints in electromagnetic output power, efficiency and power factor.

The most feasible power rating is again chosen to be 2.2 kW. The current density in the stator winding is $J_a \approx 10.4$ A/mm^2 (forced ventilation).

Table 13.2. Optimization of buried PM rotor using RSM

P_{elm}	η	minimum $\cos\phi$	g	w_M	h_M	$\cos\phi$	PM volume per pole
kW	%		mm	mm	mm		mm^3
1.5	90	–	0.3	15.21	1.21	0.717	3672
2.0	90	–	0.3	16.90	1.35	0.817	4561
2.5	90	–	0.3	20.18	1.62	0.913	6535
1.5	90	0.90	0.3	21.54	2.01	0.910	8673
2.0	90	0.90	0.3	20.87	1.52	0.900	6329
2.5	90	0.90	0.3	19.32	1.72	0.910	6627
Final result							
2.2	90	0.90	0.3	19.28	1.55	0.909	5977

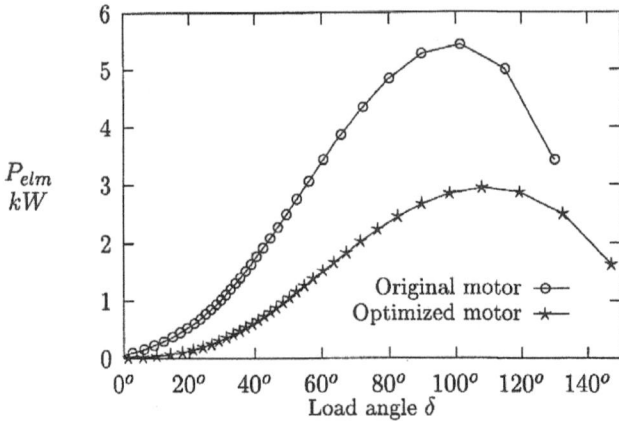

Fig. 13.6. Electromagnetic power versus load angle for buried PM motors. *Numerical example 13.2.*

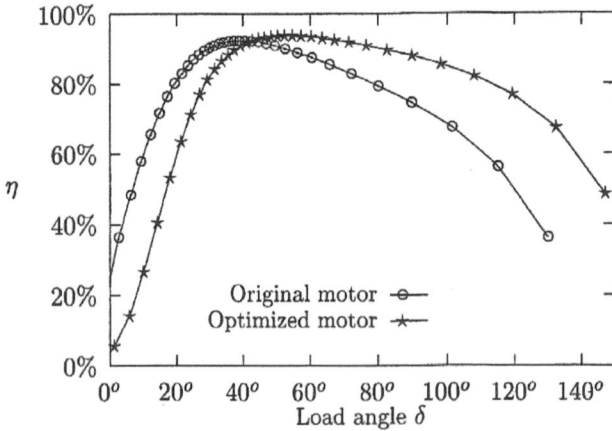

Fig. 13.7. Efficiency versus load angle for buried PM motors. *Numerical example 13.2.*

The cogging torque can be minimized in the final design by creating asymmetry of the rotor magnetic circuit. This has a marginal effect on the optimization point since it does not change any of the optimization parameters.

The performance of this optimized motor is compared with an initial buried PM motor design (Figs 13.6 and 13.7). The optimized buried PM motor uses 6627 mm^3 of PM material, while the initial design uses 16,200 mm^3 (Example 5.3). It has also superior efficiency at rated output power.

The optimization of the surface and buried PM synchronous motors both showed improved performance over their initial designs. The volume of PM material was also reduced in both designs. The buried PM motor design is considered the superior design since it uses the minimum of PM material and has a high efficiency over a wide power range (Fig. 13.7).

The RSM using the PBIL is thus seen as an appropriate method for optimization of PM synchronous motors with the aid of the FEM. This optimization technique can easily be extended for the optimum design of the whole synchronous motor.

14

Maintenance

14.1 Basic requirements for electric motors

Electric motors, both a.c. and d.c. motors, come in many shapes and sizes. Some electric motors are standardized for general-purpose applications. Other electric motors are intended for specific tasks. Very different applications and requirements have resulted in the manufacture of many different types and topologies. The customer's demand for safety, comfort, economy, clean environment and quality is another reason for the explosive growth in the variety of electric motors. The basic technical and economic requirements to electric motors can be classified as follows:

- General requirements:
    - low cost
    - simple construction
    - simple manufacture
    - high efficiency and power factor
    - low EMI and RFI level
    - long service life
    - high reliability
- Requirements depending on application and operating conditions:
    - repairability, essential for medium and large power motors
    - extended speed range and energy efficiency for motors for electric vehicles
    - low noise, essential for public life and consumer electronics
    - minimum size and mass at desired performance for airborne apparatus, handpieces and power hand tools
    - resistance to vibration and shocks, essential for transport and agricultural drives and also for airborne equipment
    - resistance to environmental effects and radiation essential for electromechanical drives operating in nuclear reactors, space vehicles, underwater vehicles and in tropics

    – explosion safety, essential for mine drives
    – low amount of gas escape, essential for electromechanical drives installed in vacuum equipment
- Additional requirements for motors used in servo drives and automatic control systems:
    – fast response
    – temperature-independent response
    – high torque at high speed
    – high overload capacity
    – stability of performance

14.2 Reliability

Reliability of an electrical machine is the probability that the machine will perform adequately for the length of time intended and under the operating environment encountered. Reliability of machines intended for a long service life is to be ensured on a reasonable level with due regard to economic factors, i.e., the best level of reliability is that obtained at minimum cost. The quantitative estimate of machine reliability is made by using *probability* and *mathematical statistics* methods.

According to reliability theory, all pieces of equipment are classified as *repairable* or *nonrepairable*, i.e., those that can be repaired upon failure and those that cannot. A *failure* (mechanical, thermal, electric, magnetic or performance degradation) is an event involving a full or partial loss of *serviceability*. Electrical machines, depending on their applications, may fall into either of these classes.

The theory of reliability usually treats failures as random events. Therefore, all the quantitative characteristics are of a probabilistic nature.

Failure density, $f(t)$. This is the unconditional probability $\Delta t f(t)$ of failure in a given time interval Δt.

Probability of trouble-free operation (reliability), $P(t)$. This is the probability that the trouble-free running time of a machine, until failure, is longer than or equal to the specified time interval $\Delta t = t_2 - t_1$, i.e.

$$P(t) = \int_{t1}^{t2} f(t)dt \tag{14.1}$$

Statistical estimation of the non-failure probability is made, provided that the machines that fail are neither repaired nor replaced by new units, by using the formula

$$P^*(t) = \frac{N_0 - n(t)}{N_0} \tag{14.2}$$

where N_0 is the number of machines at the beginning of the test (population size) and $n(t)$ is the number of machines failed within the time t. The relationship between the failure function $Q(t)$ an realiability function $P(t)$ is

$$Q(t) = 1 - P(t) \qquad (14.3)$$

Failure rate, $\lambda(t)$. This is the probability of failures of a machine per time unit Δt after a predetermined moment. It can simply be expressed as

$$\lambda(t) = \frac{n(t)}{\Delta t} \qquad (14.4)$$

where $n(t)$ is the number of machines failed within the time interval Δt. Statistical estimation of the failure rate is made from the formula

$$\lambda^*(t) = \frac{n(t)}{\Delta t N_{av}} \qquad (14.5)$$

where N_{av} is the average number of machines operating trouble-free within the observation time interval Δt.

Failure rate is expressed as the probability of failure density $f(t)$–to–the non-failure probability ratio $P(t)$, i.e.,

$$\lambda(t) = \frac{f(t)}{P(t)} \qquad (14.6)$$

A typical characteristic of the failure rate, $\lambda(t)$, obtained from the tests, is shown in Fig. 14.1. A relatively high failure rate is within the time interval $0 \leq t \leq t_1$. This interval illustrates an early-failure period (the so-called "infant mortality") within which failures occur due to manufacturing defects. Then, the failure rate is reduced rather abruptly. The interval $t_1 \leq t \leq t_2$ shows a period of normal operation. Upon the expiration of time t_2 the failure rate suddenly increases due to mechanical wear, electrical wear or deterioration of material characteristics.

Hence, the failure rate for a period of normal operation of the machine can be assumed constant, i.e., $\lambda = const$. The reliability function or probability of trouble-free operation $P(t)$ for a machine with constant failure rate λ is a negative exponential distribution with failure rate λ, i.e.,

$$P(t) = e^{-\lambda t} \qquad (14.7)$$

Statistics theory shows that the random failure rate, independent of the time, is distributed according to the exponential law. The failure density is

$$f(t) = \frac{dQ(t)}{dt} = \frac{d[1 - P(t)]}{dt} = \frac{d[1 - \exp(-\lambda t)]}{dt} = \lambda e^{-\lambda t} \qquad (14.8)$$

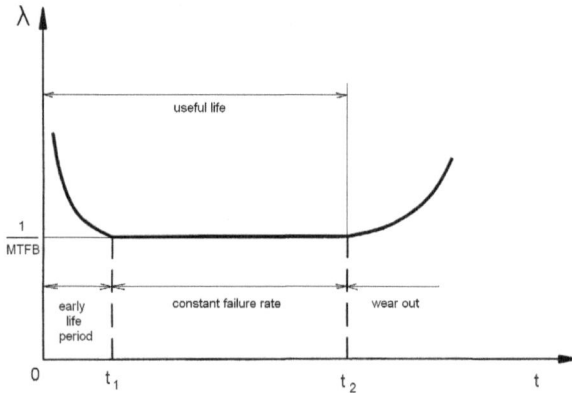

Fig. 14.1. Characteristic of failure rate $\lambda(t)$ ("bathtub" curve).

where $Q(t)$ is given by eqn (14.3) and $P(t)$ is given by eqn (14.7). Putting eqn (14.8) to eqn (14.6) the failure rate $\lambda(t) = \lambda$, which means that the motor, its parts or a system are characterized by a constant failure rate $\lambda = const$.

Mean time between failures, MTFB. The following formula is used for statistical estimation of the mean no-trouble time

$$\mathrm{MTFB}^* = t^*_{mean} = \frac{1}{N_0} \sum_{i=1}^{N_0} t_{ti} \qquad (14.9)$$

where t_{ti} is the no-trouble time of an ith specimen.

The *mean time between failures* MTFB (trouble-free operation) is the mathematical expectation of the no-trouble service time

$$\mathrm{MTFB} = t_{mean} = \int_0^\infty P(t)dt \qquad (14.10)$$

which defines an average time interval between the beginning of operation and the first failure. At $\lambda = const$, the mean time between failures

$$\mathrm{MTFB} = t_{mean} = \frac{1}{\lambda} \qquad (14.11)$$

Failure density $\lambda = const$ and MTBF for selected electromagnetic and electronic components are given in Table 14.1

Weibull distribution is a general-purpose reliability distribution used to model times to failure of mechanical and electronic components, devices, systems and material strength. The most general expression of Weibull *probability density function* (PDF) is given by the three-parameter Weibull distribution expression

Table 14.1. Failure rate $\lambda(t)$ and MTBF of electromagnetic and electronic components per hour

Machines, devices, components and systems	$\lambda(t)$, 1/h	MTFB, h
Small electrical machines	$(0.01 \ldots 8.0) \times 10^{-4}$	$1250 \ldots 1{,}000{,}000$
Transformers	$(0.0002 \ldots 0.64) \times 10^{-4}$	$15{,}000 \ldots 5 \times 10^{7}$
Resistors	$(0.0001 \ldots 0.15) \times 10^{-4}$	$67{,}000 \ldots 1 \times 10^{8}$
Semiconductor devices	$(0.0012 \ldots 5.0) \times 10^{-4}$	$2000 \ldots 8{,}333{,}000$
Intel solid state devices	0.01×10^{-4}	$1{,}000{,}000$
IC L293N *Texas Instruments*	0.09×10^{-8}	1.07×10^{9}
UPS	$(0.017 \ldots 2.5) \times 10^{-4}$	$4000 \ldots 580{,}000$
Computer HDD	0.01×10^{-4}	up to $1{,}000{,}000$
Computer CD-DVD drives	1×10^{-5}	$100{,}000$

$$f(t) = \frac{\beta}{\gamma} \left(\frac{t - \gamma}{\eta} \right)^{\beta - 1} \exp\left[- \left(\frac{t - \gamma}{\eta} \right)^{\beta} \right] \tag{14.12}$$

in which $\beta > 0$ is the shape parameter, also known as Weibull slope, $\eta > 0$ is the scale parameter (expressed in units of time) and $-\infty < \gamma < \infty$ is the location parametr. Frequently, the location parameter is not used ($\gamma = 0$) and eqn (14.12) reduces to that of two-parameter Weibull distribution. The *Weibull reliability function* is given by

$$P(t) = \exp\left[- \left(\frac{t - \gamma}{\eta} \right)^{\beta} \right] \tag{14.13}$$

Similar to eqn (14.6) the *Weibull failure rate* function is

$$\lambda(t) = \frac{f(t)}{P(t)} = \frac{\beta}{\eta} \left(\frac{t - \gamma}{\eta} \right)^{\beta - 1} \tag{14.14}$$

Weibull mean life or *mean time to failure* (MTTF) is

$$\text{MTTF} = t_{W\,mean} = \gamma + \eta \Gamma \left(\frac{1}{\beta} + 1 \right) \tag{14.15}$$

where $\Gamma(1/\beta + 1)$ is the *gamma function* defined as

$$\Gamma(n) = \int_{0}^{\infty} e^{-x} x^{n-1} dx \tag{14.16}$$

The values of the shape parameter β and MTTF for selected machines, devices and components are given in Table 14.2. While the MTFB (number of hours that pass before a component, assembly, or system fails) is a basic measure of reliability for *repairable* items, the MTTF (mean time expected until

Table 14.2. Weibull database

Machines, devices or components	Shape factor β			MTTF, h		
	Low	Typical	High	Low	Typical	High
Motors, a.c. brushless	0.5	1.2	3.0	1000	100,000	200,000
Motors, d.c. brush	0.5	1.2	3.0	100	50,000	100,000
Transformers	0.5	1.1	3.0	14,000	200,000	420,000
Solenoid valves	0.5	1.1	3.0	50,000	75,000	1,000,000
Transducers	0.5	1.0	3.0	11,000	20,000	90,000
Magnetic clutches	0.8	1.0	1.6	100,000	150,000	333,000
Ball bearing	0.7	1.3	3.5	14,000	40,000	250,000
Roller bearings	0.7	1.3	3.5	9000	50,000	125,000
Sleeve bearing	0.7	1.0	3.0	10,000	50,000	143,000
Couplings	0.8	2.0	6.0	25,000	75,000	333,000
Gears	0.5	2.0	6.0	33,000	75,000	500,000
Centrifugal pumps	0.5	1.2	3.0	1000	35,000	125,000
Coolants	0.5	1.1	2.0	11,000	15,000	33,000
Lube oils, mineral	0.5	1.1	3.0	3000	10,000	25,000
Lube oils, synthetic	0.5	1.1	3.0	33,000	50,000	250,000
Greases	0.5	1.1	3.0	7000	10,000	33,000

the first failure of a piece of equipment) is a basic measure of reliability for *nonrepairable* items.

14.3 Failures of electric motors

The service expectation of electrical machines depends on the type, applications and operating conditions. For large and medium power machines it is usually over 20 years, for general purpose small d.c. and a.c. machines up to 10 years, and for very small d.c. brush motors as for toys about 100 hours.

Electrical machines with movable parts are less reliable than semiconductor devices or static converters as, e.g., transformers (Table 14.1). It is evident from experience that most failures occur due to trouble with the mechanical parts and windings of electrical motors or changes in their material characteristics. Given below are the parts most likely to fail:

- Bearings:
    - contamination and scoring by foreign matter or dirt
    - premature fatigue due to excessive loads
    - loss in hardness, reduction of bearing capacity and deformation of balls and rings due to overheating
    - brinelling when loads exceed the elastic limit of the ring material
    - excessive wear of balls, ring, and cages due to lubricant failure
    - broken spacer or ring

- babbitt fatigue, wiping, creep and thermal ratcheting
- porosity and blisters in babbitted bearings
- Speed reduction gearboxes, if built-in:
  - cracked teeth
  - worn teeth
- Sliding contact:
  - poor contact between commutator and brushes due to wear of brushes or insufficient pressure of brush stud on brush
  - mechanical damage to brush holders
  - damage to commutator caused by brushes
- Windings:
  - broken turns or leads as result of burning due to overloads, mechanical strain due to temperature fluctuations or electrical corrosion action, especially at high humidity
  - disrupted soldered connections
  - earth or turn–to–turn fault of insulation as a result of poor electric strength, especially at severe thermal conditions and high humidity
  - transient voltage surges or spikes
- Magnetic system:
  - changes in PM performance due to high temperature, shocks, vibrations, strain and hazardous gases
  - changes in characteristics of laminations as a result of short circuits between them, electrical corrosion, etc.

In the case of d.c. PM brush motors the most vulnerable parts are the commutator and brushes. In brushless types of motors the most vulnerable parts are bearings. The condition of bearings and sliding contacts greatly depends on the speed of rotation of the rotor. Wear of most parts, commutator and brushes in particular, increases with the rotational speed, hence, the reliability of these parts and of the motor as a whole reduces.

Fig. 14.2 illustrates the MTBF t_{mean} of small d.c. brush motors as a function of angular speed $\Omega = 2\pi n$ (solid line) at which they were tested [15]. The rated speed and the time within which they were tested are assumed as unity. It means that the service life that can be guaranteed by the manufacturer depends on the motor rated speed. For fractional horsepower commutator motors the guaranteed service life is about 3000 h for the rated speed $n = 2500$ rpm and $200 \ldots 600$ h for the rated speed $n = 9000$ rpm.

A higher reliability of electric motors can be obtained by eliminating the sliding contact. The guaranteed service life of small brushless motors is minimum $10,000$ h at $12,500$ rpm [69]. Some PM brushless motors can run for almost $200,000$ h without a failure .

Reliability of motor insulation greatly depends on ambient temperature, relative humidity and temperature of the motor itself. Experiments with a number of d.c. and a.c. fractional horsepower motors have shown that failures

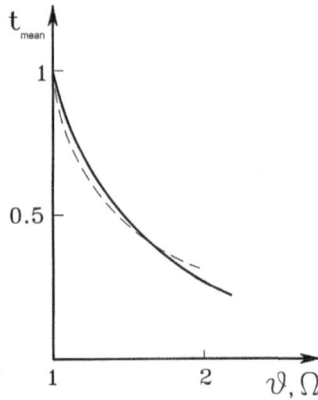

Fig. 14.2. Per unit MTFB t_{mean} of small d.c. commutator motors as a function of per unit angular speed Ω (solid line) and temperature ϑ (dash line).

due to excessive humidity and high or low temperature account for 70 to 100% of the total number of failures resulting from operating the motors under unspecified conditions. Fig. 14.2 shows the MTBF t_{mean} of fractional horsepower d.c. commutator motors as a function of ambient temperature ϑ (dash line) at which they were tested. The rated ambient temperature and the time corresponding to this temperature are assumed as unity. In addition to probability of failures, motors operating at temperatures beyond the specified limits worsen their performance characteristics.

In selecting electrical motors for a particular drive system, their thermal conditions must be given serious consideration. Small motors are commonly manufactured as totally enclosed machines. When mounted in the equipment, the frame can be joined to a metal panel or any other element to improve the conditions of heat transfer. When installing small machines in enclosed units or compartments accommodating other heat-emitting components, it is extremely essential to calculate correctly the surrounding temperature at which the machine is expected to operate. An increase in the ambient temperature causes an increase in the absolute temperature of the motor, which influences its reliability and characteristics. Even very small motors are themselves intensive heat sources. In thermal calculations, it is important to take into account their operating conditions such as duty cycle, no-load or full-load running, frequency of startings and reversals.

Reliability of electric motors may reduce over the course of service due to vibrations, shocks, and too low atmospheric pressure.

To increase the reliability of electric motors the following measures should be taken:

- optimum electromagnetic design
- effective cooling system
- robust mechanical design

- application of good quality materials
- increase of heat resistance, particularly mechanical and electrical properties of insulation
- quality assurance in manufacturing
- use machines under operating conditions as specified by the manufacturer.

Reliability of electrical motors is associated with their mechanical and electrical *endurance*. This may be determined by the service life of a motor from the beginning of its operation until its depreciation. The mechanical and electrical endurance is usually a criterion for evaluating repairable motors.

14.4 Calculation of reliability of small PM brushless motors

Probability of trouble-free operation of a small PM machine with built-in speed reduction gearbox can be calculated using the following equation

$$P(t) = P_b(t)P_g(t)P_w(t) \tag{14.17}$$

where $P_b(t)$, $P_g(t)$ and $P_w(t)$ are probabilities of trouble-free operation of bearings, gears and winding, respectively. It means that the most vulnerable parts of small PM brushless motors are their bearings, gears (if any) and the armature winding. Under normal operating conditions PMs have practically no influence on the reliability of a motor.

Probability of trouble-free operation of bearings is

$$P_b(t) = \prod_{i=1}^{l} P_{bi}(t) \tag{14.18}$$

where $P_{bi}(t)$ is probability of trouble-free operation of the ith bearing and l is the total number of bearings. The life of an individual bearing is defined as the number of revolutions (or hours at some given speed), which the bearing runs before the first evidence of fatigue develops in the material of either ring or any rolling element.

The dynamic specific load capacity of a bearing

$$C_b = Q_b(nt_t)^{0.3} \tag{14.19}$$

is a function of the equivalent load Q_b in kG, speed n in rpm and time t_t of trouble-free operation in hours also called "rating life." The equivalent load Q_b may be the actual, or the permissible, load on the bearing and depends on radial and axial loads. Under normal conditions, a bearing installed in an electric motor should work on average 77,000 h or MTFB = 77 000/(365 × 24) = 8.8 years. The failure rate is $\lambda = 1/77\ 000 = 1.3 \times 10^{-5}$ 1/h.

Probability of trouble-free operation over a period of normal operation can be found on the basis of eqn (14.13) in which $\gamma = 0$ and $\eta = T'_b$, i.e.,

$$P_{bi}(t) = \exp\left[-\left(\frac{t}{T'_b}\right)^\beta\right] \tag{14.20}$$

Parameters β and T'_b are obtained from experiments (Table 14.2). According to [200] the ratio $5.35 \leq T'_b/t_t \leq 6.84$.

If the failure rate λ_b of a bearing is independent of time, probability of trouble-free operation is expressed by eqn (14.7) in which $\lambda = 2\lambda_b$.

Probability of trouble-free operation of gears is

$$P_g(t) = \prod_{j=1}^g P_{gj}(t) \tag{14.21}$$

where $P_{gj}(t)$ is probability of trouble-free operation of jth toothed wheel and g is the number of wheels. The approximate probability of trouble-free operation P_{gj} can be calculated on the basis of eqn (14.7) in which $\lambda = \lambda_{gj}$ is the failure rate of the jth toothed wheel.

Probability of trouble-free operation of a winding is expressed by eqn (14.7) in which $\lambda = \lambda_w$. To find the failure rate λ_w of the armature windings, the average time of service expectation

$$t' = T'_w \exp[-\alpha_t(\vartheta - \vartheta_{max})] \tag{14.22}$$

which, in principle, depends on the time of service expectation of the winding. In eqn (14.22) T'_w is the mean time of service expectation for the winding at the permissible temperature ϑ_{max} for a given class of insulation (Table 14.3) and relative humidity from 40 to 60% , ϑ is the temperature of the winding under operation and α_t is the temperature coefficient of time of service expectation (Table 14.3). Time T'_w corresponds to the calendar time of service expectation, approximately 15 to 20 years or 1.314×10^5 to 1.752×10^5 h [200]. The higher the service temperature for the given class of insulation the longer the time T'_w.

Failure rate of a winding is a function of T'_w, i.e.,

$$\lambda'_w = \frac{1}{t'} = \frac{1}{T'_w} \exp[\alpha_t(\vartheta - \vartheta_{max})] = \lambda_{wT} \exp[\alpha_t(\vartheta - \vartheta_{max})] \tag{14.23}$$

where $\lambda_{wT} = 1/T'_w$ is the failure rate of a winding alone at permissible temperature ϑ_{max} and relative humidity 40 to 60%.

There are also possible failures as a result of broken or aging soldered connections between coils of the winding or terminals. Failure rate of a winding including soldered connections is [200]

Table 14.3. Classes of insulation, permissible service temperatures ϑ_{max} and temperature coefficients α_t of time of service expectation

Class of insulation	Y	A	E	B	F	H	C
ϑ_{max}, $^\circ$C	90	105	120	130	155	180	> 180
α_t, $1/^\circ$C	0.057	0.032	-	0.073	0.078	0.085	0.055

$$\lambda_w'' = \lambda_w' + m_{sc}\lambda_{sc} \qquad (14.24)$$

where λ_{sc} is the failure rate of one soldered connection and m_{sc} is the number of connections.

Reliability of a winding depends very much on the operating conditions. The influence of such external factors as humidity, overload, vibration and shocks is taken into account by the coefficient of operating conditions γ_w. The failure rate of a winding including conditions of operation is

$$\lambda_w = \gamma_w \lambda_w'' \qquad (14.25)$$

14.5 Vibration and noise

Sounds are set up by oscillating bodies of solids, liquids or gases. The oscillation is characterized by its frequency and amplitude. According to frequency, oscillations are classified as

- low frequency oscillations, $f < 5$ Hz
- infrasound, $5 < f < 20$ Hz
- audible sounds, $20 \leq f \leq 16,000$ Hz or even up to $20,000$ Hz
- ultrasound, $16,000 < f < 10^6$ Hz.

In engineering practice, low frequency oscillations of solids, below 1 kHz, are called *vibration*.

Noise is an audible sound or mixture of sound that has an unpleasant effect on human beings, disturbs their ability to think, and does not convey any useful information [98].

Part of the vibrational energy within the audible range is transformed into sound energy. There is *airborne noise* radiating directly from the vibration source and *structure-borne noise* transmitted to the surroundings via mechanical connections, couplings, base plates, supports, etc. Vibration and noise produced by electrical machines can be divided into three categories [290]:

- electromagnetic vibration and noise associated with parasitic effects due to higher space and time harmonics, phase unbalance, and magnetostrictive expansion of the core laminations
- mechanical vibration and noise associated with the mechanical assembly, in particular bearings
- aerodynamic vibration and noise associated with flow of ventilating air through or over the motor.

14.5.1 Generation and radiation of sound

The acoustic field is characterized by the acoustic pressure p and velocity v of vibrating particles. The *acoustic wave* propagation is expressed as

$$y(x,t) = A \cos \left[\omega \left(t - \frac{x}{c} \right) \right] \qquad (14.26)$$

where A is the wave amplitude, x is the direction of wave propagation, c is the *wave velocity* or *phase velocity* (for air $c = 344$ m/s), $\omega = 2\pi f$ and f is the frequency.

The *wavelength* is expressed as a function of c and f, i.e.,

$$\lambda = \frac{c}{f} = \frac{2\pi c}{\omega} \qquad \text{m} \qquad (14.27)$$

The velocity of particles (in the y direction, i.e., axis corresponding to the displacement)

$$v(x,t) = \frac{\partial y}{\partial t} = -A\omega \sin \left[\omega \left(t - \frac{x}{c} \right) \right] \qquad (14.28)$$

In a medium with linear acoustic properties the *sound pressure* p is proportional to the particle velocity v, i.e.,

$$p = \rho c v \qquad \text{N/m}^2 \qquad (14.29)$$

where p is in phase with the vector $\mathbf{v}$, ρ is the specific density of the medium (kg/m^3). For air $\rho = 1.188$ kg/m^3 at 20^0C and 1000 mbar (1 bar $= 10^5$ Pa $= 10^5$ N/m^2). The product ρc is called the *specific acoustic resistance*, i.e.,

$$\Re[\mathbf{Z}_a] = \rho c = \frac{p}{v} \qquad \text{Ns/m}^3 \qquad (14.30)$$

where $\mathbf{Z}_a$ is the complex acoustic impedance [290].

The sound pressure is characterized by the *rms* value of the pressure change caused by the sound or vibration wave [290]

$$p = \sqrt{\frac{1}{T_p} \int_0^{T_p} [p(t)]^2 dt} \qquad \text{N/m}^2 \tag{14.31}$$

Usually, there is more than one component of different frequencies, so that the *rms* value is calculated as

$$p = \sqrt{p_1^2 + p_2^2 + p_3^2 + \dots} \tag{14.32}$$

The *sound intensity* is the rate of flow of energy through the unit area perpendicular to the direction of travel of the sound [290]

$$I = \frac{1}{T_p} \int_0^{T_p} pv dt \qquad \text{W/m}^2 \tag{14.33}$$

where v is the component of speed perpendicular to the surface.

The *acoustic power* is the sound intensity over a surface S perpendicular to the direction of travel of sound [290]

$$P = \int_0^S I dS \qquad \text{W} \tag{14.34}$$

The radiated acoustic power as a function of the accoustic resistance

$$P_{ar} = \Re[\mathbf{Z}_a] v^2 \sigma(r) S = v^2 \rho c \sigma(r) S \qquad \text{W} \tag{14.35}$$

where $\sigma(r)$ is the *radiation factor* which depends on the mode number r and S is the *radiation surface*. The normal point velocity v on the surface is proportional to the frequency. Higher frequencies radiate more acoustic power than low frequencies. For a sound radiator emitting various mode numbers

$$P = \sum_r P_{ar} \tag{14.36}$$

The radiation factor for a cylindrical radiator, i.e., a frame of electrical machine is [290]

$$\sigma(r) = (kx)^2 \frac{N_r I_{r+1} - I_r N_{r+1}}{[r I_r - (kx) I_{r+1}]^2 + [r N_r - (kx) N_{r+1}]^2} \tag{14.37}$$

where N_r and N_{r+1} are Neumann functions, I_r and I_{r+1} are Bessel functions, $k = 2\pi/\lambda = \omega/c$ is the wave number and λ is the wave length according to eqn (14.27). The radiation factor $\sigma(r)$ as a function of kx, where x is the distance from the center, at $r = const$ is plotted in Fig. 14.3.

A particle of the medium under an action of a sound wave oscillates and possesses both the kinetic energy and strain (potential) energy. The variation of the *density of the kinetic energy* is

$$e_{kin} = \frac{1}{2}\rho v^2 = \frac{1}{2}\rho \omega^2 A^2 \sin^2\left(\omega t - \frac{\omega}{c}x\right) = E_{kin} \sin^2\left(\omega t - \frac{\omega}{c}x\right) \tag{14.38}$$

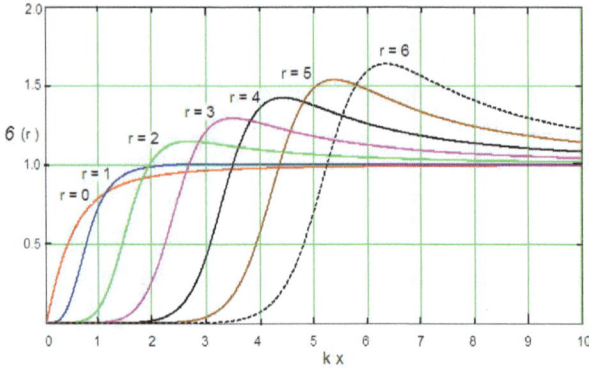

Fig. 14.3. Radiation factor curves for a cylindrical radiator and $r = 0, 1, 2, \ldots 6$.

where the average (mean) energy density

$$E_{kin} = \frac{1}{2}\rho\omega^2 A^2 \quad J/m^3 \tag{14.39}$$

The *sound power level* (SWL)

$$L_W = 10\log_{10}\frac{P}{P_0} = 10\log_{10}\frac{IS}{I_0 S_0} = L + 10\log_{10} S \qquad dB \tag{14.40}$$

where $S_0 = 1$ m^2, $P_0 = I_0 S_0 = 10^{-12}$ W is the reference sound power and L is the sound level. One decibel is expressed as

$$10\log_{10}\frac{I}{I_0} = 1 \qquad or \qquad \frac{I}{I_0} = 10^{\frac{1}{10}} = 1.2589 \tag{14.41}$$

The ratio of sound intensities of n dB is equal to 1.2589^n. The SWL of 100 dB corresponds to the power of 0.01 W, while 60 dB corresponds to the power of 10^{-6} W. Among other things, this causes a low accuracy of calculation of the SWL.

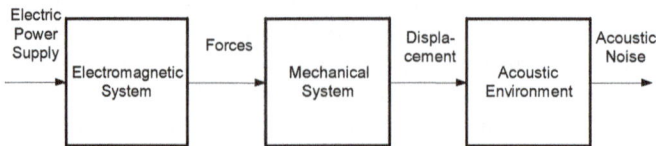

Fig. 14.4. Magneto-mechanical-acoustic system.

14.5.2 Mechanical model

Fig. 14.4 shows how the *electrical energy* is converted into *acoustic energy* in an electrical machine. The input current interacts with the magnetic field producing high-frequency forces which act on the inner stator core surface. These forces excite the stator core and frame in the corresponding frequency range and generate mechanical vibration. As a result of vibration, the surface of the stator yoke and frame displaces with frequencies corresponding to the frequencies of forces. The surrounding medium (air) is excited to vibrate too and generates acoustic noise.

By only considering the pure circumferential vibration modes of the stator core, the deflection Δd of the stator core is an inverse function of the fourth power of the force order r, i.e.,

$$\Delta d \propto \frac{1}{r^4} \tag{14.42}$$

The stator and frame assembly as a mechanical system is characterized by a distributed mass M, damping D and stiffness K. The electromagnetic force waves excite the mechanical system to generate vibration, the amplitude of which is a function of the magnitude and frequency of those forces.

The mechanical system can be simply described by a lumped parameter model with N degrees of freedom in the following matrix form

$$[M]\{\ddot{q}\} + [D]\{\dot{q}\} + [K]\{q\} = \{F(t)\} \tag{14.43}$$

where q is an $(N, 1)$ vector expressing the displacement of N degrees of freedom, $\{F(t)\}$ is the force vector applying to the degrees of freedom, $[M]$ is the mass matrix, $[D]$ is the damping matrix and $[K]$ is the stiffness matrix. Theoretically, this equation can be solved using a structural FEM package. In practice, there are problems with predictions of $[D]$ matrix for laminated materials, physical properties of materials, accuracy in calculation of forces and proper selection of force components [298].

14.5.3 Electromagnetic vibration and noise

Electromagnetic vibration and *noise* is caused by generation of electromagnetic fields. The slots, distribution of windings in slots, air gap permeance fluctuations, rotor eccentricity and phase unbalance give rise to mechanical deformations and vibration. MMF space harmonics, saturation harmonics, slot harmonics and eccentricity harmonics produce parasitic higher harmonic forces and torques. Especially radial force waves in a.c. machines, which act both on the stator and rotor, produce deformation of the magnetic circuit. If the frequency of the radial force is close to or equal to any of the natural frequencies of the machine, resonance occurs. The effects are dangerous deformation, vibration and increased noise.

Magnetostrictive noise in most electrical machines can be neglected due to low sound intensity and low frequency.

In inverter-fed motors parasitic oscillating torques are produced due to higher time harmonics. These parasitic torques are, in general, greater than oscillating torques produced by space harmonics. Moreover, the voltage ripple of the rectifier is transmitted through the intermediate circuit to the inverter and produces another kind of oscillating torque.

For the stator with frame under assumption of identical vibration of the core and frame, the *amplitude of the radial displacement* due to vibration given by eqn (14.26) can be found as

$$A_r = \frac{\pi D_{1in} L_i P_r}{K_r} \frac{1}{\sqrt{[1 - (f/f_r)^2]^2 + (2\zeta_D f/f_r)^2}} \quad \text{m} \quad (14.44)$$

where P_r is the amplitude of the radial force pressure of electromagnetic origin (5.114), K_r is the lumped stiffness of the stator core (yoke), f is the frequency of the excitation radial force density wave of the given mode, f_r is the natural frequency of a particular mode and ζ_D is the internal damping ratio of the stator (obtained from measurements). According to [311]

$$2\pi\zeta_D = 2.76 \times 10^{-5} f + 0.062 \quad (14.45)$$

The single-mode *radiated acoustic power* as a function of the radiation factor $\sigma(r)$ given by eqn (14.37) and radial vibration displacement A_r given by eqn (14.44) is expressed by eqns (14.35) and (14.36) in which $v = \omega A_r$, i.e.,

$$P = \rho c (\omega A_r)^2 \sigma(r) S \quad (14.46)$$

where for a frame with its dimensions D_f and L_f the radiation cylindrical surface $S = \pi D_f L_f$.

14.5.4 Mechanical vibration and noise

Mechanical vibration and noise is mainly due to bearings, their defects, journal ovality, sliding contacts, bent shaft, joints, rotor unbalance, etc. The rotor should be precisely balanced as it can significantly reduce the vibration. The rotor unbalance causes rotor dynamic vibration and eccentricity which in turn results in noise emission from the stator, rotor, and rotor support structure.

Both sleeve and rolling bearings are used in PM electrical machines. The sound pressure level of sleeve bearings is lower than that of rolling bearings. The vibration and noise produced by sleeve bearings depends on the roughness of sliding surfaces, lubrication, stability and whirling of the oil film in the bearing, manufacture process, quality and installation. The exciting forces are produced at frequencies $f = n$ due to rotor unbalance and/or eccentricity and $f = N_g n$ due to axial grooves where n is the speed of the rotor in rev/s and N_g is the number of grooves [290].

The noise of rolling bearings depends on the accuracy of bearing parts, mechanical resonance frequency of the outer ring, running speed, lubrication conditions, tolerances, alignment, load, temperature and presence of foreign materials. The frequency of noise due to unbalance and eccentricity is $f = n$ and frequencies of noise due to other reasons are $f \propto d_i n/(d_i + d_o)$ where d_i is the diameter of the inner contact surface and d_o is the diameter of the outer contact surface [290].

14.5.5 Aerodynamic noise

The basic source of *noise of an aerodynamic nature* is the fan. Any obstacle placed in the air stream produces a noise. In non-sealed motors, the noise of the internal fan is emitted by the vent holes. In totally enclosed motors, the noise of the external fan predominates.

The acoustic power of turbulent noise produced by a fan is [77]

$$P = k_f \rho c^3 \left(\frac{v_{bl}}{c}\right)^6 D_{bl}^2 \tag{14.47}$$

where k_f is a coefficient dependent on the fan shape, ρ is the specific density of the cooling medium, c is the velocity of sound, v_{bl} is the circumferential speed of the blade wheel and D_{bl} is the diameter of the blade wheel. The level of aerodynamic noise level due to the fan is calculated by dividing eqn (14.47) by $P_0 = 10^{-12}$ W and putting the result into eqn (14.40). For example, for $k_f = 1$, $\rho = 1.188$ kg/m^3, $c = 344$ m/s, $v_{bl} = 15$ m/s and $D_{bl} = 0.2$ m the aerodynamic noise level is $L_w = 75.9$ dB.

According to the spectral distribution of the fan noise, there is broad-band noise (100 to 10,000 Hz) and siren noise (tonal noise). The siren effect is a pure tone being produced as a result of the interaction between fan blades, rotor slots or rotor axial ventilation ducts and stationary obstacles. The frequency of the siren noise is [77, 290]

$$f_s = \nu N_{bl} n_{bl} \tag{14.48}$$

where ν are numbers corresponding to harmonic numbers, N_{bl} is the number of fan blades and n_{bl} is the circumferential speed of the fan in rev/s. Siren noise can be eliminated by increasing the distance between the fan or impeller and the stationary obstacle.

14.5.6 d.c. commutator motors

In d.c. commutator motors, noise of mechanical and aerodynamic origin is much higher than that of electromagnetic origin. The commutator and brushes are one of the major sources of noise. This noise depends on the following factors [290]:

- design of brushgear and rigidity of its mounting
- backlash between the brush holder and brush, and between the commutator and brush
- brush pressure
- materials of the brush and commutator, especially the friction coeffcient between them
- size, tolerances and commutator unbalance
- deflection of the commutator sliding surface during rotation
- condition of the commutator and brushes, especially coarseness of the sliding surfaces
- current load
- operating temperature
- environmental effects such as dust, ambient temperature, humidity.

14.5.7 PM synchronous motors

In large synchronous motors the vibration and noise of mechanical origin can exceed those of electromagnetic origin. Noise produced by small synchronous motors is mainly produced by electromagnetic effects. Mechanical natural frequencies of small motors are very high so they are poor sound generators.

The predominant amplitude of the SWL in synchronous and PM brushless machines is due to the radial force produced by interaction of the rotor poles and slotted structure of the stator [121]. The frequency and order of this force is

$$f_r = 2\mu_\lambda f \qquad\qquad r = 2|\mu_\lambda p \pm s_1| \qquad (14.49)$$

where $\mu_\lambda = \mathrm{integer}(s_1/p)$. If the frequency of this force is close to the natural frequency of the order $r = 2$, a large amplitude of the SWL is produced.

For both the induction and PM synchronous motors fed from solid state converters the most significant sound levels occur at the modulation frequency of the inverter, i.e., at the frequency of the major current harmonic with lesser contributions at multiples of that frequency. Important causes of sound generation are torque pulsations [301].

14.5.8 Reduction of noise

Electromagnetic, mechanical and aerodynamic noise can be reduced by proper motor design and maintenance.

The electromagnetic noise can be reduced from the design point of view by proper selection of the number of slots and poles, i.e., suppressing the parasitic radial forces and torque pulsations, *length–to–diameter* ratio, electromagnetic loadings, skewing of the slots, keeping the same impedances of phase windings, designing a thick stator core (yoke). The noise of electromagnetic origin is high, if circumferential orders r of radial stator deformations given by eqn (14.49) are low ($r = 1$ and $r = 2$). On the other hand, minimization of radial forces due to interaction of the rotor poles and stator slot openings does not guarantee that other noise producing harmonics of the magnetic field are suppressed.

Proper maintenance, i.e., feeding the motor with balanced voltage system, elimination of time harmonics in the inverter output voltage and selection of proper modulation frequency of the inverter has also a significant effect on the electromagnetic noise reduction.

The noise of mechanical origin can be reduced, from the design point of view, by predicting the mechanical natural frequencies, proper selection of materials, components and bearings, proper assembly, foundation, etc., and from a maintenance point of view by proper lubrication of bearings, monitoring their looseness, rotor eccentricity, commutator and brush wear, joints, couplings and rotor mechanical balance.

The aerodynamic noise can be reduced, from the design point of view, by proper selection of the number of fan blades, rotor slots and ventilation ducts and dimensions of ventilation ducts to suppress the siren effect as well as optimization of air inlets and outlets, fan cover, etc., and from a maintenance point of view by keeping the ventilation ducts and fan clean.

14.6 Condition monitoring

As an electromechanical drive becomes more complex, its cost of maintenance rises. The *condition monitoring* can predict when the electromechanical drive will break down and substantially reduce the cost of maintenance including repairing. Electric motors are the most vulnerable components of electrical drive systems (see Table 14.1) so their condition and early fault detecting is crucial.

Condition monitoring of electric motors should be done *on-line* with the aid of externally mounted sensors as, for example, current transformers, accelerometers, temperature sensors, search coils, etc. without any change of the motor construction, rearrangement of its parts or changing its rated parameters. The following methods can be employed:

- aural and visual monitoring
- measurements of operational variables
- current monitoring
- vibration monitoring
- axial flux sensing.

Aural and visual monitoring require highly skilled engineering personel. The failure is usually discovered when it is well advanced. Measurements of operational variables such as electric parameters, temperature, shaft eccentricity, etc., are simple and inexpensive but as in the case of aural and visual monitoring, there is a danger that the fault is detected too late.

The most reliable, rich in information, simple and sophisticated method is monitoring the input current, vibration or axial flux. A current transformer can be used as a current transducer. For vibration monitoring piezoelectric accelerometers or other transducers can be installed. For axial flux sensing a coil wound around the motor shaft can be installed. Because it is often inconvenient to wind coils around the shaft of a motor that is in service, a printed circuit split coil has been proposed [243].

Monitoring techniques are based on *time domain* measurements and *frequency domain* measurements.

Time domain signals can detect, e.g., mechanical damage to bearings and gears. The time signal may be averaged over a large number of periods, synchronous with the motor speed, to allow synchronous averaging. Background noise and periodic events not synchronous with the motor speed are filtered out.

A periodic time domain waveform, if passed through a narrow band-pass filter with a controllable center frequency, is converted into frequency components which appear as output peaks when the filter pass band matches the frequency components. This technique is adopted in high frequency spectral analysers. An alternative approach is to sample the time domain waveform at discrete intervals and to perform a *discrete Fourier transform* (DFT) on the sample data values to determine the resultant frequency spectra. The time necessary to compute the DFT of the time domain signal can be reduced by using a *fast Fourier transform* (FFT) calculation process which rearranges and minimizes the computational requirements of the DFT process. Modern monitoring techniques commonly use FFT.

Vibration or current spectra are often unique to a particular series of motors or even particular motors. When a motor is commissioned or when it is in a healthy state a reference spectrum is monitored which can later be compared with spectra taken in successive time intervals. This allows some conclusion and progressive motor condition to be formulated.

In analysis of frequency spectra of the input current usually the amplitude of sidebands are compared to the amplitude of the line frequency. The following problems in PM brushless motors can be detected on the basis of time domain input current measurement:

(a) unbalanced magnetic pull and air gap irregularities
(b) rotor mechanical unbalance
(c) bent shaft
(d) oval stator, rotor or bearings.

Common faults which can be detected by vibration monitoring of electric motors are [244, 290]:

(a) rotor mechanical unbalance or eccentricity, characterized by sinusoidal vibration at a frequency of once per revolution (number of revolutions per second)

(b) defects of bearings, which result in frequencies depending on the defect, bearing geometry and speed, usually 200 to 500 Hz

(c) oil whirl in bearings, characterized typically by frequencies of 0.43 to 0.48 number of revolutions per second

(d) rubbing parts, characterized by vibration frequency equal to or a multiple of the number of revolutions per second

(e) shaft misalignment, usually characterized by a frequency of twice per revolution

(f) mechanical looseness in either motor mounts or bearing end bells, which results in directional vibration with a large number of harmonics

(g) gear problems characterized by frequencies of the number of teeth per revolution, usually modulated by speed

(h) resonance (natural frequencies of shaft, machine housing or attached structures are excited by speed or speed harmonics), which results in sharp drop of vibration amplitudes with small change in speed

(i) thermal unbalance, which results in a slow change in vibration amplitude as the motor heats up

(j) loose stator laminations, which result in vibration frequencies equal to double the line frequency with frequency sidebands approximately equal to 1000 Hz

(k) unbalance line voltage, characterized by vibration frequency equal to double the line frequency.

By sensing the axial flux, many abnormal operating conditions can be identified [243], e.g. :

(a) unbalanced supply, characterized by an increase in certain even harmonics of flux spectrum proportional to the degree of unbalance

(b) stator winding interturn short circuits, characterized by a decrease in certain higher harmonics and subharmonics of flux spectrum

(c) rotor eccentricity characterized by an increase in frequency of flux spectrum equal to the line frequency and its second harmonic.

A unified analysis of the various parameter estimators and condition monitoring methods and diagnosis of electrical machines can be found in dedicated literature, e.g., in [297].

14.7 Protection

The *motor protection* depends principally on the motor importance which is a function of motor size and type of service. The bigger the motor the more expensive the electromechanical drive system is and all necessary measures should be taken to protect the motor against damage. The *main function of protection* is the detection of a fault condition, and through the opening of appropriate contactors or circuit breakers, the disconnection of the faulty item from the plant. The potential hazards considered in electric motor protection are:

(a) Phase and ground faults (short circuits between phases or phase and earth)
(b) Thermal damage from:
   – overload (excessive mechanical load)
   – locked rotor
(c) Abnormal conditions:
   – unbalanced operation
   – undervoltage and overvoltage
   – reversed phases
   – switching on the voltage while motor is still running
   – unusual environmental conditions (too high or too low temperature, pressure, humidity)
   – incomplete feeding, e.g., rupturing of a fuse in one phase
(d) Loss of excitation (in PM motors this means the total demagnetization of magnets)
(e) Operation out of synchronism (for synchronous motors only)
(f) Synchronizing out of phase (for synchronous motors only).

Most of the undesired effects cause excessive temperatures of the motor parts, in particular windings. There is an old rule of thumb that says *each 10^0C increase in the operating temperature of the winding results in a 50% loss of insulation life*.

Protective devices applied for one hazard may operate for other, e.g., overload relay can also protect against phase faults. Protection can be built in the motor controller or installed directly on the motor. Motors rated up to 600 V are usually switched by contactors or solid state devices and protected by fuses or low-voltage circuit breakers equipped with magnetic trips. Motors rated from 600 to 4800 V are switched by power circuit breakers or contactors. Motors rated from 2400 V to 13,000 V are switched by power circuit breakers.

The simplest one-time protective devices are *fuses*. As a result of excessive current, the fusible element melts, opens the circuit and disconnects the motor from the power supply.

The *thermal relay* of a "replica" type is the thermoelectromechanical apparatus which simulates, as closely as possible, the changing thermal conditions in the motor allowing the motor to operate up to the point beyond which damage would probably be caused. This relay consists of three single-phase units, each unit comprising a heater and an associated *bimetal* spiral element. The bimetal elements respond to a rise in temperature of the heaters which in turn produce movement of the contact assembly. The *single-phasing protection* is usually incorporated in the three-phase thermal simulation relay. The *short circuit protection* against short circuits in the motor winding or terminal leads is often built in the thermal relay as separate overcurrent or earth-faults elements or both.

The *stalling relay* is used in conjunction with the thermal overload and single-phasing relay. It consists of a control contactor and a thermal overload unit fitted in the same case.

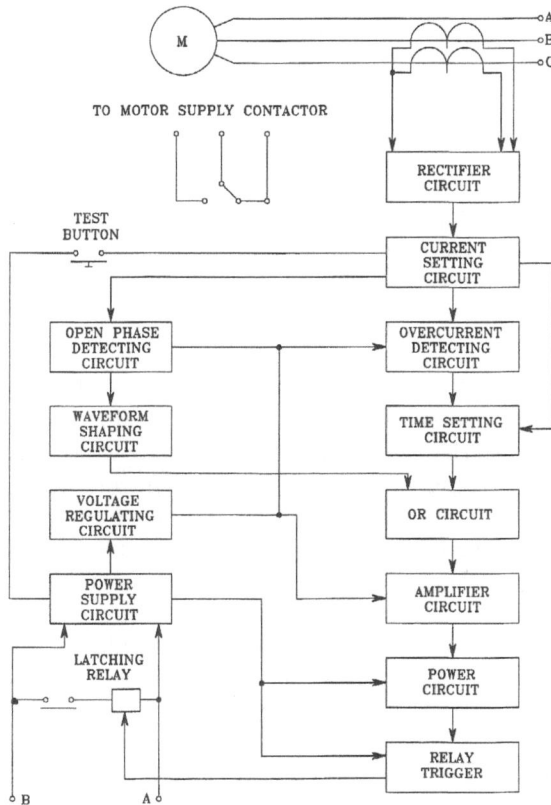

Fig. 14.5. Block diagram of a typical electronic overload relay.

An alternative to the thermal relay is an *electronic overload relay* (Fig. 14.5) using solid state devices instead of bimetallic elements [287].

For small and medium power motors with rated currents up to 25 A (or $P_{out} \leq 15$ kW), simpler and less expensive thermal trips and electromagnetic trips instead of thermal overload relays are technically and economically justified. In a *thermal trip* the motor current passing directly through a bimetal

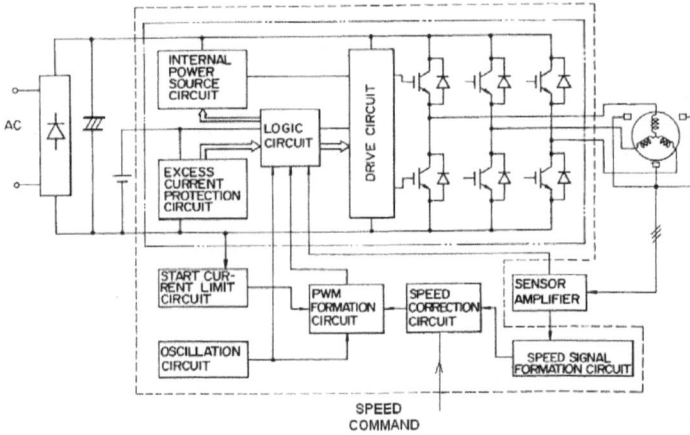

Fig. 14.6. Integrated circuit, which includes inverter for driving a PM brushless motor, drive circuits, power source circuit, circuits for controlling the speed and protection circuit for protecting the inverter from excess current according to U.S. Patent 5327064.

strip or the heater operating with a bimetal element is fed from current transformers connected to the motor power circuit. Bimetal elements operate the mechanical trip to open the motor contactor under an overload condition. The *electromagnetic trip* consists of a series-wound coil surrounding a vertical ferromagnetic plunger and an associated time-lag, i.e., oil or silicone fluid filled dashpot or air vane. The adjustable overload current lifts the plunger which opens the motor contactor. The electromagnetic trips are relatively insensitive to small overloads.

Thermistors bonded to the enamel armature conductors during manufacture are commonly referred to as *motor overheat protection*. Thermistor connections are brought out to an electronic control unit and interposing relay mounted separately for small motors and usually built into terminal boxes of motors rated above 7.5 kW. The relay is activated when the thermistors indicate the winding temperature and, indirectly, the phase currents, exceed their permissible values.

Undervoltage protection is necessary to ensure that the motor contactors or circuit breakers are tripped on a complete loss of supply, so that when the supply is restored it is not overloaded by the simultaneous starting of all the motors. Undervoltage release coils operating direct on the contactor or circuit breaker, relays or contactors with electrically held-in coils are usually used.

The state of the art in the motor protection are *microprocessor protection relays*. These advanced technology multifunction relays are programmed to provide the following functions:

- thermal overload protection with adjustable current/time curves
- overload prior alarm through separate output relay
- locked rotor and stall protection
- high-set overcurrent protection
- zero phase sequence or earth-fault protection
- negative phase sequence or phase unbalance protection
- undercurrent protection
- continuous self-supervision.

ICs for driving PM brushless motors (Fig. 6.28) have built-in protective features such as, for example, current limit ciruits, thermal shutdown, undervoltage lockout, etc. Fig. 14.6 shows an IC for driving a variable speed PM brushless motor with start current limit and excess current protection circuits according to US Patent 5327064.

14.8 Electromagnetic and radio frequency interference

Electromagnetic compatibility (EMC) is defined as the ability of all types of equipment that emit high frequency signals, frequencies higher than the fundamental supply frequency, to operate in a manner that is mutually compatible. Designers of electrical machines should be aware of the EMC specifications which affect their overall design. All electrical motors can be a source of *electromagnetic interference* (EMI) and *radio frequency interference* (RFI). RFI is an electric disturbance making undesired audio or video effects in received signals caused by current interruption in electric circuits as, e.g., sparking between a brush and commutator.

The main reasons for minimizing EMI are the high frequency noise conducted into the mains supply will be injected into other equipment, which could affect their operation, i.e., noisy signals being fed into sensitive loads such as computers and communication equipment, and the interference radiated into the atmosphere by electric and magnetic fields can interfere with various communication equipments. Three-phase electrical motors often draw currents that have frequency components that are odd integer multiples (harmonics) of the fundamental supply frequency. These harmonic currents cause increased heating and lead to a shorter lifetime of appliances.

Conducted noise emissions are suppressed using filters, usually with a passive low pass filter designed to attenuate frequencies above 10 kHz. Under nonlinear loads, such as those associated with adjustable speed drives and electronic power supplies, significant power dissipation can occur within these filters [43]. *Shield radiated emissions* can be suppressed by metallic shields and minimizing openings in the enclosures.

The filtering required to reduce high frequency emissions may be rather costly and reduce the efficiency of the electromechanical drive.

EMI regulations set the limits for conducted and radiated emissions for several classes of products. One of the most important international standards-setting organizations for commercial EMC is CISPR, the International Special Committee on Radio Interference in IEC (International Electrotechnical Commission) [284]. The European community has developed a common set of EMC requirements largely based on CISPR standards. The Federal Communication Commission (FCC) sets the limits of radiated and conducted emissions in the U.S.A. An example of a guide for electrical and electronic engineers containing complete coverage of EMI filter design is the book cited in reference [235].

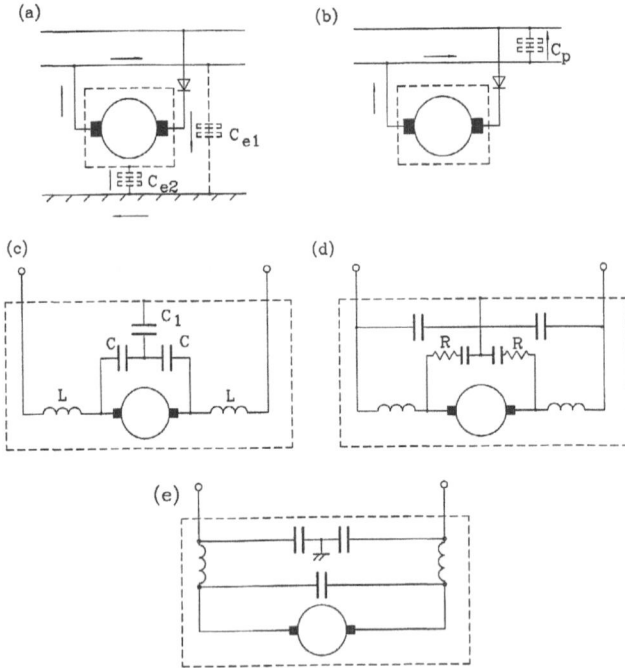

Fig. 14.7. RFI circuits and filters of brush motors: (a) asymmetrical circuit of high frequency current, (b) symmetrical circuit of high frequency current, (c), (d), (e) RFI filters.

14.8.1 Brush motors

Commutator (brush) motors are a source of serious EMI and RFI. The stronger the sparking the more intense the interferences. Wrong maintenance, dirty or worn commutator, wrong selection of brushes, unbalanced commutator, unbalanced rotor, etc., are the most serious reasons for EMI and RFI.

RFI causes clicks in radio reception being heard in the whole band of radio waves. TV reception is disturbed by change in brightness of screen and time base which in turn makes a vertical movement of image lines on the screen. EMI and RFI are emitted directly from their source and connecting leads as well from the power network from which the motor is fed via converter. They are received by radio or TV antennas. In addition, RFI can be transmitted to radio or TV sets by electric installation, or even by water and gas pipes and any metal bars.

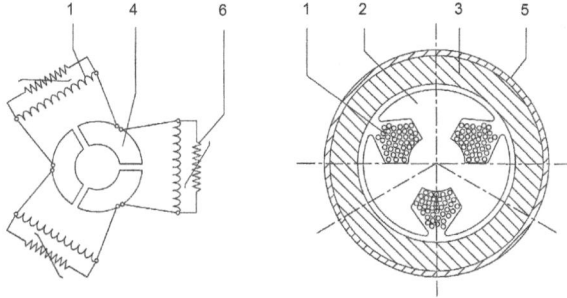

Fig. 14.8. Elimination of RFI in a three-coil d.c. PM brush motor: 1 — Δ-connected armature winding, 2 — armature stack, 3 — cylindrical PM, 4 — commutator segment, 5 — frame, 6 — resistor of RFI filter.

The interference current of high frequency flows through the electric network feeding the motor, then through capacitance C_{e1} between the feeding wire and the earth and capacitance C_{e2} between the source of interference and the earth (Fig. 14.7a). This high-frequency current, called asymmetrical high frequency current causes particularly strong interference. A grounded motor frame intensifies the level of RFI since it closes the circuit for high frequency current. Symmetrical high frequency currents are closed by capacitances C_p between wires (Fig. 14.7b).

To eliminate RFI, filters consisting of RLC elements are used (Fig. 14.7c,d,e). The resistance R is needed to damp high frequency oscillations. The choke has a high inductance L for the high frequency current, since its reactance $X_L = 2\pi fL$ increases as the frequency f and inductance L increases. The capacitive reactance $X_C = 1/(2\pi fC)$ is inversely proportional to the frequency and capacitance C. For high frequency currents the capacitive reactance is low.

Symmetrical inductances connected in series with the armature winding at both its terminals can damp oscillations more effectively than one inductance only. In series motors, field coils are used as RFI filter inductances. In the filter shown in Fig. 14.7c the high-frequency current circuit is closed by capacitors C. The capacitor C_1 is a protective capacitor against electric shock.

If the motor frame is touched by a person who is in contact with the ground, the current flowing through the human body to the earth will be limited by capacitor C_1, the capacity of which is low (about 0.005 μF) as compared with C (from 1 to 2 μF for d.c. motors).

RFI filters are effective if the connection leads between capacitors and brushes are as short as possible (less than 0.3 m). This results in minimization of interference emitted directly by the source. It is recommended that the RFI filters be built into motors. If this is not possible, all connection leads must be shielded. Resistances can effectively damp oscillations and improve the quality of RFI filter (Fig. 14.7d). A simple RFI RL filter for three-coil PM commutator motors for toys and home appliances is shown in Fig. 14.8.

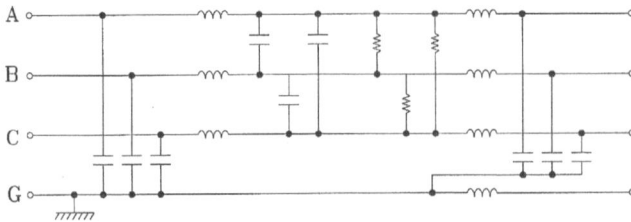

Fig. 14.9. Typical EMI filter for three-phase input.

14.8.2 Electronically commutated brushless motors

Brushless d.c. motors use inverter-based solid state converters. The current and voltage waveforms from these converters are either sinusoidal or square-wave. In both cases the converter generates the desired waveforms using PWM, switched at 8 to 20 kHz typically. When a voltage abruptly changes amplitude with respect to time, the derivative dv/dt changes produce unwanted harmonics. The nonlinear characteristics of solid-state devices worsen the situation. This accounts for the large impulse currents through the power leads, which are associated with EMI and significant voltage waveform distortion in the power system.

The reduction of electrical noise is usually done by ensuring proper grounding, avoiding extended cables from inverter to motor, twisting the cables and filtering the input power to the inverter drive.

To prevent radiated noise, the *motor ground wire* is required to be twisted or tightly bundled with the three line wires. The motor power wiring is to be kept as far away as possible from the rotor position signal wiring and any other light current wiring. A *shielded cable* is recommended for connection of the encoder or resolver with the inverter motion control section.

Placing filters on power lines to inverters not only suppresses harmonics leaving the drive, but also protects the drive from incoming high frequency signals. Fig 14.9 shows a three-phase low pass filter used for EMI/RFI filtering [284].

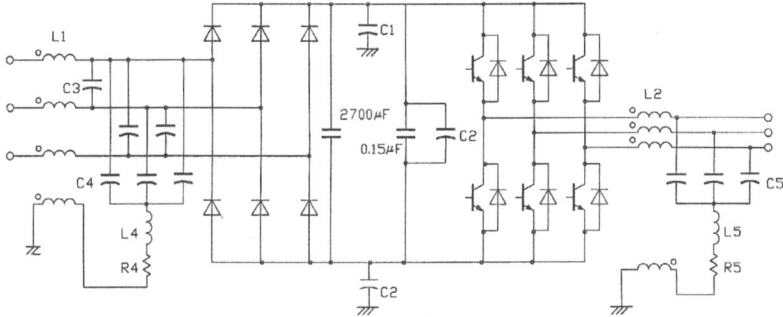

Fig. 14.10. Power circuit of a three-phase modified converter with EMI suppression components.

These line filters are bulky and increase the cost of the drive substantially. Fig 14.10 shows a low cost alternative: an inverter with EMI suppression components [315]. These include:

(a) grounding capacitance C1 from both sides of the d.c. link to the heat sink close to the switching devices which provides a physically short path for RF ground currents flowing from the switching device to the motor;
(b) Line capacitance C2 across the d.c. link, close to the switching devices which provides a low impedance for differential mode RF current flowing from switching devices such as reverse recovery current of the diodes as well as from the cable-motor load;
(c) line capacitance C3 across the a.c. power input terminals close to the diode rectifiers which serves as another shunt circuit in combination with the d.c. line capacitance for differential mode noise compensation, particularly for the noise caused by the diodes of rectifier;
(d) a common mode line inductance L1 inserted in each phase of the a.c. input power circuit of the rectifier provide a high impedance for the RF currents to the power mains;
(e) a common mode inductance L2 inserted in each phase of the a.c. output power circuit of the inverter reduce the time derivative of output mode voltages imposed on the motor, but do not affect the line to line voltages.

14.9 Lubrication

14.9.1 Bearings

In PM machines usually *rolling bearings* and *porous metal bearings* are used. In high speed PM machines magnetic, air or foil bearings are used, which will not be discussed in this section.

The stress levels in rolling bearings limit the choice of materials to those with a high yield and high creep strength. Steels have gained the widest acceptance as rolling contact materials as they represent the best compromise among the requirements and also because of economic considerations. Steels with the addition of C, Si, Mn and Cr are the most popular. To increase hardenability and operating temperature tungsten (W), vanadium (V), molybdenium (Mo) and nickel (Ni) are added. Basic methods of mounting rolling bearings for horizontal shafts are shown in Fig. 14.11.

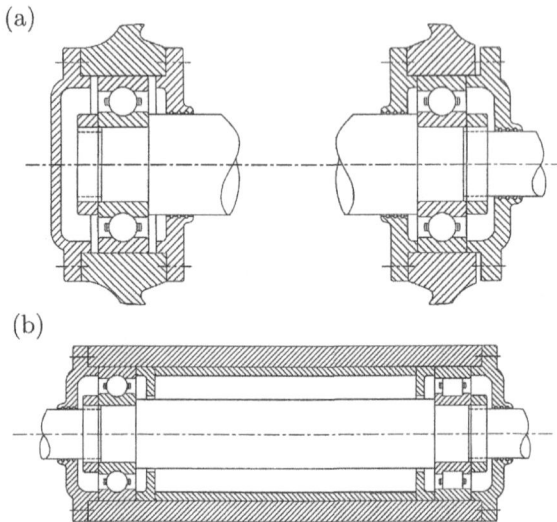

Fig. 14.11. Basic methods of installation of rolling bearing for horizontal shaft: (a) two deep groove radial ball bearings, (b) one ball bearing with one cylindrical roller bearing.

Porous metal bearings are used in small or large electric motors. The graphited tin bronze (Cu-Sn-graphite) is a general purpose alloy and gives a good balance between strength, wear resistance, conformability and ease of manufacture. Where rusting is not a problem, less expensive and stronger iron-based alloys can be used. Assemblies of self-aligning porous metal bearings with provision for additional lubrication are shown in Fig. 14.12. In most electric motor bearings the lubricant material is *oil* or *grease*.

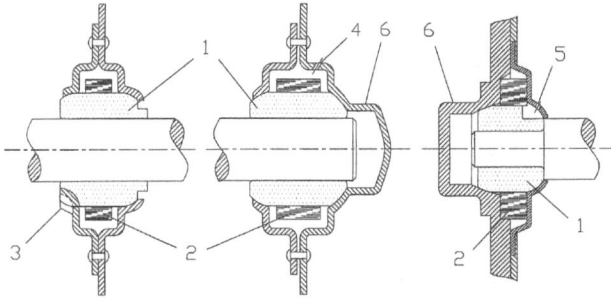

Fig. 14.12. Installation of self-aligning porous metal bearings for small, horizonal shaft motors: 1 — bearing, 2 — oil soaked felt pad, 3 — key hole, 4 — oil hole, 5 — slot to take key, 6 — end cap (may be filled with grease).

14.9.2 Lubrication of rolling bearings

Grease lubrication. Grease lubrication is generally used when rolling bearings operate at normal speeds, loads and temperatures. The bearings and housings for normal applications should be filled with grease up to 30 to 50% of the free space. Too much grease will result in overheating. The consistency, rust-inhibiting property and temperature range must be carefully considered when selecting a grease. The grease relubrication period is the same as the service life of the grease and can be estimated from the formula [226]:

$$t_g = k_b \left(\frac{14 \times 10^6}{n\sqrt{d_b}} - 4d_b \right) \qquad \text{h} \qquad (14.50)$$

where k_b is a factor depending on the type of bearings, n is speed in rpm and d_b is bearing bore diameter in mm. For spherical roller bearing and tapered roller bearings $k_b = 1$, for cylindrical and needle roller bearings $k_b = 5$ and for radial ball bearings $k_b = 10$. The amount of grease required for relubrication is [226]

$$m_g = 0.005 D_b w_b \qquad \text{g} \qquad (14.51)$$

where D_b is the outer bearing diameter and w_b is its width, both in millimeters.

The advantages of using a grease lubricant are:

- it is convenient to apply and retain the lubricant within the bearing housing
- it stays to cover and protect the highly polished surfaces even when the bearing is at rest
- it helps to form a very effective closure between shaft and housing thus preventing entry of foreign matter
- it offers freedom from lubricant contamination of the surrounding areas.

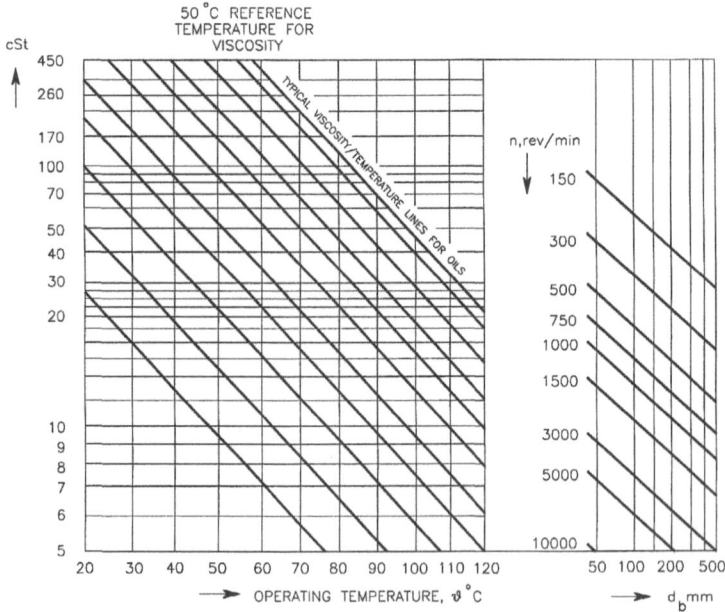

Fig. 14.13. Graphs for the selection of oil kinematic viscosities for rolling bearings versus bearing bore diameter d_b and operating temperature ϑ at constant speed n.

Oil lubrication. Oil lubrication is used when operating conditions such as speed or temperature preclude the use of grease. Ball and roller bearings must be lubricated with oil when the running speed is in excess of the recommended maximum grease speed and also when the operating temperature is over 93^0C. A guide to suitable oil kinematic viscosity for rolling bearings is presented in the form of graphs in Fig. 14.13 [226]. The unit for kinematic viscosity is 1 m^2/s or centistoke, i.e., 1 cSt $= 10^{-6}$ m^2/s. Oil viscosity is estimated on the basis of bearing bore, speed and operating temperature.

14.9.3 Lubrication of porous metal bearings

As a general recommendation, the oil in the pores should be replenished every 1000 h of use or every year, whichever is sooner. In some cases, graphs in Fig. 14.14 should be used to modify this general recommendation [226]. The lower the bearing porosity the more frequent the replenishment. The oil loss increases with the shaft velocity and bearing temperature.

Graphs in Fig. 14.15 give general guidance on the choice of oil dynamic viscosity according to load and temperature [226]. The unit for dynamic viscosity is 1 Pa s $= 1$ N/m^2 s or centipoise, i.e., 1 cP $= 10^{-3}$ kg/(ms) $= 10^{-3}$ Pa s.

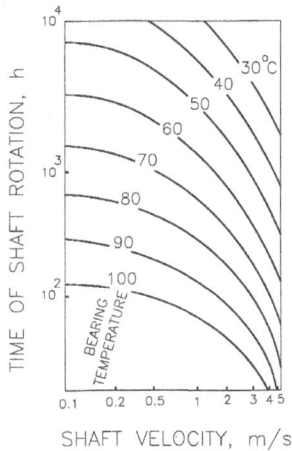

Fig. 14.14. Permissible time of shaft rotation without oil replenishment as a function of shaft velocity at constant bearing temperature.

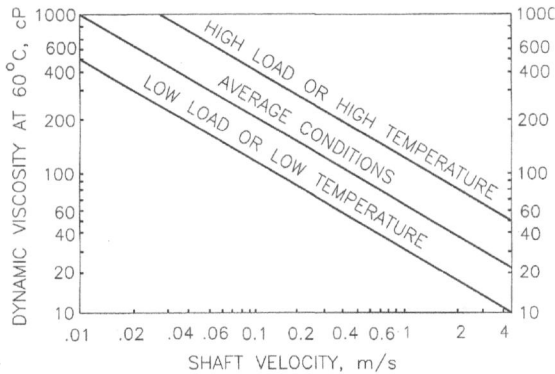

Fig. 14.15. Graphs for the selection of oil dynamic viscosity expressed in centipoises at 60^0C.

The following rules apply to the selection of lubricants [226]:

- lubricants must have high oxidation resistance
- unless otherwise specified, most standard porous metals bearings are impregnated with a highly refined and oxidation-inhibited oil with an SAE 20/30 viscosity
- oils which are not capable of being mixed with common mineral oils should not be selected
- grease should be used only to fill a blind cavity (Fig. 14.12)
- suspensions of solid lubricants should be avoided unless experience in special applications indicates otherwise
- manufacturers should be contacted for methods of re-impregnation.

Numerical examples

Numerical example 14.1

Find the probability of trouble-free operation $P(t)$ within a time period $t = 5000$ h for a three-phase, 1500 rpm PM brushless motor with a built-in speed reduction gearbox. The motor has two radial ball bearings, three toothed wheels, $m_{sc} = 9$ soldered armature winding connections and class F insulation. The MTBF for the armature winding is $T'_w = 100,000$ h, operating temperature of the winding is $\vartheta = 105^\circ$C, coefficient of the winding operation $\gamma_w = 1.5$, intensity of failures of one soldered connection $\lambda_{sc} = 0.3 \times 10^{-7}$ 1/h, probability of failure-free operation of one toothed wheel $P_{gj}(t) = 0.9986$ ($t = 5000$ h), bearing permissible load $Q_b = 45$ kG, dynamic specific load capacity of a bearing $C_b = 7170$ kG(h× rpm)$^{0.3}$ and space factor $\beta = 1.17$.

Solution

1. Bearings

The time of trouble-free operation of one bearing can be found on the basis of eqn (14.19), i.e.,

$$t_t = \frac{1}{n}\left(\frac{C_b}{Q_b}\right)^{10/3} = \frac{1}{1500}\left(\frac{7170}{45}\right)^{10/3} \approx 14,620 \text{ h}$$

Assuming $T'_b/t_t = 6.84$, the average expectation of service of the bearing is $T'_b = 6.84 \times 14,620 = 100,000$ h. Probability of trouble-free operation of one bearing according to eqn (14.20) is

$$P_{bi}(t) = \exp\left[-\left(\frac{5000}{100,000}\right)^{1.17}\right] = 0.9704$$

Probability of trouble-free operation of two the same bearings

$$P_b(t) = P_{bi}^2(t) = 0.9704^2 = 0.9417$$

2. Reduction gears

Probability of trouble-free operation of three similar toothed wheels in a reduction gear box is

$$P_g(t) = P_{gj}^3(t) = 0.9986^3 = 0.9958$$

3. Armature winding

The failure rate of the armature winding according to eqn (14.23)

$$\lambda'_w = 10^{-5}\exp[0.078(105 - 155)] = 0.2024 \times 10^{-6} \text{ 1/h}$$

where $\lambda_{wT} = 1/10^5 = 10^{-5}$ 1/h and the temperature coefficient $\alpha_t = 0.078$ 1/$^\circ$C is according to Table 14.3.

The failure rate of the armature winding including soldered connections on the basis of eqn (14.24) is

$$\lambda_w'' = 0.2024 \times 10^{-6} + 9 \times 0.03 \times 10^{-6} = 0.4724 \times 10^{-6} \text{ 1/h}$$

The failure rate of the armature winding including conditions of operation on the basis of eqn (14.25) is

$$\lambda_w = 1.5 \times 0.4724 \times 10^{-6} = 0.7086 \times 10^{-6} \text{ 1/h}$$

Probability of trouble-free operation of the winding is expressed by eqn (14.7) in which $\lambda = \lambda_w$, i.e.

$$P_w(t) = \exp(-0.7086 \times 10^{-6} \times 0.5 \times 10^4) = 0.9965$$

4. Probability of failure-free operation of the motor

Probability of failure-free operation of the motor within the time period of 5000 h is calculated using eqn (14.17), i.e.,

$$P(t) = 0.9417 \times 0.9958 \times 0.9965 = 0.9344$$

If, say, 100 such motors operate within the time period $t = 5000$ h, 7 motors out of 100 will probably not survive, i.e., 6 motors will fail due to failure of bearings, 1 motor due to failure of gears, and 1 motor due to failure of the armature winding. One out of 7 damaged motors can fail either as a result of bearings and winding failure, or bearings and gears failure, or winding and gear failure.

Numerical example 14.2

A 7.5 kW PM brushless motor has the frame diameter $D_f = 0.248$ m and frame length $L_f = 0.242$ m. The amplitude of surface vibration of a standing wave at $f = 2792.2$ Hz and mode number $r = 2$ is 0.8×10^{-8} mm. Calculate the radiated acoustic power.

Solution

The angular frequency is $\omega = 2\pi f = 2\pi \times 2792.2 = 17,543.91$ 1/s, the wave number $k = \omega/c = 17,543.91/344 = 51$ 1/m, the radius from the center $x = D_f/2 = 0.248/2 = 0.124$ m and the product $kx = 51 \times 0.124 = 6.324$. The wave velocity in the air is $c = 344$ m/s.

From Fig. 14.3 the radiation factor for $r = 2$ and $kx = 6.324$ is $\sigma \approx 1.2$. The external surface of cylindrical frame

$$S = \pi D_f L_f = \pi 0.248 \times 0.242 = 0.1885 \text{ m}^2$$

A standing surface wave can be split into two waves rotating in opposite directions with half of the amplitude of vibration, i.e., $A_r = 0.4 \times 10^{-9}$ mm [311]. The radiated acoustic power according to eqn (14.46)

$$P = \rho c (\omega A_r)^2 \sigma(r) S$$

$$= 1.188 \times 344 (17,543 \times 0.4 \times 10^{-9})^2 \times 1.2 \times 0.1885 = 4.552 \times 10^{-9} \text{ W}$$

where the air density $\rho = 1.188 \text{ kg/m}^3$ at 20^0C and 1000 mbar.

The sound level according to eqn (14.40)

$$L_w = 10 \log_{10} \frac{4.552 \times 10^{-9}}{10^{-12}} = 36.58 \text{ dB}$$

The other rotating wave of the same vibration amplitude will double the sound level, i.e.,

$$L_w = 10 \log_{10} \frac{2 \times 4.552 \times 10^{-9}}{10^{-12}} = 39.59 \text{ dB}$$

Numerical example 14.3

The bore diameter of a ball bearing is $d_b = 70$ mm, its outer diameter $D_b = 180$ mm, width $w_b = 42$ mm, operating temperature $\vartheta = 70^0$C and speed $n = 3000$ rpm. Estimate the relubrication period and amount of grease for grease lubrication and the kinematic viscosity of oil for oil lubrication.

Solution

1. Grease lubrication

The relubrication period according to eqn (14.50) is

$$t_g = 10 \left(\frac{14 \times 10^6}{3\,000\sqrt{70}} - 4 \times 70 \right) = 2\,778 \text{ h} \approx 4 \text{ months}$$

where $k_b = 10$ for radial ball bearings. The amount of grease is estimated according to eqn (14.51), i.e.,

$$m_g = 0.005 \times 180 \times 42 = 37.8 \text{ g}$$

2. Oil lubrication

The oil kinematic viscosity for $d_b = 70$ mm, $n = 3000$ rpm and $\vartheta = 70^0$C on the basis of graphs in Fig. 14.13 is 8.5 centistokes $= 8.5 \times 10^{-6} \text{ m}^2/\text{s}$.

Appendix A

Leakage Inductance of a.c. Stator Windings

A.1 Stator winding factor

The stator *winding factor* for the fundamental space harmonic $\nu = 1$

$$k_{w1} = k_{d1}k_{p1} \tag{A.1}$$

is the product of the *distribution factor* k_{d1} and *pitch factor* k_{p1} , i.e.,

$$k_{d1} = \frac{\sin(q_1\gamma/2)}{q_1 \sin(\gamma/2)} \tag{A.2}$$

$$k_{p1} = \sin\left(\frac{\pi}{2}\frac{w_c}{\tau}\right) = \sin\left(\frac{\pi}{2}\frac{w_{sl}}{Q_1}\right) \tag{A.3}$$

where $\gamma = \pi/(m_1q_1)$ for a 60° phase belt, $\gamma = 2\pi/(m_1q_1)$ for a 120° phase belt, w_c is the coil span and w_{sl} is the coil span measured in the number of slot. The *number of slots per pole* Q_1 and *number of slots per pole per phase* q_1 are defined as

$$Q_1 = \frac{s_1}{2p} \tag{A.4}$$

$$q_1 = \frac{s_1}{2pm_1} \tag{A.5}$$

In the above equations s_1 is the number of slots, $2p$ is the number of poles and m_1 is the number of phases. Putting $\gamma = \pi/(m_1q_1)$ into eqn (A.2) the distribution factor for a 60° phase belt has the following well known form

$$k_{d1} = \frac{\sin[\pi/(2m_1)]}{q_1 \sin[\pi/(2m_1q_1)]} \tag{A.6}$$

Including the stator slot skew and the effect of slot openings, the total winding factor for the fundamental space harmonic $\nu = 1$ is

(a)

(b)

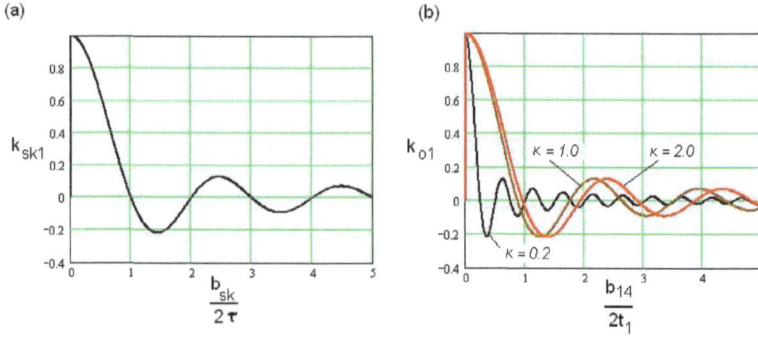

Fig. A.1. Stator skew factor k_{sk1} and slot opening factor k_{o1}: (a) k_{sk1} as a function of $b_{sk}/(2\tau)$; (b) k_{o1} as a function of $b_{14}/(2t_1)$ at $\kappa = constant$. Parameter κ appears in eqn (5.107).

$$k_{w1} = k_{d1}k_{p1}k_{sk1}k_{o1} \tag{A.7}$$

where the *skew factor*

$$k_{sk1} = \frac{\sin[\pi b_{sk}/(2\tau)]}{\pi b_{sk}/(2\tau)} = \frac{\sin[\pi p b_{sk}/(s_1 t_1)]}{\pi p b_{sk}/(s_1 t_1)} \tag{A.8}$$

and *slot opening factor*

$$k_{o1} = \frac{\sin[\pi \rho b_{14}/(2t_1)]}{\pi \rho b_{14}/(2t_1)} \tag{A.9}$$

In eqns (A.8) and (A.9) b_{sk} is the stator slot skew, b_{14} is the stator slot opening (Fig. A.2) and ρ is given by eqn (5.107). The stator slot opening factor is very small and in typical machines can be neglected, e.g., for $b_{14}/(2t_1) = 0.1$ and $\rho = 1$ the slot opening factor $k_{o1} = 0.9995$.

A.2 Slot leakage permeance

In analytical calculations of the *slot leakage permance* the saturation of the magnetic circuit due to leakage fluxes is neglected.

The coefficient of leakage permeance (specific-slot permeance) of a rectangular open slot is

- in the case of a slot totally filled with conductors

$$\lambda_{1s} = \frac{h_{11}}{3b_{11}} \tag{A.10}$$

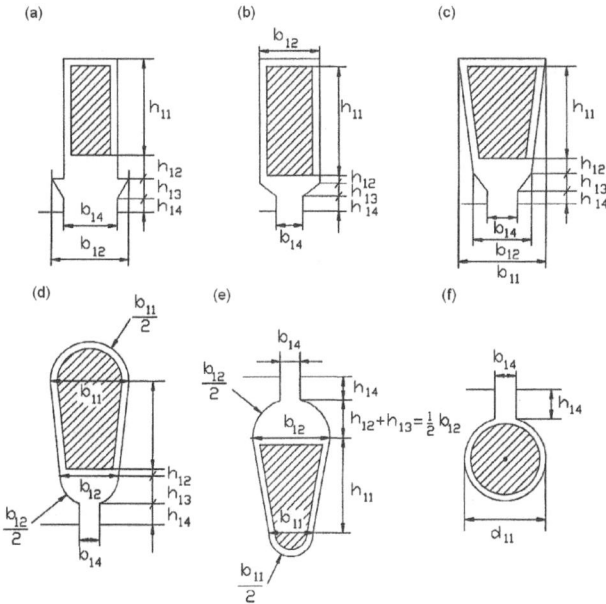

Fig. A.2. Armature slots of single layer windings: (a) open rectangular slot, (b) semi-closed rectangular slot, (c) semi-closed trapezoidal slot, (d) semi-closed oval slot, (e) semi-closed oval slot of internal stators of brushless motors or d.c. commutator motor rotors, (f) semi-closed round slot.

- in the case of an empty slot (without any conductors)

$$\lambda_{1s} = \frac{h_{11}}{b_{11}} \tag{A.11}$$

The width of the rectangular slots is b_{11} and its height is h_{11}.

The coefficients of leakage permeances of the slots shown in Fig. A.1 are

- open rectangular slot (Fig. A.2a):

$$\lambda_{1s} = \frac{h_{11}}{3b_{14}} + \frac{h_{12} + h_{14}}{b_{14}} + \frac{2h_{13}}{b_{12} + b_{14}} \tag{A.12}$$

- semi-open (semi-closed) slot (Fig. A.2b):

$$\lambda_{1s} = \frac{h_{11}}{3b_{11}} + \frac{h_{12}}{b_{11}} + + \frac{2h_{13}}{b_{11} + b_{14}} + \frac{h_{14}}{b_{14}} \tag{A.13}$$

- semi-open trapezoidal slot (Fig. A.2c):

$$\lambda_{1s} = \frac{h_{11}}{3b_{12}} k_t + \frac{h_{12}}{b_{12}} + \frac{2h_{13}}{b_{12} + b_{14}} + \frac{h_{14}}{b_{14}} \tag{A.14}$$

where

$$k_t = 3\frac{4t^2 - t^4(3 - 4\ln t) - 1}{4(t^2 - 1)^2(t - 1)}, \qquad\qquad t = \frac{b_{11}}{b_{12}} \qquad (A.15)$$

- semi-open oval slot according to Fig. A.2d or Fig. A.2e (internal stators of brushless motors or d.c. commutator motor rotors):

$$\lambda_{1s} = 0.1424 + \frac{h_{11}}{3b_{12}}k_t + \frac{h_{12}}{b_{12}} + 0.5\arcsin[\sqrt{1 - (b_{14}/b_{12})^2}] + \frac{h_{14}}{b_{14}} \quad (A.16)$$

where k_t is according to eqn (A.15).
- semi-closed round slot (Fig. A.2f):

$$\lambda_{1s} = \frac{\pi}{6} + \frac{\pi}{16\pi} + \frac{h_{24}}{b_{24}} \approx 0.623 + \frac{h_{24}}{b_{24}} \qquad (A.17)$$

The above specific-slot permeances are for single-layer windings. To obtain the specific permeances of slots containing double-layer windings, it is necessary to multiply eqns (A.12) to (A.17) by the factor

$$\frac{3w_c/\tau + 1}{4} \qquad (A.18)$$

This approach is justified if $2/3 \leq w_c/\tau \leq 1.0$.

A.3 End winding connection leakage permeance

The specific *leakage permeance of the end winding connection* (overhang) is estimated on the basis of experiments. For double-layer, low-voltage, small- and medium-power motors:

$$\lambda_{1e} \approx 0.34q_1 \left(1 - \frac{2}{\pi}\frac{w_c}{l_{1e}}\right) \qquad (A.19)$$

where l_{1e} is the length of a single end connection and the number of slots per pole per phase q_1 is defined by eqn (A.5). For cylindrical-type medium power a.c. machines

$$l_{1e} \approx (0.05p + 1.2)\frac{\pi(D_{1in} + h_{1t})}{2p}\frac{w_c}{\tau} + 0.02 \text{ m} \qquad (A.20)$$

where h_{1t} is the height of the stator tooth.

Putting $w_c/l_{1e} = 0.64$, eqn (A.19) also gives good results for single-layer windings, i.e.,

$$\lambda_{1e} \approx 0.2q_1 \qquad (A.21)$$

For double-layer, high-voltage windings,

$$\lambda_{1e} \approx 0.42 q_1 \left(1 - \frac{2}{\pi}\frac{w_c}{l_{1e}}\right) k_{w1}^2 \tag{A.22}$$

where the stator (armature) winding factor k_{w1} for the fundamental space harmonic $\nu = 1$ is given by eqn (A.1). In general,

$$\lambda_{1e} \approx 0.3 q_1 \tag{A.23}$$

for most windings.

A.4 Differential leakage permeance

The specific *permeance of the differential leakage flux* is

$$\lambda_{1d} = \frac{m_1 q_1 \tau k_{w1}^2}{\pi^2 g k_C k_{sat}} \tau_{d1} \tag{A.24}$$

where the *differential leakage factor* τ_{d1} can be calculated according to [98, 185, 264], i.e.,

$$\tau_{d1} = \frac{1}{k_{w1}^2} \sum_{\nu>1} \left(\frac{k_{w1\nu}}{\nu}\right)^2 \tag{A.25}$$

The winding factor for the higher space harmonics $\nu > 1$ is $k_{w1\nu}$. In practical calculations it is convenient to use the following formula

$$\tau_{d1} = \frac{\pi^2 (10 q_1^2 + 2)}{27} \left[\sin\left(\frac{30^0}{q_1}\right)\right]^2 - 1 \tag{A.26}$$

The Carter's coefficient is

$$k_C = \frac{t_1}{t_1 - \gamma_1 g} \tag{A.27}$$

where t_1 is the slot pitch and

$$\gamma_1 = \frac{4}{\pi} \left[\frac{b_{14}}{2g} \arctan\left(\frac{b_{14}}{2g}\right) - \ln\sqrt{1 + \left(\frac{b_{14}}{2g}\right)^2}\right] \tag{A.28}$$

The curves of the differential leakage factor τ_{d1} are plotted in Fig. A.3.

A.5 Tooth-top leakage permeance

The *tooth-top specific permeance* (between the heads of teeth) is

$$\lambda_{1t} \approx \frac{5g/b_{14}}{5 + 4g/b_{14}} \tag{A.29}$$

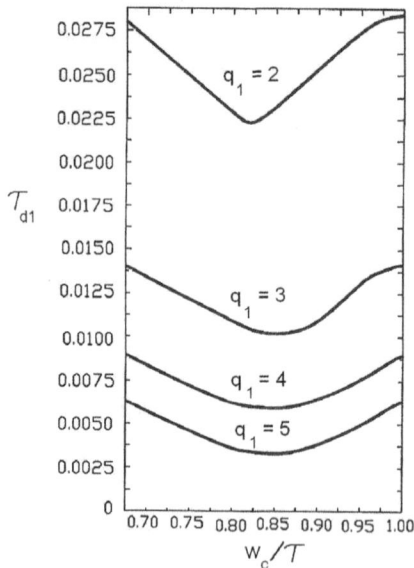

Fig. A.3. Curves of the differential leakage factor τ_{d1} plotted against w_c/τ ratio at $q_1 = constant$.

A.6 Leakage reactance per phase

The *leakage reactance* of the armature windings of a.c. machines [98, 185, 264] is

$$X_1 = 4\pi\mu_0 f \frac{N_1^2 L_i}{pq_1}\left(\lambda_{1s} + \frac{l_{1e}}{L_i}\lambda_{1e} + \lambda_{1d} + \lambda_{1t}\right) \qquad (A.30)$$

where $\mu_0 = 0.4\pi \times 10^{-6}$ H/m is the permeability of free space, q_1 is according to eqn (A.5), N_1 is the number of turns per phase, L_i is the effective length of the armature core and p is the number of pole pairs.

Appendix B

Losses in a.c. Motors

B.1 Armature winding losses

The *armature winding resistance* per phase for the d.c. current is

$$R_{1dc} = \frac{N_1 l_{1av}}{a\sigma_1 s_a} \tag{B.1}$$

where N_1 is the number of armature turns per phase, l_{1av} is the average length of turn, a is the number of parallel paths or conductors, σ_1 is the electric conductivity of the armature conductor at given temperature (for a copper conductor $\sigma_1 \approx 57 \times 10^6$ S/m at 20^0C and $\sigma_1 \approx 47 \times 10^6$ S/m at 75^0C), and s_a is the conductor cross section. The average length of the armature turn is

$$l_{1av} = 2(L_i + l_{1e}) \tag{B.2}$$

in which the length l_{1e} of a single end connection is according to eqn (A.20).

For a.c. current the armature winding resistance should be divided into the resistance R_{1b} of the winding portion located in slots (bars) and resistance R_{1e} of the end connections, i.e.,

$$R_1 = R_{1b} + R_{1e} = \frac{2N_1}{a\sigma_1 s_a}(L_i k_{1R} + l_{1e}) \approx k_{1R} R_{1dc} \tag{B.3}$$

where k_{1R} is the skin-effect coefficient for the armature resistance. For a double-layer winding and $w_c = \tau$ [192]:

$$k_{1R} = \varphi_1(\xi_1) + \left[\frac{m_{sl}^2 - 1}{3} - \left(\frac{m_{sl}}{2} \sin \frac{\gamma}{2} \right)^2 \right] \Psi_1(\xi_1) \tag{B.4}$$

where

$$\varphi_1(\xi_1) = \xi_1 \frac{\sinh 2\xi_1 + \sin 2\xi_1}{\cosh 2\xi_1 - \cos 2\xi_1} \tag{B.5}$$

$$\Psi_1(\xi_1) = 2\xi_1 \frac{\sinh \xi_1 - \sin \xi_1}{\cosh \xi_1 + \cos \xi_1} \tag{B.6}$$

$$\xi_1 = h_c \sqrt{\pi f \mu_o \sigma_1 \frac{b_{1con}}{b_{11}} \frac{L_i}{L_{1b}}} \tag{B.7}$$

and m_{sl} is the number of conductors per slot arranged above each other in two layers (this must be an even number), γ is the phase angle between the currents of the two layers, σ_1 is the electric conductivity of the primary wire, f is the input frequency, b_{1con} is the width of all the conductors in a slot, b_{11} is the slot width (Fig. A.2), h_c is the height of a conductor in the slot and L_{1b} is the length of the conductor (bar) if different from L_i. If there are n_{sl} conductors side by side at the same height of the slot, they are taken as a single conductor carrying n_{sl}–times greater current.

In general, for a three-phase winding $\gamma = 60^0$ and

$$k_{1R} = \varphi_1(\xi_1) + \left(\frac{m_{sl}^2 - 1}{3} - \frac{m_{sl}^2}{16} \right) \Psi_1(\xi_1) \tag{B.8}$$

For a chorded winding $(w_c < \tau)$ and $\gamma = 60^0$

$$k_{1R} \approx \varphi_1(\xi_1) + \left[\frac{m_{sl}^2 - 1}{3} - \frac{3(1 - w_c/\tau)}{16} m_{sl}^2 \right] \Psi_1(\xi_1) \tag{B.9}$$

The skin-effect coefficient k_{1R} for hollow conductors is given, for example, in [192]

If $m_{sl} = 1$ and $\gamma = 0$, the skin-effect coefficient $k_{1R} = \varphi_1(\xi_1)$ (as for a cage winding).

If $\gamma = 0$ the currents in all conductors are equal and

$$k_{1R} \approx \varphi_1(\xi_1) + \frac{m_{sl}^2 - 1}{3} \Psi_1(\xi_1) \tag{B.10}$$

Eqn (B.10) can also be used for calculation of additional winding losses in large d.c. commutator motors.

For small motors with round armature conductors fed from power frequencies of 50 or 60 Hz,

$$R_1 \approx R_{1dc} \tag{B.11}$$

The *armature winding losses* are

$$\Delta P_a = m_1 I_a^2 R_1 \approx m_1 I_a^2 R_{1dc} k'_{1R} \tag{B.12}$$

Since the skin effect is in only this part of the conductor which is located in the slot, the armature winding losses should be multiplied by the coefficient

$$k'_{1R} = \frac{k_{1R} + l_{1e}/L_i}{1 + l_{1e}/L_i} \tag{B.13}$$

B.2 Armature core losses

The magnetic flux in the armature (stator) core is nonsinusoidal. The rotor PM excitation system produces a trapezoidal shape of the magnetic flux density waveform. The stator windings are fed from switched d.c. sources with PWM or square wave control. The applied voltage thus contains many harmonics which are seen in the stator flux.

The *hysteresis losses* can be expressed with the aid of Richter's formula [250], i.e.

$$\Delta P_{hFe} = \epsilon \frac{f}{100} m_{Fe} \sum_{n=1}^{\infty} n[B_{mtn}^2 + B_{mrn}^2]$$

$$= \epsilon \frac{f}{100} m_{Fe} [B_{mt1}^2 + B_{mr1}^2] \eta_{dh}^2 \qquad (B.14)$$

where $\epsilon = 1.2$ to 2.0 for anisotropic laminations with 4% Si, $\epsilon = 3.8$ for isotropic laminations with 2% Si and $\epsilon = 4.4$ to 4.8 for isotropic siliconless laminations, n are the odd time harmonics, B_{mtn} and B_{mrn} are the harmonic components of the magnetic flux density in the tangential and radial (normal) directions, respectively.

The *coefficient of distortion of the magnetic flux density* for hysteresis losses is

$$\eta_{dh} = \sqrt{1 + \frac{3(B_{mt3})^2 + 3(B_{mr3})^2}{B_{mt1}^2 + B_{mr1}^2} + \frac{5(B_{mt5})^2 + 5(B_{mr5})^2}{B_{mt1}^2 + B_{mr1}^2} + \dots} \qquad (B.15)$$

For $\eta_{dh} = 1$, eqn (B.14) expresses the hysteresis losses under sinusoidal magnetic flux density.

The *eddy current losses* can be calculated using the following classical formula:

$$\Delta P_{eFe} = \frac{\pi^2}{6} \frac{\sigma_{Fe}}{\rho_{Fe}} f^2 d_{Fe}^2 m_{Fe} \sum_{n=1}^{\infty} n^2 [B_{mtn}^2 + B_{mrn}^2]$$

$$= \frac{\pi^2}{6} \frac{\sigma_{Fe}}{\rho_{Fe}} f^2 d_{Fe}^2 m_{Fe} [B_{mt1}^2 + B_{mr1}^2] \eta_{de}^2 \qquad (B.16)$$

where σ_{Fe}, d_{Fe}, ρ_{Fe} and m_{Fe} are the electric conductivity, thickness, specific density and mass of laminations, respectively. The coefficient of distortion of the magnetic flux density for eddy current losses is

$$\eta_{de} = \sqrt{1 + \frac{(3B_{mt3})^2 + (3B_{mr3})^2}{B_{mt1}^2 + B_{mr1}^2} + \frac{(5B_{mt5})^2 + (5B_{mr5})^2}{B_{mt1}^2 + B_{mr1}^2} + \dots} \qquad (B.17)$$

Note that $\eta_{de} > \eta_{dh}$, i.e., the influence of higher harmonics on eddy current losses is higher than on hysteresis losses.

Eqns (B.16) and (B.14) exclude the *excess losses* (due to magnetic anomaly) and losses due to metallurgical and manufacturing processes. Analytical and numerical methods of calculating the excess losses published so far are either unsuccessful or have limited applications.

There is a poor correlation between measured core losses and those calculated using classical methods. The losses calculated according to eqns (B.16) and (B.14) are lower than those obtained from measurements. There are errors between 25 to 75% [26]. The coefficient of additional core losses $k_{ad} > 1$ can help to obtain a better agreement of predicted and measured core losses

$$\Delta P_{Fe} = k_{ad}(\Delta P_{eFe} + \Delta P_{hFe}) \tag{B.18}$$

If the *specific core losses* are known, the stator core losses ΔP_{1Fe} can be calculated on the basis of the specific core losses and masses of teeth and yoke, i.e.

$$\Delta P_{Fe} = \Delta p_{1/50} \left(\frac{f}{50}\right)^{4/3} [k_{adt} B_{mt}^2 m_t + k_{ady} B_{my}^2 m_y] \tag{B.19}$$

where $k_{adt} > 1$ and $k_{ady} > 1$ are the factors accounting for the increase in losses due to metallurgical and manufacturing processes, $\Delta p_{1/50}$ is the specific core loss in W/kg at 1 T and 50 Hz, B_{mt} is the magnetic flux density in a tooth, B_{my} is the magnetic flux density in the yoke, m_t is the mass of the teeth, and m_y is the mass of the yoke. For teeth $k_{adt} = 1.7$ to 2.0 and for the yoke $k_{ady} = 2.4$ to 4.0 [185].

B.3 Rotor core losses

For PM synchronous and PM d.c. brushless motors the *rotor core losses* due to fundamental harmonic do not exist. The rotor core losses in PM brushless machines are due to the pulsating flux produced by the rapid changes in air gap reluctance as the rotor passes the stator teeth. These losses are negligible in surface-mounted PM motors, due to their large effective air gaps (including PM radial thicknes). These rotor losses can sometimes be significant in buried PM motors, salient pole rotors (Fig. 8.5) and surface PM motors with mild steel pole shoes. The rotor losses of a higher frequency can be obtained using, e.g., methods described in [104, 196, 230].

B.4 Core loss FEM model

The core losses within the stator and rotor are calculated using a set of finite element method (FEM) models, assuming constant rotor speed and balanced three-phase armature currents. The eddy current and hysteresis losses, within the cores, in the 2D FEM including distorted flux waveforms can be expressed by eqns (B.14) and (B.16).

The field distribution at several time intervals in the fundamental current cycle is needed to create the magnetic flux density waveforms. This is obtained by rotation of the rotor grid and phase advancement of the stator currents. From a field solution, for a particular rotor position, the magnetic flux density is calculated at each element centroid.

B.5 Losses in conductive retaining sleeves

The slot ripple losses in the retaining sleeve of the rotor can be calculated with the aid of the following simple equation [123, 261]:

$$\Delta P_{sl} = \frac{\pi^3}{2}\sigma_{sl}k_r(B_{msl}n)^2 D_{sl}^3 l_{sl}d_{sl} \ \ \text{W} \tag{B.20}$$

where n is the rotor speed in rev/s, $D_{sl} = D_{2out} - d_{sl}$, l_{sl}, d_{sl} and σ_{sl} are the mid-diameter in meters, effective length in meters, thickness in meters and the electric conductivity in S/m of the retaining sleeve, respectively. For Inconel 718 the electric conductivity $\sigma = 0.826 \times 10^6$ S/m at 20^0C. The coeficient for increasing the sleeve resistance due to tangential sleeve currents is [123]

$$k_r \approx 1 + \frac{1}{\pi}\frac{t_1}{l_{sl}} \tag{B.21}$$

The amplitude of the high frequency magnetic flux density due to slot openings (slot ripple) can be calculated according to Richter [250]

$$B_{msl} = 2\beta B_{mean} = 2\beta \frac{1}{k_C}\frac{2}{\pi}B_{mg} \tag{B.22}$$

where

$$\beta = \frac{B_{msl}}{2B_{mean}} = \frac{1 + u^2 - 2u}{2(1 + u^2)} \tag{B.23}$$

$$u = \frac{b_{14}}{2g} + \sqrt{1 + \left(\frac{b_{14}}{2g}\right)^2} \tag{B.24}$$

and B_{mean} is the mean value of the magnetic flux density in the air gap under the stator slot opening, B_{mg} is the peak value of the magnetic flux density in the air gap, b_{14} is the stator slot opening in m (see Fig. A.1), g is the air gap between the stator core and PM, and k_C is Carter's coefficient of the air gap according to eqn (A.27). Decrease in the slot opening and/or increase in the air gap reduce the slot ripple. For example, for an Inconel sleeve with its electric conductivity $\sigma_{sl} = 0.5295 \times 10^6$ S/m at 100^0C, diameter $D_{sl} = 0.1$ m, length $l_{sl} = 0.1$ m, and thickness $d_{sl} = 1.6$ mm the slot ripple losses in the sleeve are $\Delta P_{sl} = 1411$ W at $n = 100,000$ rpm, $\Delta P_{sl} = 353$ W at $n = 50,000$ rpm and $\Delta P_{sl} = 14$ W at $n = 10,000$ rpm. The following machine parameters have been assumed: $p = 2$, $s_1 = 36$, $b_{14} = 2$ mm, $g = 3.2$ mm, $B_{mg} = 0.7$ T, which give $u = 1.36$ and $B_{msl} = 0.02$ T.

Losses in the sleeve due to other harmonics can also be calculated with the aid of eqn (B.20) putting instead of B_{msl} the amplitude of the considered higher harmonic.

B.6 Losses in permanent magnets of brushless motors

The electric conductivity of sintered NdFeB magnets is from 0.6 to 0.85×10^6 S/m. The electric conductivity of SmCo magnets is from 1.1 to 1.4×10^6 S/m. Since the electric conductivity of rare earth PMs is only 4 to 9 times lower than that of a copper conductor, the *losses in conductive PMs due to higher harmonic magnetic fields* produced by the stator cannot be neglected, particularly in the case of high speed motors.

Similar to losses in a conductive retaining sleeve, the most important losses in PMs are those generated by the fundamental frequency magnetic flux due to the stator slot openings. Slot ripple losses are only in motors with slotted armature ferromagnetic cores and do not exists in slotless machines.

The slot ripple losses in PMs can be approximately estimated using eqn (B.20), in which $\sigma_{sl} = \sigma_{PM}$, $l_{sl} = l_M$, $D_{sl} = D_{2out} - h_M$ and $d_{sl} = h_M$, where σ_{PM} is the electric conductivity of PM, l_M is the axial length of PM (usually $l_M = L_i$) and h_M is the radial height of PM, i.e.,

$$\Delta P_{sl} = \frac{\pi^3}{2} \sigma_{PM} k_r (B_{msl} n)^2 (D_{2out} - h_M)^3 l_M h_M \quad \text{W} \qquad \text{(B.25)}$$

Since the external surface area of PMs is smaller than that of the rotor, eqn (B.25) should be multiplied by the factor $S_{PM}/(\pi D_{2out} l_M)$, where S_{PM} is the surface area of PMs. For surface PMs

$$S_{PM} = \alpha_i \pi D_{2out} L_i \qquad \text{(B.26)}$$

The magnetic flux density B_{msl} can be estimated on the basis of eqn (B.22) only if the rotor is not equipped with conductive retaining sleeve. Otherwise, the peak value B_{msl} will be much smaller. Eqn (B.25) may give too high value of slot ripple losses in PMs.

Power losses in PMs due to the νth higher space harmonic can be estimated using the following equation obtained from a 2D electromagnetic field distribution [122] on assumption that the relative recoil magnetic permeability $\mu_{rrec} \approx 1$, i.e.,

$$\Delta P_{PM\nu} = a_{R\nu} k_{r\nu} \frac{k_\nu^3}{\beta_\nu^2} \left(\frac{B_{m\nu}}{\mu_0 \mu_{rrec}} \right)^2 \frac{1}{\sigma_{PM}} S_{PM} \quad \text{W} \qquad \text{(B.27)}$$

where the coefficients

$$a_{R\nu} = \frac{1}{\sqrt{2}} \sqrt{\sqrt{4 + \left(\frac{\beta_\nu}{k_\nu}\right)^4} + \left(\frac{\beta_\nu}{k_\nu}\right)^2} \qquad \beta_\nu = \nu \frac{\pi}{\tau} \qquad (B.28)$$

the coefficient of attenuation of the electromagnetic field in PMs

$$k_\nu = \sqrt{\pi(1 \mp \nu) f \mu_0 \mu_{rrec} \sigma_{PM}} \qquad (B.29)$$

the edge effect coefficient

$$k_{r\nu} = 1 + \frac{1}{\nu} \frac{2}{\pi} \frac{\tau}{L_i} \qquad (B.30)$$

and τ is the stator slot pitch, σ_{PM} is the electric conductivity of PMs, $(1 \mp \nu)f$ is the frequency of the magnetic flux density component in the rotor due to the νth space harmonic, B_ν is the peak value of the νth harmonic of the magnetic flux density and S_{PM} is the active surface area (adjacent to the air gap) of all PMs.

Losses in PMs (B.25) and (B.27) can be neglected if the rotor is firnished with a current conducting retaining sleeve, because eddy currents in the conductive sleeve repell the higher harmonic fields.

B.7 Rotational losses

The *rotational* or *mechanical losses* ΔP_{rot} consist of friction losses ΔP_{fr} in bearings, windage losses ΔP_{wind} and ventilation losses ΔP_{vent}. There are many semi-empirical equations for calculating the rotational losses giving various degrees of accuracy.

The friction losses in bearings of small machines can be evaluated using the following formula

$$\Delta P_{fr} = k_{fb} m_r n \times 10^{-3} \ \text{W} \qquad (B.31)$$

where $k_{fb} = 1$ to 3 W/(kg rpm), m_r is the mass of rotor in kg and n is the speed in rpm.

The windage losses of small machines without a fan can be found as follows:

- when the speed does not exceed 6000 rpm:

$$\Delta P_{wind} \approx 2 D_{2out}^3 L_i n^3 \times 10^{-6} \ \text{W} \qquad (B.32)$$

- when the speed is higher than 15,000 rpm:

$$\Delta P_{wind} \approx 0.3 D_{2out}^5 \left(1 + 5\frac{L_i}{D_{2out}}\right) n^3 \times 10^{-6} \ \text{W} \qquad (B.33)$$

where the outer diameter D_{2out} of the rotor and effective length L_i of the core are in meters and the speed n is in rpm.

According to ABB in Switzerland, the rotational losses of 50-Hz salient-pole synchronous machines can be expressed by the formula

$$\Delta P_{rot} = \frac{1}{k_m}(D_{1in} + 0.15)^4 \sqrt{L_i} \left(\frac{n}{100}\right)^{2.5} \text{ kW} \tag{B.34}$$

where the stator inner diameter D_{1in} and effective length of the stator core are in meters, n is in rev/s and $k_m = 30$ for salient pole synchronous machines.

B.8 Windage losses in high speed motors

It is better in high speed machines to calculate the windage losses and bearing friction losses separately, including variation of the cooling medium density and its dynamic viscosity with temperature. The total windage losses

$$\Delta P_{wind} = \Delta P_a + \Delta P_{ad} + \Delta P_c \tag{B.35}$$

have three components: losses ΔP_a in the air gap due to resisting drag torque, losses ΔP_{ad} due to resisting drag torque at each flat cicular surface of the rotor and losses ΔP_c in the air gap due to axial flow of cooling medium. The losses ΔP_c exist only if the cooling medium is forced to pass through the air gap. The losses in the air gap due to resisting drag torque are

$$\Delta P_a = \pi c_f \rho \Omega^3 \frac{D_{2out}^4}{16} L_i \tag{B.36}$$

in which

- the air density as a function of temperature (^{0}C) at the atmospheric pressure 1000 mbar can approximately be expressed as

$$\rho = -10^{-8}\vartheta^3 + 10^{-5}\vartheta^2 - 0.0045\vartheta + 1.2777 \text{ kg/m}^3 \tag{B.37}$$

- friction coefficient [27]

$$c_f = 0.515\frac{[2(g - d_{sl})/D_{2out}]^{0.3}}{Re^{0.5}} \text{ if } Re < 10^4$$

$$c_f = 0.0325\frac{[2(g - d_{sl})/D_{2out}]^{0.3}}{Re^{0.2}} \text{ if } Re > 10^4 \tag{B.38}$$

- dynamic viscosity of air as a function of temperature (^{0}C) at the atmospheric pressure 1000 mbar

$$\mu_{dyn} = -2.1664 \times 10^{-11}\vartheta^2 + 4.7336 \times 10^{-8}\theta + 2 \times 10^{-5} \text{ Pa s} \tag{B.39}$$

- angular speed of the rotor $\Omega = 2\pi n$, where n is the speed of the rotor in rev/s.

In eqn (B.38) for the rotor with retaining sleeve the mechanical clearance is $g - d_{sl}$, where d_{sl} is the rotating sleeve thickness.

Reynolds number is defined by the ratio of the dynamic pressure ρv^2 to shearing stress $\mu_{dyn} v/l$ or the inertial force ρv to viscous force μ_{dyn}/l, i.e.,

$$Re = \frac{\rho v l}{\mu_{dyn}} \tag{B.40}$$

where v is the linear velocity and l is the *characteristic length*. For a pipe or duct the characteristic length is replaced by the hydraulic diameter d_h, i.e.

$$Re = \frac{\rho v d_h}{\mu_{dyn}} = \frac{\rho \Omega d_h^2}{2\mu_{dyn}} \tag{B.41}$$

The hydraulic diameter is defined as $d_h = 4A/R$ where A is the cross section area of the duct and R is the wetted perimeter of the duct. For a circular duct or pipe the hydraulic diameter d_h is the same as the geometrical diameter d, i.e., $d_h = \pi d^2/(\pi d) = d$. The hydraulic diameter of a circular tube with an inside circular tube is

$$d_h = 4\frac{0.25\pi(D_{1in}^2 - D_{2out}^2)}{\pi(D_{1in} + D_{2out})} = D_{1in} - D_{2out} \tag{B.42}$$

where D_{1in} is the inner diameter of the outside tube and D_{2out} is the outer diameter of the inside tube. This corresponds to the stator and rotor of an electrical machine. Thus, the Reynolds number for the air gap is

$$Re = \frac{\rho \Omega (D_{1in} - D_{2out})^2}{2\mu_{dyn}} \tag{B.43}$$

Losses due to resisting drag torque at each flat cicular surface of the rotor are calculated in the same way as for a rotating disc [248, 262], i.e.,

$$\Delta P_{ad} = \frac{1}{64} c_{fd} \rho \Omega^3 (D_{2out}^5 - d_{sh}^5) \tag{B.44}$$

The coefficient of friction c_{fd} for rotating disc [187] and Reynolds number Re_d [49] are, respectively

$$c_{fd} = \frac{3.87}{Re_d^{0.5}} \text{ if } Re_d < 3 \times 10^5$$

$$c_{fd} = \frac{0.146}{Re_d^{0.2}} \text{ if } Re_d > 3 \times 10^5 \tag{B.45}$$

$$Re_d = \frac{\rho \Omega D_{2out}^2}{4\mu_{dyn}} \tag{B.46}$$

Losses due to axial flow of cooling gas [248]

$$\Delta P_c = \frac{2}{3}\pi \rho v_t v_{ax} \Omega \left[(0.5D_{1in})^3 - (0.5D_{2out})^3 \right] \qquad (B.47)$$

where $v_t \approx 0.5v$ is the mean tangential linear velocity of the cooling medium (gas) due to rotation of the rotor, $v = \pi D_{2out} n$ is the surface linear velocity of the rotor and v_{ax} is the axial linear velocity of cooling medium, i.e., linear velocity air blown by a fan.

B.9 Losses due to higher time harmonics

Higher time harmonics generated by static converters produce additional losses. The higher harmonic frequency in the armature (stator) is nf. The armature winding losses, the core losses, and the stray losses are frequency-dependent. The mechanical losses do not depend on the shape of the input waveform.

The frequency-dependent losses of an inverter-fed a.c. motor are

- stator (armature) winding losses (see also Section 8.6.3):

$$\Delta P_a = \sum_{n=1}^{\infty} \Delta P_{an} = m_1 \sum_{n=1}^{\infty} I_{an}^2 R_{1n} \approx m_1 R_{1dc} \sum_{n=1}^{\infty} I_{an}^2 k_{1Rn}$$

$$= m_1 R_{1dc} I_{ar}^2 \sum_{n=1}^{\infty} \left(\frac{I_{an}}{I_{ar}} \right)^2 k_{1Rn} = \Delta P_a \sum_{n=1}^{\infty} \left(\frac{I_{an}}{I_{ar}} \right)^2 k_{1Rn} \qquad (B.48)$$

- stator (armature) core losses

$$\Delta P_{Fe} = \sum_{n=1}^{\infty} \Delta P_{Fen} = [\Delta P_{Fe}]_{n=1} \sum_{n=1}^{\infty} \left(\frac{V_{1n}}{V_{1r}} \right)^2 n^{4/3} \qquad (B.49)$$

where ΔP_a are the stator (armature) winding losses for rated d.c. current, $[\Delta P_{Fe}]_{n=1}$ are the stator core losses according to eqn (B.18) or eqn (B.19) for $n = 1$ and rated voltage, k_{1Rn} is the skin effect coefficient of the a.c. armature resistance for nf, I_{an} is the rms higher harmonic armature current, I_{ar} is the armature rated current, V_{1n} is the rms higher harmonic inverter output voltage and V_{1r} is the rms rated voltage approximately equal to the rms voltage for fundamental harmonic $n = 1$. For the index of summation $n = 1$ in eqn (B.49) the fundamental harmonic voltage $V_{1n} = V_{1r}$ and the first term under the summation symbol $(V_{1n}/V_{1r})^2 n^{4/3} = 1$.

An example of calculation of the increase in the stator winding and core losses due to higher time harmonics $n = 5, 7, 11, 13, \ldots$ for the fundamental harmonic $f = 50$ Hz and arbitrarily chosen harmonic currents I_{an} and harmonic voltages V_{1n} is given in Table B.1. The coefficient k_{1Rn} has been calculated using eqns (B.4), (B.5), (B.6) and (B.7), in which $m_{sl} = 6$, $\gamma = 60^0$, $f = nf$, $\sigma_1 = 57 \times 10^6$ S/m, $h_c = 2$ mm, $b_{1con} = 5$ mm, $b_{11} = 6.5$ mm and $L_{1b} = L_i$.

Table B.1. Increase in the stator winding and core losses due to higher time harmonics

Harmonics		Winding losses eqn (B.48)			Core losses eqn (B.49)		
n	nf Hz	I_{an}/I_{ar}	k_{1Rn}	$(I_{an}/I_{ar})^2 k_{1Rn}$	V_{1n}/V_{1r}	$n^{4/3}$	$(V_{1n}/V_{1r})^2 n^{4/3}$
1	50	1.000	1.0039	1.0039	1.000	1.00	1.0000
5	250	0.009	1.0966	0.0001	0.009	8.55	0.0007
7	350	0.023	1.1891	0.0006	0.023	13.39	0.0071
11	550	0.014	1.4654	0.0003	0.014	24.46	0.0048
13	650	0.027	1.6485	0.0012	0.027	30.57	0.0223
$\sum_n (I_{an}/I_{ar})^2 k_{1Rn}$				$= 1.0061$	$\sum_n (V_{1n}/V_{1r})^2 n^{4/3}$		$= 1.0349$

Symbols and Abbreviations

A	magnetic vector potential
A	line current density
A_r	amplitude of the radial displacement
a	number of parallel current paths of the armature winding of a.c. motors; number of pairs of parallel current paths of the armature winding of d.c. commutator motors
B	vector magnetic flux density
B	magnetic flux density
b	instantaneous value of the magnetic flux density; width of slot
b_{br}	brush shift
b_{fsk}	skew of PMs
b_p	pole shoe width
b_{sk}	skew of stator slots
C	number of commutator segments; capacitance; cost
C_c	cost of ferromagnetic core
C_0	cost of all other components independent of the shape of the machine
C_{PM}	cost of PMs
C_{sh}	cost of shaft
C_w	cost of winding
c	wave velocity; tooth width
c_{Cu}	cost of copper conductor per kg
c_E	armature constant (EMF constant)
c_{Fe}	cost of ferromagnetic core per kg
c_{PM}	cost of PMs per kg
c_{steel}	cost of steel per kg
c_T	torque constant
D	vector electric flux density
D	diameter
d_M	external diameter of PM
E	EMF, *rms* value; Young modulus
E_f	EMF per phase induced by the rotor without armature reaction

E_i internal EMF per phase

E_r resultant reactive EMF of self-induction and mutual induction of a short-circuited coil section during commutation

e instantaneous EMF; eccentricity

F force; MMF; energy functional; vector-optimization objective function

F_{exc} MMF of the rotor excitation system

F_a armature reaction MMF

f frequency

f_c frequency of cogging torque

f_r natural frequency of the rth order

$\mathcal{F}$ space and/or time distribution of the MMF

G permeance

$GCD(s_1, 2p)$ greatest common divisor of s_1 and $2p$

g air gap (mechanical clearance)

g_{My} air gap between PM and stator yoke in d.c. machines

g' equivalent air gap

g_i nonlinear inequality constraints

$\mathbf{H}$ vector magnetic field intensity

H magnetic field intensity

h height

h_i equality constraints

h_M height of the PM

I area moment of inertia; electric current; sound intensity

I_a armature d.c. or rms current

I_{ash} armature current at zero speed ("short circuit" current)

i instantaneous value of current or stepping motor current

i_a instantaneous value of armature current

$\mathbf{J}$ vector electric current density

J moment of inertia

J_a current density in the armature winding

K_r lumped stiffness

k coefficient, general symbol

k_{1R} skin effect coefficient for armature conductors

k_C Carter's coefficient

k_{ad} reaction factor in d-axis; coefficient of additional losses in armature core

k_{aq} reaction factor in q-axis

k_{d1} distribution factor for fundamental space harmonic $\nu = 1$

k_E EMF constant $k_E = c_E \Phi_f$

k_f form factor of the field excitation $k_f = B_{mg1}/B_{mg}$

k_{fault} fault-tolerant rating factor

$k_{fsk\mu}$ PM skew factor

k_i stacking factor of laminations

k_N coefficient depending on the number of manufactured machines

k_{o1} slot opening factor for fundamental space harmonic $\nu = 1$

k_{ocf}	overload capacity factor $k_{ocf} = P_{max}/P_{out}$
k_{p1}	pitch factor for fundamental space harmonic $\nu = 1$
k_{sat}	saturation factor of the magnetic circuit due to the main (linkage) magnetic flux
k_{skk}	stator slot skew factor referred to the slot (tooth) pitch t_1
$k_{sk\mu}$	stator slot skew factor referred to the pole pitch τ
k_T	torque constant $k_T = c_T\Phi_f$
k_{w1}	winding factor $k_{w1} = k_{d1}k_{p1}$ for fundamental space harmonic $\nu = 1$
L	inductance; length
$LCM(s_1, 2p)$	least common multiple of s_1 and $2p$
L_c	axial length of the interpole
L_i	armature stack effective length
L_w	sound power level
l_{1e}	length of the one-sided end connection
l_{Fe}	length of ferromagnetic yoke
l_M	axial length of PM
$\mathbf{M}$	magnetization vector
M	mutual inductance
M_b	ballistic coefficient of demagnetization
M_r	lumped mass
m	number of phases; mass
m_a	amplitude modulation index
m_f	frequency modulation index
N	number of turns per phase; number of machines
N_{cog}	number of poles–to–$GCD(s_1, 2p)$ ratio
n	rotational speed in rpm; independent variables
n_{cog}	fundamental cogging torque index
n_e	number of curvilinear squares between adjacent equipotential lines
n_0	no-load speed
n_Φ	number of curvilinear squares between adjacent flux lines
P	active power; acoustic power; probability
P_{elm}	electromagnetic power
ΔP	active power losses
$\Delta p_{1/50}$	specific core loss in W/kg at 1T and 50 Hz
p	number of pole pairs; sound pressure
p_r	radial force per unit area (magnetic pressure)
Q	electric charge, reactive power
Q_{en}	enclosed electric charge
R	resistance
R_a	armature winding resistance of d.c. commutator motors
R_1	armature winding resistance of a.c. motors
R_{br}	resistance of contact layer between brush and commutator
R_c	resistance of a coil section
Re	Reynolds number
R_{int}	interpole winding resistance

$R_{\mu M}$	permanent magnet reluctance
$R_{\mu g}$	air gap reluctance
$R_{\mu la}$	external armature leakage reluctance
r	vibration mode
S	apparent power; surface
S_M	cross section area of PM $S_M = w_M L_M$ or $S_M = b_p L_M$
s	cross section area; displacement
s_1	number of stator teeth or slots;
s_2	number of rotor teeth or slots;
s_e	estimate of the variance σ
T	torque
T_c	cogging torque
T_d	developed torque
T_{dsyn}	synchronous or synchronizing torque
T_{drel}	reluctance torque
T_0	constant of avarage component of the torque
T_p	period
T_r	periodic component of the torque
T_{sh}	shaft torque (output or load torque)
T_m	mechanical time constant
t	time; slot pitch; relative torque
t_N	normalized torque
t_r	torque ripple
V	electric voltage; volume
v	instantaneous value of electric voltage; linear velocity
v_C	commutator linear velocity
W	energy produced in outer space of PM; rate of change of the air gap energy
W_m	stored magnetic energy
w	energy per volume, J/m^3
w_M	width of PM
X	reactance
X_{ad}	d-axis armature reaction (mutual) reactance
X_{aq}	q-axis armature reaction (mutual) reactance
X_{damp}	reactance of damper
X_{sd}	d-axis synchronous reactance
X'_{sd}	d-axis transient synchronous reactance
X''_{sd}	d-axis subtransient synchronous reactance
X_{sq}	q-axis synchronous reactance
X'_{sq}	q-axis transient synchronous reactance
X''_{sq}	q-axis subtransient synchronous reactance
$\mathbf{Z}$	impedance $\mathbf{Z} = R + jX$; $\mid \mathbf{Z} \mid = Z = \sqrt{R^2 + X^2}$
z_1	number of teeth on wheel 1
z_2	number of teeth on wheel 2

α	electrical angle; control voltage–to–rated voltage ratio
α_d	angle between d-axis and y-axis
α_i	effective pole arc coefficient $\alpha_i = b_p/\tau$
β	overlap angle of pole
χ	magnetic susceptibility
γ	mechanical angle; gear ratio; form factor of demagnetization curve of PM material
γ_s	step angle of a rotary stepping motor
ΔV_{br}	voltage drop across commutation brushes
δ	power (load) angle; bias error
δ_i	inner torque angle
δl	learning rate
$\delta\epsilon$	random error
ϵ	relative eccentricity
η	efficiency
θ	rotor angular position for brushless motors
ϑ	temperature; angle between $\mathbf{I}_a$ and $\mathbf{I}_{ad}$
λ	coefficient of leakage permeance (specific leakage permeance); intensity of failures; wavelength
μ	number of the rotor μth harmonic
μ_{dyn}	dynamic viscosity
μ_0	magnetic permeability of free space $\mu_0 = 0.4\pi \times 10^{-6}$ H/m
μ_r	relative magnetic permeability
μ_{rec}	recoil magnetic permeability
μ_{rrec}	relative recoil permeability $\mu_{rrec} = \mu_{rec}/\mu_o$
ν	number of the stator νth harmonic; relative speed
ξ	coefficient of utilization; reduced height of the armature conductor
ρ	specific mass density
σ	electric conductivity
σ_f	form factor to include the saturation effect
σ_p	output coefficient
σ_r	radiation factor
τ	pole pitch
Φ	magnetic flux
Φ_f	excitation magnetic flux
Φ_l	leakage flux
ϕ	power factor angle
Ψ	flux linkage $\Psi = N\Phi$; angle between $\mathbf{I}_a$ and $\mathbf{E}_f$
Ψ_E	total electric flux
Ψ_{sd}	total flux linkage in d-axis
Ψ_{sq}	total flux linkage in q-axis
Ω	angular speed $\Omega = 2\pi n$
ω	angular frequency $\omega = 2\pi f$

Subscripts

a	armature
av	average
br	brush
c	commutation
cog	cogging
Cu	copper
d	direct axis; differential
dyn	dynamic
e	end connection; eddy-current
elm	electromagnetic
eq	equivalent
exc	excitation
ext	external
Fe	ferromagnetic
f	field
fr	friction
g	air gap
h	hysteresis
in	inner
l	leakage
l, m, n	labels of triangular element
M	magnet
m	peak value (amplitude)
n, t	normal and tangential components
out	output, outer
q	quadrature axis
r	rated; remanent
r, θ, x	cylindrical coordinate system
rel	reluctance
rhe	rheostat
rot	rotational
s	slot; synchronous; system
sat	saturation
sh	shaft
sl	sleeve; slot
st	starting
str	additional
syn	synchronous or synchronizing
t	teeth
u	useful
$vent$	ventilation
$wind$	windage
y	yoke

x, y, z cartesian coordinate system
1   primary; stator; fundamental harmonic
2   secondary; rotor

Superscripts

inc   incremental
(sq)   square wave

Abbreviations

A/D	analog to digital
ASM	additional synchronous motor
a.c.	alternating current
CAD	computer-aided design
CD	compact disk
CFRP	carbon fiber reinforced polymer
CLV	constant linear velocity
CPU	central processor unit
CSI	current source inverter
CVT	continuously variable transmission
DBO	double-beam oscilloscope
DFT	discrete Fourier transform
DSP	digital signal processor
d.c.	direct current
EIA	Energy Information Administration (Dept. of Energy, U.S.A.)
EMF	electromotive force
EMI	electromagnetic interference
EV	electric vehicle
EVOP	evolutionary operation
FDB	fluid dynamic bearing
FES	flywheel energy storage
FEM	finite element method
FFT	fast Fourier transform
GA	genetic algorithm
GCD	greatest common divisor
GFRP	glass fiber reinforced polymer
GS	generator/starter
GTO	gate turn-off (thyristor)
HDD	hard disk drive
HEV	hybrid electric vehicle
HVAC	heating, ventilating and air conditioning
IC	integrated circuit
IGBT	insulated-gate bipolar transistor
ISG	integrated starter-generator

LCM   least common multiple
LDDCM   liquid dielectric d.c. commutator motor
LIGA   litography (*Litographie*), electroforming (*Galvanoformung*) and molding (*Abformung*)
LVAD   left ventricular assist device
MEA   more electric aircraft
MEMS   micro electromechanical systems
MG   motor/starter
MMF   magnetomotive force
MRI   magnetic resonance imaging
MRSH   multiple-restart stochastic hill climbing
MTBF   mean time between failures
MTOE   million tons of oil equivalent
MTTF   mean time to failure
MVD   magnetic voltage drop
OECD   Organisation for Economic Co-operation and Development
PBIL   population-based incremental learning
PDF   probability density function
PFM   pulse frequency modulation
PLC   programmable logic controller
PM   permanent magnet
PSD   power split device
PWM   pulse width modulation
RESS   rechargeable energy storage system
RFI   radio frequency interference
RSM   response surface methodology
SA   simulated annealing
SCARA   selective compliance assembly robot arm
SRM   switched reluctance motor
SWL   sound power level
TDF   tip driven fan
TFM   transverse flux motor
TSM   tested synchronous motor
UPS   uninterruptible power supply
UV   underwater vehicles
URV   underwater robotic vehicle
VCM   voice coil motor
VSI   voltage source inverter
VSD   variable-speed drive
VVVF   variable voltage variable frequency

References

1. Ackermann B, Janssen JHH, Sottek R. New technique for reducing cogging torque in a class of brushless d.c. motors. IEE Proc Part B 139(4):315–320, 1992.
2. Afonin A, Kramarz W, Cierzniewski P. Electromechanical Energy Converters with Electronic Commutation (in Polish). Szczecin: Wyd Ucz PS, 2000.
3. Afonin A, Cierznewski P. Electronically commutated disc-type permanent magnet motors (in Russian). Int Conf on Unconventional Electromechanical and Electr Systems UEES'99. Sankt Petersburg, Russia, 1999, pp. 271–276.
4. Afonin A, Gieras JF, Szymczak P. Permanent magnet brushless motors with innovative excitation systems (invited paper). Int Conf on Unconventional Electromechanical and Electr Systems UEES'04. Alushta, Ukraine, 2004, pp. 27–38.
5. Ahmed AB, de Cachan LE. Comparison of two multidisc configurations of PM synchronous machines using an elementary approach. Int Conf on Electr Machines ICEM'94, Vol 1, Paris, France, 1994, pp. 175–180.
6. Aihara T, Toba A, Yanase T, Mashimo A, Endo K. Sensorless torque control of salient-pole synchronous motor at zero-speed operation. IEEE Trans on PE 14(1):202–208, 1999.
7. Altenbernd G, Mayer J. Starting of fractional horse-power single-phase synchronous motors with permanent magnetic rotor. Electr Drives Symp, Capri, Italy, 1990, pp. 131–137.
8. Altenbernd G, Wähner L. Comparison of fractional horse-power single-phase and three-phase synchronous motors with permanent magnetic rotor. Symp on Power Electronics, Electr Drives, Advanced Electr Motors SPEEDAM'92, Positano, Italy, 1992, pp. 379–384.
9. Andresen EC, Blöcher B, Heil J, Pfeiffer R. Permanentmagneterregter Synchronmotor mit maschinenkommutiertem Frequenzumrichter. etzArchiv (Germany) 9(12):399–402, 1987.
10. Andresen EC, Keller R. Comparing permanent magnet synchronous machines with cylindrical and salient-pole rotor for large power output drives. Int Conf on Electr Machines ICEM'94, Vol 1, Paris, France, 1994, pp. 316–321.
11. Andresen EC, Anders M. A three axis torque motor of very high steady-state and dynamic accuracy. Int Symp on Electr Power Eng, Stockholm, Sweden, 1995, pp. 304–309.

12. Andresen EC, Anders M. On the induction and force calculation of a three axis torque motor. Int Symp on Electromagn Fields ISEF'95, Thessaloniki, Greece, 1995, pp. 251–254.

13. Andresen EC, Keller R. Squirrel cage induction motor or permanent magnet synchronous motor. Symp on Power Electronics, Electr Drives, Advanced Electr Motors SPEEDAM'96, Capri, Italy, 1996.

14. Arena A, Boulougoura M, Chowdrey HS, Dario P, Harendt C, Irion KM, Kodogiannis V, Lenaerts B, Menciassi A, Puders R, Scherjon C, Turgis D. Intracorporeal Videoprobe (IVP), Medical and Care Compunetics 2, edited by L. Bos et al, Amsterdam: IOS Press, 2005, pp. 167–174.

15. Armensky EV, Falk GB. Fractional–Horsepower Electrical Machines. Moscow: Mir Publishers, 1978.

16. Arnold DP, Zana I, Herrault F, Galle P, Park JW, Das S, Lang, JH, Allen MG. Optimization of a microscale, axial-flux, permanent-magnet generator. 5th Int. Workshop Micro Nanotechnology for Power Generation and Energy Conversion Applications Power MEMS05, Tokyo, Japan, 2005, pp. 165–168.

17. Arnold DP, Das S, Park JW, Zana I, Lang JH, Allen MG. Design optimization of an 8-watt, microscale, axial flux permanent magnet generator. J of Microelectromech Microeng, 16(9):S290–S296, 2006.

18. Arshad WM, Bäckström T, Sadarangari C. Analytical design and analysis procedure for a transverse flux machine. Int Electr Machines and Drives Conf IEMDC'01, Cambridge, MA, USA, 2001, pp. 115-121.

19. Ashby MF. Material Selection in Mechanical Engineering, 3rd ed. Oxford: Butterworth-Heinemann, 2005.

20. Balagurov VA, Galtieev FF, Larionov AN. Permanent Magnet Electrical Machines (in Russian) Moscow: Energia, 1964.

21. Baluja S. Population-based incremental learning: a method for integrating genetic search based function optimization and competitive learning. Technical Report, Carnegie Mellon University, Pittsburgh, PA, USA, 1994.

22. Baudot JH. Les Machines Éléctriques en Automatique Appliqueé (in French). Paris: Dunod, 1967.

23. Bausch H. Large power variable speed a.c. machines with permanent magnets. Electr Energy Conf, Adelaide, Australia, 1987, pp. 265–271.

24. Bausch H. Large power variable speed a.c. machines with permanent magnet excitation. J of Electr and Electron Eng (Australia) 10(2):102–109, 1990.

25. Berardinis LA. Good motors get even better. Machine Design Nov 21:71–75, 1991.

26. Bertotti GA, Boglietti A, Chiampi M, Chiarabaglio D, Fiorillo F, Lazarri M. An improved estimation of iron losses in rotating electrical machines. IEEE Trans on MAG 27(6):5007–5009, 1991.

27. Bilgen E, Boulos R. Functional dependence of torque coefficient of coaxial cylinders on gap width and Reynolds numbers. Trans of ASME, J of Fluids Eng, Series I, 95(1):122-126, 1973.

28. Binns KJ, Chaaban FB, Hameed AAK. The use of buried magnets in high speed permanent magnet machines. Electr Drives Symposium EDS'90, Capri, Italy, 1990, pp. 145–149.

29. Binns KJ. Permanent magnet drives: the state of the art. Symp on Power Electronics, Electr Drives, Advanced Electr Motors SPEEDAM'94, Taormina, Italy, 1994, pp. 109–114.

30. Blackburn JL. Protective Relaying: Principles and Applications. New York: Marcel Dekker, 1987.

31. Blissenbach R, Schäfer U, Hackmann W, Henneberger G. Development of a transverse flux traction motor in a direct drive system. Int Conf on Electr Machines ICEM'00, Vol 3, Espoo, Finland, 2000, pp. 1457–1460.

32. Boglietti A, Pastorelli M, Profumo F. High speed brushless motors for spindle drives. Int Conf on Synchronous Machines SM100, Vol 3, Zürich, Switzerland, 1991, pp. 817–822.

33. Bolton MTW, Coleman RP. Electric propulsion systems — a new approach for a new millennium. Report Ministry of Defence, Bath, UK, 1999.

34. Boules N. Design optimization of permanent magnet d.c. motors. IEEE Trans on IA 26(4): 786–792, 1990.

35. Bowers B. The early history of electric motor. Philips Tech Review 35(4):77–95, 1975.

36. Bowes SR, Clark PR. Transputer-based harmonic elimination PWM control of inverter drives. IEEE Trans on IA 28(1):72–80, 1992.

37. Box GEP, Draper NR. Empirical Model-Building and Response Surfaces. New York: J Wiley and Sons, 1987.

38. Box MJ, Davies D, Swann WH. Non-Linear Optimization Techniques, 1st ed. London: Oliver and Boyd, 1969.

39. Braga G, Farini A, Manigrasso R. Synchronous drive for motorized wheels without gearbox for light rail systems and electric cars. 3rd European Power Electronic Conf EPE'91, Vol. 4, Florence, Italy, 1991, pp. 78–81.

40. Brandisky K, Belmans R, Pahner U. Optimization of a segmental PM d.c. motor using FEA — statistical experiment design method and evolution strategy. Symp on Power Electronics, Electr Drives, Advanced Electr Motors SPEEDAM'94, Taormina, Italy, 1994, pp. 7–12.

41. Brauer JR, ed. What Every Engineer Should Know about Finite Element Analysis. New York: Marcel Dekker, 1988.

42. Breton C, Bartolomé J, Benito JA, Tassinario G, Flotats I, Lu CW, Chalmers BJ. Influence of machine symmetry on reduction of cogging torque in permanent magnet brushless motors. IEEE Trans on MAG 36(5):3819–3823, 2000.

43. Briggs SJ, Savignon DJ, Krein PT, Kim MS. The effect of nonlinear loads on EMI/RFI filters. IEEE Trans on IA 31(1):184–189, 1995.

44. Broder DW, Burney DC, Moore SI, Witte SB. Media path for a small, low-cost, color thermal inkjet printer. Hewlett-Packard J:72–78, February 1994.

45. Brunsbach BJ, Henneberger G, Klepach T. Compensation of torque ripple. Int Conf on Electr Machines ICEM'98, Istanbul, Turkey, 1998, pp. 588–593.

46. Bryzek J, Petersen K, McCulley W. Micromachines on the march. IEEE Spectrum 31(5):20–31, 1994.

47. Cai W, Fulton D, Reichert K. Design of permanent magnet motors with low torque ripples: a review. Int Conf on Electr Machines ICEM'00, Vol 3, Espoo, Finland, 2000, pp. 1384–1388.

48. Campbell P. Performance of a permanent magnet axial-field d.c. machine. IEE Proc Pt B 2(4):139–144, 1979.

49. Cardone G, Astarita T, Carlomagno GM. Infrared heat transfer measurements on a rotating disk. Optical Diagnostics in Eng 1(2):1–7, 1996.

50. Caricchi F, Crescembini F, and E. Santini E. Basic principle and design criteria of axial-flux PM machines having counterrotating rotors. IEEE Trans on IA 31(5):1062–1068, 1995.

51. Caricchi F, Crescimbini F, Honorati O. Low-cost compact permanent magnet machine for adjustable-speed pump application. IEEE Trans on IA 34(1):109–116, 1998.

52. Carlson R, Lajoie-Mazenc M, Fagundes JCS. Analysis of torque ripple due to phase commutation in brushless d.c. machines. IEEE Trans on IA 28(3):632–638, 1992.

53. Carter GW. The Electromagnetic Field in its Engineering Aspects. London: Longmans, 1962.

54. Cascio AM. Modeling, analysis and testing of orthotropic stator structures. Naval Symp on Electr Machines, Newport, RI, USA, 1997, pp. 91–99.

55. Cerruto E, Consoli A, Raciti A, Testa A. Adaptive fuzzy control of high performance motion systems. Int Conf on Ind Electronics, Control, Instr and Automation IECON'92, San Diego, CA, USA, 1992, pp. 88–94.

56. Chang L. Comparison of a.c. drives for electric vehicles — a report on experts' opinion survey. IEEE AES Systems Magazine 8:7–11, 1994.

57. Changzhi S, Likui Y, Yuejun A, Xiying D. The combination of performance simulation with CAD used in ocean robot motor. Int Aegean Conf on Electr Machines and Power Electronics ACEMP'95, Kuşadasi, Turkey, 1995, pp. 692–695.

58. Chalmers BJ, Hamed SA, Baines GD. Parameters and performance of a high-field permanent magnet synchronous motor for variable-frequency operation. Proc IEE Pt B 132(3):117–124, 1985.

59. Chari MVK, Silvester PP. Analysis of turboalternator magnetic fields by finite elements. IEEE Trans on PAS 92:454–464, 1973.

60. Chari MVK, Csendes ZJ, Minnich SH, Tandon SC, Berkery J. Load characteristics of synchronous generators by the finite-element method. IEEE Trans on PAS 100(1):1–13, 1981.

61. Chen SX, Low TS, Lin H, Liu ZJ. Design trends of spindle motors for high performance hard disk drives. IEEE Trans on MAG 32(5):3848–3850, 1996.

62. Chidambaram B. Catalog-Based Customization. PhD dissertation, University of California, Berkeley, CA, USA, 1997.

63. Chillet C, Brissonneau P, Yonnet JP. Development of a water cooled permanent magnet synchronous machine. Int Conf on Synchr Machines SM100, Vol 3, Zürich, Switzerland, 1991, pp. 1094–1097.

64. Cho CP, Lee CO, Uhlman J. Modeling and simulation of a novel integrated electric motor/propulsor for underwater propulsion. Naval Symp on Electr Machines, Newport, RI, USA, 1997, pp. 38–44.

65. Chong AH, Yong KJ, Allen MG. A planar variable reluctance magnetic micromotor with fully integrated stator and coils. J Microelectromechanical Systems 2(4): 165–173, 1993.

66. Christensen GJ. Are electric handpieces an improvement? J of Amer Dental Assoc, 133(10):1433–1434, 2002.

67. Ciurys M, Dudzikowski I. Brushless d.c. motor tests (in Polish). Zeszyty Probl Komel – Maszyny Elektr 83:183–188, 2009.

68. Cohon JL. Multiobjective Programming and Planning. New York: Academic Press, 1978.

69. Coilgun research spawns mighty motors and more. Machine Design 9(Sept 24):24–25, 1993.

70. Compumotor Digiplan: Positioning Control Systems and Drives. Parker Hannifin Corporation, Rohnert Park, CA, USA, 1991.

71. Consoli A, Abela A. Transient performance of permanent magnet a.c. motor drives. IEEE Trans on IA 22(1):32–41, 1986.

72. Consoli A, Testa A. A DSP sliding mode field oriented control of an interior permanent magnet motor drive. Int Power Electronics Conf, Tokyo, Japan, 1990, pp. 296–303.

73. Consoli A, Musumeci S, Raciti A, Testa A. Sensorless vector and speed control of brushless motor drives. IEEE Trans on IE 41(1):91–96, 1994.

74. Consterdine E, Hesmondhalgh DE, Reece ABJ, Tipping D. An assessment of the power available from a permanent magnet synchronous motor which rotates at 500,000 rpm. Int Conf on Electr Machines ICEM'92, Manchester, UK, 1992, pp. 746–750.

75. Cremer R. Current status of rare-earth permanent magnets. Int Conf on Maglev and Linear Drives, Hamburg, Germany, 1988, pp. 391–398.

76. Dąbrowski M. Magnetic Fields and Circuits of Electrical Machines (in Polish). Warsaw: WNT, 1971.

77. Dąbrowski M. Construction of Electrical Machines (in Polish). Warsaw: WNT, 1977.

78. Dąbrowski M. Joint action of permanent magnets in an electrical machine (in Polish). Zeszyty Nauk Polit Pozn Elektryka 21:7–17, 1980.

79. DeGarmo EP, Black JT, Kosher RA. Materials and Processes in Manufacturing. New York: Macmillan, 1988.

80. De La Ree J, Boules N. Torque production in permanent magnet synchronous motors. IEEE Trans on IA 25(1):107–112, 1989.

81. Demenko A. Time stepping FE analysis of electric motor drives with semiconductor converters. IEEE Trans on MAG 30(5):3264–3267, 1994.

82. Demerdash NA, Hamilton HB. A simplified approach to determination of saturated synchronous reactances of large turboalternators under load. IEEE Trans on PAS 95(2):560–569, 1976.

83. Demerdash NA, Fouad FA, Nehl TW. Determination of winding inductances in ferrite type permanent magnet electric machinery by finite elements. IEEE Trans on MAG 18(6):1052–1054, 1982.

84. Demerdash NA, Hijazi TM, Arkadan AA. Computation of winding inductances of permanent magnet brushless d.c. motors with damper windings by energy perturbation. IEEE Trans EC 3(3):705–713, 1988.

85. Deodhar RP, Staton DA, Jahns TM, Miller TJE. Prediction of cogging torque using the flux-MMF diagram technique. IEEE Trans on IAS 32(3):569–576, 1996.

86. Dreyfus L. Die Theorie des Drehstrommotors mit Kurzschlussanker. Handlikar 34, Stockholm, Ingeniors Vetenkaps Akademien, 1924.

87. Dudzikowski I, Kubzdela S. Thermal problems in permanent magnet commutator motors (in Polish). 23rd Int Symp on Electr Machines SME'97, Poznan, Poland, pp. 133–138, 1997.

88. Dunkerley S. On the whirling and vibration of shafts. Proc of the Royal Soc of London 54:365–370, 1893.

89. Digital signal processing solution for permanent magnet synchronous motor: application note. Texas Instruments, USA, 1996.

90. Dote Y, Kinoshita S. Brushless Servomotors: Fundamentals and Applications. Oxford: Clarendon Press, 1990.

91. Drozdowski P. Equivalent circuit and performance characteristics of 9-phase cage induction motor. Int Conf on Electr Machines ICEM'94 Vol 1, Paris, France, 1994, pp. 118–123.

92. Drozdowski P. Some circumstances for an application of the 9-phase induction motor to the traction drive. 2nd Int Conf on Modern Supply Systems and Drives for Electr Traction, Warsaw, Poland, 1995, pp. 53–56.

93. Ede JD, Jewell GW, Atallah K, Powel DJ, Cullen JJA, Mitcham AJ. Design of a 250-kW, fault-tolerant PM generator for the more-electric aircraft. 3rd Int Energy Conv Eng Conf, San Francisco, CA, USA, 2005, AIAA 2005–5644.

94. Edwards JD, Freeman EM. MagNet 5.1 User Guide. Using the MagNet Version 5.1 Package from Infolytica. London: Infolytica, 1995.

95. Eisen HJ, Buck CW, Gillis-Smith GR, Umland JW. Mechanical design of the Mars Pathfinder Mission. 7th European Space Mechanisms and Tribology Symp ESTEC'97, Noordwijk, Netherlands, 1997, pp. 293–301.

96. Elmore WA, ed. Protective relaying: theory and applications. New York: Marcel Dekker, 1994.

97. Engström J. Design of a slotless PM motor for a screw compressor drive. 9th Int Conf on Electr Machines and Drives (Conf. Publ. No. 468) EMD'99, Canterbury, UK, 1999, pp. 154–158.

98. Engelmann RH, Middendorf WH, ed. Handbook of Electric Motors. New York: Marcel Dekker, 1995.

99. Eriksson S. Drive systems with permanent magnet synchronous motors. Automotive Engineering 2:75–81, 1995.

100. Ermolin NP. Calculations of Small Commutator Machines (in Russian). Sankt Petersburg: Energia, 1973.

101. Ertugrul N, Acarnley P. A new algorithm for sensorless operation of permanent magnet motors. IEEE Trans on IA 30(1):126–133, 1994.

102. Favre E, Cardoletti L, Jufer M. Permanent magnet synchronous motors: a comprehensive approach to cogging torque suppression. IEEE Trans on IA 29(6):1141–1149, 1993.

103. Ferreira da Luz MV, Batistela NJ, Sadowski N, Carlson R, Bastos JPA. Calculation of losses in induction motors using the finite element method. Int Conf on Electr Machines ICEM'2000 Vol 3, Espoo, Finland, 2000, pp. 1512–1515.

104. Fiorillo F, Novikov A. An approach to power losses in magnetic laminations under nonsinusoidal induction waveform. IEEE Trans on MAG 26(5):2904–2910, 1990.

105. Fitzgerald AE, Kingsley C. Electric Machinery, 2nd ed. New York: McGraw-Hill, 1961.

106. Fletcher R. Practical Methods of Optimization, 2nd ed. New York: J Wiley and Sons, 1987.

107. Fox RL. Optimization Methods for Engineering Design. London: Addison-Wesley, 1971.

108. Fouad FA, Nehl TW, Demerdash NA. Magnetic field modeling of permanent magnet type electronically operated synchronous machines using finite elements. IEEE Trans on PAS 100(9):4125–4133, 1981.

109. Fracchia M, Sciutto G. Cycloconverter drives for ship propulsion. Symp on Power Electronics, Electr Drives, Advanced Electr Motors SPEEDAM'94, Taormina, Italy, 1994, pp. 255–260.

110. Freise W. Jordan H. Einsertige magnetische Zugkräfte in Drehstrommaschinen. ETZ: Elektrische Zeitschrift, Ausgabe A 83:299–303, 1962.

111. Fuchs EF, Erdélyi EA. Determination of waterwheel alternator steady-state reactances from flux plots. IEEE Trans on PAS 91:2510–2527, 1972.

112. Furlani EP. Computing the field in permanent magnet axial-field motors. IEEE Trans on MAG 30(5):3660–3663, 1994.

113. Gair S, Eastham JF, Profumo F. Permanent magnet brushless d.c. drives for electric vehicles. Int Aeagean Conf on Electr Machines and Power Electronics ACEMP'95, Kuşadasi, Turkey, 1995, pp. 638–643.

114. Gerald CF, Wheatley PO. Applied Numerical Analysis. London: Addison-Wesley, 1989.

115. Gieras JF. Performance calculation for small d.c. motors with segmental permanent magnets. Trans. of SA IEE 82(1):14–21, 1991.

116. Gieras JF, Moos EE, Wing M. Calculation of cross MMF of armature winding for permanent magnet d.c. motors. Proc of South African Universities Power Eng Conf SAUPEC'91, Johannesburg, South Africa, 1991, pp. 273–279.

117. Gieras JF, Wing M. The comparative analysis of small three-phase motors with cage, cylindrical steel, salient pole, and permanent magnet rotors. Symp of Power Electronics, Electr Drives, Advanced Electr Motors SPEEDAM'94, Taormina, Italy, 1994, pp. 59–64.

118. Gieras JF, Kileff I, Wing M. Investigation into an electronically-commutated d.c. motor with NdFe permanent magnets. MELECON'94, Vol 2, Antalya, Turkey, 1994, pp. 845–848.

119. Gieras JF, Wing M. Design of synchronous motors with rare-earth surface permanent magnets. Int Conf on Electr Machines ICEM'94, Vol 1, Paris, France, 1994, pp. 159–164.

120. Gieras JF, Santini E, Wing M. Calculation of synchronous reactances of small permanent magnet alternating-current motors: comparison of analytical approach and finite element method with measuremets. IEEE Trans on MAG 34(5):3712–3720, 1998.

121. Gieras JF, Wang C, Lai JC. Noise of Polyphase Electric Motors. Boca Raton: CRC Taylor & Francis, 2006.

122. Gieras JF, Wang RJ, Kamper MJ. Axial Flux Permanent Magnet Brushless Machines, 2nd ed. London: Springer, 2008.

123. Gieras JF, Koenig AC, Vanek LD. Calculation of eddy current losses in conductive sleeves of synchronous machines. Int Conf on Electr Machines ICEM'08, Vilamoura, Portugal, 1998, paper ID 1061.

124. Gieras JF. Advancements in Electric Machines. London: Springer, 2008.

125. Gilbert W. De Magnete, Magneticisque Corporibus et de Magno Magnete Tellure (On the Magnet, Magnetic Bodies and on the Great Magnet the Earth). London: 1600 (translated 1893 by Mottelay PF, Dover Books).

126. Glinka T. Electrical Micromachines with Permanent Magnet Excitation (in Polish). Gliwice (Poland): Silesian Techn University, 1995.

127. Glinka T, Grzenik R, Mołoń Z. Drive system of a wheelchair (in Polish). 2nd Int Conf on Modern Supply Systems and Drives for Electr Traction, Warsaw, Poland, 1995, pp. 101–105.

128. Gogolewski Z, Gabryś W. Direct Current Machines (in Polish). Warsaw: PWT, 1960.

129. Goldemberg C, Lobosco OS. Prototype for a large converter-fed permanent magnet motor. Symp on Power Electronics, Electr Drives, Advanced Electr Motors SPEEDAM'92, Positano, Italy, 1992, pp. 93–98.

130. Gottvald A, Preis K, Magele C, Biro O, Savini A. Global optimization methods for computational electromagnetics. IEEE Trans on MAG 28(2):1537–1540, 1992.

131. Greenwood R. Automotive and Aircraft Electricity. Toronto: Sir Isaac Pitman, 1969.

132. Grumbrecht P, Shehata MA. Comparative study on different high power variable speed drives with permanent magnet synchronous motors. Electr Drives Symp EDS'90, Capri, Italy, 1990, pp. 151–156.

133. Hague B. The principles of electromagnetism applied to electrical machines. New York: Dover Publications, 1962.

134. Hakala H. Integration of motor and hoisting machine changes the elevator business. Int Conf on Electr Machines ICEM'00 Vol 3, Espoo, Finland, 2000, pp. 1242–1245.

135. Halbach K. Design of permanent multipole magnets with oriented rare earth cobalt material. Nuclear Instruments and Methods, 169:1–10, 1980.

136. Halbach K. Physical and optical properties of rare earth cobalt magnets. Nuclear Instruments and Methods 187:109–117, 1981.

137. Halbach K. Application of permanent magnets in accelerators and electron storage rings. J Appl Physics 57:3605–3608, 1985.

138. Hanitsch R, Belmans R, Stephan R. Small axial flux motor with permanent magnet excitation and etched air gap winding. IEEE Trans on MAG 30(2):592–594, 1994.

139. Hanitsch R. Microactuators and micromotors — technologies and characteristics. Int Conf on Electr Machines ICEM'94, Vol 1, Paris, France, 1994, pp. 20–27.

140. Hanitsch R. Microactuators and micromotors. Int Aeagean Conf on Electr Machines and Power Electronics ACEMP'95, Kuşadasi, Turkey, 1995, pp. 119–128.

141. Hanselman DC. Effect of skew, pole count and slot count on brushless motor radial force, cogging torque and back EMF. IEE Proc Part B 144(5):325–330, 1997.

142. Hanselman DC. Brushless Permanent-Magnet Motor Design, 2nd ed. Cranston, RI: The Writers' Collective, 2003.

143. Hardware interfacing to the TMS320C25, Texas Instruments.

144. Heller B, Hamata V. Harmonic Field Effect in Induction Machines. Prague: Academia (Czechoslovak Academy of Sciences), 1977.

145. Henneberger G, Schustek S. Wirtz R. Inverter-fed three-phase drive for hybrid vehicle applications. Int Conf on Electr Machines ICEM'88 Vol 2, Pisa, Italy, 1988, pp. 293–298.

146. Henneberger G, Bork M. Development of a new transverse flux motor. IEE Colloquium on New Topologies of PM Machines, London, UK, 1997, pp. 1/1–1/6.

147. Henneberger G, Viorel IA, Blissenbach R. Single-sided transverse flux motors. Int Conf Power Electronics and Motion Control EPE-PEMC'00, Vol 1, Košice, Slovakia, 2000, pp. 19–26.

148. Hendershot JH, Miller TJE. Design of Brushless Permanent Magnet Motors. Oxford: Clarendon Press, 1994.

149. Hesmondhalgh DE, Tipping D. Slotless construction for small synchronous motors using samarium cobalt magnets. IEE Proc Pt B 129(5):251–261, 1982.

150. Holland JH. Adaption in Natural and Artificial Systems, 3rd ed. New York: Bradford Books, 1994.

151. Honorati O, Solero L, Caricchi F, Crescimbini F. Comparison of motor drive arrangements for single-phase PM motors. Int Conf on Electr Machines ICEM'98 Vol 2, Istanbul, Turkey, 1998, pp. 1261–1266.

152. Honsinger VB. Performance of polyphase permanent magnet machines. IEEE Trans on PAS 99(4):1510–1516, 1980.

153. Honsinger VB. Permanent magnet machines: asynchronous operation. IEEE Trans on PAS 99(4):1503–1509, 1980.

154. Horber R. Permanent-magnet steppers edge into servo territory. Machine Design, 12:99–102, 1987.

155. Hrabovcová V, Brślica V. Equivalent circuit parameters of disc synchronous motors with PMs. Electr Drives and Power Electronics Symp EDPE'92, Košice, Slovakia, 1992, pp. 348–353.

156. Huang DR, Fan CY, Wang SJ, Pan HP, Ying TF, Chao CM, Lean EG. A new type single-phase spindle motor for HDD and DVD. IEEE Trans on MAG 35(2): 839–844, 1999.

157. Hughes A, Lawrenson PJ. Electromagnetic damping in stepping motors. Proc IEE 122(8):819–824, 1975.

158. International Energy Outlook 2008. Report No DOE/EIA-0484(2008), www.eia.doe.gov/oiaf/ieo

159. Ishikawa T, Slemon G. A method of reducing ripple torque in permanent magnet motors without skewing. IEEE Trans on MAG 29(3):2028–2033, 1993.

160. Ivanuskin VA, Sarapulov FN, Szymczak P. Structural Simulation of Electromechanical Systems and their Elements (in Russian). Szczecin (Poland): Wyd Ucz PS, 2000.

161. Jabbar MA, Tan TS, Binns KJ. Recent developments in disk drive spindle motors. Int Conf on Electr Machines ICEM'92, Vol 2, Manchester, UK, 1992, pp. 381–385.

162. Jabbar MA, Tan TS, Yuen WY. Some design aspects of spindle motors for computer disk drives. J Inst Electr Eng (Singapore) 32(1):75–83, 1992.

163. Jabbar MA. Torque requirement in a disc-drive spindle motor. Int Power Eng Conf IPEC'95, Vol 2, Singapore, 1995, pp. 596–600.

164. Jahns TM. Torque production in permanent-magnet synchronous motor drives with rectangular current excitation. IEEE Trans on IA 20(4):803–813, 1984.

165. Jahns TM, Kliman GB, Neumann TW. Interior PM synchronous motors for adjustable-speed drives. IEEE Trans on IA 22(4):738–747, 1986.

166. Jahns TM. Motion control with permanent magnet a.c. machines. Proc IEEE 82(8):1241–1252, 1994.

167. Jones BL, Brown JE. Electrical variable-speed drives. IEE Proc Pt A 131(7):516–558, 1987.

168. Kamiya M. Development of traction drive motors for the Toyota hybrid system. IEEJ Trans on IA (Japan) 126(4):473–479, 2006.

169. Kaneyuki K, Koyama M. Motor-drive control technology for electric vehicles. Mitsubishi Electric Advance (3):17–19, 1997.

170. Kasper M. Shape optimization by evolution strategy. IEEE Trans on MAG 28(2):1556–1560, 1992.

171. Kawashima K, Shimada A. Spindle motors for machine tools. Mitsubishi Electric Advance, 2003,Sept, pp. 17-19.

172. Kenjo T, Nagamori S. Permanent Magnet and Brushless d.c. Motors. Oxford: Clarendon Press, 1985.
173. Kenjo T. Power Electronics for the Microprocessor Era. Oxford: OUP, 1990.
174. Kenjo T. Stepping Motors and their Microprocessor Control. Oxford: Clarendon Press, 1990.
175. Kenjo T. Electric Motors and their Control. Oxford: OUP, 1991.
176. Kiley J, Tolikas M. Design of a 28 hp, 47,000 rpm permanent magnet motor for rooftop air conditioning. www.satcon.com
177. King RD, Haefner KB, Salasoo L, Koegl RA. Hybrid electric transit bus pollutes less, conserves fuel. IEEE Spectrum 32(7): 26–31, 1995.
178. Kleen S, Ehrfeld W, Michel F, Nienhaus M, Stölting HD. Penny-motor: A family of novel ultraflat electromagnetic micromotors. Int Conf Actuator'00, Bremen, Germany, 193–196, 2000.
179. Klein FN, Kenyon ME. Permanent magnet d.c. motors design criteria and operation advantages. IEEE Trans on IA 20(6):1525–1531, 1984.
180. Klug L. Axial field a.c. servomotor. Electr Drives and Power Electronics Symp EDPE'90, Košice, Slovakia, 1990, pp. 154–159.
181. Klug L. Synchronous servo motor with a disk rotor (in Czech). Elektrotechnický Obzor 80(1–2):13–17, 1991.
182. Klug L, Guba R. Disc rotor a.c. servo motor drive. Electr Drives and Power Electronics Symp EDPE'92, Košice, Slovakia, 1992, pp. 341–344.
183. Koh CS. Magnetic pole shape optimization of permanent magnet motor for reduction of cogging torque. IEEE Trans on MAG, 33(2):1822–1827, 1997.
184. Korane KJ. Replacing the human heart. Machine Design 9(Nov 7):100–105, 1991.
185. Kostenko M, Piotrovsky L. Electrical Machines. Vol 1: Direct Current Machines and Transformers. Vol 2: Alternating Current Machines. Moscow: Mir Publishers, 1974.
186. Kozłowski HS, Turowski E. Induction Motors: Design, Construction, Manufacturing (in Polish). Warsaw: WNT, 1961.
187. Kreith F. Convection heat transfer in rotating systems. Advances in Heat Transfer Vol. 1. New York: Academic Press, 1968, pp. 129–251.
188. Kumada M, Iwashita Y, Aoki M, Sugiyama E. The strongest permanent dipole magnets. Particle Accelerator Conf 2003, pp. 1993–1995.
189. Kumar P, Bauer P. Improved analytical model of a permanent-magnet brushless d.c. motor. IEEE Trans on MAG 44(10):2299–2309, 2008.
190. Kurihara K, Rahman, MA. High efficiency line-start permanent magnet motor. IEEE Trans on IAS, 40(3):789–796, 2004.
191. Kurtzman GM. Electric handpieces: an overview of current technology. Inside Dentistry February 2007, pp. 88–90.
192. Lammeraner J, Štafl M. Eddy Currents. London: Iliffe Books, 1964.
193. Lange A, Canders WR, Laube F, Mosebach H. Comparison of different drive systems for a 75 kW electrical vehicles drive. Int Conf on Electr Machines ICEM'00, Vol 3, Espoo, Finland, 2000, pp. 1308–1312.
194. Lange A, Canders WR, Mosebach H. Investigation of iron losses of soft magnetic powder components for electrical machines, Int Conf on Electr Machines ICEM'00, Vol 3, Espoo, Finland, 2000, pp. 1521–1525.
195. Larminie J, Lowry J. Electric Vehicle Technology, New York: J Wiley and Sons, 2003.

196. Lavers JD, Biringer PP. Prediction of core losses for high flux densities and distorted flux waveforms. IEEE Trans on MAG 12(6):1053–1055, 1976.
197. Li T, Slemon G. Reduction of cogging torque in permanent magnet motors. IEEE Trans on MAG 24(6):2901–2903, 1988.
198. Linear Interface ICs. Device Data Vol 1. Motorola, 1993.
199. Lobosco OS, Jordao RG. Armature reaction of large converter-fed permanent magnet motor. Int Conf on Electr Machines ICEM'94 Vol 1, Paris, France, 1994, pp. 154–158.
200. Lodochnikov EA, Serov AB, Vialykh VG. Some problems of reliability calculation of small electric motors with built-in gears (in Russian). Reliability and Quality of Small Electrical Machines. St Petersburg: Nauka (Academy of Sciences of USSR), 1971, pp. 29–42.
201. Low TS, Jabbar MA, Rahman MA. Permanent-magnet motors for brushless operation. IEEE Trans on IA, 26(1):124–129, 1990.
202. Lowther DA, Silvester PP. Computer-Aided Design in Magnetics. Berlin: Springer Verlag, 1986.
203. Lukaniszyn M, Wróbel R, Mendrela A, Drzewoski R. Towards optimisation of the disc-type brushless d.c. motor by changing the stator core structure. Int Conf on Electr Machines ICEM'00, Vol 3, Espoo, Finland, 2000, pp. 1357-1360.
204. Magnetfabrik Schramberg GmbH & Co, Schramberg–Sulgen, 1989.
205. Magureanu R, Kreindler L, Giuclea D, Boghiu D. Optimal DSP control of brushless d.c. servosystems. Symp on Power Electronics, Electr Drives, Advanced Electr Motors SPEEDAM'94, Taormina, Italy, 1994, pp. 121–126.
206. Mallinson JC. One-sided fluxes — A magnetic curiosity? IEEE Trans on MAG, 9(4):678–682, 1973.
207. Marinescu M, Marinescu N. Numerical computation of torques in permanent magnet motors by Maxwell stress and energy method. IEEE Trans on MAG 24(1):463–466, 1988.
208. Marshall SV, Skitek GG. Electromagnetic Concepts and Applications. Englewood Cliffs: Prentice-Hall, 1987.
209. Maxon Motors. Sachseln, Switzerland: Interelectric AG, 1991/92.
210. Matsuoka K, Kondou K. Development of wheel mounted direct drive traction motor. RTRI Report (Kokubunji-shi, Tokyo, Japan) 10(5):37–44, 1996.
211. Mayer J. A big market for micromotors. Design News 3(March 27):182, 1995.
212. McNaught C. Running smoothly — making motors more efficient. IEE Review 39(2): 89–91, 1993.
213. Mecrow BC, Jack AG, Atkinson DJ, Green S, Atkinson GJ, King A, Green B. Design and testing of a 4 phase fault tolerant permanent magnet machine for an engine fuel pump. Int Electr Machines and Drives Conf IEMDC'03, Madison, WI, USA, 2007, pp. 1301-1307.
214. Mellara B, Santini E. FEM computation and optimization of L_d and L_q in disc PM machines. 2nd Int Workshop on Electr and Magn Fields, Leuven, Belgium, 1994, Paper No. 89.
215. Merrill FW. Permanent magnet excited synchronous motors. AIEE Trans 72 Part III(June):581–585, 1953.
216. Mhango LMC. Advantages of brushless d.c. motor high-speed aerospace drives. Int Conf on Synchronous Machines SM100, Zürich, Switzerland, 1991, 829–833.
217. Miller I, Freund JE. Probability and Statistics for Engineers. 3rd ed. Englewood Cliffs: Prentice-Hall, 1977.

218. Miller TJE. Brushless Permanent-Magnet and Reluctance Motor Drives. Oxford: Clarendon Press, 1989.

219. Miniature motors. Portescap: A Danaher Motion Company, West Chester, PA, USA, www.portescap.com

220. Mitcham AJ, Bolton MTW. The transverse flux motor: a new approach to naval propulsion. Naval Symp on Electr Machines, Newport, RI, USA, 1997, pp. 1–8.

221. Megaperm 40L. Vacuumschmelze GmbH & Co. KG, Hanau, Germany, www.vacuumschmelze.com

222. Mohan N, Undeland TM, Robbins WP. Power Electronics Converters Applications and Design. New York: J Wiley and Sons, 1989.

223. Mongeau P. High torque/high power density permanent magnet motors. Naval Symp on Electr Machines, Newport, RI, USA, 1997, pp. 9–16.

224. Morimoto S, Takeda Y, Hatanaka K, Tong Y, Hirasa T. Design and control system of inverter driven permanent magnet synchronous motors for high torque operation. IEEE Trans on IA 29(6):1150–1155, 1993.

225. Nasar SA, Boldea I, Unnewehr LE. Permanent Magnet, Reluctance, and Self-Synchronous Motors. Boca Raton: CRC Press, 1993.

226. Neale MJ, ed. Bearings: A Tribology Handbook. Oxford: Butterworth-Heinemann, 1993.

227. Nehl TW, Fouad FA, Demerdash NA. Determination of saturated values of rotating machinery incremental and apparent inductances by an energy perturbation method. IEEE Trans on PAS 101(12):4441–4451, 1982.

228. Nehl TW, Pawlak AM, Boules NM. ANTIC85: A general purpose finite element package for computer aided design and analysis of electromagnetic devices. IEEE Trans MAG 24(1):358–361, 1988.

229. Neves CGC, Carslon R, Sadowski, N, Bastos JPA, Soeiro NS, Gerges SNY. Calculation of electromechanic-mechanic-acoustic behavior of a switched reluctance motor. IEEE Trans on MAG 36(4):1364–1367, 2000.

230. Newbury RA. Prediction of losses in silicon steel from distorted waveforms. IEEE Trans on MAG 14(4):263–268, 1978.

231. Norman HM. Induction motor locked saturation curves. AIEE Trans Electr Eng (April):536–541, 1934.

232. Odor F, Mohr A. Two-component magnets for d.c. motors. IEEE Trans on MAG 13(5):1161–1162, 1977.

233. Osin IL, Kolesnikov VP, Yuferov FM. Permanent Magnet Synchronous Micromotors (in Russian). Moscow: Energia, 1976.

234. Oyama J, Higuchi T, Abe T, Shigematsu K, Yang X, Matsuo E. A trial production of small size ultra-high speed drive system. Dept of Electr and Electronic Eng, Nagasaki University, Japan.

235. Ozenbaugh RL. EMI Filter Design. New York: Marcel Dekker, 1995.

236. Parker RJ. Advances in Permanent Magnetism. New York: J Wiley and Sons, 1990.

237. Patterson D, Spée R. The design and development of an axial flux permanent magnet brushless d.c. motor for wheel drive in a solar powered vehicle. IEEE Trans on Ind Appl 31(5): 1054–1061, 1995.

238. Patterson ML, Haselby RD, Kemplin RM. Speed, precision, and smoothness characterize four-color plotter pen drive system. Hewlett Packard Journal 29(1):13–19, 1977.

239. Pavlik D, Garg VK, Repp JR, Weiss J. A finite element technique for calculating the magnet sizes and inductances of permanent magnet machines. IEEE Trans EC 3(1):116–122, 1988.

240. Pawlak AM, Graber DW, Eckhard DC. Magnetic power steering assist system – Magnasteer. SAE Paper 940867, 1994.

241. Pawlak AM. Magnets in modern automative applications. Gorham Conf on Permanent Magnet Systems, Atlanta, GA, USA, 2000.

242. Pawlak AM. Sensors and Actuators in Mechatronics: Design and Applications. Boca Raton: CRC Taylor & Francis, 2006.

243. Penman J, Dey MN, Tait AJ, Bryan WE. Condition monitoring of electrical drives. IEE Proc Pt B 133(3):142–148, 1986.

244. Pichler M, Tranter J. Computer based techniques for predictive maintenance of rotating machinery. Electr Energy Conf, Adelaide, Australia, 1987, pp. 226–226.

245. Pillay P, Krishnan R. An investigation into the torque behavior of a brushless d.c. motor. IEEE IAS Annual Meeting, New York, 1988, pp. 201–208.

246. Pillay P, Krishnan R. Modelling, simulation, and analysis of permanent magnet motor drives, Part 1 and 2. IEEE Trans on IA 25(2):265–279, 1989.

247. Pillay P, Krishman R. Application characteristics of permanent magnet synchronous and brushless d.c. motors for servo drives. IEEE Trans IA, 27:986–996, 1991.

248. Polkowski JW. Turbulent flow between coaxial cylinders with the inner cylinder rotating. ASME Trans 106:128–135, 1984.

249. Power IC's databook. National Semiconductor, 1993.

250. Richter R. Elektrische Machinen, Band I. 3 Auflage. Basel: Birkhäuser Verlag, 1967.

251. Radulescu MM, Oriold A, Muresan P. Microcontroller-based sensorless driving of a small electronically-commutated permanent-magnet motor. Electromotion 2(4):188–192, 1995.

252. Rahman MA, Osheiba AM. Performance of large line-start permanent magnet synchronous motor. IEEE Trans on EC 5(1):211–217, 1990.

253. Rahman MA, Zhou P. Determination of saturated parameters of PM motors using loading magnetic fields. IEEE Trans on MAG 27(5):3947–3950, 1991.

254. Rajashekara K, Kawamura A, Matsuse K, ed. Sensorless Control of a.c. Motor Drives. New York: IEEE Press, 1996.

255. Ramsden VS, Nguyen HT. Brushless d.c. motors using neodymium iron boron permanent magnets. Electr Energy Conf, Adelaide, Australia, 1987, pp. 22–27.

256. Ramsden VS, Holliday WM, Dunlop JB. Design of a hand-held motor using a rare-earth permanent magnet rotor and glassy-metal stator. Int Conf on Electr Machines ICEM'92, Vol 2, Manchester, UK, 1992, pp. 376–380.

257. Ramsden VS, Mecrow BC, Lovatt HC. Design of an in wheel motor for a solar-powered electric vehicle. Eighth Int Conf on Electr Machines and Drives (Conf. Publ. No. 444) EMD'97, Cambridge, UK, 1997, pp. 234–238.

258. Rao J.S.: Rotor Dynamics, 3rd ed. New Delhi: New Age Int Publishers, 1983.

259. Renyuan T, Shiyou Y. Combined strategy of improved simulated annealing and genetic algorithm for inverse problem. 10th Conf on the Computation of Electromagn Fields COMPUMAG'95, Berlin, Germany, 1995, pp. 196–197.

260. RF System Lab, Nagano-shi, Japan, www.rfsystemlab.com

261. Robinson RC, Rowe I, Donelan LE. The calculations of can losses in canned motors. AIEE Trans on PAS Part III (June):312-315, 1957.

262. Saari J, Arkkio A. Losses in high speed asynchronous motors. Int Conf on Elec Machines ICEM'94, Vol 3, Paris, France, 1994, pp. 704–708.
263. Say MG, Taylor ED. Direct Current Machines. London: Pitman, 1980.
264. Say MG. Alternating Current Machines. Singapore: ELBS with Longman, 1992.
265. Seely S. Electromechanical Energy Conversion. New York: McGraw-Hill, 1962.
266. Servax drives. Landert-Motoren AG, Bülach, Switzerland, www.servax.com
267. SGS–Thomson motion control applications manual. 1991.
268. Schiferl RF, Colby RS, Novotny DW. Efficiency considerations in permanent magnet synchronous motor drives. Electr Energy Conf, Adelaide, Australia, 1987, pp. 286–291.
269. Sato E. Permanent magnet synchronous motor drives for hybrid electric vehicles. IEEJ Trans on EEE (Japan) 2(2):162–168, 2007.
270. Shin-Etsu Co Ltd. Permanent magnet motor with low cogging torque by simulated magnetic field analysis. Shin-Etsu Magnetic Materials Research Center, Takefu-shi, Fukui Prefecture, Japan, 1999.
271. Shingo K, Kubo K, Katsu T, Hata Y. Development of electric motors for the Toyota hybrid vehicle Prius. 17th Int Elec Vehicle Symp EVS–17, Montreal, Canada, 2000.
272. Sidelnikov B, Szymczak P. Areas of application and appraisal of control methods for converter-fed disk motors (in Polish). Prace Nauk. IMNiPE, Technical University of Wroclaw, 48: Studies and Research 20:182–191, 2000.
273. Silvester PP, Ferrari RL. Finite Elements for Electrical Engineers, 2nd ed. Cambridge (UK): CUP, 1990.
274. Sobczyk TJ, Węgiel T. Investigation of the steady-state performance of a brushless d.c. motor with permanent magnets. Int Conf on Elec Machines ICEM'98, Istanbul, Turkey, 1998, pp. 1196–1201.
275. Sokira TJ, Jaffe W. Brushless d.c. Motors — Electronic Commutation and Controls. Blue Ridge Summit, PA: Tab Books, 1990.
276. Soyk KH. The MEP motor with permanent magnet excitation for ship propulsion. Electr Drives Symp EDS'90, Capri, Italy, 1990, pp. 235–238.
277. Spooner E, Chalmers B, El-Missiry MM. A compact brushless d.c. machine. Electr Drives Symp EDS'90, Capri, Italy, 1990, pp. 239–243.
278. Stefani P, Zandla G. Cruise liners diesel electric propulsion. Cyclo- or synchroconverter? The shipyard opinion. Int Symp on Ship and Shipping Research Vol 2, Genoa, Italy, 1992, pp. 6.5.1–6.5.32.
279. Stemme O, Wolf P. Principles and properties of highly dynamic d.c. miniature motors, 2nd ed. Sachseln (Switzerland): Interelectric AG, 1994.
280. Stiebler M. Design criteria for large permanent magnet synchronous machines. Int Conf on Elec Machines ICEM'00, Vol 3, Espoo, Finland, 2000, pp. 1261–1264.
281. Strauss F. Synchronous machines with rotating permanent magnet fields. Part II: Magnetic and electric design considerations. AIEE Trans on PAS 71(1):887–893, 1952.
282. Szeląg A. Numerical method for determining parameters of permanent magnet synchronous motor, 12th Symp. on Electromagn Phenomena in Nonlinear Circuits, Poznań, Poland, 1991, pp. 331–335.
283. Takahashi T, Koganezawa T, Su G, Ohyama K. A super high speed PM motor drive system by a quasi-current source inverter. IEEE Trans on IA 30(3):683–690, 1994.

284. Tarter RE. Solid-State Power Conversion Handbook. New York: J Wiley and Sons, 1993.

285. Teppan W, Protas E. Simulation, finite element calculations and measurements on a single phase permanent magnet synchronous motor. Int Aegean Conf on Electr Machines and Power Electronics ACEMP'95, Kuşadasi, Turkey, 1995, pp. 609–613.

286. Teshigahara A, Watanabe M, Kawahara N, Ohtsuka Y, Hattori T. Performance of a 7-mm microfabricated car. J of Micromechanical Systems 4(2):76–80, 1995.

287. The Electricity Council. Power system protection, Chapter 14. Stevenage and New York: Peter Peregrinus, 1990.

288. The European variable-speed drive market. Frost & Sullivan Market Intelligence, 1993.

289. The smallest drive system in the world, Faulhaber, Schönaich, Germany, 2004, www.faulhaber.com

290. Timar PL, Fazekas A, Kiss J, Miklos A, Yang SJ. Noise and Vibration of Electrical Machines. Amsterdam: Elsevier, 1989.

291. Tian Y, Chen KW, Lin CY. An attempt of constructing a PM synchronous motor drive using single chip TMS32020-based controller. Int Conf on Synchronous Machines SM100 Vol 2, Zürich, Switzerland, 1991, pp. 440–444.

292. Toader S. Combined analytical and finite element approach to the field harmonics and magnetic forces in synchronous machines. Int Conf on Electr Machines ICEM'94 Vol 3, Paris, 1994, pp. 583–588.

293. Tokarev BP, Morozkin VP, Todos PI. Direct Current Motors for Underwater Technology (in Russian). Moscow: Energia, 1977.

294. Turowski J. Technical Electrodynamics (in Polish), 2nd ed. Warsaw: WNT, 1993.

295. Vacoflux 48 – Vacoflux 50 – Vacodur 50 – Vacoflux 17. Vacuumschmelze GmbH & Co. KG, Hanau, Germany, www.vacuumschmelze.com

296. Varga JS. A breakthrough in axial induction and synchronous machines. Int Conf on Elec Machines ICEM'92 Vol 3, Manchester, UK, 1992, pp. 1107–1111.

297. Vas P. Parameter Estimation, Condition Monitoring, and Diagnosis of Electrical Machines. Oxford: OUP, 1993.

298. Verdyck D, Belmans RJM. An acoustic model for a permanent magnet machine: modal shapes and magnetic forces. IEEE Trans on IA 30(6):1625–1631, 1994.

299. Voldek AI. Electrical Machines (in Russian). St Petersburg: Energia, 1974.

300. Wagner B, Kreutzer M, Benecke W. Permanent magnet micromotors on silicon substrates. J of Microelectromechanical Systems 2(1):23–29, 1993.

301. Wallace AK, Spée R, Martin LG. Current harmonics and acoustic noise in a.c. adjustable-speed drives. IEEE Trans on IA 26(2):267–273, 1990.

302. Wang R, Demerdash NA. Comparison of load performance and other parameters of extra high speed modified Lundell alternators. IEEE Trans on EC 7(2):342–352, 1992.

303. Warlimont H. Opening address. 9th Int Workshop on Rare-Earth Permanent Magnets, Bad Soden, Germany, 1987.

304. Weh H. Permanentmagneterregte Synchronmaschinen höher Krafdichte nach dem Transversalflusskonzept. etz Archiv 10(5):143–149, 1988.

305. Weh H. Linear electromagnetic drives in traffic systems and industry. 1st Int Symp on Linear Drives for Ind Appl LDIA'95, Nagasaki, Japan, 1995, pp. 1–8.

306. Weimer, JA. The role of electric machines and drives in the more electric aircraft. Int Electr Machines and Drives Conf IEMDC'03, Madison, WI, USA, 2003, pp. 11–15.

307. Wheatley CT. Drives. 5th European Power Electronics Conf EPE'93 Vol 1, Brighton, UK, 1993, pp. 33–39.

308. Wing M, Gieras JF. Calculation of the steady state performance for small commutator permanent magnet d.c. motors: classical and finite element approaches. IEEE Trans on MAG 28(5):2067–2071, 1992.

309. Wróbel T. Stepping Motors (in Polish), Warsaw: WNT, 1993.

310. Yamamoto K, Shinohara K. Comparison between space vector modulation and subharmonic methods for current harmonics of DSP-based permanent-magnet a.c. servo motor drive system. IEE Proc – Electr Power Appl 143(2):151–156, 1996.

311. Yang S.J. Low-Noise Electrical Motors. Oxford: Clarendon Press, 1981.

312. Yuh J. Learning control for underwater robotic vehicles. Control Systems 14(2):39–46, 1994.

313. Zawilak T. Minimization of higher harmonics in line-start permanent magnet synchronous motors (in Polish). Prace Naukowe IMNiPE, Technical University of Wroclaw, Poland, 62(28): 251–258, 2008.

314. Zhang Z, Profumo, Tenconi A. Axial flux interior PM synchronous motors for electric vehicle drives. Symp on Power Electronics, Electr Drives, Advanced Electr Motors SPEEDAM'94, Taormina, Italy, 1994, pp. 323–328.

315. Zhong Z, Lipo TA. Improvements in EMC performance of inverter-fed motor drives. IEEE Trans on IA 31(6):1247–1256, 1995.

316. Zhu ZQ, Howe D. Analytical prediction of the cogging torque in radial field permanent magnet brushless motors. IEEE Trans on MAG 28(2):1371–1374, 1992.

317. Zhu ZQ. Recent development of Halbach permanent magnet machines and applications. Power Conversion Conf PCC'07, Nagaoya, Japan, 2007, pp. K9–K16.

Index

PEFC Certified

This product is
from sustainably
managed forests
and controlled
sources

PEFC™
PEFC/16-33-415

www.pefc.org

This book is made of chain-of-custody materials; FSC materials for the cover and PEFC materials for the text pages.

#0109 - 160516 - C0 - 234/156/34 [36] - CB - 9781420064407